To: Beth,
My best wishes on
career!
C. I. [illegible]

AF789786

Plant Bacteriology

Clarence I. Kado
Professor Emeritus
University of California, Davis

APS PRESS
The American Phytopathological Society
St. Paul, Minnesota U.S.A.

APS PRESS thanks the following sponsor for its financial support of this book.

AgBioChem, Inc., Los Molinos, CA

The cover images were used, by permission, as detailed below. **Top:** Meg McGrath, pumpkin with bacterial wilt caused by *Erwinia tracheiphila*; **Middle row, left:** Margery Daughtrey, geranium with bacterial blight caused by *Xanthomonas campestris* pv. *pelargonii*; **middle:** Steven T. Koike, rose with crown gall caused by *Agrobacterium tumefaciens*; **right:** Steven T. Koike, broccoli with bacterial blight caused by *Pseudomonas syringae* pv. *alisalensis*; **Bottom row, left:** Meg McGrath, tomato with bacterial speck, spot, and canker caused by *Pseudomonas syringae* pv. *tomato*, *Xanthomonas campestris* pv. *vesicatoria*, and *Clavibacter michiganensis* subsp. *michiganensis*, respectively; **middle:** Meg McGrath, snap bean with bacterial blight caused by *Xanthomonas campestris* pv. *phaseoli*; **right:** Steven T. Koike, manzanita with bacterial leaf spot caused by a *Xanthomonas* sp.

Library of Congress Control Number: 2010908304
International Standard Book Number: 978-0-89054-388-7

Printed in the United States of America on acid-free paper

The American Phytopathological Society
3340 Pilot Knob Road
St. Paul, Minnesota 55121, U.S.A.

Preface

Our knowledge of plant bacteriology has now reached a stage where it would be useful to integrate the concepts and ideas into book form. The purpose of this book is thus to provide basic background information, including a brief background of important historical events that gave birth to the field, and to collate and show recent advances in plant bacteriology. Plant bacteriology is clearly a rapidly advancing field, so various aspects of the discussions in the book will be, of course, overtaken by ongoing research. Nonetheless, although important problems remain to be solved, much of our present conceptual framework will remain valid, with results from future work adding to changes in detail rather than to basic revision. The sequencing of the genomes of every known bacterial plant pathogen will certainly make changes in the ever-fluctuating bacterial nomenclature and taxonomy.

In its organization, I have devoted the first part of the book to a brief historical account of microbiology, which led to the field of plant bacteriology, my thoughts and ideas on how bacterial pathogens evolved, and how these pathogens are currently classified (albeit in flux and awaiting advances provided by genomics and gene functions). These three chapters are followed by descriptions of the types of symptoms caused by bacteria, including illustrative diseases. Each type of symptom is presented with a detailed example of a causal agent and its characteristics, diagnostics, and mechanisms of virulence and pathogenicity. It is not my intention to cover *all* plant diseases caused by bacterial pathogens as would appear in a compendium of bacterial diseases. Instead, some key examples of diseases and their pathogens are described and discussed. This is followed by an extended discussion on the molecular mechanisms of virulence. This chapter is followed by a presentation of the concepts of epidemiology and disease control. The book is concluded by a chapter on the future of plant bacteriology and the developments that we hopefully expect in the near future.

I express my gratitude to the staff of APS PRESS in making this book possible, to Margery Daughtrey for her time and dedication as editor-in-chief, and especially to Anne Alvarez who kept encouraging me to continue with my efforts on the book by her endearing qualities and acts of kindness. Her nice suggestions and work as senior editor were also greatly appreciated. I also thank David Coplin, Dean Gabriel, Arun Chatterjee, Margaret Essenberg, and six anonymous reviewers, who dedicated their valuable time and expertise in reviewing this book. I also thank the following colleagues who provided photographs and technical advice: R. Michael Davis, Norman W. Schaad, George W. Sundin, Diane A. Cuppels, Isaac Barash, Susanne B. von Bodman, Martin Romantschuk, Carol A. Ishimaru, Erh-Min Lai, Dennis Gross, Carlos F. Gonzalez, Susan Hirano, Chris Schardl, Kay Ryugo, Teresa Coutinho, Hong Lin, Melodie Putnam, Jeff Hall, and others, with my apologies, whom I may have overlooked. I am very grateful to the National Institutes of Health for providing continuous support for 30 years of my career in plant bacteriology. I am also indebted to the American Cancer Society for providing valuable support in launching my career in 1968, to the USDA for additional financial assistance, and to the National Science Foundation for supporting various phases of crown gall research in my laboratory as well as the Fallen Leaf Lake Conferences, which are dedicated to plant microbiology and genetics. I am extremely grateful to the Toyota and Dole Foods corporations for generously supporting the latter parts of my research endeavors in biotechnology and on the pink disease of pineapple. My plant bacteriologist colleagues and I are all very grateful to the federal granting agencies for their continued support of much of the excellent research described in this book.

Clarence I. Kado

Contents

Plant Bacteriology

Historical Development of Plant Bacteriology

1

Before microscopes made microbes visible, written records documented the occurrence of various maladies of humans, animals, and plants. The roots of bacteriology can be found throughout the course of early history, but this field started to blossom only in the late sixteenth century, when microbiological equipment and techniques began to be developed. The specialized field of plant bacteriology was not fully initiated until the late nineteenth century, when bacteria were first understood to cause plant diseases. This chapter provides a categorical overview on how plant bacteriology started. It is not intended to provide a comprehensive review of the development of this field, because various detailed historical aspects of plant bacteriology have been well covered elsewhere (Smith, 1896; Whetzel, 1918; Dowson, 1957; Stapp, 1961; Ainsworth, 1980; Starr, 1984; Goto, 1990).

The Observational Period

The history of agriculture begins with the Fertile Crescent, where agriculture started before 7000 B.C. (Diamond, 1999) and flourished by 3100 B.C. during the early Nubian culture and Egyptian dynastic period (Gaballa, 2002). Diseases of man and plants were recorded during Egyptian and biblical times when the Fertile Crescent was the region of food production (Diamond, 1999). The cultivation and pruning of grapevines were clearly recorded in 2700 B.C. in hieroglyphics (McGovern et al., 1997). Plant diseases, such as blasting and mildew, are mentioned in the Bible, Deuteronomy 28:22 and I Kings 8:37, and famines resulting from drought and crop destruction by locusts are also recorded. The major crops of this period in the Fertile Crescent were emmer wheat, einkorn wheat, barley, pea, chickpea, lentil, muskmelon, olive, grape, fig, and date (Diamond, 1999; Gaballa, 2002).

Credited for his keen and accurate records of his observations, Hippocrates (ca. 460–370 B.C.) initiated scientific inquiry by declaring that there are natural explanations for all observable events in disease. In 430 B.C., Thucydides reported a plague of plants and man, while in 320 B.C., Cleidemus described diseases of fig, olive, and grapevine. In his treatise *Historia Plantarum*, written about 320 B.C., Theophrastus of Eresus, a student of Aristotle, observed,

> "The olive, in addition to having worms, produces also a 'knot' … it resembles the effect of sun-scorch." (*Historia Plantarum*, Book IV, Chapter 14)

At that time, the lack of scientific terms hindered the accurate description of the diseased parts of the plants. With regard to the etiology of disease, Theophrastus believed that man's cultivation of plants made the plants susceptible to disease. Indeed, pruning, as then practiced, made olive trees vulnerable to the later-known olive knot pathogen, *Pseudomonas syringae* pv. *savastanoi*. Like all the observers during this period, Theophrastus believed in spontaneous generation. Gaius

Plinius Secundus (A.D. 23–79) noticed diseased plants in his garden but attributed the condition to the evil influence of the stars. The occurrence of diseased plants was noticed almost fortuitously because comparisons between healthy and diseased plants were rarely considered. As late as 1866, Ernst Hallier was convinced that bacteria developed into fungi, which had no direct influence on disease, and proposed that the true nature of disease of plants was the result of bad soil and unnatural growing conditions—thinking that followed from that of Theophrastus.

Era of the Miasmatic Concept

The miasmatic period was characterized by the widespread belief that noxious effluvium emanated from putrescent matter and floated into the air, especially in night mists. However, this belief was not held universally. It was Marcus Terentius Varro (116–27 B.C.) who described in *De Re Rustica* that tiny animals spread disease:

> "Precautions must also be taken in the neighborhood of swamps ... because there are bred certain minute creatures which cannot be seen by the eyes, which float in the air and enter the body through the mouth and nose and there cause serious diseases." (Hooper, W. D., and Ash, H. B. 1934. Book I, section 12 in: De Re Rustica by Varro. English translation. Loeb Classical Library.)

This is one of the first instances in which some smaller unseeable form of life is proposed to be the cause of disease.

A number of years after Varro's declaration, Lucius Junius Moderatus Columella (A.D. 4 to ca. 70) echoed this idea about the association of disease with marsh land:

> "... it sends forth plagues of swimming and crawling things deprived of their winter moisture and infected with poison by the mud and decaying filth, from which are often contracted mysterious diseases whose causes are even beyond the understanding of physicians ..." (Ash, H. B. 1941. Book I, section 5 in: De Re Rustica of Columella. English translation. Loeb Classical Library.)

From this time on, there was a definite transitional period headed toward modernistic views. There were those who continued to believe in the Miasmatic Theory of Disease, however, even as late as the 1600s–1800s. For example, in 1631, Peter Lauremberg believed that certain stars, such as those making up the constellation Orion, exert an especially injurious influence from which the so-called secret evils arise, including rust, carbuncle, and mildew. In 1690, Heinrich Hesse believed that, if one grafted trees when the moon lay in the sign of the Crab or the Scorpion, the trees would develop canker. Indeed, biodynamic planting today retains elements of this belief. Even later, in 1773, Johann B. Zallinger believed that fungi were just abnormal parts of the plant. Although lauded for his astute classification and accurate description of a number of fungi growing on diseased tissue, Franz J. F. Meyen was convinced in 1837 that cereal smut was caused by stagnation of the sap and excessive fertilization. In 1833, Franz J. A. N. Unger, an army surgeon who spent much of his time on botany and plant pathology, strongly held that disease organisms originated from the host tissues.

Era of the Contagion Concept

In 1530, a physician and poet named Girolamo Fracastoro accurately described the symptoms of syphilis in one of his poems, which recounted the legend of the shepherd Syphilis who had caught the disease. Although it was not known at that time that syphilis is caused by a bacterium (which was observed 345 years later by Edwin Klebs), Fracastoro wrote about this disease and also about the diseases of plants. In his first of three books, he observed that diseases among fruit trees are communicable by contact and the disease is due to some contagion, the contagion being a seed of disease. He also proposed that contagions are host specific. Although he was not aware of the actual causes of disease, his observations and proposals were very advanced for his time. They were forgotten for nearly 350 years! Fracastoro is best known as the Father of the Germ Theory of Disease.

Escalation of the Microbial Theory of Disease

As with all natural sciences, major breakthroughs are made when novel methods and instruments are invented. The Microbial Theory of Disease gathered strength with the development of the compound microscope around 1595 by Zacharias Janssen and his father Hans. Antonie van Leeuwenhoek (1632–1723) made many keen observations using single-lens microscopes that he made himself by hand (Fig. 1-1). The microscope allowed Athanasius Kircher in 1656 to presumably see small worms in the blood from victims of the black plague. Although his observations were probably due to an overactive imagination, they led others to develop more powerful magnifying instruments, which led to the development of the light microscope and to subsequent seminal discoveries. Clearly, this single instrument led to the escalation of the Microbial Theory of Disease.

Robert Hooke (1635–1703) and Louis Joblot (1645–1723) used improved compound light microscopes and recorded observing many minute objects, tissue sections, and organisms (Pirie, 1964; Lechevalier, 1976). They confirmed many of the observations made earlier by van Leeuwenhoek (Porter, 1976).

For his description of microorganisms, including accurate descriptions of spirochetes, cocci, and bacilliform bacteria, van Leeuwenhoek was given the historical

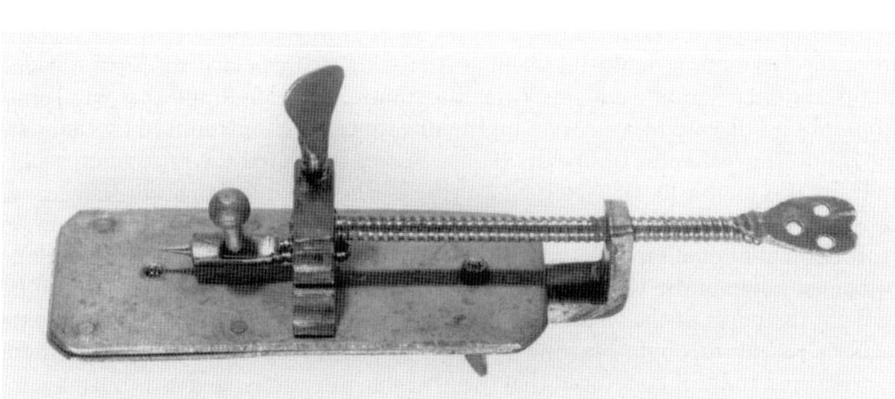

Fig. 1-1. An operational single-lens Leeuwenhoek microscope purchased at the Rijksmuseum, Leiden. (Photo by Jeff Hall)

title of Father of Bacteriology. These and many early observations made with the early microscopes eventually led to the experimental demonstration that bacteria indeed are the etiological organisms of disease and set forth the Theory of Infectious Microorganisms.

Establishment of the Theory of Infectious Microorganisms

The invention of the compound microscope and the development of sterile media for culturing microorganisms were instrumental in allowing the development of the Theory of Infectious Microorganisms. In 1774, a Danish entomologist, Johann C. Fabricius, initiated the idea that fungi associated with the lesions on a plant are actually distinct organisms rather than a part or result of abnormal plant development. Such a postulate was not well received, because many scientists remained advocates of the Theory of Spontaneous Generation of Microorganisms.

The Theory of Spontaneous Generation of Microorganisms was crushed by the ingenious experiments conducted by van Leeuwenhoek, Lazzaro Spallanzani (1729–1799), Louis Pasteur (1822–1895), and Robert Koch (1843–1910). Spallanzani cleverly used closed and open-mouth vessels containing infusions made from seeds, such as kidney bean, vetch, buckwheat, barley, maize, mallow, and beet, that were heated for various times in boiling water (Doetsch, 1976). After incubation at room temperature for several days, and using both single-lens and compound microscopes, Spallanzani saw no "animalcules" in the infusions from the closed vessels, but many such organisms were found swimming in the infusions from the open-mouth vessels. They were also seen in infusions made from the ash of seeds.

Using culture media, it was Koch who clearly demonstrated that bacteria are the etiological organisms of anthrax, tuberculosis, and cholera (Penn and Dworkin, 1976). The astute observations and inventive experiments by Rev. Miles Joseph Berkeley in 1845 definitely silenced further arguments by spontaneous generationists when he clearly showed that a mold was responsible for the blight of potatoes in Ireland (Berkeley, 1846). Julius Kühn in 1861 extended Berkeley's observations to diseases caused by fungi. About the same time period, Heinrich Anton de Bary (1831–1888), a German botanist, described many different fungi and devoted some time to bacteria as well.

These were also the times of discovery of plant viruses, in particular the agent that causes tobacco mosaic, a disease first named "mosaic disease of tobacco". In 1876, the German agricultural chemist Adolf Mayer inspected diseased tobacco plants and then set out to identify the infectious entity. He found that emulsions from ground diseased leaves were highly infectious when introduced by capillary tubes into the leaf veins of healthy plants. Mayer inspected the juice microscopically but was unable to see bacteria; nevertheless, he concluded that the disease was caused by a bacterium that remained to be identified. In 1892, Dmitrij J. Iwanowski reported that filtered sap of leaves from diseased plants still retained the infectious principle; however, he was not convinced that his results were valid because he thought the pores of the Chamberland filter that he used may have allowed some bacteria to leak through. Martinus Willem Beijerinck (1851–1931) repeated these experiments and was convinced that the filterable infectious agent was not a bacterium because the infectious agent diffused rapidly through agar. He called the infectious agent causing the mosaic disease "contagium vivum fluidum" (contagious living fluid or fluid infectious principle). In 1897, Friedrich Loeffler and Paul Frosch used kieselguhr filters to exclude bacteria but allow passage of the infectious agent that causes the foot-and-mouth disease of cattle. After repeated experimentation, including the use of *Bacillus fluorescens* (now *Pseudomonas fluorescens*) as the control to verify complete exclusion of bacteria by the filtration, Loeffler and Frosch concluded that the causal agent is not a bacterium but a very tiny organism capable of replication and possibly akin to those agents that cause smallpox, cowpox, and measles.

Birth of Plant Bacteriology

Bacteria were first recognized as causal organisms of plant disease in 1878 when Thomas Jonathan Burrill proposed that bacteria may cause fire blight of pear and apple. Using the microscope, Burrill saw swarms of minute particles resembling bacteria in the mucilaginous fluid derived from the inner phloem of recently blighted limbs of apple trees. No such fluid was present in healthy trees and the minute particles were also absent. He thus proposed that bacteria may be the cause of fire blight. He presented this observation and reported his findings to the Board of Trustees of the Illinois Industrial University (now the University of Illinois, Urbana). Because the pure culture technique was only in an early stage of development in the laboratory of Koch in Germany, Burrill was unaware of this modern microbiological method. Although personally convinced of the bacterial etiology of fire blight, he was unable to convince skeptics. As the pure culture technique became known in the United States, bacteria from blighted twigs were identified as the cause of fire blight (see Chapter 5 for further historical background on fire blight). In 1885, Joseph C. Arthur, a botanist at New York's Geneva Experiment Station, published convincing proof that fire blight is caused by a bacterium. Aware of the pure culture technique established in Koch's laboratory, Arthur isolated in pure culture a bacterium derived from diseased tissue and, with the pure culture, induced the typical disease in healthy trees.

A yellow bacterium, which was later identified to be *Xanthomonas hyacinthi*, was first isolated by J. H. Wakker in 1883 and shown to be the cause of hyacinth yellows disease. In 1893, L. H. Pammel found that bacteriosis of cabbage is caused by a yellow bacterium, *Bacillus campestris*, a name that was later changed to *Xanthomonas campestris* pv. *campestris*.

Frediano Cavara (1857–1929), an Italian botanist who was the director of the Royal Botanical Gardens at the University of Naples, had for some time keenly observed galls on grapevines. He recorded the successful isolation of a bacterium from the galls and demonstrated that it caused an identical gall disease on healthy grapevines (Cavara, 1897). This bacterium was later identified by Erwin F. Smith, who named the organism *Bacterium tumefaciens* (Smith and Townsend, 1907), which was changed to *Phytomonas tumefaciens*, a name that was again changed to *Agrobacterium tumefaciens*, as we know it presently. In the reports by Cavara (Cavara, 1897), it is clear that Smith, who visited Cavara in 1904, was shown the bacterial cause of the crown gall disease. Although Smith and C. O. Townsend (Smith and Townsend, 1907) are cited as the discoverers of the bacterial nature of crown gall, credit should be given to Cavara for the initial discovery and demonstration of the bacterial etiology of crown gall. Smith and Townsend actually confirmed the earlier finding by Cavara. It is apparent that Smith became intrigued with this disease and used it as a paradigm of cancer in animals and humans (Smith, 1911).

Despite the increasing number of reports that implicated bacteria as causal organisms of disease, established scientists in Europe refused to accept the evidence

Fig. 1-2. Erwin F. Smith. (From National Archives and Records Administration)

and maintained that bacteria were simply secondary effects of disorders caused by environmental conditions. In a long controversy with Emil Fischer, Smith compellingly argued that all the previous evidence, along with his own observations and experimental findings, proved that these bacteria are the cause, not the result, of disease. A series of animated exchanges took place between Fischer and Smith from 1897 to 1901 (Fischer and Smith, 1981), and the eventual acceptance of Smith's reasoning and proofs marked the beginning of a new era for plant bacteriology.

Smith expanded his work with studies of a number of plant-pathogenic bacteria, including *Bacillus carotovorus* (changed to *Erwinia carotovora* subsp. *carotovora*; now *Pectobacterium carotovorum* subsp. *carotovorum*), *Xanthomonas campestris*, and *Pseudomonas* (now *Ralstonia*) *solanacearum*. It was L. R. Jones, in 1901, who showed that the soft rot of vegetables is caused by *Bacillus carotovorous*. Likewise, olive knot was shown by Luigi S. Savastano, in 1887–1889, to be caused by a bacterium. In 1896, Smith postulated that there are in all probability as many bacterial diseases of plants as there are of animals (Smith, 1896). As late as 1920, Smith predicted that there will be bacterial diseases found on plants of every plant family. Because of the breadth and magnitude of his work and numerous contributions in the science of plant bacteria, Smith is appropriately named the Father of Plant Bacteriology (Fig. 1-2).

During the same era, Jakob Eriksson, in 1896, proposed that stripe rust on wheat overwintered as naked protoplasm mingled with that of the host. His vision of a pathogen lacking a cell wall was realized in the 1960s when phytoplasmas (then called mycoplasmalike organisms, or MLOs) were discovered. Because of his ideas, confirmed only in recent times, Eriksson is credited as the Father of Mycoplasmas.

The Development of Culture Media

Although the development of the compound light microscope was one of the great scientific breakthroughs in microbiology, the development of synthetic media to permit the growth of bacteria as a single-membered culture was equally a major step in microbiology.

The culture of microorganisms occurred long before it was recognized that beer and wine are products of microbial fermentation. Indeed, beer and wine qualify as some of the first means of growing microorganisms in relatively single-membered cultures. It was van Leeuwenhoek in 1680 who used extracts from boiled oats, from seeds, and from powdered pepper to grow microorganisms in his home laboratory. In 1769, Spallanzani also used extracts from seeds and even included some gravy to grow microorganisms in liquid culture. Such approaches were modified to include beef extracts; Max Schutze in 1836 and Theodor Schwann in 1837 employed beef broth as liquid media. It was H. G. F. Schröeder and Tn. von Dusch in 1854 who introduced the use of wool fibers to allow air to pass into culture flasks. Later, cotton fibers were used, and various slip caps made of metal or plastic serve the same purpose today.

By 1863, Pasteur refined these media by turning to the extract of yeast cells, hydrolyzed casein from milk, and urine. All of these developments were focused on liquid media. To purify cultures, the technique of serial transfer was developed by Klebs in 1873 and the method of serial dilution was cleverly devised by J. Lister in 1878.

The development of solid media can find its origins in the use of bread: V. Sette observed red spots on old bread and used bread to grow bacteria. Vittadini in 1852 first used gelatin to grow fungi, while Sergius Winogradsky in 1890 used silica gel for isolating denitrifying bacteria. Prior to this time, J. Schroeter in 1872 used mixtures of starch, flour, and egg white as solid substrates for growing bacteria. In 1881, Koch was using slices of potato; but learning of Vittadini's gelatin via Oscar Brefeld in 1882, he converted to the use of gelatin. This led to the development of the streak dilution and the pour plate techniques for purifying mixed cultures of bacteria. The use of gelatin, however, was limited to the winter and early spring months because warm conditions caused the gelatin to liquefy. Moreover, gelatin-liquefying bacteria caused problems. A dramatic breakthrough came from the discovery of agar and its excellent physical properties, including durability. The use of agar was developed serendipitously in 1882, when Walther Hesse, an assistant to Koch, recognized the usefulness of agar, which his wife, Lina (Angelina), used in making jam, jellies, and puddings at home (Gröschel, 1992). Agar remains the mainstay of modern microbiology.

Another assistant to Koch was a displaced country physician, R. J. Petri, who recommended that Koch replace bell jars with glass-walled dishes with lids on them. This innovation remains essentially the same today, albeit plastic has replaced the glass "petri dishes".

Following the development of the pure culture technique and of solid media, Loeffler painstakingly developed selective media, one of which is still used today in clinical microbiology. Loeffler also developed a number of stains, including a stain for recognizing bacterial flagella.

The Development of Bacteriological Stains

Along with the development of light microscopy, a number of efforts were made to increase the contrast of the objects viewed. The stains were actually borrowed from the dyes used to stain textiles in the clothing industry and, later, from those used to stain histological sections. The histologists H. R. Goeppert and G. Cohn in

1849 and F. Bohmer in 1865 used carmine to stain tissue specimens. It was Herman Hoffmann, a botanist and mycologist, who first attempted to stain bacteria using carmine and fuchsin. This attempt was extended in 1871 by Carl Weigert, who used methyl violet. Then, in 1877, Koch used a mixture of dyes that included methyl violet, basic fuchsin, and aniline brown. He was the first to use heat and alcohol to fix bacteria to glass slides. Using Loeffler's approach, Koch was able to stain flagella with a mixture of chromic acid and hematoxylin.

Along with his notable work on the chemotherapeutic techniques for tuberculosis and syphilis, Paul Ehrlich (1854–1915) should be credited for developing the differential staining technique in 1877. F. Neilsen in 1880 used sulfuric acid as a fixative, while F. Ziehl in 1900 used phenol. The differential staining technique of Ehrlich was noted by Hans Christian Gram, who found that Ehrlich's aniline-water-gentian violet, when mixed on the slide with a solution of potassium iodide, formed a complex that was retained by certain bacteria but not by other bacteria when washed with alcohol. This observation in 1884 led to the technique of differential staining that is commonly known as the Gram stain and remains widely used as a benchmark in bacterial classification.

The Development of Sterilization

Pertinent to the advent of the pure culture technique was the concurrent development of sterilization methods. John Tyndall (1820–1893), a British physicist, developed a method of sterilizing broth media using periodic heat treatments. The idea was to boil the medium for 5 min and allow it to cool overnight in a closed vessel. The medium was then boiled again for 5 min and again allowed to cool overnight. This cycle of heating and cooling was repeated three or four times, yielding a final sterile medium in a closed container. Tyndall observed that the treated medium showed no growth of microorganisms over extended periods. The power of the method reflects the present knowledge that bacterial spores germinate after heat shock and cooling and that such germinated spores are highly susceptible to heat. Tyndall astutely realized that vegetatively active microbes are highly susceptible to heat. Quite interestingly, being a keen observer, Tyndall also had noted the inhibition of bacterial growth by *Penicillium* species more than 50 years before the same observations were made by Sir Alexander Fleming.

Meanwhile, in 1881, Koch determined that steam at 100°C is not always able to sterilize media. C. Chamberland showed in 1880 that high-pressure steam is essential for this process. He also devised the candle filter in 1884, using unglazed porcelain to filter away contaminating microbes from unsterilized media. In 1891, H. Nordtmeyer developed the Berkefeld filter to sterilize broth. This filter was made of kieselguhr, a dolomite that was mined in Hanover, Germany. Nordtmeyer named the filter after the owner of the dolomite mine. These early filters, including those made of asbestos, have long been replaced by filters made from cellulose acetate/nitrate mixtures, from nylon, and from highly durable Teflon and Kevlar synthetic polymers of modern times.

First Descriptions of Insects Being Vectors of Plant Pathogens

Hired by the United States Department of Agriculture to help control the fire blight disease, Merton B. Waite discovered in 1891 that the fire blight bacterium overwinters in the cankers and appears in the ooze material during the spring

(Waite, 1896). It was noted that insects carry the ooze from blossom to blossom, prompting Waite to propose that insects are transmitting the fire blight organism from tree to tree. This is the first realization that an insect can be a vector of a plant pathogen.

Not knowing that the serious Anaheim disease of grapevines in southern California was caused by a bacterium transmitted by leafhoppers, Newton D. Pierce in 1885 investigated this disease extensively—the disease is presently known as Pierce's disease—and concluded that it is caused by a minute unidentifiable microbe. In 1940 and later, it was shown that this disease and alfalfa dwarf disease can be transmitted to healthy grapevine seedlings by grafting (Hewitt, 1945) and by leafhoppers (Houston et al., 1940; Hewitt et al., 1946). Because a number of viruses were known to be transmissible by grafting and by insects, Pierce's disease at that time was believed by many pathologists to be caused by a virus. This notion was countered by the claim that Pierce's disease is caused by a Gram-positive, rod-shaped bacterium (Auger et al., 1974), which was later shown to be *Lactobacillus hordniae*, a new species that is associated with the leafhopper vector *Hordnia circellata* (Baker) and is not the cause of Pierce's disease (Latorre-Guzman et al., 1977). Intrigued by the novel notion that Pierce's disease might be caused by a bacterium, Davis and colleagues successfully isolated the causal bacterium (Davis et al., 1978). This organism is presently known as *Xylella fastidiosa*, a relative of *Xanthomonas* species with Gram-negative, nonmotile, rod-shaped, and xylem-associated features (Wells et al., 1987). These more recent findings support the earlier claim by Pierce that the etiological organism of the disease is likely to be a bacterium. An important clue that a bacterium might be the cause of Pierce's disease was overlooked in the beautiful histological sections of the xylem-bearing tissues from Pierce's diseased grapevines published in 1948 by Katherine Esau (Esau, 1948). What were interpreted as curious nuclei in tyloses in the xylem at the time appear in hindsight to be bacterial cells (Esau, 1948 [Plate 10]).

Summary

A survey from ancient times to the present day shows that the advances in bacteriology and subsequently plant bacteriology occurred in small increments until novel technical advances, such as the invention of the microscope, steam sterilization, and culture media, were made. Many of these technical advances were the result of the industrial revolution, but the creative genius of many of our pioneers in microbiology clearly allowed them to make lasting marks on the science. Their major advances are summarized in Figure 1-3.

Epilogue

The growth of plant bacteriology continues to be concurrent with the development of new techniques and instruments. Once bacteria were found to be plant disease-causing organisms, many of the techniques used in promoting the fields of medical microbiology and molecular microbiology were quickly adopted by plant bacteriologists. In reverse fashion, the field of virology began with the discovery of *Tobacco mosaic virus*, its isolation and purification, and with the subsequent purification and characterization of a number of plant viruses. Virology of animals and man followed after knowledge of viruses infecting plants. Significant funding was generated (ca. 1950) for the biomedical field with the goal of curbing the polio

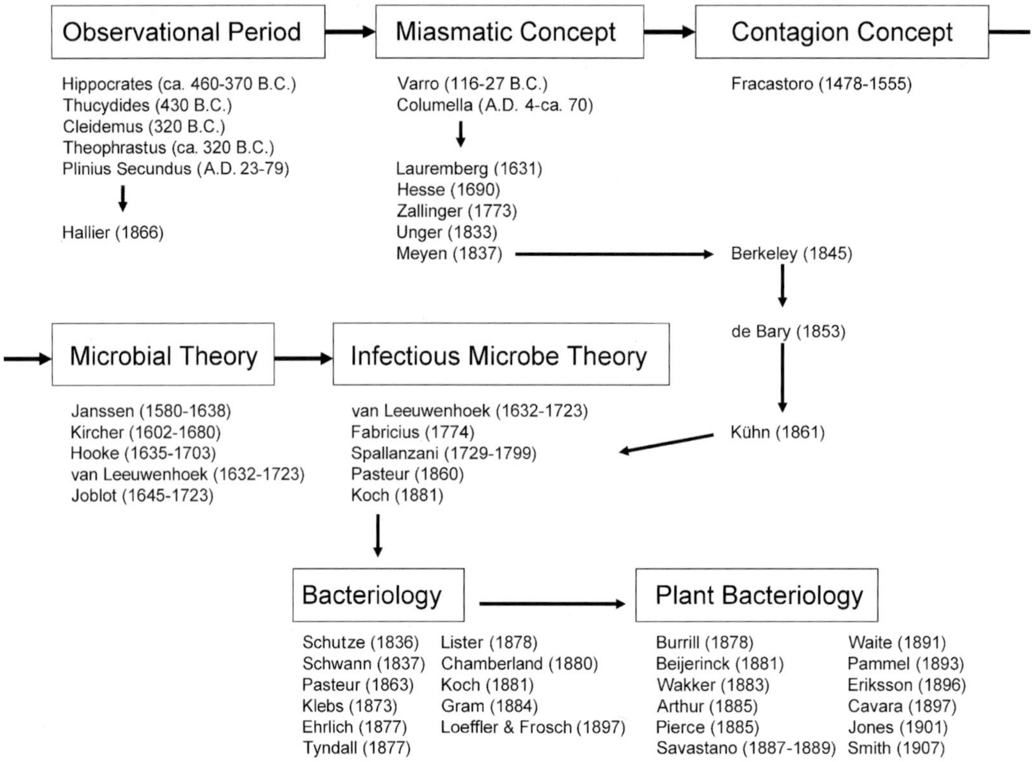

Fig. 1-3. Flow diagram of the historical development of plant bacteriology.

epidemic and rising cancer rates. A part of this funding was the original driving force for investigating *Agrobacterium tumefaciens*, which was considered a cancer paradigm of plants (Smith and Townsend, 1907; Smith, 1911).

Knowledge in the field of plant bacteriology has progressed from the initial descriptive aspects to today's investigations of fundamental mechanisms. Many of the developments in plant bacteriology, such as the understanding of insects vectoring pathogens, the biology of the Ti plasmid, and the type III and type IV secretion systems, have aided clinical microbiology. The research efforts made in plant bacteriology therefore should continue to be funded not only by agricultural agencies but also by the biomedical granting agencies, such as the National Institutes of Health in the United States, the National Research Council of Canada, and the European granting agencies. The history of microbiology has shown that many of the novel findings dovetail nicely; the subfields of microbiology are not mutually exclusive.

REFERENCES

Ainsworth, G. C. 1980. Introduction to the History of Plant Pathology. Cambridge University Press, New York.

Auger, J. G., Shalla, T. A., and Kado, C. I. 1974. Pierce's disease of grapevines: Evidence for a bacterial etiology. Science 184:1375-1377.

Berkeley, M. J. 1846. Observations, botanical and physiological, on the potato murrain. J. R. Hortic. Soc. Lond. 1:9-34.

Cavara, F. 1897. Eziologia di alcune malattie de piante coltivate. Staz. Sper. Agr. Ital. 30:482-509.

Davis, M. J., Purcell, A. H., and Thompson, S. V. 1978. Pierce's disease of grapevines: Isolation of the causal bacterium. Science 199:75-77.

Diamond, J. M. 1999. Guns, Germs and Steel: The Fates of Human Societies. W. W. Norton, New York.

Doetsch, R. N. 1976. Lazzaro Spallanzani's Opuscoli of 1776. Bacteriol. Rev. 40:270-275.

Dowson, W. J. 1957. Plant Diseases Due To Bacteria, 2nd ed. Cambridge University Press, Cambridge.

Esau, K. 1948. Anatomic effects of the viruses of Pierce's disease and phony peach. Hilgardia 18:423-482.

Fischer, A., and Smith, E. F. 1981. The Fischer-Smith Controversy: Are There Bacterial Diseases of Plants? Phytopathological Classics No. 13. C. L. Campbell, translator. American Phytopathological Society, St. Paul, MN.

Gaballa, G. A. 2002. The History and Culture of Nubia. Ministry of Culture, Higher Council for Antiquities, Museums' Sector, Cairo, Egypt.

Goto, M. 1990. Fundamentals of Bacterial Plant Pathology. Academic Press, San Diego, CA.

Gröschel, D. H. M. 1992. Walther and Angelina Hesse—Early contributors to bacteriology. (A translation of Wolfgang Hesse.) ASM News 58:425-428.

Hewitt, W. B. 1945. A graft-transmissible mosaic disease of grapevine. Phytopathology 35:940-942.

Hewitt, W. B., Houston, B. R., Frazier, N. W., and Freitag, J. H. 1946. Leafhopper transmission of the virus causing Pierce's disease of grape and dwarf of alfalfa. Phytopathology 36:117-128.

Houston, B. R., Frazier, N. W., and Hewitt, W. B. 1942. Leaf-hopper transmission of the alfalfa dwarf virus. (Abstr.) Phytopathology 32:10.

Latorre-Guzman, B. A., Kado, C. I., and Kunkee, R. E. 1977. *Lactobacillus hordniae*, a new species from the leafhopper (*Hordnia circellata*). Int. J. Syst. Bacteriol. 27:362-370.

Lechevalier, H. 1976. Louis Joblot and his microscopes. Bacteriol. Rev. 40:241-258.

McGovern, P. E., Fleming, S. J., and Katz, S. H., eds. 1997. The Origins and Ancient History of Wine. Routledge, Taylor & Francis, New York.

Penn, M., and Dworkin, M. 1976. Robert Koch and two visions of microbiology. Bacteriol. Rev. 40:276-283.

Pirie, N. W. 1964. The Leeuwenhoek lecture, 1963, the size of small organisms. Proc. R. Soc. B 160:149-166.

Porter, J. R. 1976. Antony van Leeuwenhoek: Tercentenary of his discovery of bacteria. Bacteriol. Rev. 40:260-269.

Smith, E. F. 1896. The bacterial diseases of plants: A critical review of the present state of our knowledge. Am. Nat. 30:797-804.

Smith, E. F. 1911. Crown-gall and sarcoma. U.S. Dep. Agric. Bur. Plant Ind. Circ. 85:1-4.

Smith, E. F., and Townsend, C. O. 1907. A plant-tumor of bacterial origin. Science 25:671-673.

Stapp, C. 1961. Bacterial Plant Pathogens. Oxford University Press, Oxford.

Starr, M. P. 1984. Landmarks in the development of phytobacteriology. Annu. Rev. Phytopathol. 22:169-188.

Waite, M. B. 1896. Cause and prevention of pear blight. U.S. Dep. Agric. Yearb. 1895:295-300.

Wells, J. M., Raju, B. C., Hung, H.-Y., Weisburg, W. G., Mandelco-Paul, L., and Brenner, D. J. 1987. *Xylella fastidiosa* gen. nov., sp. nov.: Gram-negative, xylem-limited, fastidious plant bacteria related to *Xanthomonas* spp. Int. J. Syst. Bacteriol. 37:136-143.

Whetzel, H. H. 1918. An Outline of the History of Phytopathology. W. B. Saunders, Philadelphia.

2 Evolution of Bacterial Pathogens

The association of bacteria with plants is long standing. The initial intimate association probably began with the appearance of photosynthetic cyanobacteria as endophytes of a heterotrophic eukaryote that evolved into the first primitive plants (Delwiche and Palmer, 1997; McFadden, 1999; Moreira et al., 2000). Microbial and ecological diversity then led to microbe–plant interactions between free-living bacteria and plants. These bacteria were capable of living off a number of different types of substrates shed by the plants. By their nutritional diversity, these bacteria, particularly the chemolithotrophs, had no specific ecological habitat. Such bacterial types are often classified as saprophytic bacteria, owing to their ability to consume a wide range of inorganic and nonliving organic substrates.

Through evolutionary steps, involving chance associations and acquisition of metabolic networks through horizontal gene transfer events in response to changing environments (Pal et al., 2005), some saprophytic bacteria gradually developed a preference for plants as a niche to escape competitors and predators. In the beginning, these bacteria evolved into epiphytic types, choosing the plant environment as a more favorable niche rather than those in water and soil, where the microbial competition for space and food remained extremely fierce. Some of these epiphytic bacteria evolved to develop highly specialized features to increase their efficiency in occupying plant surfaces. The abundance of fissures and crevices of the cuticular layer, composed of cutin (cross-linked hydroxyl fatty acids) bounded by a layer of wax on its surface and a layer of pectin bound to the cellulose of the cell wall (Eglinton and Hamilton, 1967), provided excellent sites for bacterial colonization. Natural exposure to ultraviolet radiation in sunlight may be an evolutionary pressure toward high guanine-plus-cytosine (GC) content in the bacterial chromosome (Singer and Ames, 1970). Thus, a number of epiphytes have high GC compositions, between 60 and 70 mol %. Those bacteria with GC contents lower than these mole percentages have other means of escaping the deleterious effects of ultraviolet radiation. Many of these bacteria are pigmented, and it is the pigment that serves as a screen against the harmful ultraviolet rays (Niyogi et al., 1997). As epiphytic bacteria became more and more specialized, their habitat, in turn, became considerably narrow and yet provided a niche sufficient for competitive advantages. In the modern era, epiphytes are specialized bacteria whose sustained presence on the host plant is not harmful to either party.

Long-term and constant associations of epiphytic bacteria and plants have led some of them to become mutually beneficial by exchanging requirements for growth and maintenance to ensure the health of both partners. This type of mutualism is exemplified by *Janthinobacterium lividum* (formerly *Chromobacterium lividum*), whose niche is the *Ardisia* plant (Rodrigues-Pereira et al., 1972). These purple-pigmented bacteria are ectosymbionts whose beneficial contributions are in the form of phytohormones (cytokinins) that are elaborated by these bacteria to stimulate and promote plant growth as well as seed germination (Rodrigues-Pereira et al., 1972; Tsavkelova et al., 2007). In turn, there is sufficient substrate on the plant surface to provide the food base for the ectosymbiont. These ectosymbionts have retained the purple insoluble pigmentation, presumably as protection against harmful ultraviolet light.

By providing nutritional and growth-promoting factors (e.g., cytokinins) to the plant, ectosymbionts have the added advantage of a protective habitat created by abnormal growths (fasciations and phyllodies) produced by the cytokinins. Ectosymbionts are known to defend their marine ciliate host against predators, as is the case with *Euplotidium* species against predation by *Litonotus lamella* (Ehrenberg) Schewiakoff (Petroni et al., 2000). Host specificity conferred on the bacteria adds to this advantage. By being host specific, the competitive advantage of the ectosymbiont over epiphytes is substantial.

The long-term interactions between ectosymbionts and their hosts have led to endophytic and endosymbiotic cohabitations among bacteria and plants. For example, ectosymbionts of *Euplotidium* species have evolved into an episymbiotic association between the ectosymbiont and the host (Petroni et al., 2000). It also appears that bacteria aided by twitching motility via type IV pili lead to the establishment of microcolonies of an *Azoarcus* sp. on the root surface and as endosymbionts in the roots of rice plants (Bőhm et al., 2007). The endosymbiont colonizes mainly in the apoplast of the root cortex and not inside living plant cells (Reinhold-Hurek et al., 2007). Foliar parts (phyllosphere and phylloplane) can also harbor endophytes. Natural bacterial colonization of cotton and sweet corn leaves by endophytic bacteria has been observed (Misaghi and Donndelinger, 1990; McInroy and Kloepper, 1995). When population growth favors a given species of bacteria, the transient establishment of endophytic associations can be promoted (Misaghi and Donndelinger, 1990; Mahaffee et al., 1994; Quadt-Hallmann et al., 1995). The leaf cavity of the mosquito fern, *Azolla caroliniana* Willd., contains the nitrogen-fixing endosymbiont *Anabaena azollae*. This interrelationship appears essential for the symbiosis between the fern and the cyanobacterium (Gates et al., 1980).

Evolutionary selective pressures, such as continued stressful competition for food in a given niche, select for bacterial strains that possess efficient escape mechanisms for their survival. Thus, naturally occurring endosymbionts, such as highly motile members of the genus *Phyllobacterium*, are sheltered in specific protective niches within their specific plant hosts. Members of the families *Rhizobiaceae*, *Phyllobacteriaceae*, and *Bradyrhizobiaceae* often favor the rhizosphere and rhizoplane and are intimately associated with plant roots in the form of nodules and with leaves in the form of galls. In the case of *Sinorhizobium* species, this intimacy extends to the establishment of intracellular infections that is mediated by means of an unusually modified lipid A (Ferguson et al., 2005). *Sinorhizobium meliloti* traverses down infection threads in root hairs of the alfalfa host (*Medicago sativa* L.) and forms compartments called symbiosomes, in which bacterial cells differentiate into morphologically distinct Y-shaped bacteroids that fix atmospheric nitrogen to ammonia for the plant and in turn receive nutrients from the plant.

Endosymbionts are highly evolved to promote the growth of their plant hosts and at the same time create a niche that precludes the habitation of foreign bacteria. These intimate associations require numerous genes to fix N_2 and reduce it to ammonium ions, neutralize toxic plant phenols, and produce structures such as infection threads culminating in nodule or gall formation, which serve as unique homes for these bacteria. These genes are frequently harbored on large plasmids (e.g., Sym plasmids), suggesting that these genes were obtained and sequestered in more recent evolutionary times by horizontal gene transfer.

Shifts in the environment can perturb the existing endosymbiotic association to gradually convert it into a pathogenic association with the plant. In this case, the plant no longer benefits from the association, and bacteria remain as long as the pathogenic association continues to benefit the survival and perpetuation of the pathogen. Here, the host provides a unique niche to the pathogen that allows it to circumvent competing microorganisms and to replicate to sufficient numbers to be transmitted to a new susceptible host. Also, the pathogen has evolved to circumvent host defense barriers.

In nature, there is usually some base equilibrium in pathogenic associations with the host plant so that the host is not killed outright. Disruption of this equilibrium results in severe reactions by the host to the pathogen. A classic example in which the base equilibrium has been drastically changed is the generation of crop plants from wild-type plants through plant breeding and domestication. The wild-type plant is highly tolerant of the pathogen but the corresponding domesticated crop plant is highly susceptible to the same pathogen. Thus, the series of pandemics that have occurred throughout historical times simply reflect the vulnerability of the crop plant to a pathogen that long existed in equilibrium with the wild-type plant host.

The overall transition from a saprophytic association to a pathogenic association between a bacterium and its host plant is shown in Figure 2-1. The gradual shift of the environment to adverse conditions for growth (physical, chemical, and genetic alterations) can cause a shift of the pathogen into an innocuous endosymbiont and vice versa. A more rapid shift from epiphyte and endophyte to a pathogen is evidenced by the evolution of *Pantoea agglomerans*, a common epiphyte and endophyte, into the gall-forming pathogen *Pantoea agglomerans* pv. *gypsophilae* (Barash and Manulis-Sasson, 2007).

The highly developed interrelationship of a pathogen with its host plant is exemplified by crown gall and hairy root. Crown gall is caused by *Agrobacterium tumefaciens* and hairy root is caused by *Agrobacterium rhizogenes*. (Note: A proposal was made to change the name *Agrobacterium* to *Rhizobium* based on nongenomic evidence.) *Agrobacterium tumefaciens* and *Agrobacterium rhizogenes* have created highly specific niches for themselves by converting normal plant tissue into galls and hairy galls, respectively, by the transfer of specific genes into the host cell and host chromosomes (see Chapter 8 for information on bacteria causing abnormal growths). These specific genes encode the biosynthesis of auxin, cytokinin, and a

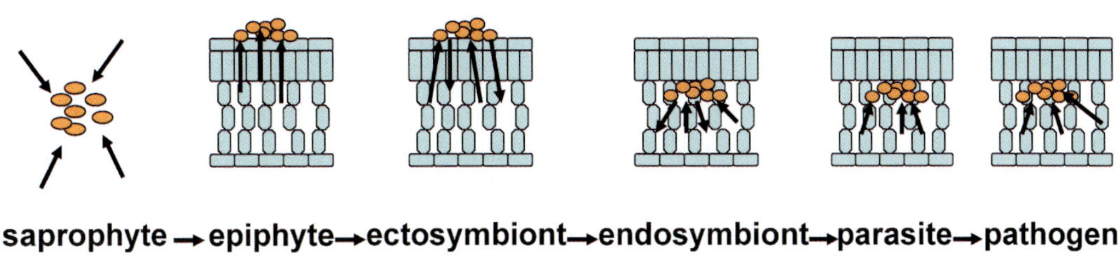

Fig. 2-1. Transition of microbe–plant interactions over evolutionary time that led to endosymbiotic, parasitic, and pathogenic relationships as seen today. Vertical arrows indicate some beneficial attributes provided by either the associated bacterium or the invaded plant cell or both.

histone-binding protein for abnormal appendage formation and encode enzymes to synthesize a specific food base composed of arginine or other basic amino acids conjugated with a keto organic acid to form unusual compounds called opines. Opines are not usually catabolized by bacteria other than *Agrobacterium tumefaciens* or *Agrobacterium rhizogenes*. This type of substrate specificity provides a highly competitive advantage over other competing bacteria that are unable to utilize opines. Since *Agrobacterium tumefaciens* and *Agrobacterium rhizogenes* cleverly genetically engineer the host plant to provide a specific compound for nutritional utilization, and a niche in the form of galls or hairy galls, this plant–microbe association epitomizes one of the highest levels of pathogen–host interactions.

The microbe–plant interactions are dynamic associations. Any imbalance in the environmental conditions that would perturb the association may favor one partner of the association over the other. That is, environmental changes can disturb the nature of an association between the plant and its associated microbial symbiont. When this occurs, either the habitat becomes unfit for the microbe or the plant harboring the microbe becomes susceptible to the advances of the microbe with the creation of a pathogenic relationship. Plant pathologists have routinely employed cultural practices as one means of controlling the advance of the pathogen. Such practices are used to favor the crop plant over the pathogen by shifting the environmental conditions in one manner or another to make the crop unfit for habitation by the pathogen. Cultural adjustments in fertilization schedules, watering regimes, calcium (gypsum) applications, and soil pH, as well as aeration and pruning, are some of the methods used for this purpose. For example, gypsum is routinely applied to young pineapple plants to reduce the incidence of pink disease (Anthony Hepton, Dole Foods, Inc., personal communication). The interaction between the pathogen and its host plant remains dynamic since the evolutionary pressure to remain successful in a given niche is enormous. The transition from the saprophytic way of life among microbes to a highly specialized host-specific relationship is a sophisticated means of escaping the constant competition among microbes in the soil and on plant surfaces. Such transitions have culminated in the present collection of plant pathogens that are classified in the bacterial phylum *Proteobacteria*: the class *Alphaproteobacteria* containing the genera *Sphingomonas*, *Agrobacterium*, and *Phyllobacterium*; the class *Betaproteobacteria* containing the genera *Burkholderia*, *Ralstonia*, and *Acidovorax*; and the class *Gammaproteobacteria* containing the genera *Xanthomonas*, *Xylella*, *Pseudomonas*, *Brenneria*, *Dickeya*, *Erwinia*, *Pantoea*, *Pectobacterium*, *Salmonella*, *Serratia*, and *Tatumella*. In the phylum *Actinobacteria*, class *Actinobacteria*, and family *Microbacteriaceae* are the genera *Clavibacter*, *Curtobacterium*, *Leifsonia*, and *Rathayibacter*; in the family *Nocardiaceae* is the genus *Rhodococcus*; and in the family *Streptomycetaceae* is the genus *Streptomyces*. Moreover, in the phylum *Firmicutes* and class *Mollicutes* is the genus *Spiroplasma*. *Spiroplasma* species are considered highly evolved through genome size amelioration since they are nutritionally fastidious, requiring specific substrates that they cannot produce on their own (Ochman and Lawrence, 1996; Lawrence, 2002; Pal et al., 2005; Comer et al., 2007). Spiroplasmas have limited genome size and must rely on the acquisition of genetic information for adaptation in the host as well as in the insect vector (Melcher and Fletcher, 1999). The mechanisms of genetic flux of these plant pathogens have evolved to adapt to differing environments of the host. Acquisition of genes for environment adaptation through lateral gene transfer is one of these mechanisms (Lawrence, 2002).

General Characteristics of Plant-Pathogenic Bacteria

Bacterial plant pathogens have evolved to grow well in their plant hosts, often in specific hosts as well as in a specific plant part, such as blossoms, leaves, roots, or storage organs (e.g., tubers, corms). The general characteristics of these pathogens are that the majority of them are non-spore-forming, rod-shaped organisms. Only *Streptomyces* species form spores and are filamentous. Depending on their cell wall structure, the bacterial plant pathogens can be either Gram negative or Gram positive. Most plant-pathogenic bacteria are nutritionally nonfastidious and have growth temperature optima around 23–28°C. A few pathogens, such as *Ralstonia solanacearum* and *Burkholderia cepacia*, can grow at 42°C. Plant-pathogenic bacteria can be either strictly aerobic or facultatively anaerobic. Unlike most mammalian pathogens, plant-pathogenic bacteria infect intercellularly. Genetic studies have shown the involvement of numerous genes in virulence and pathogenesis. A number of plant-pathogenic bacteria harbor virulence and pathogenicity genes on an extrachromosomal element (plasmid), whereas other pathogens bear virulence and pathogenicity genes in their chromosomes. The observed conservation of virulence and pathogenicity genes in pathogenicity islands strongly suggests interchangeable movement of these genes across species boundaries via lateral gene transfer mechanisms (Hacker and Kaper, 2000; Syvanen and Kado, 2002). Many of these genes provide an inherent ability of the pathogen to grow in its host, often a specific host, and are clearly essential for successful colonization of the host. Efforts to control and eradicate plant diseases are constantly being challenged by bacterial pathogens developing new survival strategies. Mutator phenotypes that are defective in methyl-directed DNA mismatch repair (such as a mutation in the *mutS* gene) might play a major role in pathogen adaptation since the mutator allele enhances mutation and genetic recombination during mating between *Escherichia coli* and *Salmonella typhimurium* by up to 1,000-fold (LeClerc et al., 1996). Recombination enhancement has been observed in a number of different bacterial species (Matic et al., 2002). Recombination increases adaptability by increasing genetic variability (Crow and Kimura, 1965). The importance of mutator alleles for the evolution of pathogens is directly linked to their particular lifestyle. The foliar pathogen *Xanthomonas vesicatoria*, for example, harbors a *hrp*-type pathogenicity island that encodes an extracellular filamentous "Hrp" pilus (see Chapter 10). Comparative sequence analyses have revealed that the Hrp pilus structure has undergone evolutionary adaptation to avoid detection by the plant defense system (Weber and Koebnik, 2006).

In summation, successful pathogen colonization of the host depends on how well the pathogen can develop novel survival mechanisms. Acquisition of novel genes through lateral gene transfer is a powerful means for pathogen development (Kado, 2009). The generation, through selective mutational events, of novel and functional genes whose products enhance successful colonization can also facilitate pathogen fitness. Whatever the mechanism, genetic changes that are advantageous to pathogen survival occur under selection pressures put on the pathogen by various applied control measures implemented in farming practices.

REFERENCES

Barash, I., and Manulis-Sasson, S. 2007. Virulence mechanisms and host specificity of gall-forming *Pantoea agglomerans*. Trends Microbiol. 15:538-545.

Bőhm, M., Hurek, T., and Reinhold-Hurek, B. 2007. Twitching motility is essential for endophytic rice colonization by the N_2-fixing endophyte *Azoarcus* sp. strain BH72. Mol. Plant-Microbe Interact. 20:526-533.

Comer, J., Fletcher, J., Davis, R. E., and Melcher, U. 2007. Evolution of the *Spiroplasma* P58 multigene family. Biochem. Genet. 45:25-32.

Crow, J. F., and Kimura, M. 1965. Evolution in sexual and asexual populations. Am. Nat. 99:439-450.

Delwiche, C. F., and Palmer, J. D. 1997. The origin of plastids and their spread via secondary symbiosis. Pages 53-86 in: Origins of Algae and Their Plastids. D. Bhattacharya, ed. Springer-Verlag, New York.

Eglinton, G., and Hamilton, R. J. 1967. Leaf epicuticular waxes. Science 156:1322-1335.

Ferguson, G. P., Datta, A., Carlson, R. W., and Walker, G. C. 2005. Importance of unusually modified lipid A in *Sinorhizobium* stress resistance and legume symbiosis. Mol. Microbiol. 56:68-80.

Gates, J. E., Fisher, R. W., and Chandler, R. A. 1980. The occurrence of coryneform bacteria in the leaf cavity of *Azolla*. Arch. Microbiol. 127:163-165.

Hacker, J., and Kaper, J. B. 2000. Pathogenicity islands and the evolution of microbes. Annu. Rev. Microbiol. 54:641-679.

Kado, C. I. 2009. Horizontal gene transfer: Sustaining pathogenicity and optimizing host-pathogen interactions. Mol. Plant Pathol. 10:143-150.

Lawrence, J. G. 2002. The dynamics of bacterial genomes. Pages 95-110 in: Horizontal Gene Transfer, 2nd ed. M. Syvanen and C. I. Kado, eds. Academic Press, San Diego, CA.

LeClerc, J. E., Li, B., Payne, W. L., and Cebula, T. A. 1996. High mutation frequencies among *Escherichia coli* and *Salmonella* pathogens. Science 274:1208-1211.

Mahaffee, W. F., Kloepper, J. W., van Vuurde, J. W. L., Wolf, J. M., and van den Brink, M. 1994. Endophytic colonization of *Phaseolus vulgaris* by *Pseudomonas fluorescens* strain 89B-27 and *Enterobacter asburiae* strain JM22. Page 180 in: Improving Plant Productivity with Rhizosphere Bacteria. M. H. Ryder, P. M. Stephens, and G. D. Bowen, eds. CSIRO Press, Adelaide, Australia.

Matic, I., Tenaillon, O., Lecointre, G., Darlu, P., Radman, M., Taddei, F., and Denamur, E. 2002. Horizontal transfer of mismatch repair genes and the variable speed of bacterial evolution. Pages 147-155 in: Horizontal Gene Transfer, 2nd ed. M. Syvanen and C. I. Kado, eds. Academic Press, San Diego, CA.

McFadden, G. I. 1999. Endosymbiosis and evolution of the plant cell. Curr. Opin. Plant Biol. 2:513-519.

McInroy, J. A., and Kloepper, J. W. 1995. Survey of indigenous bacterial endophytes from cotton and sweet corn. Plant Soil 173:1-6.

Melcher, U., and Fletcher, J. 1999. Genetic variation in *Spiroplasma citri*. Eur. J. Plant Pathol. 105:519-533.

Misaghi, I. J., and Donndelinger, C. R. 1990. Endophytic bacteria in symptom-free cotton plants. Phytopathology 80:808-811.

Moreira, D., Le Guyader, H., and Philippe, H. 2000. The origin of red algae and the evolution of chloroplasts. Nature 405:69-72.

Niyogi, K. K., Björkman, O., and Grossman, A. R. 1997. The roles of specific xanthophylls in photoprotection. Proc. Natl. Acad. Sci. U.S.A. 94:14162-14167.

Ochman, H., and Lawrence, J. G. 1996. Phylogenetics and the amelioration of bacterial genomes. Pages 1649-1654 in: *Escherichia coli* and *Salmonella typhimurium*: Cellular and Molecular Biology. F. C. Neidhardt, J. L. Ingraham, B. K. Low, B. Magasanik, M. Schaechter, and E. Umbarger, eds. American Society for Microbiology, Washington, DC.

Pal, C., Papp, B., and Lercher, M. J. 2005. Adaptive evolution of bacterial metabolic networks by horizontal gene transfer. Nat. Genet. 37:1372-1375.

Petroni, G., Spring, S., Schleifer, K.-H., Verni, F., and Rosati, G. 2000. Defensive extrusive ectosymbionts of *Euplotidium* (*Ciliophora*) that contain microtubule-like structures are bacteria related to *Verrucomicrobia*. Proc. Natl. Acad. Sci. U.S.A. 97:1813-1817.

Quadt-Hallmann, A., Kloepper, J. W., Rodriguez-Kábana, R., and Sikora, R. A. 1995. Immunological studies on the detection and localization of plant-associated bacteria. (Abstr.) Phytopathology 85:1136.

Reinhold-Hurek, B., Krause, A., Leyser, B., Miché, L., and Hurek, T. 2007. The rice apoplast as a habitat for endophytic N_2-fixing bacteria. Pages 427-443 in: The Apoplast of Higher Plants: Compartment of Storage, Transport and Reactions. B. Sattelmacher and W. J. Horst, eds. Springer, Dordrecht, the Netherlands.

Rodrigues-Pereira, A. S., Houwen, P. J. W., Deurenberg-Vos, H. W. J., and Pey, E. B. F. 1972. Cytokinins and the bacterial symbiosis of *Ardisia* species. Z. Pflanzenphysiol. 68:170-177.

Singer, C. E., and Ames, B. N. 1970. Sunlight ultraviolet and bacterial DNA base ratios. Science 170:822-826.

Syvanen, M., and Kado, C. I., eds. 2002. Horizontal Gene Transfer, 2nd ed. Academic Press, San Diego, CA.

Tsavkelova, E. A., Cherdyntseva, T. A., Klimova, S. Y., Shestakov, A. I., Botina, S. G., and Netrusov, A. I. 2007. Orchid-associated bacteria produce indole-3-acetic acid, promote seed germination and increase their microbial yield response to exogenous auxin. Arch. Microbiol. 188:655-664.

Weber, E., and Koebnik, R. 2006. Positive selection of the *hrp* pilin HrpE of the plant pathogen *Xanthomonas*. J. Bacteriol. 188:1405-1410.

Classification of Plant-Pathogenic Bacteria

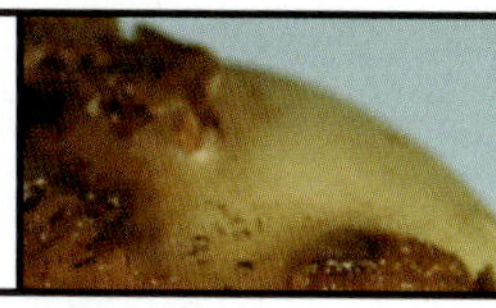

3

A taxon is a group of related organisms that is delineated by phenotypic and genotypic similarities, such as a genus. A species is the taxonomic unit below genus and above subspecies that represents the basic unit in biological classification. Changes in the names of species and other taxa occur when new classifications are proposed and the new taxa are named. Plant-pathogenic bacteria are currently classified according to taxonomic criteria employing qualitative and quantitative physiological, biochemical, and molecular methods, the last of which includes sequence comparisons of ribosomal genes (primarily 16S small subunit rDNA). Underlying this system of classification is the basic presumption that the combined data reflect the whole organism and provide the best estimate of actual phylogenetic relationships. To date, generally accepted phylogenies of bacterial plant pathogens have not been reached using these approaches. New technologies used to evaluate bacterial diversity result in changes in our understanding of bacterial phylogenies that result in novel classification proposals of new taxa or altered taxa. Confusion caused by changes in bacterial nomenclature will continue as long as the classifications remain in flux.

Genotyping (whole-genome sequence comparisons) is the most powerful procedure that will likely stabilize the taxonomy of bacteria. The new generation of high-throughput DNA sequencing technologies is supplanting current array-based technologies (Shendure and Ji, 2008). High-throughput DNA sequencing allows us to follow the evolution of plant-pathogenic traits, such as bacterial host range and hypersensitive response (HR) elicitation, in real time. The new technologies promise to uncover diverse novel genes that are currently inaccessible and will integrate biological information to higher levels than currently possible. Current phylogenetic approaches support a more objective and comprehensive species concept than do phenetic approaches (Vinatzer and Bull, 2009). However, comprehensive modern phenetic approaches (e.g., sequence alignment of nucleic acids and proteins and similarity in enzymatic function) clearly delimit and rank taxa on the basis of overall similarity in weighted characters, reflecting the geneology of the included species and revealing lateral gene transfer events (Brown et al., 2002; Kado, 2009). Furthemore, genotyping techniques are less prone to subjective error and provide a fast, reliable, and stable classification. The advent of rapid and economical genome sequencing will result in the availability of many complete genomes for comparative analysis (Shendure et al., 2004). These data will provide novel insights into the evolutionary ties between various plant-pathogenic and nonpathogenic bacteria and result in the reclassification of many plant pathogens based on evolutionary alignments.

Current taxonomic classifications utilize DNA–DNA hybridization values, guanine-plus-cytosine (GC) content in mole percent (mol %), 16S rDNA sequences, and to a lesser extent, intergenic spacer sequences (16S-23S) and sequence comparisons of a few characterized genes (such as ATPase, the β subunit of F_1F_0 elongation factor Tu, *recA*, and *gyrA*) to generate new taxonomic proposals. Genome and plasmid sequence analyses have revealed complete profiles of gene families and identified pathogenicity islands and specific genes or gene sets that proved more useful and reliable than physiological and biochemical tests. Previous molecular techniques, such as mapping repetitive extragenic palindromic (REP) and enterobacterial repetitive intergenic consensus (ERIC) sequences (de Bruijn, 1992) and random amplified polymorphic DNA (RAPD) sequences (Welsh and McClelland, 1990; Williams et al., 1993), were used as a limited means of distinguishing similarities and differences between strains of a given species. The growing number of whole-genome sequences available provides the needed data to analyze evolutionary trends and define the age of each genus or bacterial family by using molecular clocks (Novichkov et al., 2004) and protein clocks (Adkins and Li, 2002). Moreover, microbial genomics should provide a sound means for the rapid identification of bacterial species as well as the development of a solid phylogenic tree.

For the purposes of this text, the classification and associated nomenclature presented are primarily based on that of Bergey's Manual Trust (www.bergeys.org). Phylogenetic relationships among bacteria are inferred from the 16S rDNA sequences used to develop the new tripartite universal tree of life (Bacteria [formally Eubacteria], Archaea [formally Archaebacteria], and Eucarya) (Woese et al., 1990). Additional classification proposals are discussed for some organisms in which significant restructuring has been proposed. There are legitimately published synonyms for some of the plant-pathogenic taxa that can be easily found in the list presented in the Names of Plant Pathogenic Bacteria, which is updated periodically (Young et al., 2004; www.isppweb.org/names_bacterial_revised.asp), and in lists of all bacteria (www.bacterio.cict.fr/; www.bergeys.org).

There are 24 phyla listed under the domain *Bacteria*. Most phyla (BI to BXI) list no plant-pathogenic bacteria. The majority of plant-pathogenic bacteria are Gram-negative, rod-shaped, and non-spore-forming species in phylum BXII, *Proteobacteria*. The phylum *Proteobacteria* is divided into five classes: alpha, beta, gamma, delta, and epsilon. Plant-pathogenic bacteria are present in genera in the alpha, beta, and gamma classes. The class *Alphaproteobacteria* contains the genera *Sphingomonas* and *Agrobacterium* that list plant-pathogenic species. The genera *Rhizobium*, *Sinorhizobium*, *Allorhizobium*, *Mesorhizobium*, *Bradyrhizobium*, *Azorhizobium*, and *Phyllobacterium* each contain species that interact with plants but are usually nitrogen-fixing endosymbionts. The class *Betaproteobacteria* contains the families *Burkholderiaceae*, *Ralstoniaceae*, and *Comamonadaceae*, containing plant pathogens in the genera *Burkholderia*, *Ralstonia*, and *Acidovorax*, respectively. The class *Gammaproteobacteria* encompasses the families *Xanthomonadaceae*, *Pseudomonadaceae*, and *Enterobacteriaceae*, which contain plant pathogens in the genera *Xanthomonas*, *Xylella*, *Pseudomonas*, *Brenneria*, *Dickeya*, *Erwinia*, *Pantoea*, *Pectobacterium*, *Salmonella*, *Serratia*, and *Tatumella*. Sequence analysis of the 16S rDNA of a novel nonculturable bacterium that causes marginal chlorosis of strawberry has been named '*Candidatus* Phlomobacter fragariae'. This organism is detected in whiteflies but not in leafhoppers or psyllids (Foissac et al., 2000) and is restricted to the phloem of its host plant. Sequence analysis of the 16S rDNA indicates that this bacterium is a member of the gamma-3 subdivision of *Proteobacteria* (Zreik et al., 1998).

Some plant-pathogenic bacteria are obligate parasites that cannot be cultured in vitro. These organisms do not synthesize cell walls and are unable to retain the crystal violet-iodine complex of the Gram stain and therefore always react as

Gram-negative organisms. They are classified in the phylum BXIII, *Firmicutes*. The plant pathogens are found within the class *Mollicutes*, order II, family *Spiroplasmataceae*. Two genera in this family, *Spiroplasma* and Phytoplasma, contain plant pathogens.

Gram-positive bacterial plant pathogens are found in several subclasses within the class *Actinobacteria* in the phylum BXIV (also called *Actinobacteria*). Subclass V contains four families, *Micrococcaceae*, *Microbacteriaceae*, *Nocardiaceae*, and *Streptomycetaceae*, which have plant-pathogenic members in the genera *Arthrobacter*, *Clavibacter*, *Curtobacterium*, *Leifsonia*, *Rathayibacter*, *Rhodococcus*, and *Streptomyces*. Genera containing plant pathogens are listed in Table 3-1.

As indicated above, the current taxonomic classification, which is based primarily on 16S rDNA sequence comparisons, will undoubtedly change as whole-genome sequence data accumulate and become the criteria for speciation. When using polymerase chain reaction (PCR) amplification of 16S rDNA, precautionary measures must be taken to avoid the formation of artifacts that can lead to

Table 3-1. Families and Genera in the Domain *Bacteria* Containing Plant-Pathogenic Species[a]

Phylum	Class	Family	Genus
BXII	*Alphaproteobacteria*	*Rhizobiaceae*	*Agrobacterium*
		Sphingomonadaceae	*Sphingomonas*
		Phyllobacteriaceae	'*Candidatus* Liberibacter'[b]
BXII	*Betaproteobacteria*	*Burkholderiaceae*	*Burkholderia*
		Ralstoniaceae	*Ralstonia*
		Comamonadaceae	*Acidovorax*
BXII	*Gammaproteobacteria*	*Xanthomonadaceae*	*Xanthomonas* *Xylella*
		Pseudomonadaceae	*Pseudomonas* *Rhizobacter* *Xylophilus*
		Enterobacteriaceae	*Brenneria* *Dickeya* *Erwinia* *Pantoea* *Pectobacterium* *Salmonella* *Serratia* *Tatumella* '*Candidatus* Phlomobacter'[c]
BXIII	*Mollicutes*	*Spiroplasmataceae*	*Spiroplasma* '*Candidatus* Phytoplasma'
BXIV	*Actinobacteria*	*Micrococcaceae*	*Arthrobacter*
		Microbacteriaceae	*Clavibacter* *Curtobacterium* *Leifsonia* *Rathayibacter*
		Nocardiaceae	*Rhodococcus*
		Streptomycetaceae	*Streptomyces*

[a] Other genera in which proposals have placed plant-pathogenic bacteria include *Acetobacter, Bacillus, Clostridium, Enterobacter, Gluconobacter, Herbaspirillum, Janthinobacterium, Nocardia, Rhizobium*, and *Samsonia*.

[b] New genus proposed by Jagoueix et al., 1994.

[c] New genus proposed by Zreik et al., 1998.

erroneous conclusions, especially when using universal primers (Osborne et al., 2005). Comparative sequence analyses using PCR amplification of REP, ERIC, and so-called highly conserved (BOX) sequences have helped delineate various genera and species (Rademaker et al., 2005), but such analyses reflect only bits of the chromosomal signature. However, even using multilocus sequence typing (MLST, one of the most advanced techniques being used to estimate phylogeny) to delineate the taxonomy of plant-pathogenic bacteria is not sufficient. Shuffling of taxa and species names will continue until whole-genome sequences of many strains (including the type and pathotypes) of all species and pathovars of interest become available. Because taxonomy is a researchable subject, it will remain in a transitional state, changing as knowledge is accumulated.

PHYLUM BXII: *PROTEOBACTERIA*

Class *Alphaproteobacteria*

Rhizobiaceae

The family *Rhizobiaceae* is composed of eight genera, *Agrobacterium*, *Allorhizobium*, *Carbophilus*, *Chelatobacter*, *Kaistia*, *Rhizobium*, and *Ensifer*/*Sinorhizobium* (*Ensifer* and *Sinorhizobium* are considered synonyms), as members of the class *Alphaproteobacteria*. The genera *Agrobacterium*, *Rhizobium*, and *Sinorhizobium* contain species that can induce hypertrophic and hyperplastic growth of plant cells. *Agrobacterium* species cause crown gall and hairy root on many plant species. They are best known for their close association with plant roots and the soil surrounding the root environment. To avoid direct competition with other microorganisms in the soil or plant surface environments, *Rhizobium*, *Sinorhizobium*, and *Agrobacterium* species have evolved to colonize their respective hosts and settle into a distinctive protected niche that provides a unique competitive advantage over other microbes. These infections lead to abnormal plant growth, culminating in the formation of nodules on the roots, in the case of *Rhizobium* and *Sinorhizobium* species, or in tumors or hairy roots, in the case of *Agrobacterium* species.

This family of bacteria is characterized by Gram-negative and non-spore-forming rods that are obligately aerobic and nutritionally nonfastidious, enabling growth on a wide variety of carbon sources (succinic acid is one of the best utilized for optimal growth). *Agrobacterium* species possess flagella that are arranged circumthecally. Unlike peritrichous flagella, circumthecal flagella are anchored around the first third of the bacilliform bacterial cell (Fig. 3-1). *Agrobacterium* species are able to catabolize opines and also use opines, such as octopine and agrocinopine, for conjugative transfer of the Ti plasmid. Many *Agrobacterium* species that contain octopine Ti plasmids also produce bionic acid (3-ketolactose).

Rhizobium and *Sinorhizobium*

Rhizobium and *Sinorhizobium* species interact with the roots of legumes, producing overgrowths in the form of nodules, within which atmospheric dinitrogen is converted to ammonia. Nodule development begins when these bacteria bind to and cause deformation and curling of the root hairs of the legume. The rhizobia continue to invade the root hair through the establishment of an infection thread

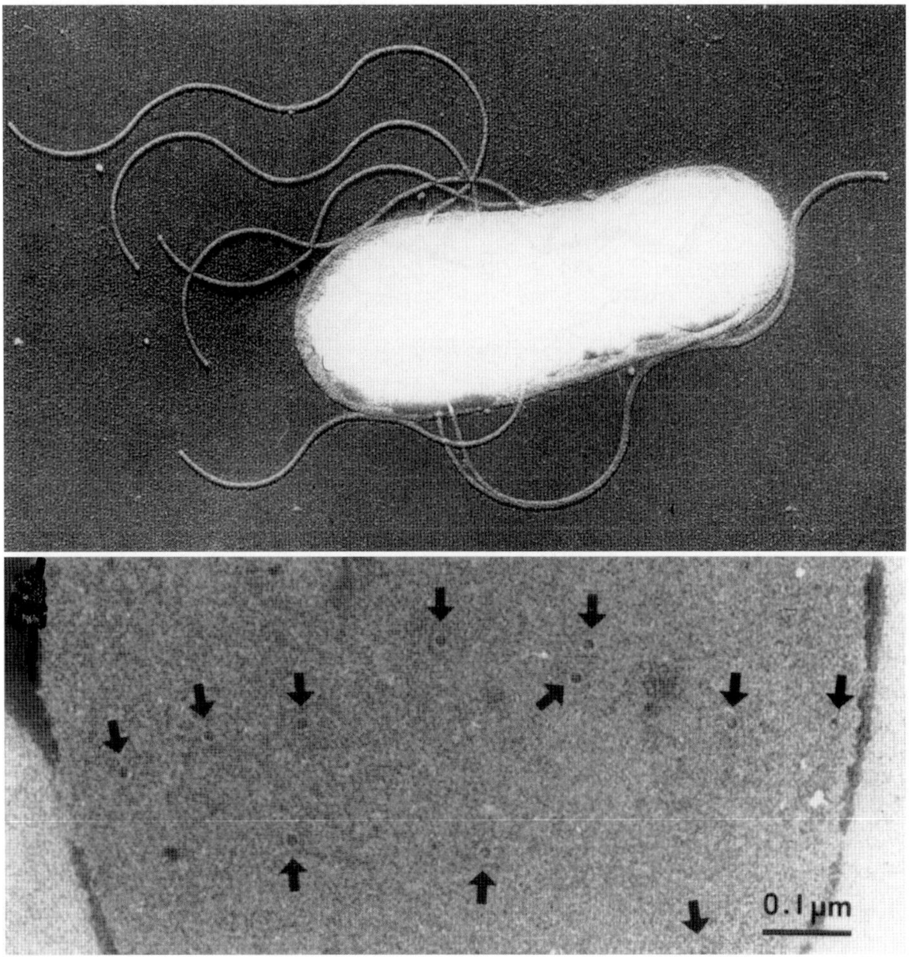

Fig. 3-1. **Top,** Electron micrograph of a rod-shaped *Agrobacterium tumefaciens* cell with circumthecally arranged flagella. **Bottom,** Flagella anchor sites (arrows) arranged around the cylinder of the cell. (Courtesy Shuhei Tanaka; reprinted from Kado, 1992, with kind permission of Springer Science and Business Media)

that becomes surrounded by dividing cortical cells. The complex of bacterial and host cortical cells differentiates into bacteroids that fix atmospheric nitrogen. This symbiotic interaction is triggered by the exchange of chemical signals elaborated by the rhizobial cells and root tissues. Legume roots secrete various flavonoids, some of which regulate the expression of nodulation genes (*nod*, *nol*, and *noe*) in the infecting rhizobial cells. Flavonoids perceived by the *Rhizobium* (or *Sinorhizobium*) cell activate the expression of the *nodD* gene whose product transcriptionally activates the *nod* genes. Specific genes (*nodA*, *nodB*, *nodC*, *nodE*, *nodH*, *nodP*, *nodQ*, and *nodS*) in the rhizobial bacteria encode enzymes that are required for the biosyntheses of nodulation factors (Nod factors), and *nol* and *noe* genes are involved in the secretion of Nod factors that facilitate nodular entry via the infection threads formed in the root hairs. Nod factors of *Rhizobium leguminosarum* and *Sinorhizobium meliloti* are lipochitin oligosaccharides with a chitinlike sugar backbone and an

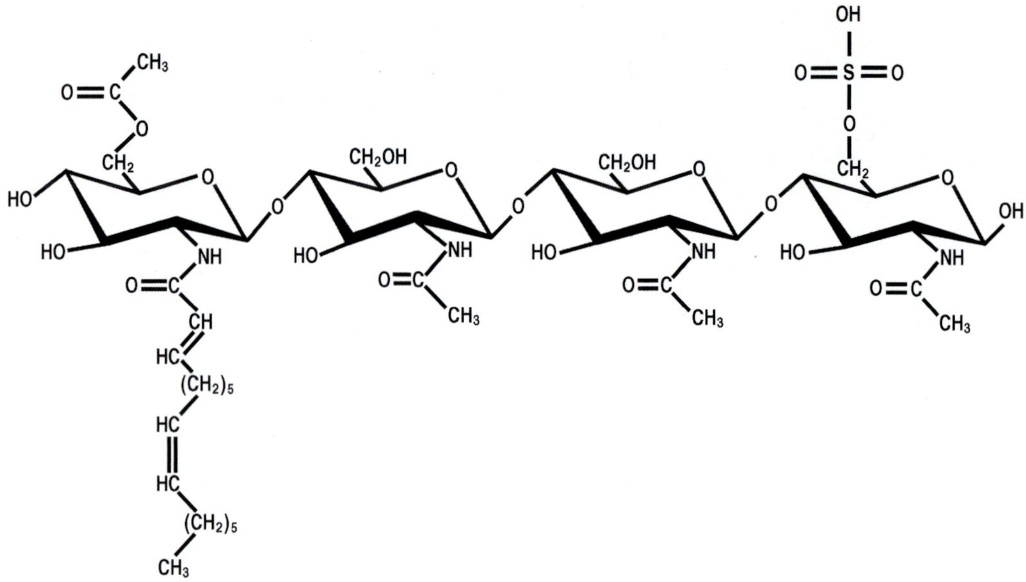

Fig. 3-2. Nod factor composed of four glucosamine moieties with the first glucosamine unit linked to a 16-carbon–acylated chain and an acyl group. A sulfate group is linked to the terminal glucosamine residue. The degree of saturation and the presence of the sulfate group determine the type of Nod factor produced by *Sinorhizobium meliloti.*

Table 3-2. Plant-Associated and Plant-Pathogenic Members of the Genus *Agrobacterium*

Species	GC (mol %)	Pathogenic Phenotype	Location of Virulence Genes and Oncogenes
A. tumefaciens	58	Crown gall	200-kb Ti plasmid
A. rhizogenes	59	Hairy root	220-kb Ri plasmid
A. rubi	58	Cane gall	200-kb Ti plasmid
A. vitis	59	Vine gall	280-kb Ti plasmid
A. radiobacter	58	Nonpathogenic	180- to 220-kb Cryptic plasmid with no oncogenes

N-linked long-chain fatty acid (N-acylated oligomers of *N*-acetyl-D-glucosamine containing various substitutions) (Fig. 3-2). These Nod factors can initiate nodulation at picomolar concentrations.

The *nod* genes conferring root nodule induction and nitrogen fixation are contained on large plasmids known as Sym plasmids, ranging in molecular length from 0.14 to 1.7 megabases (Mb). Since the main chromosome of *Sinorhizobium meliloti* is 3.4 Mb long, the larger Sym plasmids constitute nearly half of the bacterium's genome. Some Sym plasmids, such as pSym5 and pRL1JI, are conjugative, while others, such as pSym1, are nonconjugative (Johnston et al., 1978; Hooykaas et al., 1982).

Agrobacterium

The genus *Agrobacterium* comprises pathogens that cause hyperplastic diseases in many different plants. *Agrobacterium tumefaciens* has a wide host range and has been experimentally demonstrated to cause crown gall tumors in plants representing 93 plant families. *Agrobacterium radiobacter* represents the naturally occurring nonpathogenic member of the genus. Members that cause abnormal growths in the form of non-self-limiting tumors and proliferation of numerous roots are listed in Table 3-2. The classification of the genus *Agrobacterium* and many of its species has recently been reevaluated. Pathogenicity in the genus *Agrobacterium* is mediated by plasmid-borne genes that are interchangeable among species. Young et al. (2001) proposed to amend the genus *Rhizobium* to include all *Agrobacterium* species. Although these names have appeared in the List of Prokaryotic Names with Standing in Nomenclature (LPSN) (www.bacterio.cict.fr/), few researchers use the proposed names in order to avoid confusion (Farrand et al., 2003).

It should be noted that there are other species related to those of the genus *Agrobacterium* occurring in marine environments that are not included on the LPSN yet possess some genetic and phenotypic traits that show distant relatedness to *Agrobacterium* species. It has been proposed that the marine *Agrobacterium* species be reclassified into the genera *Stappia*, *Ruegeria*, and *Ahrensia* (Uchino et al., 1998). Whole-genome sequence-based analyses will be required to accurately reclassify these marine species.

Sphingomonadaceae

Sphingomonas

Sphingomonas species are Gram-negative rods that can be either motile (with a single polar flagellum) or nonmotile, are nonfermentative, and produce yellow colonies. These species can synthesize 3-ketolactose from lactose and produce β-galactosidase. They all contain large quantities of a unique sphingoglycolipid containing a long-chain base, dihydrosphingosin, with 2-hydroxymyristic acid side chains. Ubiquinone-10 and glucoronosyl ceramide are also present.

The genus *Sphingomonas* contains plant pathogens, including *Sphingomonas suberifaciens* (illegitimate basonym *Rhizomonas suberifaciens*), a root colonizer that causes necrotic lesions on roots, resulting in brown hypertrophic overgrowths that appear like cork along the lesions (Table 3-3). The disease appropriately is called "corky root" (Fig. 3-3). *Sphingomonas suberifaciens* was originally classified as a member of the proposed genus *Rhizomonas* (van Bruggen et al., 1990a), but the genus *Rhizomonas* was illegitimate and the pathogen subsequently was reclassified as a member of the genus *Sphingomonas* (Yabuuchi et al., 1999). Other species of the genus *Sphingomonas* and related genera may also cause corky root (van Bruggen and Jochimsen, 1992). *Sphingomonas suberifaciens* is soilborne and

Table 3-3. Species of the Genus *Sphingomonas*

Species	Principal Hosts	Disease or Habitat
S. asaccharolytica	Rose (*Rosa* spp.)	Rhizoplane inhabitant
S. mali	Apple (*Malus* spp.)	Rhizoplane inhabitant
S. melonis	Spanish melon (*Cucumis melo* L. var. *inodorus* H. Jacq.)	Brown spot
S. pruni	Peach (*Prunus persica* (L.) Batsch)	Rhizoplane inhabitant
S. rosa	Rose (*Rosa* spp.)	Hairy root
S. suberifaciens	Asteraceae members	Hypertrophy of root surfaces

appears to be a natural resident on the roots of wild lettuce (*Lactuca serriola* L.), common sowthistle (*Sonchus oleraceus* L.), prickly sowthistle (*Sonchus asper* (L.) Hill) (van Bruggen et al., 1990b), and many other weed species (Barak et al., 2001).

The host range of *Sphingomonas suberifaciens* appears to be limited to the family Asteraceae, subfamily Liguiliflorae, and tribe Cichoriae, but it has been found associated with the roots of nonhosts, such as barley, broadbean, and rye (van Bruggen et al., 1990b).

A pathogen of Spanish melon that causes brown spots on the fruits and not on the leaves has been named *Sphingomonas melonis* (Buonaurio et al., 2002). Based on phenotypic profiling and 16S rDNA sequence comparisons, *Sphingomonas melonis* is phylogenetically related to nonpathogenic *Sphingomonas mali*, *Sphingomonas pruni*, and *Sphingomonas asaccharolytica*, which are rhizoplane inhabitants of *Rosa* species (rose), *Psychotria nairobiensis* Bremek., *Ardisia crispa* (Thunb.) DC., *Prunus persica* (L.) Batsch (peach), and *Malus* × *domestica* Borkh. (apple) (Takeuchi et al., 1995).

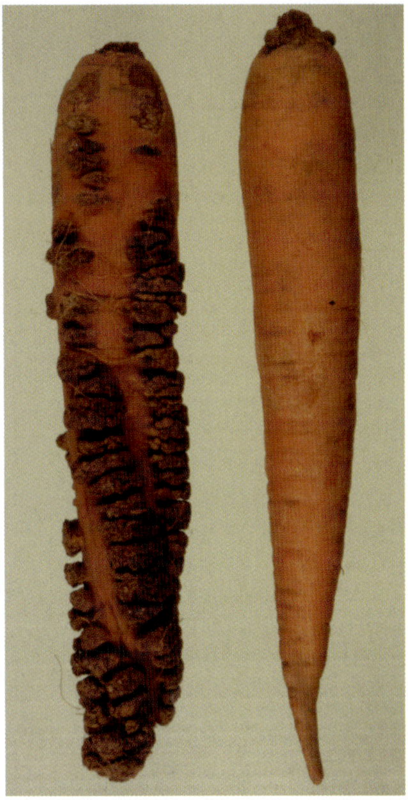

Fig. 3-3. Corky root symptoms on carrot root caused by *Sphingomonas suberifaciens*.

Phyllobacteriaceae

'*Candidatus* Liberibacter'

Members of the genus '*Candidatus* Liberibacter' are Gram-negative, unculturable, and phloem-limited bacteria-like organisms. '*Candidatus* Liberibacter americanus' is associated with citrus huanglongbing (citrus greening) in South America (see Chapter 4 for symptoms). '*Candidatus* Liberibacter asiaticus' and '*Candidatus* Liberibacter africanus' are associated with citrus greening in Asia and in Africa, respectively. These bacteria are vectored by psyllids as is a novel '*Candidatus* Liberibacter' strain that causes the zebra chip disease of potato (Lin et al., 2009). This latter strain infects various solanaceous crops. Infected tubers display swollen lenticels and vascular browning. When processed as potato chips, brown and dark stripes in a zebralike pattern appear, rendering them unmarketable (Fig. 3-4).

Fig. 3-4. Zebra chip disease symptoms of processed potato chips. (Courtesy Hong Lin)

Class *Betaproteobacteria*

Burkholderiaceae

Burkholderia

The genus *Burkholderia*, named after Walter H. Burkholder of Cornell University, contains 20 species: *B. andropogonis*, *B. caribensis*, *B. caryophylli*, *B. cepacia*, *B. cocovenenans*, *B. gladioli*, *B. glathei*, *B. glumae*, *B. graminis*, *B. kururiensis*, *B. mallei*, *B. multivorans*, *B. norimbergensis* (Wittker et al., 1997), *B. phenazinium*, *B. plantarii*, *B. pseudomallei*, *B. pyrrocinia*, *B. thailandensis*, *B. vandii*, and *B. vietnamiensis*. The plant pathogens are members of the species *B. andropogonis*, *B. caryophylli*, *B. cepacia*, *B. gladioli*, *B. glumae*, and *B. plantarii* (Table 3-4).

The taxonomy of *Burkholderia cepacia* and its relatives is complex and the group is often referred to as the *Burkholderia cepacia* complex (Coenye et al., 2001). *Burkholderia cepacia* genomovar I members cause sour skin of onion bulbs, a disease characterized by slimy, light brown decay of the body and neck of the bulb (Burkholder, 1950). Colonies are nonpigmented. *Burkholderia cepacia* produces catalase and five distinct classes of peritrichous pili. Optimal growth temperatures for *Burkholderia cepacia* are slightly higher (30–35°C) than for many other plant-pathogenic species, and it can grow at 41°C. Other species with the ability to grow at mammalian physiological temperatures (37°C) have been identified as *Burkholderia cenocepacia* (formerly *Burkholderia cepacia* genomovar III; Vandamme et al., 2003) in human patients with cystic fibrosis and chronic granulomatous disease. Some strains of *Burkholderia cepacia* are used in the biological control of plant-pathogenic fungi and oomycetes.

Plant-pathogenic *Burkholderia* species accumulate poly-β-hydroxybutyric acid as reserve material and do not produce diffusible fluorescent pigments. Members of this group are either vascular necrogens or macerative types. They all elaborate catalase and can catabolize citrate and inositol but not inulin.

Table 3-4. Plant-Pathogenic Members of the Genus *Burkholderia*

Species	GC (mol %)	Principal Hosts	Diseases and Pathogenic Factor
B. andropogonis	59–61.3	Grasses (*Andropogon* spp.); Sudan grass (*Sorghum vulgare* Pers.); clover (*Trifolium* spp.)	Bacterial leaf spot; produces rhizobitoxine (= L-2-amino-4-(2 amino-3-hydroxypropoxy)-*trans*-but-3-enoic acid) toxin, an inhibitor of homocysteine biosynthesis
B. caryophylli	64.6–65.3	Carnation (*Dianthus caryophyllus* L.)	Bacterial wilt; 1.8-kDa phytotoxic polar compound
B. cepacia	66.6–67.4	Onion (*Allium cepa* L.); garlic (*Allium sativum* L.)	Sour skin, bulb rot; endopolygalacturonase encoded by plasmid-borne *pehA* gene
B. gladioli pv. *agaricicola*	57.8	Mushroom (*Agaricus bitorquis* (Quél.) Sacc.)	Soft rot; produces pectate lyase
B. gladioli pv. *allicola*	68–69	Onion (*Allium cepa* L.)	Slippery skin, bulb rot; produces pectate lyase
B. gladioli pv. *gladioli*	67.9–69.8	Gladioli (*Gladiolus hortulanus* Bailey and *Gladiolus gandavensis* Van Houtte)	Basal leaf rot, corm soft rot, leaf spot; produces pectate lyase
B. glumae	69.6	Rice (*Oryza sativa* L.)	Bacterial grain rot, seedling rot; produces toxoflavin phytotoxin
B. plantarii	64.8	Rice (*Oryza sativa* L.)	Seedling blight; produces tropolone phytotoxin

Ralstoniaceae

Ralstonia

Ralstonia solanacearum (Yabuuchi et al., 1995) (formerly *Pseudomonas solanacearum* and *Burkholderia solanacearum*) is a member of the family *Ralstoniaceae* (Table 3-5). *Ralstonia solanacearum* is the causal organism of a vascular wilt disease that affects a wide range of species (encompassing more than 35 genera in 50 plant families, e.g., Asteraceae, Casuarinaceae, Cruciferae, Geraniaceae, Leguminosae, Musaceae, Solanaceae, Tropaeolaceae, Urticaceae, Verbenaceae, and Zingiberaceae [see Chapter 7 for details]). *Ralstonia solanacearum*, a xylem vessel colonizer, produces cell wall-degrading enzymes, such as a 43-kDa endoglucanase, polygalacturonases of 52 (PglA), 80 (PglB), and 80 kDa (PglC), a 42-kDa pectin methylesterase, and a 28-kDa exocellular protein, all regulated by a complex sensory network, including extracellular polysaccharide production (Schell, 2000). Several virulence factors are expressed by 3-hydroxypalmitic acid methyl ester, a volatile extracellular factor produced by *Ralstonia solanacearum* (Flavier et al., 1997) and by quorum-sensing systems (see Chapter 10 for details on quorum sensing). Ethylene, auxin, and cytokinin are also produced.

Generally, *Ralstonia solanacearum* is prevalent in the tropical, subtropical, and warm temperate regions (Hayward, 1991). A few strains of this species have been found in colder regions. The taxonomy of *Ralstonia solanacearum* is complex; two subspecific classification systems have developed independently. The term "race" as used in relation to *Ralstonia solanacearum* is equivalent to the more broadly used term "pathovar" in other plant-pathogenic bacteria. Three host specificity-based "races" of the former *Pseudomonas solanacearum* were designated (Buddenhagen et al., 1962). Race 1 affects tobacco, tomato, solanaceous weeds, and certain diploid bananas. Race 2 causes wilt in triploid bananas and *Heliconia* species. Race 3 attacks potato, tomato, and geranium and is weakly virulent on other solanaceous hosts. Later, races 4 and 5 were added (Buddenhagen and Kelman, 1964; He et al., 1983). Four biotypes were established based on carbohydrate and hexose alcohol utilization profiles (Hayward, 1964). Recently, a third subspecific classification has been proposed. Genetic data based on internal transcribed spacer (ITS) sequence analysis have been used to restructure the species into four phylotypes, which do not correspond exactly with races and biotypes (Fegan and Prior, 2005; Fegan and Prior, 2006).

Ralstonia solanacearum is a Gram-negative rod with polar flagella. *Ralstonia solanacearum* isolates lack fluorescent, phenazine, or carotenoid pigments. Colonies appearing on tetrazolium agar medium (Kelman, 1954) are irregularly round or elliptical, smooth, and fluidal. Usually the white colony displays a red formazan pigment in its center. On media containing tyrosine, many solanaceous-derived

Table 3-5. Plant-Pathogenic Members of the Genus *Ralstonia*

Species	Hosts	Disease and Pathogenic Factors
R. solanacearum	Numerous in at least 50 plant families	Vascular occlusion causing browning, chlorosis, stunting, wilting; a 3.7-Mb chromosome and a 2.1-Mb megaplasmid contain virulence genes
R. solanacearum complex	Banana (*Musa* spp.)	Blood disease, Indonesia
R. syzygii	Clove (*Syzygium aromaticum* (L.) Merr. & L. M. Perry)	Sumatra disease of clove

isolates produce a brown-black, soluble pigment (melanin), turning the agar medium dark brown to black. On sucrose-peptone–containing medium, poly-β-hydroxybutyric acid is accumulated in most fluidal isolates. The reserve granules (poly-β-hydroxybutyrate aggregates) can be stained with Sudan Black B (Hayward, 1960).

Comamonadaceae

Acidovorax

There are seven plant-pathogenic species of *Acidovorax* (recently proposed to be reclassified into the genus *Mesothermophilus* [Schaad et al., 2008a,b]) (Table 3-6). Ecologically, *Acidovorax* members are mainly soil inhabitants or endosymbionts of earthworms. Members of the genus *Acidovorax* are Gram-negative, rod-shaped cells bearing a single polar flagellum. *Acidovorax avenae* and related strains do not contain 3-hydroxyoctanoic acid, which is found in saprophytic and endosymbiotic *Acidovorax* species (Schaad et al., 2008a). *Acidovorax avenae* subsp. *avenae* causes leaf blight on members of the Poaceae family. *Acidovorax avenae* subsp. *cattleyae* causes leaf spot on orchids, namely, *Cattleya*, *Odontoglossum*, *Phalaenopsis*, *Dendrobium*, and likely other genera. *Acidovorax oryzae* causes leaf spot on rice and is apparently locally systemic. *Acidovorax anthurii* causes necrotic lesions that develop into leaf spot along veins and leaf margins of anthurium (Gardan et al., 2000). *Acidovorax konjaci* causes leaf spot on konjac (*Amorphophallus konjac* K. Koch) *Acidovorax avenae* subsp. *citrulli* causes water-soaked lesions known as bacterial fruit blotch on watermelon, canteloupe, and other cucurbits. This bacterium is seed transmitted. *Acidovorax valerianellae* causes black leaf spot on lamb's lettuce (*Valerianella locusta* (L.) Laterr.). Its GC content is 65.5 mol % and it produces catalase, urease, and cytochrome oxidase (Gardan et al., 2003). Lamb's lettuce is grown commercially in western France and central coastal California.

Table 3-6. Plant-Pathogenic Members of the Genus *Acidovorax*

Species	Hosts	Sites of Infection	Diseases
A. anthurii	Flamingo flower (*Anthurium* spp.); dumbcane (*Dieffenbachia seguine* (Jacq.) Schott)	Invades close to veins and leaf margins; systemic	Black and gray leaf spot; blotch
A. avenae subsp. *avenae*	Poaceae members (e.g., oat, corn, sugarcane, millet); Leguminosae member (soybean)	Invades veinal region; systemic; potentially seedborne	Stripe and stalk rot of corn; leaf blight of oat; stunting of soybean
A. avenae subsp. *cattleyae*	Orchidaceae members (*Cattleya* spp., *Phalaenopsis* spp., *Dendrobium* spp.)	Invades injured areas, particularly during bulb dividing and handling	Leaf spot
A. avenae subsp. *citrulli*	Cucurbitaceae members (e.g., watermelon, cantaloupe, pumpkin)	Colonizes blossoms, infests seeds; suspected to be seedborne	Fruit blotch; leaf spot
A. konjaci	Konjac (*Amorphophallus konjac* K. Koch)	Invades leaves, petioles, roots, and corms; overwinters in corms	Leaf spot
A. oryzae	Rice (*Oryza sativa* L.)	Invasive along veins; seedborne	Stripe; leaf spot
A. valerianellae	Lamb's lettuce (*Valerianella locusta* (L.) Laterr.)	Cotyledons and leaves	Water-soaking; black leaf spot

Class *Gammaproteobacteria*

Xanthomonadaceae

This family contains nine genera, of which two (*Xanthomonas* and *Xylella*) contain plant pathogens.

Xanthomonas

Plant-pathogenic members of the family *Xanthomonadaceae* produce single polar flagella (*Xylella* species are an exception) and yellow pigments, known as xanthomonadins. These yellow pigments have been characterized as brominated aryl-

Table 3-7. *Xanthomonas axonopodis* Pathovars and Associated Diseases

Pathovar	Hosts	Primary Diseases and Virulence Factors
alfalfae	Alfalfa/lucerne (*Medicago sativa* L.); common bean (*Phaseolus vulgaris* L.); pea (*Lathyrus* spp.)	Leaf spot
alli	Welsh onion (*Allium fistulosum* L.)	Leaf spot and streaks, tip dieback, stunting; seedborne
aurantifolii	Mexican lime/dwarf citrus (*Citrus* × *aurantiifolia* (Christm.) Swingle); weaker virulence on other *Citrus* spp.	Leaf spot, water-soaking of fruit followed by spotting; produces virulence effector Pth
axonopodis	Carpet grass (*Axonopus scoparius* (Flügge) Hitchc. and other *Axonopus* spp.)	Leaf spot, gummosis of pasture grass
bauhiniae	Orchid tree (*Bauhinia racemosa* Lam.)	Leaf spot
begoniae	Begonia (*Begonia* spp.) (Rex begonias are highly susceptible)	Leaf spot, infects only interveinal tissues, never veins, petioles, or stems; forms irregular coalesced spots
betlicola	Betel pepper (*Piper betle* L.), long pepper (*Piper longum* L.), and a wild pepper (*Piper hookeri* Miq.)	Leaf spot
biophyti	Life plant (*Biophytum sensitivum* (L.) DC.)	Leaf spot
cajani	Pigeonpea (*Cajanus cajan* (L.) Millsp.)	Leaf spot
cassavae	Cassava (*Manihot esculenta* Crantz)	Leaf spot
cassiae	Cassia (*Cassia tora* L.); pea (*Pisum sativum* L.)	Leaf spot
citri	Citrus (various *Citrus* spp.)	Water-soaking, leaf spot, raised pustules; PthA effector present
clitoriae	Butterfly pea (*Clitoria biflora* Dalz.)	Leaf spot
coracanae	African millet (*Eleusine coracana* (L.) Gaertn.)	Leaf spot
cyamopsidis	Guar (*Cyamopsis tetragonoloba* (L.) Taub.)	Leaf spot
desmodii	Tick trefoil (*Desmodium diffusum* DC.)	Leaf spot
desmodiigangetici	Asumat (*Desmodium gangeticum* (L.) DC.)	Leaf spot
desmodiirotundifolii	Prostrate tick trefoil (*Desmodium rotundifolium* DC.)	Irregular leaf spots with yellow halo
dieffenbachiae	Dieffenbachia/dumbcane (*Dieffenbachia seguine* (Jacq.) Schott); flamingo flower (*Anthurium andraeanum* Linden); philodendron (*Philodendron* spp.)	Leaf spot, leaf blight; invades leaf margins along veins with systemic invasion all the way into stems
erythrinae	Coral tree (*Erythrina variegata* L.)	Leaf spot, localized water-soaking
fascicularis	Hoof (*Corchorus fascicularis* Lam.)	Leaf spot
glycines	Soybean (*Glycine max* (L.) Merr.); red vine (*Brunnichia cirrhosa* Gaertn.); horse gram (*Macrotyloma uniflorum* (Lam.) Verdc.); common bean (*Phaseolus vulgaris* L.)	Leaf spot ranging from yellow to brown, raised brown pustules; its pathogenicity island is composed of nine *hrp*, nine *hrc*, and eight *hpa* genes

(continued)

polyenes (Andrewes et al., 1976). On sugar-containing medium, xanthomonads produce copious exopolysaccharides known as "xanthan gum" and appear as shiny, yellow, mucoid colonies (except *Xanthomonas axonopodis* pv. *manihotis*). They do not produce poly-β-hydroxybutyric acid. Most species produce acid slowly from sucrose. Most species can grow in media containing 0.5% NaCl, but not on DL-tartrate media, and are stimulated in growth by methionine, glutamate, and nicotinic acid. Growth is inhibited by 0.1% triphenyltetrazolium chloride.

The taxonomy of the plant-pathogenic xanthomonads remains in flux. There are numerous species of the genus *Xanthomonas*, whose members are unable to produce nitrate reductase but all produce catalase. Within the species *Xanthomonas axonopodis* (Table 3-7) and *Xanthomonas campestris* (Table 3-8), there are numerous

Table 3-7. *Xanthomonas axonopodis* Pathovars and Associated Diseases *(continued)*

Pathovar	Hosts	Primary Diseases and Virulence Factors
khayae	Mahogany (*Khaya senegalensis* (Desr.) A. Juss.)	Leaf spot
lespedezae	Bush clover/Japanese clover (*Lespedeza* spp.)	Leaf spot
maculifolliigardeniae	Gardenia (*Gardenia* spp.)	Leaf spot
malvacearum	Cotton (*Gossypium hirsutum* L.); karpasa (*Gossypium indicum* Medik.); kapok tree (*Ceiba pentandra* (L.) Gaertn.)	Angular leaf spot, boll rot; seedborne on lint; invasive along vascular tissues
manihotis	Cassava (*Manihot esculenta* Crantz)	Leaf spot, tuber rot; secretes virulence effector pectate lyase
martyniicola	Bombay presidency (*Martynia diandra* Gloxin)	Leaf spot
melhusii	Teak (*Tectona grandis* L.)	Leaf spot
nakataecorchori	Jute (*Corchorus acutangulus* Lam.)	Leaf spot
patelii	Brown hemp/sunn hemp (*Crotalaria juncea* L.)	Leaf spot
pedalii	Bara gokhru (*Pedalium murex* L.)	Leaf spot
phaseoli	Common bean (*Phaseolus vulgaris* L.); hyacinth bean (*Dolichos lablab* L.)	Leaf spot with yellow margins, water-soaking of stems, greasy spots on seed pods
phyllanthi	Bahupatra/bhuin amia (Bengali)/bhu amla (Hindi) (*Phyllanthus niruri* L.)	Leaf spot
physalidicola	Chinese lantern (*Physalis alkekengi* L.)	Leaf spot
poinsettiicola	Poinsettia (*Euphorbia pulcherrima* Willd. ex Klotzsch); cassava (*Manihot esculenta* Crantz)	Leaf spot, scabby leaf spot
punicae	Pomegranate (*Punica granatum* L.)	Leaf spot, red-brown angular lesions
rhynchosiae	Rhynchosia (*Rhynchosia* spp.)	Leaf spot
ricini	Castor bean (*Ricinus communis* L.)	Leaf spot
sesbaniae	Sesban (*Sesbania sesban* (L.) Merr.)	Leaf spot
tamarindi	Tamarind (*Tamarindus indica* L.)	Leaf spot
vasculorum	Sugarcane (*Saccharum officinarum* L.)	Red leaf stripe
vignaeradiatae	Mung bean (*Vigna radiata* (L.) R. Wilczek)	Leaf spot
vignicola	Cowpea (*Vigna unguiculata* (L.) Walp.); Japanese red bean/adzuki (*Vigna angularis* (Willd.) Ohwi & H. Ohashi)	Leaf spot
vitians	Lettuce (*Lactuca sativa* L.)	Leaf spot, water-soaked brown leaf lesions turn black with time; lesions expand along leaf veins; 22–25°C optimum for infection

Table 3-8. *Xanthomonas campestris* Pathovars and Associated Diseases

Pathovar	Hosts	Primary Disease and Virulence Effectors
aberrans	Cauliflower (*Brassica oleracea* L. var. *botrytis* L.)	Leaf spot
alangii	Ankola (*Alangium salviifolium* (L. f.) Wangerin)	Leaf spot
amaranthicola	Green amaranth (*Amaranthus viridis* L.)	Leaf spot
amorphophalli	Elephant yam (*Amorphophallus campanulatus* (Roxb.) Blume ex Decne.)	Leaf spot
armoraciae	Horseradish (*Armoracia rusticana* G. Gaertn., B. Mey. & Scherb.)	Leaf spot, coalesce into angular spot
aracearum	Taro (*Colocasia esculenta* (L.) Schott)	Leaf spot
arecae	Betel nut (*Areca catechu* L.)	Leaf spot, leaf stripe
argemones	Mexican prickly poppy (*Argemone mexicana* L.)	Leaf spot
arracaciae	Arracacha (*Arracacia xanthorrhiza* Bancr.)	Watery leaf spot
asclepiadis	Milkweed (*Asclepias syriaca* L.)	Leaf spot
azadirachtae	Neem tree (*Azadirachta indica* A. Juss.)	Leaf spot
badrii	Cocklebur (*Xanthium strumarium* L.); pea (*Pisum sativum* L.)	Leaf spot; produces pectate lyase
barbareae	Winter cress (*Barbarea vulgaris* R. Br.)	Leaf spot
betae	Beet (*Beta vulgaris* L.)	Leaf spot
bilvae	Indian bael (*Aegle marmelos* (L.) Corrêa)	Leaf spot with water-soaked lesions
blepharidis	*Blepharis boerhaviaefolia* Pers.	Leaf spot
boerhaaviae	Wine flower (*Boerhaavia repens* L.)	Leaf spot
brunneivaginae	Rice (*Oryza sativa* L.)	Sheath necrosis, browning
campestris	Cruciferae members (*Brassica* spp.)	Angular leaf spot; moves systemically in vascular tissue; produces pectate lyase
cannabis	Hemp (*Cannabis sativa* L.)	Leaf spot
cannae	Canna (*Canna indica* L.)	Leaf spot, blight
carissae	Caranda (*Carissa carandas* L.)	Leaf spot
centellae	Asian pennywort (*Centella asiatica* (L.) Urb.)	Leaf spot
citrumelo	Sweet orange (*Citrus* × *sinensis* (L.) Osbeck)	Leaf spot; produces virulence effector PthA
clerodendri	Agnimantha (*Clerodendron phlomoides* L.)	Leaf spot
convolvuli	Small bindweed (*Convolvulus arvensis* L.)	Leaf spot
coriandri	Coriander (*Coriandrum sativum* L.)	Leaf spot
daturae	Hindu datura/downy thorn apple (*Datura metel* L.)	Leaf spot
durantae	Golden dewdrop (*Duranta repens* L.)	Leaf spot
esculenti	Okra (*Hibiscus esculentus* L.)	Leaf spot
eucalypti	Eucalyptus/lemon-scented gum tree (*Eucalyptus citriodora* Hook.)	Leaf spot
euphorbiae	Wild spurge (*Euphorbia acalyphoides* Hochst. ex Boiss.)	Leaf spot
fici	Pipal tree (*Ficus religiosa* L.); weeping fig (*Ficus benjamina* L.)	Leaf spot
guizotiae	Ramtilla (*Guizotia abyssinica* (L. f.) Cass.)	Leaf spot
gummisudans	Corn flag (*Gladiolus communis* L.)	Leaf streak, water-soaked lesions that turn brown and coalesce into streaks
heliotropii	Heliotrope (*Heliotropium* spp.)	Leaf spot
incanae	Stock (*Matthiola incana* (L.) R. Br.)	Pith necrosis
ionidii	*Ionidium heterophyllum* Vent.	Leaf spot
lantanae	Weeping lantana (*Lantana camara* L.)	Leaf spot
laureliae	Pukatea (*Laurelia novae-zelandiae* A. Cunn.)	Leaf spot
lawsoniae	Henna (*Lawsonia inermis* L.); loosestrife (*Ammania multiflora* Roxb.)	Leaf spot
leeana	*Leea edgeworthii* Santapau	Leaf blight
leersiae	Southern cutgrass (*Leersia hexandra* Sw.)	Leaf spot, leaf streak

(continued)

Table 3-8. *Xanthomonas campestris* Pathovars and Associated Diseases *(continued)*

Pathovar	Hosts	Primary Disease and Virulence Effectors
malloti	Akamegashiwa/Japanese mallotus (*Mallotus japonicus* (L. f.) Müll. Arg.)	Dark brown leaf spot
mangiferaeindicae	Cashew (*Anacardium occidentale* L.); mango (*Mangifera indica* L.); hog plum (*Spondias mangifera* Willd.)	Leaf spot
merremiae	Kidney leaf morning glory (*Merremia gangetica* (L.) Cufod.)	Leaf spot
mirabilis	Four o'clock (*Mirabilis jalapa* L.)	Leaf spot, leaf blight
musacearum	Ethiopian banana/red banana (*Ensete ventricosum* (Welw.) Cheesman)	Leaf spot, leaf streak
nigromaculans	Zinnia (*Zinnia elegans* Jacq.)	Leaf spot
obscurae	Obscure morning glory (*Ipomoea obscura* (L.) Ker Gawl.)	Brown leaf spot
olitorii	Jew's mallow/jute/Nalta jute/tossa jute/molokhiya (*Corchorus olitorius* L.)	Leaf spot
papavericola	Prickly poppy (*Argemone* spp.); Shirley poppy (*Papaver rhoeas* L.)	Leaf spot
parthenii	Parthenium (*Parthenium hysterophorus* L.)	Leaf spot, blight
passiflorae	Wing-stemmed passion flower (*Passiflora alata* Curtis); passion fruit (*Passiflora edulis* Sims)	Leaf spot, greasy systemic necrosis
paulliniae	Guarana (*Paullinia cupana* Kunth)	Leaf spot
pennamericanum	Fodder pearl millet (*Pennisetum americanum* (L.) Leeke)	Leaf spot
phormiicola	New Zealand flax/harakeke (*Phormium tenax* J. R. Forst. & G. Forst.)	Leaf spot, blotch, greasy lesions on pod
physalidis	Physalis (*Physalis minima* L.)	Leaf spot
plantaginis	Ribgrass (*Plantago lanceolata* L.)	Leaf spot, leaf stripe
raphani	Garden radish (*Raphanus sativus* L.)	Leaf spot
sesami	Sesame (*Sesamum indicum* L.)	Leaf spot, stem blight
spermacoces	Button weed (*Spermacoce hispida* L.)	Leaf spot
syngonii	Mexican creeping vine/arrowhead (*Syngonium podophyllum* Schott)	Leaf spot
tardicrescens	Wall iris (*Iris tectorum* Maxim.)	Leaf spot, dark green streaks that become brown
thespesiae	Portia tree (*Thespesia populnea* (L.) Sol. ex Correa)	Leaf spot
thirumalacharii	Pilose triumfetta (*Triumfetta pilosa* Roth)	Leaf spot
tribuli	Puncture vine (*Tribulus terrestris* L.)	Leaf spot
trichodesmae	Camel bush (*Trichodesma zeylanicum* (Burm. f.) R. Br.)	Leaf spot
uppalii	Lavender moonvine (*Ipomoea muricata* (L.) Jacq.)	Leaf spot, blight
vernoniae	Iron weed (*Vernonia cinerea* (L.) Less.)	Leaf spot
viegasii	Golden shrimp plant (*Pachystachys lutea* Nees)	Leaf spot
viticola	Grapevine (*Vitis* spp.)	Angular leaf spot
vitiscarnosae	Grapevine (*Vitis carnosa* Wall. ex Laws.)	Leaf spot
vitistrifoliae	Grapevine (*Vitis trifolia* L.)	Leaf spot
vitiswoodrowii	Sand grape (*Vitis woodrowii* Stapf)	Leaf spot
zantedeschiae	Calla (*Zantedeschia aethiopica* (L.) Spreng.)	Leaf spot originating from water-soaked spot
zingibericola	Japanese ginger (*Zingiber mioga* (Thunb.) Roscoe)	Leaf spot
zinniae	Zinnia/youth-and-age (*Zinnia elegans* Jacq.)	Angular leaf spot, flower spot

pathovars, the number of which changes regularly because of the importance of these pathogens and the subsequent prioritization of research on these groups. The taxonomy of many pathovars of *Xanthomonas campestris* is obscure because they are inadequately described or taxonomic research did not include them (Alvarez and Louws, 2001). The classification of plant-pathogenic xanthomonads and adherent nomenclature will continue to change as whole-genome sequencing data are accumulated and analyzed. Species of plant-pathogenic xanthomonads other than *Xanthomonas campestris* and *Xanthomonas axonopodis* are listed in Table 3-9. Some species and their pathovars were placed in Table 3-9 according to current listings (see "Bacterial Nomenclature up-to-date" at Bergey's Manual Trust: www.bergeys.org) *Xanthomonas* species reclassified in 1995 (Vauterin et al., 1995) are subject to change.

Table 3-9. *Xanthomonas* Species and Pathovars Other than *Xanthomonas axonopodis* and *Xanthomonas campestris*

Species	Hosts	Primary Disease and Virulence Effector
X. albilineans	Sugarcane (*Saccharum officinarum* L.)	Leaf scald; produces albicidin that inhibits plastid development by inhibiting DNA replication in proplastids
X. arboricola pv. *celebensis*	Dwarf banana (*Musa acuminata* Colla cv. Dwarf Cavendish)	Leaf spot
X. arboricola pv. *corylina*	Filbert/hazel nut (*Corylus americana* Marshall)	Leaf spot, twig dieback, canker, floret dieback
X. arboricola pv. *juglandis*	English walnut (*Juglans regia* L.)	Leaf spot, twig necrosis, yellow creamy internal nut
X. arboricola pv. *populi*	Poplar (*Populus* spp.)	Black and brown lesions, necrosis
X. arboricola pv. *pruni*	Apricot (*Prunus armeniaca* L.)	Leaf spot, red-brown angular lesions
X. bromi	California brome (*Bromus carinatus* Hook. & Arn.)	Leaf streak
X. cassavae[a]	Cassava (*Manihot* spp.)	Leaf spot
X. citri pv. *aurantifolii*	Citrus (*Citrus* spp.)	South American strains B and C causing leaf spot and canker
X. citri pv. *citri*	Rutaceae cultivated species (primarily *Citrus* spp., *Fortunella* spp., *Poncirus* spp.)	Leaf spots, canker, raised brown pustules; Asian strains A and A* cause canker and raised brown pustules; produces a *pthA* gene product
X. cucurbitae	Cucumber (*Cucumis sativus* L.); pumpkin (*Cucurbita maxima* Duchesne); watermelon (*Citrullus vulgaris* Schrad.)	Leaf spot, angular fruit crack, greasy fruit lesions
X. fragariae	Strawberry (*Fragaria chiloensis* (L.) Mill.)	Leaf spot, spots coalesce into angular areas; produces phytotoxic protease
X. hortorum pv. *carotae*[a]	Carrot (*Daucus carota* L.)	Watery leaf spot; produces pectate lyase
X. hortorum pv. *hederae*	English ivy (*Hedera helix* L.); Australian umbrella tree (*Schefflera actinophylla* (Endl.) Harms)	Leaf spot
X. hortorum pv. *pelargonii*	Meadow cranesbill (*Geranium pratense* L. cv. Splish Splash); Storkbill (*Pelargonium* spp.)	Leaf spot, stem necrosis, stem rot; produces pectate lyase
X. hortorum pv. *taraxaci*	Russian dandelion (*Taraxacum kok-saghyz* L. E. Rodin)	Leaf spot
X. hyacinthi	Garden hyacinth (*Hyacinthus orientalis* L.)	Leaf spot
X. melonis	Watermelon (*Citrullus vulgaris* Schrad.)	Leaf spot, soft rot; produces pectate lyase

(continued)

[a] Listed in current Bergey's Manual Trust Taxonomic Outlines (www.bergeys.org).
[b] Synonym: *X. euvesicatoria* with strain variations (Jones et al., 2004).

That many *Xanthomonas campestris* pathovars show distinct host specificities strongly suggests that they are genetically distinct, at least in their host-specifying genes. Whole-genome sequencing, which is rapidly progressing toward high-speed mega-nucleotide sequencing, will clearly delineate differences and similarities that exist among these pathovars. Genome plasticity and pathogenicity of *Xanthomonas campestris* pv. *vesicatoria* (elevated synonym: *Xanthomonas*

Table 3-9. *Xanthomonas* Species and Pathovars Other than *Xanthomonas axonopodis* and *Xanthomonas campestris* (continued)

Species	Hosts	Primary Disease and Virulence Effector
X. oryzae pv. *oryzae*	Rice (*Oryza sativa* L.)	Leaf blight, kresek (wilt), chlorosis; produces an *avrXa7* gene product via type III secretory pathway regulated by *hrpX* (see Chapter 10)
X. oryzae pv. *oryzicola*	Rice (*Oryza sativa* L.)	Leaf streak, translucent lesions
X. pisi	Pea (*Pisum sativum* L.)	Leaf spot
X. populi	Poplar (*Populus* spp.); poplar hybrids (*Populus deltoides* W. Bartram ex Marshall and other hybrids in the sections *Aigeiros*, *Leuce*, and *Tacamahaca*)	Canker
X. sacchari	Sugarcane (*Saccharum officinarum* L.)	Leaf spot
X. theicola	Tea (*Camellia sinensis* (L.) Kuntze)	Leaf spot
X. translucens pv. *arrhenatheri*	Oatgrass (*Arrhenatherum elatius* (L.) P. Beauv. ex J. Presl & C. Presl)	Leaf spot, leaf stripe
X. translucens pv. *cerealis*	Brome grass (*Bromus inermis* Leyss.)	Leaf spot, leaf streak
X. translucens pv. *graminis*	Brome grass (*Bromus* spp.); cocksfoot (*Dactylis glomerata* L.); ryegrass (*Lolium perenne* L.)	Leaf stripe
X. translucens pv. *phlei*	Timothy grass (*Phleum pratense* L.)	Leaf stripe
X. translucens pv. *phleipratensis*	Timothy grass (*Phleum pratense* L.)	Brown stripes with yellow halos, brown decay, copious gum exudates
X. translucens pv. *poae*	Rough bluegrass (*Poa trivialis* L.)	Leaf stripe
X. translucens pv. *secalis*	Rye (*Secale cereale* L.)	Leaf stripe
X. translucens pv. *translucens*	Barley (*Hordeum vulgare* L.); wheat (*Triticum aestivum* L.)	Leaf stripe
X. translucens pv. *undulosa*	Wheat (*Triticum aestivum* L.)	Leaf stripe
X. vasicola	Banana (*Musa* spp.); enset (*Ensete* spp.)	Wilt of male buds, premature ripening, fruit decay
X. vasicola pv. *holcicola*	Sorghum (*Sorghum bicolor* (L.) Moench subsp. *bicolor* (L.) Moench); Sudan grass (*Sorghum bicolor* (L.) Moench subsp. *drummondii* (Nees ex Steud.) de Wet & Harlan); switch grass (*Panicum* spp.); millet (*Pennisetum glaucum* (L.) R. Br.)	Streaky spot, red, dark brown, black leaf stripes
X. vasicola pv. *vasculorum*	Royal palm (*Roystonea regia* (Kunth) O. F. Cook); sugarcane (*Saccharum officinarum* L.)	Leaf streak, blight
X. vesicatoria[b]	Pepper (*Capsicum* spp.); nightshade family members, including tomato (*Lycopersicon esculentum* Mill. var. *esculentum* Mill.)	Leaf spot, stem necrosis; produces *AvrBs2*, *AvrRxv*, etc. virulence effectors via type III secretory pathway regulated by *hrpX*

Table 3-10. Characteristics of *Xylella fastidiosa* Subspecies

Subspecies[a]	Primary Hosts	Vector, Disease, and Pathogenicity Factor
piercei	Grapevine (*Vitis vinifera* L.); numerous wild plants (more than 100 species)	Harbored in the blue-green sharpshooter (*Graphocephala atropunctata* (Signoret)) and the glassy-winged sharpshooter (*Homalodisca coagulata* (Say)) leafhopper vectors; cause of leafhopper-vectored Pierce's disease of grapevines; produces haemolysin III (XF0175), which belongs to an uncharacterized protein family, and four others (XF0668, XF1011, XF2407, XF2759), which belong to the RTX toxin family that contains tandemly repeated glycine-rich nonapeptide motifs at the C-terminal domain (Simpson et al., 2000); one of these is closely related to a bacteriocin, an RTX toxin also found in the plant bacterium *Rhizobium leguminosarum*
multiplex	Peach (*Prunus persica* (L.) Batsch); plum (*Prunus domestica* L.); almond (*Prunus dulcis* (Mill.) D. A. Webb); elm (*Ulmus* spp.); pigeon grape (*Vitis aestivalis* Michx.); sycamore (*Platanus* spp.)	Vectored by the same leafhoppers as *Xylella fastidiosa* subsp. *piercei*; leaf scorch, veinal chlorosis, wilt, dwarfing
pauca	Citrus (*Citrus* spp.)	Vectored by the same leafhoppers as *Xylella fastidiosa* subsp. *piercei*; veinal chlorosis; small fruit size

[a] Subspecies names as of 2004 (Schaad et al., 2004).

euvesicatoria [Jones et al., 2004]) was revealed through sequencing of its entire genome (Thieme et al., 2005). With the accumulation of novel genomic information on the complex genetic organization conferring host specificity of each pathovar, there is strong motivation to either convert many of the pathovars into distinct species or consolidate them into a few species as revealed by lateral gene transfers that are known to take place between distantly related microorganisms (Syvanen and Kado, 2002).

Xylella

Xylella species have been recently reclassified (Schaad et al., 2004). Continuing genomic sequence analyses of *Xylella fastidiosa* subspecies have revealed strain differences due to perhaps the lateral transfer of genes by temperate phages (Simpson et al., 2000; Monteiro-Vitorello et al., 2005; Doddapaneni et al., 2006). One of these temperate phages has been isolated and propagated on *Xylella fastidiosa*. Genomic analysis of phage Xfas53 revealed interesting degenerate tandem direct repeats that may encode DNA-binding proteins (Summer et al., 2010). Cells are nonpigmented, nonmotile, and nonflagellated. They produce catalase but not cytochrome oxidase. For growth, *Xylella* species require a medium containing glutamine, peptone, and serum albumin. *Xylella fastidiosa* cells are limited to the xylem of many different plants, frequently as asymptomatic infections (Table 3-10; see Chapter 9).

Pseudomonadaceae

There are 15 genera listed as members of the family *Pseudomonadaceae*. Three of these genera (*Pseudomonas*, *Rhizobacter*, and *Xylophilus*) contain plant pathogens. The genus *Rhizobacter* contains one species (*Rhizobacter dauci*) that causes galls on carrot.

Pseudomonas

Members of the family *Pseudomonadaceae* have been found in virtually every habitat in which they have been sought. They have been recovered from the deepest parts of the oceans to the soils of the highest mountains as well as from plants and animals. Their highly cosmopolitan nature is an indication of strong survival attributes in the members of this family (Table 3-11). These attributes include their ability to grow utilizing a wide variety of carbon sources. In addition, the genomic GC content averages about 65 mol %, which enables the DNA of these bacteria to survive the damaging effects of ultraviolet light emitted by the sun better than bacteria with genomes of GC content in the range of 50 mol % (e.g., members of the family *Enterobacteriaceae*).

Pseudomonas species have features in common that distinguish them from other bacteria. They are Gram negative and rod shaped with monotrichous or multitrichous flagella attached polarly. They are strictly aerobic and produce catalase. Most members of the family *Pseudomonadaceae* can utilize more than one type of carbon source and require no external source of vitamins. They can grow on organic substrates but do not ferment carbohydrates. They are nonspore forming. Except for many of the pathogens in the species *Pseudomonas syringae* and epiphytic *Frateuria* species, cytochrome oxidase is produced. Although strains of the genus *Frateuria* are commonly associated with raspberry plants, their association is mainly epiphytic and they are apparently unable to interact in an opportunistic pathogenic manner. Members of the genus *Pseudomonas* produce diffusible pigments in iron-free medium and also produce catalase and nitrate reductase (Azegame et al., 1988). Some species are pathogens of humans, animals, and plants. Three genera within the family *Pseudomonadaceae* include plant-pathogenic species.

Pseudomonas syringae sensu lato is a heterogeneous species including a broad variety of pathogenic forms (pathovars) and epiphytes (Table 3-12). This species

Table 3-11. Distinguishing Features of Genera of the Family *Pseudomonadaceae*

Genus	Key Features	Habitat
Frateria	Yellow or orange colonies; produces acid and hydrogen sulfide	Epiphytic on Japanese raspberry (*Rubus parvifolius* L.); lily (*Lilium auratum* Lindl.)
Pseudomonas[a]	Produces diffusible pigments in iron-free medium; produces catalase and nitrate reductase	Cosmopolitan; some species are pathogens of humans, animals, and plants
Rhizobacter[a]	Yellowish white or white colonies; tends to aggregate and flocculate in broth culture; can utilize starch and dextrin	Causes galls on carrot (*Daucus carota* L.)
Xylophilus[a]	Cells are pigmented and motile with a single polar flagellum; produces catalase but not cytochrome oxidase; grows best on medium containing yeast extract (1%), D-galactose (2%), and calcium carbonate (2%), on which a diffusible brown pigment is produced	Associated with woody parts of the grapevine (*Vitis* spp.), causing necrosis culminating in canker formation
Zoogloea	Cells aggregate into flocculant sediments due to extracellular fibrillar acidic polysaccharide material; cells accumulate poly-β-hydroxybutyric acid on rich medium; can utilize a wide variety of sugars, alcohols, and amino acids; most strains produce cytochrome oxidase, protease, and urease	Organically enriched aquatic environments, such as sewage

[a] These genera contain plant-pathogenic species.

Table 3-12. *Pseudomonas syringae* Pathovars, Their Hosts and Diseases

Pathovar	Principal Hosts	Disease and Virulence Effectors
aceris	Big leaf maple (*Acer macrophyllum* Pursh)	Leaf blight, veinal necrosis
aesculi	Indian horse chestnut (*Aesculus indica* Colebr.)	Leaf blight, veinal necrosis, stem canker
alisalensis	Arugula (*Eruca vesicaria* (L.) Cav. subsp. *sativa* (Mill.) Thell.)	Angular, water-soaked leaf spots that turn brown with age
antirrhini	Snapdragon (*Antirrhinum majus* L.)	Leaf margin blight, seedling blight
apii	Celery (*Apium graveolens* L.)	Blight, leaf spot, water-soaked stem lesions
aptata	Sugar beet and garden beet (*Beta vulgaris* L.); lettuce (*Lactuca sativa* L.); nasturtium (*Tropaeolum majus* L.)	Blight, leaf spot; contains *syrD* gene that encodes lipodepsinonapeptide phytotoxin; produces a peptide siderophore
atrofaciens	Wheat (*Triticum vulgare* Vill.)	Basal glume rot, water-soaked green to translucent lesions; contains *syrD* gene that encodes lipodepsinonapeptide phytotoxin
atropurpurea	Smooth brome grass (*Bromus inermis* Leyss.); rye (*Secale cereale* L.)	Leaf halo blight; causes diffuse chlorosis, cell wall thickening, hypertrophy; tan spots with greasy margins; produces coronatine (dicyclic coronafacic-aminoethylcyclopropyl amino acid) and coronamic acid polyketide phytotoxins
berberidis	Japanese barberry (*Berberis thunbergii* DC.)	Blight, leaf spot
cannabina	Hemp (*Cannabis sativa* L.); common bean (*Phaseolus vulgaris* L.); broadbean (*Vicia faba* L.)	Bacteriosis
ciccaronei	Carob tree (*Ceratonia siliqua* L.)	Leaf spot, localized necrosis
coriandricola	Coriander (*Coriandrum sativum* L.)	Seed decay, umbel blight
daphniphylli	Himeyuzuriha (*Daphniphyllum teijsmanni* Zoll.)	Irregular galls on trunks and twigs (Ogimi et al., 1990)
delphinii	New Zealand mahogany (*Dysoxylum spectabile* (G. Forst.) Hook. f.)	Leaf black spot
dendropanacis	Kakuremino (*Dendropanax trifidus* (Thunb.) Makino)	Induces bacterial galls
dysoxyli	New Zealand mahogany (*Dysoxylum spectabile* (G. Forst.) Hook. f.)	Greasy leaf spot
eriobotryae	Loquat (*Eriobotrya japonica* (Thunb.) Lindl.)	Black rot of loquat fruit, leaf spot
garcae	Coffee (*Coffea arabica* L.)	Greasy leaf spot; produces tabtoxin (2-serine tabtoxin or tabtoxinine-β-lactam)
glycinea	Soybean (*Glycine max* (L.) Merr.)	Blight, brown angular leaf spots with yellow or white margin; produces coronatine (dicyclic coronafacic-aminoethylcyclopropyl amino acid) and coronamic acid polyketide phytotoxins
helianthi	Sunflower (*Helianthus annuus* L.)	Leaf spot, stem necrosis
hibisci	Hibiscus (*Hibiscus mutabilis* L.)	Irregular leaf spot, stem necrosis
japonica	Rye (*Secale cereale* L.); barley (*Hordeum vulgare* L.); wheat (*Triticum vulgare* Vill.)	Black node, necrosis
lachrymans	Cucumber (*Cucumis sativus* L.)	Angular leaf spot, soft rot, premature fruit abscission; infestation of seed
lapsa	Corn (*Zea mays* L.); sugarcane (*Saccharum officinarum* L.)	Stalk rot
maculicola	Cruciferae members (including *Arabidopsis* spp.)	Causes diffuse chlorosis, cell wall thickening, hypertrophy; produces coronatine (dicyclic coronafacic-aminoethylcyclopropyl amino acid) and coronamic acid polyketide phytotoxins
mellea	Tobacco (*Nicotiana tabacum* L.)	Leaf spot
mori	Mulberry (*Morus alba* L.)	Veinal necrosis, blight

(continued)

Table 3-12. *Pseudomonas syringae* Pathovars, Their Hosts and Diseases *(continued)*

Pathovar	Principal Hosts	Disease and Virulence Effectors
morsprunorum	Cherry (*Prunus avium* L.); plum (*Prunus domestica* L.); various *Prunus* spp.	Causes diffuse chlorosis, cell wall thickening, hypertrophy; produces coronatine (dicyclic coronafacic-aminoethylcyclopropyl amino acid) and coronamic acid polyketides phytotoxin
myricae	Yamamomo/Chinese bayberry (*Myrica rubra* Siebold & Zucc.)	Bacterial gall
oryzae	Rice (*Oryza sativa* L.)	Leaf spot
panici	Broomcorn millet (*Panicum miliaceum* L.)	Leaf stripe
papulans	European crabapple (*Malus sylvestris* (L.) Mill.)	Midvein necrosis, blister spot
passiflorae	Passion fruit (*Passiflora edulis* Sims)	Leaf spot with greasy margins, coalescing into brown-black greasy spots
persicae	Peach (*Prunus persica* (L.) Batsch)	Leaf necrosis, fruit necrosis
phaseolicola	Common bean (*Phaseolus vulgaris* L.)	Halo blight, brown angular leaf spots with chlorotic halos; seeds infected; produces phaseolotoxin (N^{δ}-[*N'*-sulfo-diaminophosphinyl]-ornithyl-alanyl-homoarginine)
philadelphi	Mock orange (*Philadelphus coronarius* L.)	Leaf spot
photiniae	Japanese photinia (*Photinia glabra* (Thunb.) Maxim.)	Leaf spot
pisi	Pea (*Pisum sativum* L.)	Blight; brown angular leaf spots; pod necrotic, circular water-soaked spots
porri	Leek (*Allium porrum* L.)	Blight, watery yellow lesions, leaf curling, water-soaking of flower stems, leaf tip browning
primulae	Primrose (*Primula polyantha* Mill.)	Leaf necrosis, leaf spot
ribicola	Golden currant (*Ribes aureum* Pursh)	Leaf spot, defoliation
sesami	Sesame (*Sesamum indicum* L.)	Leaf spot
striafaciens	Oat (*Avena sativa* L.); barley (*Hordeum vulgare* L.)	Leaf and stem stripe
syringae	Lilac (*Syringa vulgaris* L.); *Prunus* spp. (apricot, peach, cherry); *Citrus* spp. (lemon, orange); pear (*Pyrus* spp.); Gramineae members	Leaf spots, stem lesions, streaking, cankers, gummosis; produces syringomycin (a cyclic lipodepsinonapeptide), syringotoxin, syringostatin, and pseudomycin A; citrus isolates produce syringotoxin B (Fukuchi et al., 1990); survives epiphytically on leaf and stem surfaces
tabaci	Tobacco (*Nicotiana tabacum* L.)	Yellow-green halo leaf spots with brown island, lesions coalesce into irregular patterns; produces tabtoxin (= tabtoxinine-β-lactam)
tagetis	Marigold (*Tagetes erecta* L.)	Angular leaf spot, tip chlorosis; produces tagetitoxin (= a substituted 9-oxa-3-thiabicyclononane)
theae	Tea (*Camellia sinensis* (L.) Kuntze)	Shoot blight, leaf stripe, dieback
tomato	Tomato (*Lycopersicon esculentum* Mill. var. *esculentum* Mill.)	Leaf and fruit speck disease, small dark brown to black spots with narrow yellow halo; produces coronatine (dicyclic coronafacic-aminoethylcyclopropyl amino acid) and coronamic acid polyketide phytotoxins
tremae	Gunpowder tree (*Trema orientalis* (L.) Blume)	Leaf spot
ulmi	Elm (*Ulmus* spp.)	Leaf spot
viburni	European cranberrybush (*Viburnum opulus* L.); Sargent viburnum (*Viburnum sargentii* Koehne); doublefile viburnum (*Viburnum plicatum* Thunb. var. *tomentosum* (Thunb.) Miq.)	Water-soaked lesions, foliage deformation, shoot necrosis
zizaniae	Wild rice (*Zizania palustri* L. and *Zizania aquatica* L.)	Leaf streak

is likely to be divided into several species. To date, nine separate clades have been identified in analyses that included DNA-DNA homology (Gardan et al., 1999). Many of the pathovars of *Pseudomonas syringae* are fluorescent or produce plant toxins, including syringomycin, coronatine, tabtoxin, and tagetitoxin, or both (DiGiorgio et al., 1997; Bare et al., 1999; Bender et al., 1999; Dalla Serra et al., 1999). Besides toxins that facilitate invasion of the host, the inherent type III secretion systems are used by *Pseudomonas syringae* pathovars to suppress or circumvent host defense responses (see Chapter 10 for details).

Within the family *Pseudomonadaceae*, there are five genera as describe in Table 3-11. Three of these genera (in bold type) contain plant-pathogenic species.

Table 3-13. Plant-Pathogenic *Pseudomonas* Species Other than *Pseudomonas syringae*

Species	Primary Hosts	Effects on Host and Other Traits
P. agarici	Table mushroom (*Agaricus bisporus* (J. E. Lange) Singer)	Drippy gill disease, water-soaked dark brown on gills
P. amygdali	Almond (*Prunus amygdalus* Batsch)	Hyperplastic canker, shoot swelling, slow progressive decline of tree
P. asplenii	Bird's nest fern (*Asplenium nidus* L.)	Water-soaked, translucent, coalescing leaf spots; frond necrosis; low-molecular-weight, heat-labile phytotoxin
P. avellanae	Hazel nut tree (*Corylus avellana* L.)	Systemic canker, decline
P. cannabina	Hemp (*Cannabis sativa* L.)	Leaf blight
P. caricapapayae	Papaya (*Carica papaya* L.)	Angular leaf spot, water-soaked leaf spot
P. cichorii	Lettuce (*Lactuca sativa* L.); cabbage (*Brassica oleracea* L. var. *capitata* L.); bok choy/pak choi (*Brassica rapa* L. subsp. *chinensis* (L.) Hanelt)	Irregular brown-black coalescing leaf lesions leading to total rot; produces a 2-kDa peptide "chichorin" phytotoxin
P. corrugata	Tomato (*Lycopersicon esculentum* Mill. var. *esculentum* Mill.); pepper (*Capsicum frutescens* L.)	Leaf tip chlorosis, stem necrosis; produces two hypersensitive response (HR) peptide elicitors
P. ficuserectae	Japanese fig (*Ficus erecta* Thunb.)	Leaf spot
P. fluorescens	Celery (*Apium graveolens* L.); broccoli and cauliflower (*Brassica oleracea* L. var. *botrytis* L.)	Water-soaked lesions, soft rot; produces biosurfactant
P. fuscovaginae	Rice (*Oryza sativa* L.); other Gramineae members	Sheath rot, leaf sheath necrosis, inhibition of panicle emergence; produces fuscopeptins (amphiphilic lipodepsinonapeptides)
P. marginalis (syn. *P. fluorescens* group)	Wide host range; rhubarb (*Rheum* spp.); lettuce (*Lactuca sativa* L.)	Soft rot, leaf spot, water-soaking
P. rubrilineans	Sugarcane (*Saccharum officinarum* L.)	Red stripe, top rot
P. rubrisubalbicans	Sugarcane (*Saccharum officinarum* L.); sorghum (*Sorghum bicolor* (L.) Moench)	Mottled stripe, red stripes with white margins, red crusty deposit
P. savastanoi pv. *fraxini*[a]	Ash (*Fraxinus excelsior* L.)	Ash gall; produces indoleacetic acid and cytokinins
P. savastanoi pv. *nerii*[a]	Oleander (*Nerium oleander* L.)	Gall disease; produces indoleacetic acid and cytokinins
P. savastanoi pv. *savastanoi*[a]	Olive (*Olea europaea* L.); oleander (*Nerium oleander* L.); privet (*Ligustrum* sp.)	Olive knot; produces indoleacetic acid and cytokinins
P. tolaasii	Mushroom (*Agaricus bisporus* (J. E. Lange) Singer and *Agaricus bitorquis* (Quél.) Sacc.)	Brown blotch; produces tolaasin (lipodepsinonapeptide)
P. tremae[b]	Gunpowder tree (*Trema orientalis* (L.) Blume)	Gall disease
P. viridiflava	Bean (*Phaseolus* spp.)	Leaf watery rot, stem necrosis; opportunistic weak pathogen

[a] *Pseudomonas savastanoi* pathovars after Gardan et al., 1992.
[b] *Pseudomonas tremae* described in Gardan et al., 1999.

The diverse phenotype and habitat of pseudomonads are depicted in this table. *Pseudomonas syringae* sensu lato is a heterogenous species that includes pathovars and epiphytes (Table 3-12).

Individual plant-pathogenic species of the genus *Pseudomonas* other than *Pseudomonas syringae* pathovars are listed in Table 3-13.

Rhizobacter

This genus lists only one plant pathogen, *Rhizobacter dauci*. *Rhizobacter dauci* causes galls on carrot roots (Goto and Kuwata, 1988). Cells are motile with a polar flagellum but can be nonmotile, suggesting regulation of motility. This species can grow aerobically on complex carbohydrates, such as starch and dextrin. The GC content ranges between 67 and 71 mol %. Colonies are yellowish white or white. Cells tend to aggregate and flocculate in broth culture. This pathogen can utilize starch and dextrin.

Xylophilus

Xylophilus ampelinus is the type species of this genus. This species is found associated with woody parts of grapevine. The species spreads through the xylem vessels and causes severe necrosis and canker of grapevines (*Vitis vinifera* cultivars). Cells are nonspore formers, pigmented, and motile with a single polar flagellum. The colonies are slow growing, requiring at least 10 days at 25°C. On nutrient agar, colonies range from 0.1 to 0.6 mm in diameter and are yellow, but the pigment is not a xanthomonadin (the type usually found in *Xanthomonas* species). A brown soluble pigment is produced on agar medium containing yeast extract (1%), galactose (2%), and calcium carbonate (2%). They produce catalase but not cytochrome oxidase.

Enterobacteriaceae

The family *Enterobacteriaceae* comprises a large number of related bacteria grouped into 41 genera, eight of which contain plant pathogens (in bold, Table 3-14). Distinguishing features of this family are as follows: Gram-negative, facultatively anaerobic rods with peritrichous flagellation that reduce nitrate to nitrite and produce catalase but not cytochrome oxidase. These bacteria reside in or on various habitats, ranging from soil and water to plants, cold- and warm-blooded animals, and humans. They can be pathogens of marine and fresh-water animals and plants and terrestrial animals, humans, and plants. Genera in bold in Table 3-14 include plant pathogens.

Of the 41 genera, the genera *Brenneria*, *Dickeya*, *Erwinia*, *Pantoea*, *Pectobacterium*, *Salmonella*, *Serratia*, and *Tatumella* contain plant-pathogenic species (Table 3-15). Members of these genera are best characterized by their motile rods that produce acid from glucose and sucrose with little if any gas production in stab culture (e.g., in glucose minimal agar medium [GM] agar, which contains per liter 10 g of glucose, 2 g of peptone, 1 g of yeast extract, 5 g of sodium chloride, 2 g of dipotassium phosphate, 3 ml of 1% bromthymol blue, and 4 g of agar). They do not release cytochrome oxidase and most of them produce nitrate reductase. Most species do not produce arginine-, lysine-, and ornithine-decarboxylases and rarely do

Table 3-14. Habitats and Traits of Genera of the Family *Enterobacteriaceae*

Genus	Habitats	Traits
Alterococcus	Intertidal marine hot spring	Moderate thermophile; produces agarase that dissolves agar
Arsenophonus	Pea aphid (*Acyrthosiphon pisum* (Harris))	Endosymbiont
Brenneria[a]	Plants	Necrogenic semisystemic pathogens of walnut, willow, alder
Buchnera	Pea aphid (*Acyrthosiphon pisum* (Harris))	Endosymbiont
Budvicia	Sulfur-rich water	Aquaphilic; produces hydrogen sulfide
Buttiauxella	Water, soil, snails	Saprophyte
Calymmatobacterium	Genitals	Sexually transmitted pathogen
Cedecea	Skin, respiratory tract, wounds	Transient opportunistic pathogen
Citrobacter	Human intestinal tract	Saprophyte
Dickeya[a]	Plants	Pathogen of plants
Edwardsiella	Reptiles, amphibian intestinal tract	Pathogen of salmonid fish; resident of amphibians and reptiles
Enterobacter	Soil, water, dairy products	Mainly saprophytes; some strains are opportunistic pathogens of urinary tract and nosocomial infections; bacteremia
Erwinia[a]	Plants, phyllosphere, rhizosphere	Pathogen of plants
Escherichia	Mammal and human intestinal tracts	Saprophyte of lower intestine; some strains toxigenic
Ewingella	Respiratory tract, blood, wounds	Weakly opportunistic pathogen
Hafnia	Soil, water, sewage, dairy products	Commensalistic saprophyte
Klebsiella	Soil, water, respiratory tract	Opportunistic pathogen; some fix nitrogen
Kluyvera	Soil, sewage, intestinal tract	Commensalistic saprophyte; some pathogenic
Leclercia	Blood, sputum, urine, plants	Commensalistic saprophyte
Leminorella	Urine, stool	Commensalistic saprophyte
Moellerella	Stool, water	Commensalistic saprophyte
Morganella	Stool, respiratory and urinary tracts, wounds	Opportunistic pathogen of humans and mammals
Obesumbacterium	Brewery contaminant	Saprophyte
Pantoea[a]	Plants	Pathogen of plants; some epiphytic
Pectobacterium[a]	Plants	Pathogen of plants; causes soft rots
Photorhabdus	Endomopathogenic nematodes	Endosymbiont; bioluminescent
Plesiomonas	Water, fresh water fish, shell fish	Opportunistic intestinal pathogen of humans
Pragia	Water, stool	Saprophyte
Proteus	Soil, sewage, urinary tract	Opportunistic pathogen of humans and mammals
Providencia	Wounds, burns, urinary tract, stool	Opportunistic pathogen of humans; commensalistic on penguins
Rahnella	Fresh water	Saprophyte
Saccharobacter	Sugarcane	Produces ethanol from sucrose
Salmonella[a]	Intestinal tract, some plants	Pathogenic on humans and mammals; survives in the vascular system of tomato; causes water-soaked lesions
Serratia[a]	Soil, water, respiratory tract, some plants	Opportunistic pathogen of humans; one strain of the species *S. marcescens* causes wilt of Cucurbitaceae members
Shigella	Lower intestine	Pathogen of humans; invasive
Sodalis	Tsetse fly (*Glossina morsitans morsitans* Westwood)	Secondary endosymbiont
Tatumella[a]	Respiratory tract, some plants	Commensalistic saprophyte; pathogen of pineapple
Trabulsiella	Water, lower intestine	Opportunistic pathogen of humans

(continued)

[a] These genera contain plant-pathogenic species.

Table 3-14. Habitats and Traits of Genera of the Family *Enterobacteriaceae* (continued)

Genus	Habitats	Traits
Wigglesworthia	Mycetocyte of tsetse fly (*Glossina morsitans morsitans* Westwood)	Primary endosymbionts
Xenorhabdus	Nematode intestinal tract	Entomotoxigenic
Yersinia	Rodents, fleas	Pathogen of humans; cause of black and bubonic plagues
Yokenella	Urine, sputum, insect intestinal tract	Saprophyte

Table 3-15. Taxa of Plant-Pathogenic Species of the Family *Enterobacteriaceae*

Genus and Species	Disease Common Name	Hosts	Disease and Virulence Effectors
Brenneria alni	Bacterial canker	Alder (*Alnus* spp.)	Necrosis
Brenneria nigrifluens	Shallow canker	Juglandaceae members; English walnut (*Juglans regia* L.)	Necrosis, grease spots on bark
Brenneria quercina	Drippy nut	Coast live oak (*Quercus agrifolia* Née); interior live oak (*Quercus wislizeni* DC.) in California; holly oak (*Quercus ilex* L.) and Pyrenean oak (*Quercus pyrenaica* Willd.) in Spain	Infections of insect punctures results in bacterial ooze and browning, necrotic cankers of inner bark tissues
Brenneria rubrifaciens	Deep bark canker	Juglandaceae members; English walnut (*Juglans regia* L.)	Necrosis, reddish brown ooze from cracks in bark
Brenneria salicis	Bacterial wilt	Cricket bat willow (*Salix alba* L. var. *caerulea* (Sm.) Sm.)	Foliage reddening and yellowing followed by wilt, internal bark stained dark brown
Dickeya chrysanthemi pv. *chrysanthemi*	Soft rot and necrosis (semitropical regions and greenhouses); sudden decline of date palm	Wide host range; date palm (*Phoenix dactylifera* L.)	Soft rot, flagging, stem necrosis
Dickeya chrysanthemi pv. *parthenii*	Soft rot	Guayule (*Parthenium argentatum* A. Gray)	Stem water-soaking, soft rot
Dickeya dadantii	Soft rot	Banana (*Musa acuminata* Colla); euphorbia (*Euphorbia* spp.); corn (*Zea mays* L.); pineapple (*Ananas comosus* (L.) Merr.); pelargonium (*Pelargonium* spp.); philodendron (*Philodendron* spp.); potato (*Solanum tuberosum* L.); African violet (*Saintpaulia* spp.)	Soft rot of many crop plants; contains the insecticidal *cyt*-like genes; pathogen of the pea aphid (*Acyrthosiphon pisum* (Harris)) (Grenier et al., 2006)
Dickeya dianthicola	Soft rot	Artichoke (*Cynara scolymus* L.); chicory (*Cichorium intybus* L.); dahlia (*Dahlia* spp.); carnation (*Dianthus caryophyllus* L.); tomato (*Lycopersicon esculentum* Mill. var. *esculentum* Mill.); iris (*Iris* spp.); kalanchoe (*Kalanchoë* spp.)	Stem rot, basal rot

(continued)

[a] From Brady et al., 2009.
[b] From Hardt and Galan, 1997; Cooley et al., 2003.
[c] Renamed and reclassified from *Pantoea citrea* to *Tatumella citrea* (Brady et al., 2010).

Table 3-15. Taxa of Plant-Pathogenic Species of the Family *Enterobacteriaceae (continued)*

Genus and Species	Disease Common Name	Hosts	Disease and Virulence Effectors
Dickeya dieffenbachiae	Leaf rot, stem rot	Dieffenbachia (*Dieffenbachia* spp.)	Soft water-soaked sunken lesions on stems, systemic; plants are stunted; leaves show yellow margins
Dickeya paradisiaca	Soft rot	Plantain (*Musa paradisiaca* L.)	Stalk rot
Dickeya zeae	Soft rot	Corn (*Zea mays* L.); rice (*Oryza sativa* L.); pineapple (*Ananas comosus* (L.) Merr.); banana (*Musa acuminata* Colla); tobacco (*Nicotiana tabacum* L.)	Soft rot; inhibits rice germination; virulence genes activated by quorum sensing (Hussain et al., 2008)
Enterobacter cancerogenus	Bacterial canker	Poplar (*Populus* spp.)	Canker
Enterobacter cloacae subsp. *dissolvens*	Bacterial stalk rot, caulescent rot	Corn (*Zea mays* L.)	Water-soaked lengthy lesions, soft rot
Erwinia amylovora	Fireblight	Rosaceae members	Necrosis, contains *hrpN* and *hrp* genes (see Chapter 5 for details)
Erwinia cacticida	Soft rot, fruit rot	Saguaro (*Carnegiea gigantea* (Engelm.) Britton & Rose); organ pipe cactus (*Stenocereus thurberi* (Engelm.) Buxb.); senita (*Pachycereus schottii* (Engelm.) D. R. Hunt); pricklypear (*Opuntia* spp.)	Soft rot of leaves and fruit
Erwinia mallotivora	Bacterial spot	Akamegashiwa/Japanese mallotus (*Mallotus japonicus* (L. f.) Müll. Arg.)	Leaf spot
Erwinia papayae	Bacterial canker	Papaya (*Carica papaya* L.)	Firm water-soaked lesions, sometimes greasy, on stems; angular leaf spot
Erwinia persicina	Leaf spot	Common bean (*Phaseolus vulgaris* L.)	Coalescing leaf spots
Erwinia psidii	Soft rot	Guava (*Psidium guajava* L.)	Water-soaked lesions followed by soft rot of foliage and infected fruit; virulence activated by quorum sensing (see Chapter 10 for details on quorum sensing)
Erwinia rhapontici	Soft rot	Rhubarb (*Rheum rhaponticum* L.); hyacinth (*Hyacinthus* spp.)	Stem soft rot; yellow-browning of hyacinth bulbs; produces water-soluble red pigment proferrorosamine A; converts sucrose into isomaltulose
Erwinia tracheiphila	Vascular wilt	Cucurbitaceae members; corn (*Zea mays* L.); goldenrod (*Solidago nemoralis* Aiton and *Solidago altissima* L.); Johnsongrass (*Sorghum halepense* (L.) Pers.)	Invades xylem vessels, causes drooping of foliage, general wilt, and collapse of vines; pathogen is transmitted and spread by the striped cucumber beetle (*Acalymma vittata* (Fabricius)) and spotted cucumber beetle (*Diabrotica undecimpunctata* Mannerheim)

(continued)

Table 3-15. Taxa of Plant-Pathogenic Species of the Family *Enterobacteriaceae* (continued)

Genus and Species	Disease Common Name	Hosts	Disease and Virulence Effectors
Pantoea agglomerans pv. *gypsophilae*	Galls	Gypsophila (*Gypsophila* spp.)	Galls produced on exposed cuts following harvest (see Chapter 8 for details)
Pantoea agglomerans pv. *millettiae*	Galls	Japanese wisteria (*Wisteria floribunda* (Willd.) DC.); Chinese wisteria (*Wisteria sinensis* (Sims) DC.)	Galls produced on stems; contains *crt* genes of the carotenoid pathway
Pantoea ananas	Soft rot	Rice (*Oryza sativa* L.); banana (*Musa acuminata* Colla); pineapple (*Ananas comosus* (L.) Merr.)	Soft rot; epiphyte, some strains cause soft rot of pineapple fruit; some grow on *Puccinia graminis* Pers.:Pers. uredia
Pantoea ananatis[a]	Internal fruit rot; bacterial blight and dieback; brown stalk rot	Pineapple (*Ananas comosus* (L.) Merr.); gum (*Eucalyptus* spp.); corn (*Zea mays* L.); wheat (*Triticum* spp.); onion (*Allium cepa* L.)	Fruit blotch, dark browning of internal tissue, dieback of *Eucalyptus* spp., internal stalk rot, slimy head of wheat, soft center rot of onions
Pantoea ananatis pv. *ananatis*	Soft rot	Pineapple (*Ananas comosus* (L.) Merr.); gum (*Eucalyptus* spp.)	Soft rot, internal rot of pineapple fruit, blight of twigs and leaves, stunting of *Eucalyptus* spp.
Pantoea ananatis pv. *uredovora*	Soft rot	*Puccinia graminis* Pers.:Pers. urediniospores	Grows on urediniospores; contains the *crt* genes in carotenoid biosynthesis; strain 20D3 from University of California-Davis was the original source of *crt* genes for "golden rice"
Pantoea stewartii subsp. *indologenes*	Stewart's wilt, leaf spot	Foxtail millet (*Setaria italica* (L.) P. Beauv.); pearl millet (*Pennisetum glaucum* (L.) R. Br.); guar/cluster bean (*Cyamopsis tetragonoloba* (L.) Taub.)	Wilt, invasive; blue leaf spotter of cluster bean
Pantoea stewartii subsp. *stewartii*	Stewart's wilt of corn	Corn (*Zea mays* L.); corn flea beetle (*Chaetocnema pulicaria* Melsh.)	Wilt
Pectobacterium atrosepticum	Tuber soft rot, blackleg (temperate regions)	Potato (*Solanum tuberosum* L.)	Soft rot
Pectobacterium betavasculorum	Bacterial vascular necrosis, soft rot	Sugar beet (*Beta vulgaris* L.)	Soft rot
Pectobacterium cacticida	Bacterial necrosis, soft rot	Pricklypear (*Opuntia* spp.)	Soft rot
Pectobacterium carotovorum subsp. *carotovorum*	Soft rot, blackleg (temperate and semitropical regions)	Vegetables; tubers (e.g., potato [*Solanum tuberosum* L.]); corms	Soft rot, progressive masceration
Pectobacterium carotovorum subsp. *odoriferum*	Odorous soft rot	Chicory (*Cichorium intybus* L.)	Soft rot

(continued)

Table 3-15. Taxa of Plant-Pathogenic Species of the Family *Enterobacteriaceae* (continued)

Genus and Species	Disease Common Name	Hosts	Disease and Virulence Effectors
Pectobacterium cypripedii	Leaf decay, stem decay	Cypripedium orchids (*Cypripedium* spp.)	Soft rot
Pectobacterium rhapontici	Crown rot, foot rot	Rhubarb (*Rheum rhaponticum* L.)	Soft rot
Pectobacterium wasabiae	Crown rot	Japanese horseradish (*Wasabia japonica* (Miq.) Matsum.)	Soft rot
Salmonella enterica[b]	Systemic colonization	Thale cress (*Arabidopsis thaliana* (L.) Heynh.)	Invades plant tissues, secretes AvrA virulence effector that is homologous to AvrRxv of *Xanthomonas vesicatoria*
Serratia marcescens	Yellow vine wilt	Cucurbitaceae members (e.g., watermelon, pumpkin, cantaloupe, squash)	Water-soaked lesions followed by wilt; squash bug (*Anasa tristis* De Geer) vector; phloem preference
Serratia proteamaculans	Wilt	California poplar (*Populus trichocarpa* Torr. & A. Gray)	Plant-growth-promoting endophyte; opportunistic human pathogen causing pneumonia; produces ethylene, 1-aminocyclopropane-1-carboxylic acid deaminase, indoleacetic acid, and cytokinins
Tatumella citrea[c]	Pink disease	Pineapple (*Ananas comosus* (L.) Merr.)	Pink-red discoloration of inner fruit tissues; discoloration enhanced by canning of fruit

they produce urease. Most species are nutritionally nonfastidious, with the exception that *Erwinia amylovora* requires nicotinic acid. Most species produce glycogen rather than poly-β-hydroxybutyric acid as their cellular reserve material. *Brenneria*, *Dickeya*, *Erwinia*, *Pantoea*, and *Pectobacterium* species cause a wide spectrum of diseases, ranging from necrosis to soft rots. One member of the genus *Serratia*, *Serratia marcescens*, has recently entered the list of enteric group plant pathogens. A strain of *Serratia marcescens* causes wilt of cucurbits and is vectored by a squash bug, while *Serratia proteamaculans* causes disease on *Protea cynaroides* (L.) L. The genera *Salmonella* and *Tatumella* both now contain a plant pathogen.

On the basis of biochemical and physiological characters, together with 16S rDNA sequence comparisons, the epiphytic and pathogenic yellow bacteria formerly classified as *Erwinia* species have been grouped into the genus *Pantoea*. Certain members of the genus *Pantoea* are normally epiphytic, while others are opportunistic pathogens. It remains unclear if moving the yellow *Erwinia* species to the genus *Pantoea* can be justified based on evolutionary traits revealed through whole-genome sequence comparisons within the family *Enterobacteriaceae*. As it currently stands, the genus *Pantoea* has members with both plant and soil associations.

PHYLUM BXIII: *FIRMICUTES*

Class *Mollicutes*

Members of the class *Mollicutes* are phenotypically distinct from other bacteria because of their lack of a cell wall and small cell size. Mollicutes live as commensals and, in many arthropods, perhaps as endosymbionts. They possess the following distinctive features: they are highly pleomorphic because of the lack of a cell wall and hence they are Gram negative; they can vary in shape from spherical to branched filaments of uniform diameter with measurements of 125–250 nm in diameter to 10–150 µm long, with no flagella. The above features characterize members of the families *Entomoplasmataceae* and *Spiroplasmataceae*, with *Spiroplasma* cells displaying helical morphology (Fig. 3-5; Table 3-16). They are mainly facultative anaerobes and require sterols and fatty acids for growth. They are resistant to rifampin and β-lactam antibiotics but are sensitive to tetracycline and tylosin. On agar medium, the colonies have a raised center much like that of a sunny side up fried egg (without the yellow color). Growth is slow. Many other members are nonculturable at present. The genome size of mollicutes is about one-third the size of that of *Escherichia coli*. Thus, some members contain genomes of less than 600 kb, while *Spiroplasma* species have genomes from 940 to 2,300 kb. Their GC content ranges from 25 to 31 mol %. The reduced genome size may be due to the loss of genes no longer needed for residence in a specific niche, such as the phloem, or in specific association with insects, as seen with some plant-associated mollicutes.

Mycoplasmataceae

Mycoplasma

Mycoplasma species infect mammals, reptiles, fish, arthropods, humans, and plants. Originally, mycoplasmalike organisms (MLOs) were grouped here, but those that infect plants were later given the term "phytoplasmas" because they were shown to be genetically distinct. More than 200 *Mycoplasma* species have been listed. Their GC content ranges between 23 and 40 mol %. Their optimum growth temperature is 37°C. The sequencing of the entire genomes of some *Mycoplasma* species has provided new insights on the evolutionary relationships of these mollicutes to one another.

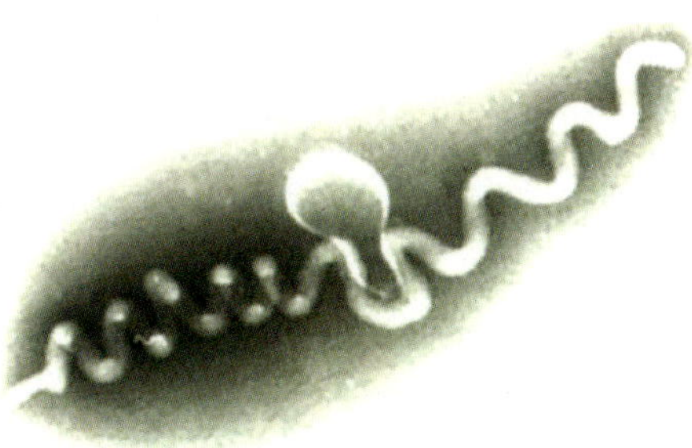

Fig. 3-5. Electron micrograph of *Spiroplasma citri*. (Courtesy Robert E. Davis)

Table 3-16. Families and Genera of the Orders *Mycoplasmatales* and *Entomoplasmatales*

Family	Genus	Habitat
Mycoplasmataceae	*Mycoplasma*	Animals, humans
Entomoplasmataceae	*Entomoplasma*	Insects, plants
	Mesoplasma	Insects, plants
Spiroplasmataceae	*Spiroplasma*	Insects, plants
	'Candidatus Phytoplasma'	Insects, plants

Entomoplasmataceae

This family contains the genera *Entomoplasma* and *Mesoplasma*. Currently, there are no plant-pathogenic species classified in this family. Recent genome sequence data suggest that some members of the genus '*Candidatus* Phytoplasma' may be related to this family.

Spiroplasmataceae

Spiroplasma

Spiroplasma species vary in size and shape. In rich media they are usually helical and become branched at later stages of growth. They are motile, showing twitching and rotary movements in broth medium. They are facultative, wall-less anaerobes and produce acid from glucose. Cholesterol is required for growth. They are sensitive to erythromycin and tetracycline but not to ampicillin. Certain species colonize bees, ticks, wasps, beetles, mosquitoes, leafhoppers, and certain dicots. Those that infect plants cause a yellowing disease vectored by leafhoppers that feed on the phloem. Plant-infecting species, such as *Spiroplasma citri*, harbor multiple plasmids, some of which play a maintenance role in the vector.

Plant pathogens are grouped in the family *Spiroplasmataceae* under the genus *Spiroplasma*. Table 3-17 lists currently known plant pathogens.

Phytoplasma

Phytoplasmas are nonculturable, wall-free bacteria with pleomorphic morphology that commensally reside in plant phloem-sucking insects, such as leaf- and planthoppers. They are easily transmitted to the phloem of a wide range of plants, the result of which manifests in both symptomless and symptom-producing diseases. Phylogenetic sequence analyses of 16S rDNA demonstrated that phytoplasmas constitute a monophyletic group within the class *Mollicutes*, probably at genus level (Gundersen et al., 1994; Seemüller et al., 1994). Based on 16S rDNA sequence comparisons, about 15 groups (putative species) of the genus Phytoplasma have been classified (Lee et al., 1993; Namba et al., 1993; Gundersen et al., 1994; Seemüller et al., 1994; Schneider et al., 1995) and additional members have been named since. Because of their strong associations with phoem-feeding insects and the harsh reactions when introduced into plants, phytoplasmas may have evolved with the insects that harbor them. Although these organisms cannot be named in accordance with the Bacteriological Code, the category *Candidatus* was proposed to provide a way to refer to and record new taxa that could not be cultured and whose description was based primarily on DNA sequence analysis. *Candidatus* species are defined on the basis of habitat and nucleic acid sequence data. By 2008, 25 species of '*Candidatus* Phytoplasma' had been named (Table 3-18). Complete genome sequence analyses

Table 3-17. Plant-Pathogenic *Spiroplasma* Species

Species	Hosts	Disease
S. citri	Many dicots; beet leafhopper (*Circulifer tenellus* (Baker))	Stubborn disease of citrus
S. kunkelii	Corn (*Zea mays* L.); leafhoppers	Corn stunt, horseradish brittle root
S. phoeniceum	Periwinkle (*Catharanthus roseus* (L.) G. Don)	Periwinkle stunt

Table 3-18. Plant-Pathogenic '*Candidatus* Phytoplasma' Species

Species	Hosts	Diseases and Pathogenic Factor
'*Candidatus* Phytoplasma allocasuarinae' Marcone, Gibb, Streten & Schneider 2004	Slaty she-oak (*Allocasuarina muelleriana* (Miq.) L. A. S. Johnson)	Yellows, branch dieback
'*Candidatus* Phytoplasma americanum' Lee, Bottner, Secor, & Rivera-Varas 2006	Potato (*Solanum tuberosum* L.)	Leaves fringed with purple coloration, wilt of upper foliage followed by complete collapse of the potato plant
'*Candidatus* Phytoplasma asteris' Lee, Gundersen-Rindal, Davis, Bottner, Marcone & Seemüller 2004	Many hosts (members of at least 60 plant families, which include Alliaceae, Anacardiaceae, Apocynaceae, Asteraceae, Cruciferae, Geraniaceae, Gramineae, Pedaliaceae, Rosaceae, Rutaceae, Solanaceae, Verbenaceae [see list in Table 8-4])	Aster yellows
'*Candidatus* Phytoplasma aurantifolia' Zreik, Carle, Bové & Garnier 1995	Acid lime (*Citrus* × *aurantiifolia* (Christm.) Swingle)	Witches'-broom, branch dieback
'*Candidatus* Phytoplasma australasia' White, Blackall, Scott & Walsh 1998	Papaya (*Carica papaya* L.)	Yellow crinkle
'*Candidatus* Phytoplasma australiense' Davis, Dally, Gundersen, Lee & Habili 1997	Papaya (*Carica papaya* L.); hibiscus (*Hibiscus* sp.); potato (*Solanum tuberosum* L.)	Dieback, hibiscus witches'-broom, rolling and purple leaves, internal tuber discoloration
'*Candidatus* Phytoplasma brasiliense' Montano, Davis, Dally, Hogenhout, Pimentel & Brioso 2001	Hibiscus (*Hibiscus rosa-sinensis* L.)	Witches'-broom characterized by excessive axillary branching, abnormally small leaves, and deformed flowers
'*Candidatus* Phytoplasma caricae' Arocha, López, Piñol, Fernández, Picornell, Almeida, Palenzuela, Wilson & Jones 2005	Papaya (*Carica papaya* L.)	Papaya bunchy top, twisted and deformed upper leaves coupled with slight yellowing; spread by the papaya leafhopper (*Empoasca papayae* Oman)
'*Candidatus* Phytoplasma castaneae' Jung, Sawayanagi, Kakizawa, Nishigawa, Miyata, Oshima, Ugaki, Lee, Hibi & Namba 2002	Japanese chestnut (*Castanea crenata* Siebold & Zucc.)	Chestnut witches'-broom
'*Candidatus* Phytoplasma cynodontis' Marcone, Schneider & Seemüller 2004	Bermuda grass (*Cynodon dactylon* (L.) Pers.)	Bermuda grass white leaf curl, bushy growth, shortened stolons and rhizomes, small leaves, stunting with eventual death
'*Candidatus* Phytoplasma fragariae' Valiunas, Staniulis & Davis 2006	Strawberry (*Fragaria* × *ananassa* Duchesne)	General stunting and yellowing
'*Candidatus* Phytoplasma fraxini' Griffiths, Sinclair, Smart & Davis 1999	Ash (*Fraxinus* spp.); lilac (*Syringa* spp.)	Ash yellows, lilac witches'-broom
'*Candidatus* Phytoplasma graminis' Arocha, López, Piñol, Fernández, Picornell, Almeida, Palenzuela, Wilson & Jones 2005	Sugarcane (*Saccharum officinarum* L.)	Yellow-leaf disease
'*Candidatus* Phytoplasma japonicum' Sawayanagi, Horikoshi, Kanehira, Shinohara, Bertaccini, Cousin, Hiruki & Namba 1999	Hydrangea/hortensia (*Hydrangea* spp.)	Hydrangea phyllody

(continued)

Table 3-18. Plant-Pathogenic '*Candidatus* Phytoplasma' Species *(continued)*

Species	Hosts	Diseases and Pathogenic Factor
'*Candidatus* Phytoplasma mali' Seemüller & Schneider 2004	Apple (*Malus* × *domestica* Borkh.); cherry (*Prunus avium* L.); apricot (*Prunus armeniaca* L.); plum (*Prunus domestica* L.)	Bushy leaflets, floral and phloem necrosis, wilting, plant decline
'*Candidatus* Phytoplasma oryzae' Jung, Sawayanagi, Wongkaew, Kakizawa, Nishigawa, Wei, Oshima, Miyata, Ugaki, Hibi & Namba 2003	Rice (*Oryza sativa* L.)	Yellow dwarf, yellowing with stunted growth; vectored by leafhoppers (*Nephotettix cincticeps* (Uhler), *Nephotettix nigropictus* (Stål), and *Nephotettix virescens* (Distant))
'*Candidatus* Phytoplasma phoenicium' Verdin, Salar, Danet, Choueiri, Jreijiri, El Zammar, Gélie, Bové & Garnier 2003	Periwinkle (*Vinca major* L.)	Periwinkle yellows, dwarfing, multiple leaflets
'*Candidatus* Phytoplasma pini' Schneider, Torres, Martín, Schröder, Behnke & Seemüller 2005	Scots pine (*Pinus sylvestris* L.); Aleppo pine (*Pinus halepensis* Mill.)	Shoot proliferation, distortion of needles (dwarfed needles, twisted and yellowish needles)
'*Candidatus* Phytoplasma prunorum' Seemüller & Schneider 2004	*Prunus* spp. (e.g., apricot and peach)	European stone fruit yellows; vectored by the apricot psyllid (*Cacopsylla pruni* Scopoli)
'*Candidatus* Phytoplasma pyri' Seemüller & Schneider 2004	Pear (*Pyrus communis* L.)	General decline of tree known as pear decline, which is transmitted by the pear psylla (*Cacopsylla pyri* L.)
'*Candidatus* Phytoplasma rhamni' Marcone, Gibb, Streten & Schneider 2004	Common buckthorn (*Rhamnus cathartica* L.)	Buckthorn witches'-broom phytoplasma
'*Candidatus* Phytoplasma spartii' Marcone, Gibb, Streten & Schneider 2004	Spanish broom (*Spartium junceum* L.)	Spartium witches'-broom
'*Candidatus* Phytoplasma trifolii' Hiruki & Wang 2004	Clover (*Trifolium* spp.); bell pepper (*Capsicum annuum* L.); tomato (*Lycopersicon esculentum* Mill. var. *esculentum* Mill.)	Clover proliferation disease, stunting, yellowing, proliferation of lateral buds, hypertrophic calyxes, virescent flowers
'*Candidatus* Phytoplasma ulmi' Lee, Martini, Marcone & Zhu 2004	Elm (*Ulmus americana* L., *Ulmus minor* Mill., *Ulmus rubra* Muhl., *Ulmus alata* Michx., *Ulmus serotina* Sarg., *Ulmus crassifolia* Nutt., *Ulmus chenmoui* W. C. Cheng)	Elm yellows, witches'-broom, stunting, general decline
'*Candidatus* Phytoplasma ziziphi' Jung, Sawayanagi, Kakizawa, Nishigawa, Wei, Oshima, Miyata, Ugaki, Hibi & Namba 2003	Common jujube (*Ziziphus jujuba* Mill.)	Jujube witches'-broom, rosetting

of the phytoplasmas may place them in the family *Entomoplasmataceae* rather than the family *Spiroplasmataceae*. Phylogenetic studies should eventually provide strong evidence for determining whether the first hosts of phytoplasmas were insects or plants.

'*Candidatus* Phytoplasma' species represents a large number of nonculturable bacteria that are grouped into the aster yellows and western X types.

Aster Yellows Group. The plant diseases caused by unculturable phytoplasmas were analyzed by 16S rDNA and restriction fragment length polymorphism (RFLP) comparisons. Phylogenetic analysis of the unculturable phytoplasmas, which are classified as the aster yellows group, showed that the Phytoplasma clade can be divided into at least 20 subclades. In the near future, the subclades may be delineated at the species level. The aster yellows group comprises more than 100 isolates. They appear relatively homogeneous, based on 16S rDNA sequence analysis. Whole-genome sequence analyses will elucidate whether there are distinctive differences within the isolates.

X-Disease Group. The western X phytoplasma infects stone fruit (peaches, sweet and sour cherries) by proliferating in the phloem of infected plants. The organism is transmitted into the phloem of plants by phloem-feeding insects. Although western X is nonculturable, modern technologies have allowed molecular analysis of its genome. An average genome size of a member of the western X group ranges between 600 and 720 kb.

PHYLUM BXIV: *ACTINOBACTERIA* PHY. NOV.

Class *Actinobacteria*

Micrococcaceae

Members of the Gram-positive family *Micrococcaceae* generally grow as spherical cells that are usually nonmotile. They are facultatively anaerobic or primarily aerobic, depending on whether they are fermentative or respiratory in their metabolism. Only one genus of this family contains plant-pathogenic species.

Arthrobacter

Cells of *Arthrobacter* species exhibit a unique morphological change by appearing, when freshly grown, as Gram-negative, elongated, irregularly shaped rods that later change into Gram-positive cocci as the culture reaches the stationary phase. When transferred into fresh medium, the cells return to the Gram-negative, elongated form. This transition suggests that major alterations take place in the cell peptidoglycan layer (containing the diamino acid lysine) as the cell matures. Most species reside in soils and readily use a wide range of organic substrates. They are nutritionally nonfastidious and obligate aerobes. The colonies of more than half of the *Arthrobacter* species listed are yellow (nondiffusible, acetone-soluble pigment). Catalase and deoxyribonuclease are produced. The optimum growth temperature is 25°C but *Arthrobacter* species can grow at wide temperature ranges (10 to 35°C). The GC content ranges between 55 and 66 mol %. *Arthrobacter ilicis* is apparently the sole unconfirmed plant-pathogenic member of this genus (only isolated once). Its colonies are yellow when grown on rich medium; it is nutritionally nonfastidious. Its GC content is 61.5 mol %. Because the species provides important natural products, sequencing of its entire genome should provide more insights on the presence of plant virulence genes.

Microbacteriaceae

Members of the family *Microbacteriaceae* are Gram-positive bacteria forming irregular, non-spore-forming cells classified into 11 genera (plant-pathogenic genera are in bold type in Table 3-19). The habitats of members of this family vary considerably. Some species are found in soil, some in fresh water, and others as epiphytes on the phyllosphere of grasses and some dicots. The metabolism is primarily aerobic.

Table 3-19. Characteristics of Genera of the Family *Microbacteriaceae*

Genus	GC (mol %)	Native Habitats
Agrococcus	74	Sandstone surfaces of ancient buildings
Agromyces	70–73	Forest, mangrove rhizospheres
Aureobacterium	67–70	Milk products, soil, insects
Clavibacter[a]	66–68	Plant pathogens
Cryobacterium	66–67	Filamentous psychrophile
Curtobacterium[a]	68–75	Plant pathogens
Frigoribacterium	71	Soil, air-duct dust, animal shed floor, psychrophile (5–10°C)
Leifsonia[a]	66–73	Bermuda grass
Leucobacter	66.2	Unknown; found as a contaminant
Microbacterium	69–75	Dairy products, sewage, insects
Rathayibacter[a]	63–72	Cereal grasses; vectored by nematode (*Anguina* spp.); plant pathogen
Subtercola	64.4–67.8	Cold groundwater, Finland, psychrophile

[a] These genera contain plant-pathogenic species.

The cells can be very small (<0.1 µm) and form slightly curved rods that are usually linked irregularly. The cell wall peptidoglycan is characterized by a cross-linkage between the α-carboxyl group of D-glutamic acid in position 2 of the peptide subunit and the C-terminal D-alanine of an adjacent subunit. Many cells are found in pairs or in multiples. The binary form of cell division is also distinctive; they frequently form short rods that are in multiple linear, angular, palisade arrangements that are characteristic of cells dividing by the so-called "snapping division" mechanism. The GC content ranges from 59 to 75 mol %. Plant pathogens share this trait less commonly. Plant pathogens are sensitive to 0.1% triphenyl-tetrazolium chloride in culture media but, at a lower concentration (0.001%), this compound is commonly used to differentiate colonies. Their Gram staining behavior varies depending on the strain, culture medium, and age of the culture. All members of the family *Microbacteriaceae* produce catalase. Interestingly, the plant-pathogenic members do not hydrolyze starch. Plant-pathogenic members also do not reduce nitrate to nitrite but catabolize D-arabinose. D-Arabinose, galactose, and meso-diaminopimelic acid constitute much of their cell wall peptidoglycan. The colonies formed on rich agar medium display a variety of colors, including shades of yellow and orange. These pigments are mainly carotenoids, whose concentration is dependent on culture media and temperature of growth. Thus, cultures growing on rich complex media and at suboptimal growth temperatures frequently produce intense color. The plant-pathogenic genera include *Clavibacter*, *Curtobacterium*, *Leifsonia*, and *Rathayibacter*. The plant-pathogenic members of *Clavibacter* and *Curtobacterium* can be classified into motile and nonmotile types and can frequently, but not always, be distinguished by colony color when cultured on rich agar medium (Table 3-20).

Clavibacter

Nonmotile cells freshly grown on agar medium are circular, smooth, opaque to white, and change to light yellow with age. Colonies therefore can be white, yellow, orange, or pink, depending on the species, strain, medium, and age. The insoluble bipyridyl pigment indigoidine, which imparts a deep dark blue color mixed in with the yellowish colony, is produced by some subspecies (*Clavibacter michiganensis* subsp. *insidiosus*). The yellow pigments are carotenoids. Most species require complex medium, providing B vitamins and amino acids. The cells produce catalase and

Table 3-20. Properties of *Curtobacterium* and *Clavibacter* Species[a]

Species	Diseases	Host	Colony Color
Motile			
Curtobacterium flaccumfaciens pv. *betae*	Leaf silvering	Sugar beet and garden beet (*Beta vulgaris* L.)	Yellow
Curtobacterium flaccumfaciens pv. *beticola*[b]	Large leaf spots and root degeneration	Sugar beet (*Beta vulgaris* L.)	Cream, pale yellow
Curtobacterium flaccumfaciens pv. *flaccumfaciens*	Wilt	Common bean (*Phaseolus vulgaris* L.)	Yellow
Curtobacterium flaccumfaciens pv. *ilicis*[c]	Leaf blight	American holly (*Ilex opaca* Aiton)	Yellow
Curtobacterium flaccumfaciens pv. *oortii*	Leaf spot, bulb canker	Tulip (*Tulipa* spp.)	Yellow
Curtobacterium flaccumfaciens pv. *poinsettiae*	Canker	Poinsettia (*Euphorbia pulcherrima* Willd. ex Klotzsch)	Orange
Nonmotile			
Clavibacter michiganensis subsp. *insidiosus*	Wilt	Alfalfa (*Medicago falcata* L.)	Cream-yellow, indigo produced with age
Clavibacter michiganensis subsp. *michiganensis*	Leaf and stem necrosis, wilt, canker	Tomato (*Lycopersicon esculentum* Mill. var. *esculentum* Mill.)	Yellow
Clavibacter michiganensis subsp. *nebraskensis*	Leaf spots, freckles, wilt	Corn (*Zea mays* L.)	Orange
Clavibacter michiganensis subsp. *sepedonicus*	Wilt, ring rot	Potato (*Solanum tuberosum* L.)	Cream

[a] Note: *Corynebacterium agropyri*, *Corynebacterium humiferum*, *Corynebacterium humuli*, and *Corynebacterium hypertrophicans* are not valid species names.
[b] From Chen et al., 2007.
[c] A nonaccepted name as synonym of *Arthrobacter ilicis* (Tindall, 2008).

some acid when grown on carbohydrates other than rhamnose or ribose. The cell wall peptidoglycan contains 2,4-diaminobutyric acid as the dibasic cross-linking acid. The GC content ranges around 70 to 72 mol %.

Curtobacterium

Curtobacterium species produce elongated, unbranched cells with slight bends when septated during cell division. The cells are usually motile. A high GC content between 66 and 71 mol % is reflected by its genome. Cells produce catalase and nuclease and weakly produce acid from glucose, fructose, rhamnose, and other simple sugars. Organic acids, such as acetic and succinic acids, can be utilized as sole carbon sources. Their cell wall contains D-ornithine as the diamino acid of the peptide bridge. *Curtobacterium flaccumfaciens* pv. *flaccumfaciens* var. *violaceum* produces a soluble purple pigment, while *Curtobacterium flaccumfaciens* var. *aurantiacum* produces a soluble orange pigment.

Leifsonia

Leifsonia is a relatively new genus that accommodates commensal bacteria found in *Poa annua* L. root gall, induced by the nematode *Subanguina radicicola* (Greeff) Paramonov. The plant pathogens *Leifsonia xyli* subsp. *xyli* and *Leifsonia xyli* subsp. *cynodontis* are reclassified members originally classified in the genus *Clavibacter*. Members of the genus *Leifsonia* are characterized by the coryneform rod morphology and the presence of 2,4-diaminobutyric acid as a peptide bridge component in the peptidoglycan. For growth, cells require hemin chloride and bovine serum albumin. The GC content ranges from 66 to 73 mol %. There are two pathogenic subspecies of *Leifsonia xyli* and several environmental isolates in different species. *Leifsonia xyli* subsp. *cynodontis* causes stunting of Bermuda grass (*Cynodon dactylon* (L.) Pers.) and corn (*Zea mays* L.). It harbors a 51-kb cryptic pCXC100 plasmid with a *parB* gene (Yin et al., 2006). *Leifsonia xyli* subsp. *xyli* causes ratoon stunt disease of sugarcane (*Saccharum* interspecific hybrids).

Rathayibacter

The classification of this genus was based on the unique L-diaminobutyric acid linkage in the peptidoglycan cell wall and on fatty acid composition. These features reflect only a small portion of the genome of members of the genus *Rathayibacter* and therefore reclassification of this genus will take place as additional whole-genome sequences become available. *Rathayibacter* members produce yellow colonies on rich medium. They are transmitted by nematodes. *Rathayibacter toxicus* causes slimy head on Wimmera ryegrass (*Lolium rigidum* Gaudin), Pacific bent grass (*Agrostis avenacea* J. F. Gmel.), and rabbitfoot grass (*Polypogon monspeliensis* (L.) Desf.). In addition to being a plant pathogen, it produces toxins poisonous to grazing animals. *Rathayibacter rathayi* causes slimy head on orchard grass (*Dactylis glomerata* L.). *Rathayibacter iranicus* and *Rathayibacter tritici* cause gumming and yellow slime disease of wheat (*Triticum aestivum* L.), respectively. *Rathayibacter tritici* was previously named *Clavibacter tritici*.

Nocardiaceae

Members of the family *Nocardiaceae* produce filamentous mycelial structures, either aerial or retained in the growth substrate. They can be rod shaped or form coccoidal cells that are nonmotile. *Nocardia* species can form aerial conidal chains. *Rhodococcus* species are coccoidal, forming filaments showing elementary branching that are easily fragmented into short rods and hyphal filaments. Both *Nocardia* and *Rhodococcus* species reproduce by mycelial growth and fragmentation. Pink to red colonies are usually produced on rich medium. *Rhodococcus fascians* is the sole plant pathogen of this genus. Its colonies produce a characteristic flesh-colored orange-pink color. *Nocardia* species reside in the soil and can be pathogens of animals, including humans. Likewise, *Rhodococcus* members have been found in the soil, as well as in marine sediments and herbivore dung. Members of the genus infest the intestinal tract of carp (*Cyprinus carpio* L.) and the intestinal tract of reduvid bug (*Rhodnius prolixus* Stål). *Rhodococcus fascians* causes fasciation of leaf and flower primordia and leafy galls on stems through the production of auxin and cytokinins. Hosts include pea (*Pisum sativum* L.), marigold (*Tagetes patula* L. and *Tagetes erecta* L.), chrysanthemum (*Chrysanthemum* × *morifolium* Ramat.), and stock (*Matthiola incana* (L.) W. T. Aiton), but galls have been experimentally induced by this pathogen on many hosts.

Streptomycetaceae

Members of the family *Streptomycetaceae* are characterized by the production of extensive branched primary mycelium with secondary aerial or substrate mycelium.

Table 3-21. Plant-Pathogenic *Streptomyces* Species

Species	Principal Host	Diseases and Pathogenic Factor
S. acidiscabies	Potato (*Solanum tuberosum* L.)	Acidiphilic scab
S. caviscabies	Potato (*Solanum tuberosum* L.)	Deep-pitted lesions
S. ipomoeae	Sweet potato (*Ipomoea batatas* (L.) Lam.)	Pitting, rot
S. scabies	Potato (*Solanum tuberosum* L.)	Severe scab, russet scab; produces phytoxins thaxtomin A and B
S. setonii	Potato (*Solanum tuberosum* L.)	Mild common scab
S. turgidiscabies	Potato (*Solanum tuberosom* L.), causes necrosis of Arabidopsis (*Arabidopsis* spp.), tobacco (*Nicotiana* spp.), and radish (*Raphanus* spp.)	Severe scab, necrosis; produces thaxtomin, a toxic nitrosodipeptide

Branched hyphae are partly transformed into chains of conidia (arthrospores). All genera listed in Table 3-21 contain peptidoglycan LL-diaminopimelic acid in their cell wall. Their peptidoglycan contains an interpeptide bridge composed of a glycine residue. The group possesses genomes with a high GC content. Many species produce antimicrobial compounds that are exploited commercially.

Streptomyces

Streptomyces species do not fragment into rods or coccoidal cells from hyphae. This genus contains the scab-producing plant pathogens that attack primarily potato cultivars (Fig. 3-6). They produce spores from terminal hyphae. The gray spores are attached as long spiral chains; they fragment and disperse when dry. *Streptomyces scabies* produces melanin on medium containing tyrosine. *Streptomyces acidiscabies* causes symptoms similar to those caused by *Streptomyces scabies* but tolerates more acidic soils and grows at pH 4.5. *Streptomyces caviscabies* causes deep-pitted scab. *Streptomyces aureofaciens* causes russet scab. *Streptomyces turgidiscabies* contains a 660-kb pathogenicity island. *Streptomyces acidiscabies*, *Streptomyces scabies*, and *Streptomyces turgidiscabies* also contain a novel *nec1* gene that encodes a secreted necrogenic protein (Joshi et al., 2007). Table 3-21 lists the plant-pathogenic members.

Fig. 3-6. Severe scab symptoms on a potato tuber. (Courtesy Madhumita V. Joshi)

REFERENCES

Adkins, R. M., and Li, W.-H. 2002. Dating the age of the last common ancestor of all living organisms with a protein clock. Pages 363-370 in: Horizontal Gene Transfer, 2nd ed. M. Syvanen and C. I. Kado, eds. Academic Press, San Diego, CA.

Alvarez, A. M., and Louws, F. J. 2001. Xanthomonads. Pages 1208-1210 in: Encyclopedia of Plant Pathology. O. C. Maloy and T. D. Murray, eds. John Wiley & Sons., Inc., New York.

Andrewes, A. G., Jenkins, C. L., Starr, M. P., Shepherd, J., and Hope, H. 1976. Structure of xanthomonadin I, a novel dibrominated aryl-polyene pigment produced by the bacterium *Xanthomonas juglandis*. Tetrahedron Lett. 45:4023-4024.

Azegame, K., Nihiyama, K., and Kato, H. 1988. Effect of iron limitation on "*Pseudomonas plantarii*" growth and tropolone and protein production. Appl. Environ. Microbiol. 54:844-847.

Barak, J. D., Koike, S. T., and Gilbertson, R. L. 2001. Role of crop debris and weeds in the epidemiology of bacterial leaf spot of lettuce in California. Plant Dis. 85:169-178.

Bare, S., Coiro, V. M., Scaloni, A., Di Nola, A., Paci, M., Segre, A. L., and Ballio, A. 1999. Conformations in solution of the fuscopeptids. Phytotoxic metabolites of *Pseudomons fuscovaginae*. Eur. J. Biochem. 266:484-492.

Bender, C. L., Alarcon-Chaidez, F., and Gross, D. C. 1999. *Pseudomonas syringae* phytotoxins: Mode of action, regulation, and biosynthesis by peptide and polyketide synthetases. Microbiol. Mol. Biol. Rev. 63:266-292.

Brady, C. L., Venter, S. N., Cleenwerck, I., Engelbeen, K., Vancanneyt, M., Swings, J., and Coutinho, T. A. 2009. *Pantoea vagans* sp. nov., *Pantoea eucalypti* sp. nov., *Pantoea deleyi* sp. nov. and *Pantoea anthophila* sp. nov. Int. J. Syst. Evol. Microbiol. 59:2339-2345.

Brady, C. L., Venter, S. N., Cleenwerck, I., Vandemeulebroecke, K., de Vos, P., and Coutinho, T. A. 2010. Transfer of *Pantoea citrea*, *Pantoea punctata* and *Pantoea terrea* to the genus *Tatumella* emend. as *Tatumella citrea* comb. nov., *Tatumella punctata* comb. nov. and *Tatumella terrea* comb. nov. and description of *Tatumella morbirosei* sp. nov. Int. J. Syst. Evol. Microbiol. 60:484-494.

Brown, J. R., Italia, M. J., Douady, C., and Stanhope, M. J. 2002. Horizontal gene transfer and the universal tree of life. Pages 305-361 in: Horizontal Gene Transfer, 2nd ed. M. Syvanen and C. I. Kado, eds. Academic Press, San Diego, CA.

Buddenhagen, I. W., and Kelman, A. 1964. Biological and physiological aspects of bacterial wilt caused by *Pseudomonas solanacearum*. Annu. Rev. Phytopathol. 2:203-230.

Buddenhagen, I., Sequeira, L., and Kelman, A. 1962. Designation of races in *Pseudomonas solanacearum*. (Abstr.) Phytopathology 52:726.

Buonaurio, R., Stravato, V. M., Kosako, Y., Fujiwara, N., Naka, T., Kobayashi, K., Cappelli, C., and Yauuchi, E. 2002. *Sphingomonas melonis* sp. nov., a novel pathogen that causes brown spots on yellow Spanish melon fruits. Int. J. Syst. Evol. Microbiol. 52:2081-2987.

Burkholder, W. H. 1950. Sour skin, a bacterial rot of onion bulbs. Phytopathology 40:115-117.

Chen, Y.-F., Yin, Y.-N., Zhang, X.-M., and Guo, J.-H. 2007. *Curtobacterium flaccumfaciens* pv. *beticola*, a new pathovar of pathogens in sugar beet. Plant Dis. 91:677-684.

Coenye, T., Vandamme, P., Govan, J. R. W., and Lipuma, J. J. 2001. Taxonomy and identification of the *Burkholderia cepacia* complex. J. Clin. Microbiol. 39:3427-3436.

Cooley, M. B., Miller, W. G., and Mandrell, R. E. 2003. Colonization of *Arabidopsis thaliana* with *Salmonella enterica* and enterohemorrhagic *Escherichia coli* O157:H7 and competition by *Enterobacter asburiae*. Appl. Environ. Microbiol. 69:4915-4926.

Dalla Serra, M., Fagiuoli, G., Nordera, P., Bernhart, I., Della Volpe, C., Di Giorgio, D., Ballio, A., and Menestrina, G. 1999. The interaction of lipodepsipeptide toxins from *Pseudomonas syringae* pv. *syringae* with biological and model membranes: A comparison of syringotoxin, syringomycin, and two syringopeptins. Mol. Plant-Microbe Interact. 12:391-400.

de Bruijn, F. J. 1992. Use of repetitive (repetitive extragenic palindromic and enterobacterial repetitive intergenic consensus) sequences and the polymerase chain reaction to fingerprint the genomes of *Rhizobium meliloti* isolates and other soil bacteria. Appl. Environ. Microbiol. 60:2286-2295.

DiGiorgio, D., Comoni, L., Marchiafava, C., and Ballio, A. 1997. Biological activities of pseudomycin A, a lipodepsinonapeptide from *Pseudomonas syringae* MSU 16H. Phytochemistry 45:1385-1391.

Doddapaneni, H., Yao, J., Lin, H., Walker, M. A., and Civerolo, E. I. 2006. Analysis of the genome-wide variations among multiple strains of the plant pathogenic bacterium *Xylella fastidiosa*. BMC Genomics 7:225.

Farrand, S. K., van Verkum, P. B., and Oger, P. 2003. *Agrobacterium* is a definable genus of the family *Rhizobiaceae*. Int. J. Syst. Evol. Microbiol. 53:1681-1687.

Fegan, M., and Prior, P. 2005. How complex is the "*Ralstonia solanacearum* species complex". Pages 449-461 in: Bacterial Wilt Disease and the *Ralstonia solanacearum* Species Complex.

C. Allen, P. Prior, and A. C. Hayward, eds. American Phytopathological Society, St. Paul, MN.

Fegan, M., and Prior, P. 2006. Diverse members of the *Ralstonia solanacearum* species complex cause bacterial wilts of banana. Aust. Plant Pathol. 35:93-101.

Flavier, A. B., Clough, S. J., Schell, M. A., and Denny, T. P. 1997. Identification of 3-hydroxypalmitic acid methyl ester as a novel autoregulator controlling virulence in *Ralstonia solanacearum*. Mol. Microbiol. 26:251-269.

Foissac, X., Danet, J.-L., Zreik, L., Gandar, J., Nourrisseau, J.-G., Bove, J.-M., and Garnier, M. 2000. Cloning of the *spoT* gene of "*Candidatus* Phlomobacter fragariae" and development of a PCR-restriction fragment length polymorphism assay for detection of the bacterium in insects. Appl. Environ. Microbiol. 66:3474-3480.

Fukuchi, N., Isogai, A., Nakayama, J., and Suzuki, A. 1990. Structure of syringotoxin B, a phytotoxin produced by citrus isolates of *Pseudomonas syringae* pv. *syringae*. Agric. Biol. Chem. 54:3377-3379.

Gardan, L., Bollet, C., Abu Ghorrah, M., Grimont, F., and Grimont, P. A. D. 1992. DNA relatedness among the pathovar strains of *Pseudomonas syringae* subsp. *savastanoi* Janse (1982) and proposal of *Pseudomonas savastanoi* sp. nov. Int. J. Syst. Bacteriol. 42:606-612.

Gardan, L., Shafik, H., Belouin, S., Broch, R., Grimont, F., and Grimont, P. A. D. 1999. DNA relatedness among the pathovars of *Pseudomonas syringae* and description of *Pseudomonas tremae* sp. nov. and *Pseudomonas cannabina* sp. nov. (ex Sutic and Dowson 1959). Int. J. Syst. Bacteriol. 49:469-478.

Gardan, L., Dauga, C., Prior, P., Gillis, M., and Saddler, G. S. 2000. *Acidovorax anthurii* sp. nov., a new phytopathogenic bacterium which causes bacterial leaf-spot of anthurium. Int. J. Syst. Evol. Microbiol. 50:235-246.

Gardan, L., Stead, D. E., Dauga, C., and Gillis, M. 2003. *Acidovorax valerianella* sp. nov., a novel pathogen of lamb's lettuce [*Valerianella locusta* (L.) Laterr.]. Int. J. Syst. Evol. Microbiol. 53:795-800.

Goto, M., and Kuwata, H. 1988. *Rhizobacter daucus* gen. nov., sp. nov., the causal agent of carrot bacterial gall. Int. J. Syst. Bacteriol. 38:233-239.

Grenier, A.-M., Dupont, G., Pagès, S., Condemine, G., and Rahbé, Y. 2006. The phytopathogen *Dickeya dadantii* (*Erwinia chrysanthemi* 3937) is a pathogen of the pea aphid. Appl. Environ. Microbiol. 72:1956-1965.

Gundersen, D. E., Lee, I.-M., Rehner, S. A., Davis, R. E., and Kingsbury, D. T. 1994. Phylogeny of mycoplasmalike organisms (phytoplasmas): A basis for their classification. J. Bacteriol. 176:5244-5254.

Hardt, W.-D., and Galan, J. E. 1997. A secreted *Salmonella* protein with homology to an avirulence determinant of plant pathogenic bacteria. Proc. Natl. Acad. Sci. U.S.A. 94:9887-9892.

Hayward, A. C. 1960. A method for characterizing *Pseudomonas solanacearum*. Nature 186:405-406.

Hayward, A. C. 1964. Characteristics of *Pseudomonas solanacearum*. J. Appl. Bacteriol. 27:265-277.

Hayward, A. C. 1991. Biology and epidemiology of bacterial wilt caused by *Pseudomonas solanacearum*. Annu. Rev. Phytopathol. 29:65-87.

He, L. Y., Sequeira, L., and Kelman, A. 1983. Characteristics of strains of *Pseudomonas solanacearum* from China. Plant Dis. 67:1357-1361.

Hooykaas, P. J. J., Snijdewint, F. G., and Schilperoort, R. A. 1982. Identification of the Sym plasmid of *Rhizobium leguminosarum* strain 1001 and its transfer to and expression in other rhizobia and *Agrobacterium tumefaciens*. Plasmid 8:73-82.

Hussain, M. B. B. M., Zhang, H.-B., Xu, J.-L., Liu, Q., Jiang, Z., and Zhang, L.-H. 2008. The acyl-homoserine lactone-type quorum-sensing system modulates cell motility and virulence of *Erwinia chrysanthemi* pv. zeae. J. Bacteriol. 190:1045-1053.

Jagoueix, S., Bove, J. M., and Garnier, M. 1994. The phloem-limited bacterium of greening disease of citrus is a member of the alpha subdivision of the Proteobacteria. Int. J. Syst. Bacteriol. 44:379-386.

Johnston, A. W., Bibb, M. J., and Beringer, J. E. 1978. Tryptophan genes in *Rhizobium*—Their organization and their transfer to other bacterial genera. Mol. Gen. Genet. 165:323-330.

Jones, J. B., Lacy, G. H., Bouzar, H., Stall, R. E., and Schaad, N. W. 2004. Reclassification of the xanthomonads associated with bacterial spot disease of tomato and pepper. Syst. Appl. Microbiol. 27:755-762.

Joshi, M., Rong, X., Moll, S., Kers, J., Franco, C., and Loria, R. 2007. *Streptomyces turgidiscabies* secretes a novel virulence protein, Nec1, which facilitates infection. Mol. Plant-Microbe Interact. 20:599-608.

Kado, C. I. 1992. Plant pathogenic bacteria. Pages 659-674 in The Prokaryotes, 2nd ed. Vol. I. A. Balows, H. G. Truper, M. Dworkin, W. Harder, and K.-H. Schleifer, eds. Springer-Verlag, New York.

Kado, C. I. 2009. Horizontal gene transfer: Sustaining pathogenicity and optimizing host-pathogen interactions. Mol. Plant Pathol. 10:143-150.

Kelman, A. 1954. The relationship of pathogenicity in *Pseudomonas solanacearum* to colony appearance on a tetrazolium medium. Phytopathology 44:693-695.

Lee, I.-M., Hammond, R. W., Davis, R. E., and Gundersen, D. E. 1993. Universal amplification and analysis of pathogen 16S rDNA for classification and identification of mycoplasmalike organisms. Phytopathology 83:834-842.

Lin, H., Doddapaneni, H., Munyaneza, J. E., Civerolo, E. L., Sengoda, V. G., Buchman, J. L., and Stenger, D. C. 2009. Molecular characterization and phylogenetic analysis of 16S rRNA from a new "*Candidatus* Liberibacter" strain associated with zebra chip disease of potato (*Solanum tuberosum* L.) and the potato psyllid (*Bactericera cockerelli* Sulc). J. Plant Pathol. 91:215-219.

Monteiro-Vitorello, C. B., de Oliveira, M. C., Zerillo, M. M., Varani, A. M., Civerolo, E., and Van Sloys, M. A. 2005. *Xylella* and *Xanthomonas* mobil'omics. OMICS 9:146-159.

Namba, S., Oyaizu, H., Kato, S., Iwanawi, S., and Tsuchizaki, T. 1993. Phylogenetic diversity of phytopathogenic mycoplasmalike organisms. Int. J. Syst. Bacteriol. 43:461-467.

Novichkov, P. S., Omelchenko, M. V., Gelfand, M. S., Mironov, A. A., Wolf, Y. I., and Koonin, E. V. 2004. Genome-wide molecular clock and horizontal gene transfer in bacterial evolution. J. Bacteriol. 186:6575-6585.

Ogimi, C., Kubo, Y., Higuchi, H., and Takikawa, Y. 1990. Bacterial gall diseases of himeyuzuriha (*Daphniphyllum teijsmanni* Z.) caused by *Pseudomonas syringae* pv. *daphniphylii* pv. nov. J. Jpn. For. Soc. 72:17-22.

Osborne, C. A., Galic, M., Sangwan, P., and Janssen, P. H. 2005. PCR-generated artifact from 16S rRNA gene-specific primers. FEMS Microbiol. Lett. 248:183-187.

Rademaker, J. L. W., Louws, F. J., Schultz, M. H., Rossbach, U., Vauterin, L., Swings, J., and de Bruijn, F. J. 2005. A comprehensive species to strain taxonomic framework for *Xanthomonas*. Phytopathology 95:1098-1111.

Schaad, N. W., Postnikova, E., Lacy, G., Fatmi, M., and Chang, C.-J. 2004. *Xylella fastidiosa* subspecies: *X. fastidiosa* subsp. *piercei*, subsp. nov., *X. fastidiosa* subsp. *multiplex* subsp. nov., and *X. fastidiosa* subsp. *pauca* subsp. nov. Syst. Appl. Microbiol. 27:290-300.

Schaad, N. W., Postnikova, E., Sechler, A., Claflin, L. E., Vidaver, A. K., Jones, J. B., Agarkova, I., Ignatov, A., Dickstein, E., and Ramundo, B. A. 2008a. Reclassification of subspecies of *Acidovorax avenae* as *A. avenae* (Manns 1905) emend., *A. cattleyae* (Pavarino, 1911) comb. nov., *A. citrulli* (Schaad et al., 1978) comb. nov., and proposal of *A. oryzae* sp. nov. Syst. Appl. Microbiol. 31:434-446.

Schaad, N. W., Postnikova, E. N., Claflin, L. E., Vidaver, A. K., Agarkova, I., Sechler, A. J., Ignatov, A., and Ramundo, B. A. 2008b. Transfer of *Acidovorax avenae* subspecies to *Mesothermophilus avenae*. Syst. Appl. Microbiol. 31:434-446.

Schell, M. A. 2000. Control of virulence and pathogenicity genes of *Ralstonia solanacearum* by an elaborate sensory network. Annu. Rev. Phytopathol. 38:263-292.

Schneider, B., Seemüller, E., Smath, C. D., and Kirkpatrick, B. C. 1995. Phylogenetic classification of plant pathogenic mycoplasma-like organisms or phytoplasmas. Pages 369-380 in: Molecular and Diagnostic Procedures in Mycoplasmology, Vol. 1. S. Razin and J. G. Tully, eds. Academic Press, San Diego, CA.

Seemüller, E., Schneider, B., Maurer, R., Ahrens, U., Daire, X., Kison, H., Lorenz, K.-H., Firrao, G., Avinent, L., Sears, B. B., and Stackebrandt, E. 1994. Phylogenetic classification of phytopathogenic mollicutes by sequence analysis of 16S ribosomal DNA. Int. J. Syst. Bacteriol. 44:440-446.

Shendure, J., and Ji, H. 2008. Next-generation DNA sequencing. Nat. Biotechnol. 26:1135-1145.

Shendure, J., Mitra, R. D., Varma, C., and Church, G. M. 2004. Advanced sequencing tehnologies: Methods and goals. Nat. Rev. Genet. 5:335-344.

Simpson, A. J. G., Reinach, F. C., Arruda, P., Abreu, F. A., Acencio, M., Alvarenga, R., Alves, L. M. C., Araya, J. E., Baia, G. S., Baptista, C. S., Barros, M. H., Bonaccorsi, E. D., Bordin, S., Bové, J. M., Briones, M. R. S., Bueno, M. R. P., Camargo, A. A., Camargo, L. E. A., Carraro, D. M., Carrer, H., Colauto, N. B., Colombo, C., Costa, F. F., Costa, M. C. R., Costa-Neto, C. M., Coutinho, L. L., Cristofani, M., Dias-Neto, E., Docena, C., El-Dorry, H., Facincani, A. P., Ferreira, A. J. S., Ferreira, V. C. A., Ferro, J. A., Fraga, J. S., França, S. C., Franco, M. C., Frohme, M., Furlan, L. R., Garnier, M., Goldman, G. H., Goldman, M. H. S., Gomes, S. L., Gruber, A., Ho, P. L., Hoheisel, J. D., Junqueira, M. L., Kemper, E. L., Kitajima, J. P., Krieger, J. E., Kuramae, E. E., Laigret, F., Lambais, M. R., Leite, L. C. C., Lemos, E. G. M., Lemos, M. V. F., Lopes, S. A., Lopes, C. R., Machado, J. A., Machado, M. A., Madeira, A. M. B. N., Madeira, H. M. F., Marino, C. L., Marques, M. V., Martins, E. A. L., Martins, E. M. F., Matsukuma, A. Y., Menck, C. F. M., Miracca, E. C., Miyaki, C. Y., Monteiro-Vitorello, C. B., Moon, D. H., Nagai, M. A., Nascimento, A. L. T. O., Netto, L. E. S., Nhani, A., Jr., Nobrega, F. G., Nunes, L. R., Oliveira, M. A., de Oliveira, M. C., de Oliveira, R. C., Palmieri, D. A., Paris, A., Peixoto, B. R., Pereira, G. A. G., Pereira, H. A., Jr., Pesquero, J. B., Quaggio, R. B., Roberto, P. G., Rodrigues, V., de M. Rosa, A. J., de Rosa, V. E., Jr., de Sá, R. G., Santelli, R. V., Sawasaki, H. E., da Silva, A. C. R., da Silva, A. M., da Silva, F. R., Silva, W. A., da Silveira, J. F., Silvestri, M. L. Z., Siqueira, W. J., de Souza, A. A., de Souza, A. P., Terenzi, M. F., Truffi, D., Tsai, S. M., Tsuhako, M. H., Vallada, H., Van Sluys, M. A., Verjovski-Almeida, S., Vettore, A. L., Zago, M. A., Zatz, M., Meidanis, J., and Setubal, J. C. 2000. The genome sequence of the plant pathogen *Xylella fastidiosa*. Nature 406:151-157.

Summer, E. J., Enderle, C. J., Ahern, S. J., Gill, J. J., Torres, C. P., Appel, D. N., Black, M. C., Young, R., and Gonzalez, C. F. 2010. Genomic and biological analysis of phage Xfas53 and related prophages of *Xylella fastidiosa*. J. Bacteriol. 192:179-190.

Syvanen, M., and Kado, C. I. 2002. Horizontal Gene Transfer, 2nd ed. Academic Press, San Diego, CA.

Takeuchi, M., Sakane, T., Yanagi, M., Yamasato, K., Hamana, K., and Yokota, A. 1995. Taxonomic study of bacteria isolated from plants: Proposal of *Sphingomonas rosa* sp. nov., *Sphingomonas pruni* sp. nov., *Sphingomonas asaccharolytica* sp. nov., and *Sphingomonas mali* sp. nov. Int. J. Syst. Bacteriol. 45:334-341.

Thieme, F., Koebnik, R., Bekel, T., Berger, C., Boch, J., Büttner, D., Caldana, C., Gaigalat, L., Goesmann, A., Kay, S., Kirchner, O., Lanz, C., Linke, B., McHardy, A. C., Meyer, F., Mittenhuber, G., Nies, D. H., Niesbach-Klösgen, U., Patschkowski, T., Rückert, C., Rupp, O., Schneiker, S., Schuster, S. C., Vorhölter, F.-J., Weber, E., Pühler, A., Bonas, U., Bartels, D., and Kaiser, O. 2005. Insights into genome plasticity and pathogenicity of the plant pathogenic bacterium *Xanthomonas campestris* pv. *vesicatoria* revealed by the complete genome sequence. J. Bacteriol. 187:7254-7266.

Tindall, B. J. 2008. *Corynebacterium ilicis* is typified by ICMP 2608 = ICPB CI144, *Arthrobacter ilicis* is typified by DSM 20138 = ATCC 14264 = NCPPB 1228 and the two are not homotypic synonyms, and clarification of the authorship of these two species. Opinion 87. Int. J. Syst. Evol. Microbiol. 58:1976-1978.

Uchino, Y., Hirota, A., Yokota, A., and Sugiyama, J. 1998. Reclassification of marine *Agrobacterium* species: Proposals of *Stappia stellulata* gen. nov., comb. nov., *Stappia aggregata* sp. nov., nom. rev., *Ruegeria atlantica* gen. nov., comb. nov., *Ruegeria gelatinovora* comb. nov., *Ruegeria algicola* comb. nov., and *Ahrensia kieliense* gen. nov., sp. nov., nom. rev. J. Gen. Appl. Microbiol. 44:201-210.

van Bruggen, A. H. C., and Jochimsen, K. N. 1992. First report of *Rhizomonas* sp. causing corky root of lettuce in Europe. Neth. J. Plant Pathol. 98:45-56.

van Bruggen, A. H. C., Jochimsen, K. N., and Brown, P. R. 1990a. *Rhizomonas suberifaciens* gen. nov., sp. nov., the causal agent of corky root of lettuce. Int. J. Syst. Bacteriol. 40:175-188.

van Bruggen, A. H. C., Brown, P. R., and Jochimsen, K. N. 1990b. Host range of *Rhizomonas suberifaciens*, the causal agent of corky root of lettuce. Plant Dis. 74:581-584.

Vandamme, P., Holmes, B., Coenye, T., Goris, J., Mahenthiralingam, E., LiPuma, J. J., and Govan, J. R. 2003. *Burkholderia cenocepacia* sp. nov.—A new twist to an old story. Res. Microbiol. 154:91-96.

Vauterin, L., Hoste, B., Kersters, K., and Swings, J. 1995. Reclassification of *Xanthomonas*. Int. J. Syst. Bacteriol. 45:472-489.

Vinatzer, B. A., and Bull, C. T. 2009. The impact of genomic approaches on our understanding of diversity and taxonomy of plant pathogenic bacteria. Pages 37-61 in: Plant Pathogenic Bacteria: Genomics and Molecular Biology. R. W. Jackson, ed. Caister Academic Press, Norfolk, United Kingdom.

Welsh, J., and McClelland, M. 1990. Fingerprinting genomes using PCR with arbitrary primers. Nucleic Acids Res. 18:7213-7218.

Williams, J. G. K., Hanafey, M. K., Rafalski, J. A., and Tingey, S. V. 1993. Genetic analysis using random amplified polymorphic DNA markers. Methods Enzymol. 218:704-740.

Wittker, R., Ludwig, W., Peiffer, S., and Kleiner, D. 1997. Isolation and characterization of *Burkholderia norimbergensis* sp. nov., a mildly alkaliphilic sulfur oxidizer. Syst. Appl. Microbiol. 20:549-553.

Woese, C. R., Kandler, O., and Wheelis, M. L. 1990. Towards a natural system of organisms: Proposal for the domains Archaea, Bacteria and Eucarya. Proc. Natl. Acad. Sci. U.S.A. 87:4576-4579.

Yabuuchi, E., Kosako, Y., Yano, I., Hotta, H., and Nishiuchi, Y. 1995. Transfer of two *Burkholderia* and an *Alcaligenes* species to *Ralstonia* gen. nov.: Proposal of *Ralstonia pickettii* (Ralston, Palleroni and Doudoroff 1973) comb. nov., *Ralstonia solanacearum* (Smith 1896) com. nov. and *Ralstonia eutropha* (Davis 1969) comb. nov. Microbiol. Immunol. 39:897-904.

Yabuuchi, E., Kosako, Y., Naka, T., Suzuki, S., and Yano, I. 1999. Proposal of *Sphingomonas suberifaciens* (van Bruggen, Jochimsen and Brown 1990) comb. nov., *Sphingomonas natatoria* (Sly 1985) comb. nov., *Sphingomonas ursincola* (Yurkov et al. 1997) comb. nov., and emendation of the genus *Sphingomonas*. Microbiol. Immunol. 43:339-349.

Yin, P., Li, T.-Y., Xie, M.-H., Jiang, L., and Zhang, Y. 2006. A type Ib ParB protein involved in plasmid partitioning in a gram-positive bacterium. J. Bacteriol. 188:8103-8108.

Young, J. M., Kuykendall, L. D., Martînez-Remero, E., Kerr, A., and Sawada, H. 2001. A revision of *Rhizobium* Frank 1889, with an emended description of the genus, and the inclusion of all species of *Agrobacterium* Conn 1942 and *Allorhizobium undicola* de Lajudie *et al.* 1998 as new combinations: *Rhizobium radiobacter, R. rhizogenes, R. rubi, R. undicola*, and *R. vitis*. Int. J. Syst. Evol. Microbiol. 51:89-103.

Young, J. M., Bull, C. T., De Boer, S. H., Firrao, G., Saddler, G. E., Stead, D. E., and Takikawa, Y. 2004. Names of plant pathogenic bacteria, 1864-2004. www.isppweb.org/names_bacterial_revised.asp.

Zreik, L., Bove, J. M., and Garnier, M. 1998. Phylogenetic characterization of the bacterium-like organism associated with marginal chlorosis of strawberry and proposition of a *Candidatus* taxon for the organism '*Candidatus* Phlomobacter fragariae'. Int. J. Syst. Bacteriol. 48:257-261.

Symptoms Caused by Plant-Pathogenic Bacteria

4

Although the host may affect symptom expression, there are four basic types of symptoms associated with bacterial diseases of plants (Billing, 1987). These symptom groupings are necrosis, vascular wilt, soft rot, and overgrowths.

Numerous bacterial species kill cells, resulting in rapid necrosis (death) of tissues, which is the most commonly encountered bacterial disease symptom. Necrosis may affect only small areas or may be so extensive that it leads to death of the plant host. Transitional changes in symptoms also lead to tissue necrosis. For example, angular leaf spot of cotton, caused by *Xanthomonas axonopodis* pv. *malvacearum*, starts out with water-soaked lesions on the leaves. The lesions gradually become necrotic and delimited by the veins. Infected cotton bolls also develop water-soaked lesions followed by necrosis.

Other bacterial species can cause vascular wilt, most often typified by the drooping of foliage due to the occlusion of vascular tissue, which inhibits movement of water and vascular fluid. Depending on the severity of the infection, water-deprived tissues may become necrotic, leading to defoliation and the complete collapse of the plant. When the phloem is infected, vascular wilt symptoms often include yellowing of the foliage and stunting of the whole plant. These phloem-affecting vascular wilt diseases are called yellows and are caused by phytoplasmas, bacteria-like prokaryotes without a cell wall.

The third symptom type results from the dissolution of plant tissues by macerating enzymes and the subsequent release of cellular fluids. This symptom is commonly known as soft rot and occurs most often in fleshy fruit, vegetables, and ornamentals. It is frequently accompanied by a foul odor produced by secondary bacteria growing on the diseased tissue.

The fourth type of symptom is typified by abnormal overgrowths of infected leaves, stems, trunks, or roots. These overgrowths often take the form of galls (plant tumors), leafy galls, fasciation (multiple fused, flattened stems), or phyllody (conversion of flower parts to leaflike structures).

It should be stressed here that there are many variations and even overlaps within these basic symptom types because infection results in many different observable host responses. For example, necrotic symptoms may occur as leaf spots, leaf streaks, leaf scorch, blights, or cankers. Likewise, vascular wilt symptoms may vary from flagging and wilt to stunting and plant death, all derived from water deprivation. Soft rots will take on different appearances depending upon the host and the portion of the host affected. Among the abnormal growth symptoms, there can be a wide range of sizes and distributions of malformed plant parts. Examples of diseases with characteristic symptoms are summarized in Table 4-1. Pathogens that cause each of these symptom types are detailed in the following chapters.

Sometimes diseases remain latent (symptom free) until environmental factors trigger symptom expression or elicit unusual effects that are nondeleterious to the plant but economically undesirable (see Chapter 9 for details).

Table 4-1. Examples of Diseases Showing Variations in the Four Basic Symptom Types[a]

Symptom	Disease	Pathogen
Leaf spot/streak/stripe (necrosis type)	Bacterial spot of tomato and pepper	*Xanthomonas axonopodis* pathovars
	Bacterial speck of tomato	*Pseudomonas syringae* pv. *tomato*
	Leaf scald of sugar cane	*Xanthomonas albilineans*
	Leaf streak of rice	*Xanthomonas oryzae* pv. *oryzicola*
	Bacterial streak of maize	*Burkholderia andropogonis*
	Black chaff of wheat	*Xanthomonas translucens* pv. *translucens*
Blight (necrosis type)	Fire blight of apple and pear	*Erwinia amylovora*
	Black rot of crucifers	*Xanthomonas campestris*
	Bacterial blight of anthurium and other aroids	*Xanthomonas axonopodis* pv. *dieffenbachiae*
	Bacterial blight of rice	*Xanthomonas oryzae* pv. *oryzae*
	Halo blight of bean	*Pseudomonas syringae* pv. *phaseolicola*
	Angular leaf spot/blight of cotton	*Xanthomonas axonopodis* pv. *malvacearum*
	Walnut blight	*Xanthomonas arboricola* pv. *juglandis*
Scorch/decline (necrosis type)	Pierce's disease of grape	*Xylella fastidiosa* subsp. *piercei*
	Almond leaf scorch	*Xylella fastidiosa* subsp. *piercei*
	Bacterial leaf scorch of elm, oak, ash, and *Prunus* spp.	*Xylella fastidiosa* subsp. *multiplex*
Canker (necrosis type)	Bacterial canker of tomato	*Clavibacter michiganensis* subsp. *michiganensis*
	Citrus canker	*Xanthomonas citri* pv. *citri*
	Bacterial canker of lilac and stone fruits	*Pseudomonas syringae* pv. *syringae*
	Deep bark canker of English walnut	*Brenneria rubrifaciens*
	Shallow bark canker of English walnut	*Brenneria nigrifluens*
Foliar streak and defoliation (necrosis type)	Citrus greening, huanglongbing	'*Candidatus* Liberibacter asiaticus'
Linear streaking of shoots and canker formation at shoot base (necrosis type)	Grapevine canker	*Xylophilus ampelinus*
Scab (necrosis type)	Potato scab	*Streptomyces acidiscabies*
Wilt (wilt type)	Bacterial wilt of solanaceous crops and many other hosts	*Ralstonia solanacearum*
	Cucumber wilt	*Erwinia tracheiphila*
	Bacterial wilt of bean	*Curtobacterium flaccumfaciens* pv. *flaccumfaciens*
	Stewart's wilt of corn	*Pantoea stewartii*
Yellows (wilt type)	Citrus stubborn disease	*Spiroplasma citri*
	Aster yellows of Asteraceae members and other families, such as Apiaceae	'*Candidatus* Phytoplasma asteris'
	Lethal yellowing of coconut palm	'*Candidatus* Phytoplasma palmae'
Galls, leafy galls, fasciation, and phyllody (tissue overgrowth type)	Crown gall	*Agrobacterium tumefaciens*
	Hairy root	*Agrobacterium rhizogenes*
	Olive knot	*Pseudomonas savastanoi* pv. *savastanoi*
	Oleander knot	*Pseudomonas savastanoi* pv. *nerii*
	Bacterial gall of *Gypsophila* spp.	*Pantoea agglomerans* pv. *gypsophilae*
	Clover phyllody	'*Candidatus* Phytoplasma asteris' group member
	Ash yellows and lilac witches'-broom	'*Candidatus* Phytoplasma fraxini'
	Periwinkle bunch disease	*Xylella fastidiosa* subsp. *piercei*
	Maize bushy stunt	'*Candidatus* Phytoplasma asteris' group member
	Ratoon stunt of sugarcane	*Leifsonia xyli* subsp. *xyli*
	Fasciation (leafy galls) of phlox, geranium, and many other hosts	*Rhodococcus fascians*

(continued)

[a] Diseases and disease categories in this table were provided by Anne Alvarez.

Table 4-1. Examples of Diseases Showing Variations in the Four Basic Symptom Types[a] *(continued)*

Symptom	Disease	Pathogen
Rot (soft rot type)	Soft rots of vegetable and ornamental crops	*Pectobacterium carotovorum* subsp. *carotovorum*
	Blackleg of potato	*Pectobacterium atrosepticum*
	Soft rot and blight of ornamental and vegetable crops	*Dickeya chrysanthemi* pv. *chrysanthemi*
	Watermelon fruit blotch	*Acidovorax avenae* subsp. *citrulli*
	Basal glume rot of wheat	*Pseudomonas syringae* pv. *atrofaciens*
	Kernel discoloration of rice	*Pseudomonas fuscovaginae*
	Ring rot of potato	*Clavibacter michiganensis* subsp. *sepedonicus*

Necrosis

Leaf spots, stem cankers, and blights are rapid necrotic reactions characterized by the progressive killing of tissues as the pathogen invades its host. There are a large number of bacteria, particularly those belonging to the genus *Xanthomonas* and the species *Pseudomonas syringae*, that cause these symptoms (see Tables 3-7, 3-12, and 4-1).

The plant-pathogenic bacteria that cause rapid necrosis contain genes whose products elicit a rapid tissue killing response in their host and progressively kill plant tissues surrounding the area of invasion. Other plant pathogens, such as members of the genus *Agrobacterium*, usually do not cause rapid necrosis of the host tissue. They can, however, induce a necrotic reaction when experimentally inoculated into grapevine explants (Deng et al., 1995; Herlache et al., 2001; Argun et al., 2002). *Agrobacterium* species apparently possess normally silent *hrp* genes that elicit the hypersensitive response when mutated (Kado, 1994; Herlache et al., 2001). Opportunistic plant-pathogenic bacteria that may reside in the intercellular spaces of their hosts can also elicit necrosis when the plant host is stressed or shocked by severe changes in the environment, such as exposure to freezing temperatures, excess fertilization, or a deficiency or excess of water (see Chapter 9).

Although rapid necrosis in the form of localized cell death can be induced experimentally by injecting large (10^7 CFU/ml or higher) numbers of certain members of the families *Pseudomonadaceae*, *Burkholderiaceae*, *Xanthomonadaceae*, and *Enterobacteriaceae* into nonhost species, it is clear that the necrotic reaction caused by this treatment is purely artificial and does not truly reflect the natural circumstances in which infection causes rapid necrosis. The rapid necrotizing reaction produced by artificial injections of cultured bacteria into nonhosts is called the hypersensitive response (HR). The ability to elicit this defense response is commonly used as a diagnostic test in the identification of certain *Pseudomonas* and *Xanthomonas* species.

Fire blight, a disease primarily of the Rosaceae family of plants caused by *Erwinia amylovora,* is an example of a rapid necrotizing disease. A member of the family *Enterobacteriaceae*, *Erwinia amylovora* has exquisite transport and secretion systems that deliver pathogenic and virulence factors into the host cells (see Chapter 5 for details). Rapid invasion by this pathogen through leaf wounds, stomata, and floral stigmata leads to the production of 2,5-dihydrophenylalanine, amylovoran, levan, and a desferrioxamine-like siderophore. The cumulative effects of these compounds are the disruption of cell integrity in petioles and stem tissues, culminating in necrosis that appears as the blackening of the affected host tissues (Fig. 4-1). Amylovoran is an exopolysaccharide elaborated during the invasion of vessel elements (primarily the xylem). Levan (a linear fructan of 2,6 β-linked fructofuranose residues) and amylovoran are thought to plug the xylem elements and cause the leakage of tissue fluids that are forced to the stem surface in the form of ooze droplets (bacteria-laden droplets of sap appearing on the plant surface) (Fig. 4-2). The above compounds contribute to the necrosis of tissues below the epidermal layer as *Erwinia amylovora* moves systemically (Fig. 4-1).

Another example of a necrogenic pathogen is *Pseudomonas syringae* pv. *syringae*, which causes bacterial canker of stone fruit trees (see Chapter 5 for details). There are both nonpathogenic (epiphytes) and pathogenic forms of this pathovar (Latorre and Jones, 1979; Hirano and Upper, 1990; Gardan et al., 1999; Schaad et al., 2000). Both strain types are able to induce the HR in tobacco (Latorre and Jones, 1979; Hirano and Upper, 1990). Wounds caused by wind-blown sand and leaf scars provide entry for the pathogen. Like *Erwinia amylovora*, *Pseudomonas syringae* pv. *syringae* is equipped with both transport and secretion systems to introduce pathogenicity and virulence factors and toxins into the host cell (see Chapter 5 for

Fig. 4-1. Subepidermal necrotic symptoms caused by the invasion of *Erwinia amylovora*, which moved systemically through the flower and down the petiole into the branch.

Fig. 4-2. Fire blight symptoms on apple with secreted ooze (green arrows).

Fig. 4-3. Necrotic symptoms of a young peach tree (left) and an established apricot tree (right) systemically infected by *Pseudomonas syringae* pv. *syringae*. The dark brown regions (exposed) demark the killed tissues of the trunk. Extensive tissue killing results in the girdling and death of the tree.

details). Initial intercellular infection by *Pseudomonas syringae* pv. *syringae* is followed by the rapid systemic invasion and killing of tissues surrounding the invaded areas (Fig. 4-3).

Soft Rot (Maceration)

Bacterial soft rots are the result of dissolution (maceration) of cells and tissues by degradative enzymes produced and exported by the pathogen. Many of these enzymes are expressed only when the pathogen interacts with the host plant. Pectin lyases, together with pectate (polygalacturonic acid) lyases, play a key role in maceration, as do proteases and lipases and, in some cases, cellulases. Collectively, these enzymes provide oligomers in the form of oligosaccharides, simple sugars, peptides, and single amino acids that can be imported and consumed by the bacteria. Often, soft rots attract opportunistic pathogens that contribute to tissue maceration and may produce a range of foul odors while the primary pathogen continues its invasion of host tissue.

Soft rot-causing bacterial species are members of the families *Enterobacteriaceae*, *Pseudomonadaceae*, and *Burkholderiaceae*. *Pectobacterium carotovorum* subsp. *carotovorum* is the prototype species for soft rot-causing bacteria. The symptoms caused by this pathogen are generally similar to those of other soft rots. Watery depressions form on host tissues. Soft rot is promoted by microaerophilic conditions, high moisture, and warm temperatures. When such conditions subside, disease progression ceases and macerated tissue dries out (Fig. 4-4).

Fig. 4-4. Soft rot injury of carrot roots caused by *Pectobacterium carotovorum* subsp. *carotovorum*. If left in the soil, the carrot roots would have been totally macerated. (Courtesy R. Michael Davis)

Fig. 4-5. Onion foliar symptoms of soft rot caused by *Pectobacterium carotovorum* subsp. *carotovorum*. (Reprinted from Mohan, 2008)

Pectobacterium carotovorum subsp. *carotovorum* also causes bulb rot of onions. The pathogen usually enters through small wounds and favors the neck of the onion bulb because of the high level of moisture present at the base of each leaf. Infection leads to degradation of the structural tissues of the plant, causing wilt and collapse of the leaves and followed by death (Fig. 4-5). Warm and humid conditions favor infection.

Pectobacterium atrosepticum causes blackleg of potato. Besides causing soft rot of the potato tuber, *Pectobacterium atrosepticum* infects the crown of the potato plant and invades all of the major stems, resulting in the total collapse of the plant (Fig. 4-6). Invaded stems appear dark brown to black. This disease is favored by mild to cool climatic conditions and high moisture levels.

Dickeya chrysanthemi pv. *chrysanthemi* causes soft rot of a number of crops, including pineapple (Fig. 4-7) and chrysanthemum. (*Erwinia chrysanthemi* pathovars have recently been reclassified into various *Dickeya* species, and not all former *Erwinia chrysanthemi* strains are now *Dickeya chrysanthemi*.) As with diseases caused by *Pectobacterium carotovorum* subsp. *carotovorum*, enzymes released by the pathogen progressively degrade tissues in advance of the bacterial invasion of the host. The action of cell wall-, pectin-, protein-, lipid-, and carbohydrate-degrading enzymes released by the pathogen is sufficient to cause the complete collapse of the plant.

It should be pointed out that saprophytic bacteria may also produce a number of pectin lyases, cellulases, proteases, and lipases. For example, some strains of the normally saprophytic bacterium *Pseudomonas fluorescens* constitutively produce pectin lyases and cellulases and are able to cause soft rots in vegetables, such as broccoli, cauliflower, lettuce, and celery, after harvest (Zucker and Hankin, 1970). In plant-pathogenic bacteria, the expression of these enzymes is induced rather than constitutive. *Pseudomonas fluorescens* strains are considered opportunistic pathogens and can cause disease under plant stress conditions, such as refrigera-

Fig. 4-6. Blackleg of potato cv. Russet Burbank caused by *Pectobacterium atrosepticum*.

Fig 4-7. Soft rot (heart rot) of pineapple in Polomolok, Mindanao, Philippines, caused by *Dickeya chrysanthemi* pv. *chrysanthemi* (previously *Erwinia chrysanthemi*). The basal rot also causes chlorosis of the lower leaves.

tor temperatures that favor psychrophilic *Pseudomonas fluorescens* (Hernandez-Anguiano et al., 2004). Initial infection causes water-soaking symptoms, followed by progressive but localized maceration.

Vascular Wilts (Bacterial Wilts and Yellows Diseases)

Bacterial wilts are characterized by the obstruction of the vascular system of stems, leaves, and flowers that eventually culminates in either total collapse of herbaceous plants or withering of the foliar parts of trees. Examination of a cross section of a stem infected by a wilt-causing bacterium usually reveals creamy to yellow bacteria-laden slime or ooze. The wilt of the host plant is caused by the progressive accumulation of slime or ooze produced by the invading organism that blocks the vascular elements used for nutrients and water conduction. Vascular wilts described as bacterial wilts are caused by both Gram-negative and Gram-positive bacteria found mainly in the families *Ralstoniaceae*, *Enterobacteriaceae*, and *Microbacteriaceae*.

Ralstonia solanacearum, a member of the family *Ralstoniaceae*, causes wilt of a wide range of host plants (see Chapter 7 for a detailed host range). The disease known as Southern wilt of geranium is characterized by the early flagging or drooping of the foliar parts followed by yellowing of the leaf margins and eventual plant collapse. In infected potatoes and tomatoes, wilting occurs suddenly, with water-soaking of the pith tissue. As the infection progresses, the pith turns brown. Wilting, yellowing, and stunting of tomato or potato plants in the absence of other symptoms is characteristic of infection by *Ralstonia solanacearum*. A different race of *Ralstonia solanacearum* also causes bacterial wilts of banana known as Moko disease, Bugtok disease, and blood disease.

Pantoea stewartii subsp. *stewartii*, a member of the family *Enterobacteriaceae*, causes Stewart's wilt of corn (sweet corn is highly susceptible). Water-soaked lesions is the primary symptom. Leaves appear pale green and display long yellow streaks with wavy margins paralleling the leaf veins (Fig. 4-8 left). The pith of the infected plant has brown cavities. Infection can occur throughout the life of the plant, including the pretasseling stage. The pathogen is transmitted by the corn flea beetle, *Chaetocnema pulicaria* Melsheimer, a small, shiny black beetle with strong hind legs that are used for jumping (Fig. 4-8 right).

Another member of the family *Enterobacteriaceae*, *Erwinia tracheiphila*, causes bacterial wilt of cucurbits, such as cucumber, muskmelon, squash, pumpkin, white gourd, wild gourd, other wild cucurbits, and occasionally watermelon. The crops most severely affected by wilt are cucumber and muskmelon.

Erwinia tracheiphila is spread primarily by the striped cucumber beetle, *Acalymma vittata* (Fabr.), and the spotted cucumber beetle, *Diabrotica undecimpunctata howardi* Barber. After infection, dull green patches form on the leaves followed by progressive wilting, starting with the flagging of the leaves and petioles and then the lateral shoots. The entire plant eventually collapses.

Curtobacterium flaccumfaciens pv. *flaccumfaciens*, a member of the Gram-positive *Microbacteriaceae* family, causes vascular wilt of beans. Invasion and proliferation of this seedborne bacterium in the vascular bundles of the host result in rapid drooping of the leaves and total collapse of the plant, particularly young seedlings. On established older plants, golden to yellow-brown, irregularly shaped lesions form and defoliation may occur. Depending on the strain of *Curtobacterium*

Fig. 4-8. Leaf symptoms of Stewart's wilt caused by *Pantoea stewartii* (left) and the flea beetle vector of the disease (right). (Left: Courtesy Jim Stack, reprinted, by permission, from Stack, 2009. Right: Courtesy Dennis Calvin, reprinted, by permission, from Pennsylvania State University.)

flaccumfaciens pv. *flaccumfaciens*, the seed may display either a purple or yellow-orange pigmentation beneath its coat.

Clavibacter michiganensis subsp. *insidiosus* causes bacterial wilt of alfalfa by colonizing the vascular system and disrupting water transport of the host. Infected plants appear yellow and frequently are stunted (Fig. 4-9). After mowing, they produce smaller leaves than do their healthy counterparts. Water stress initiates severe wilting and eventual death. Examination of the vascular region of the taproot reveals yellow banding of the invaded regions. A yellow ring demarking the vascular bundles is often seen in root cross sections. The disease is often found in fields or sections of fields with poor drainage, suggesting that water is important for the pathogen to proliferate and infect. The pathogen gains entry through wounds, such as those caused by the stem nematode, *Ditylenchus dipsaci* (Kühn) Filipjev.

Clavibacter michiganensis subsp. *sepedonicus* causes a wilt disease of potato called ring rot. The pathogen invades the vascular system of the tuber, producing cream-colored bacterial exudates. When infected tubers are cross cut at the stolon end and squeezed, the cream to yellow exudates becomes obvious (Fig. 4-10). Leaves of infected plants droop and are off-color; they often yellow at the margins and may curl inward onto themselves. The pathogen systemically invades the vascular tissues and proliferates copiously so that bacterial ooze can be squeezed out from a cut end of an infected stem. Certain cultivars, such as Russet Burbank and Nooksack,

Fig. 4-9. Dwarfed alfalfa caused by *Clavibacter michiganensis* subsp. *insidiosus* shown on the left of a typical healthy alfalfa plant. (Courtesy R. Michael Davis)

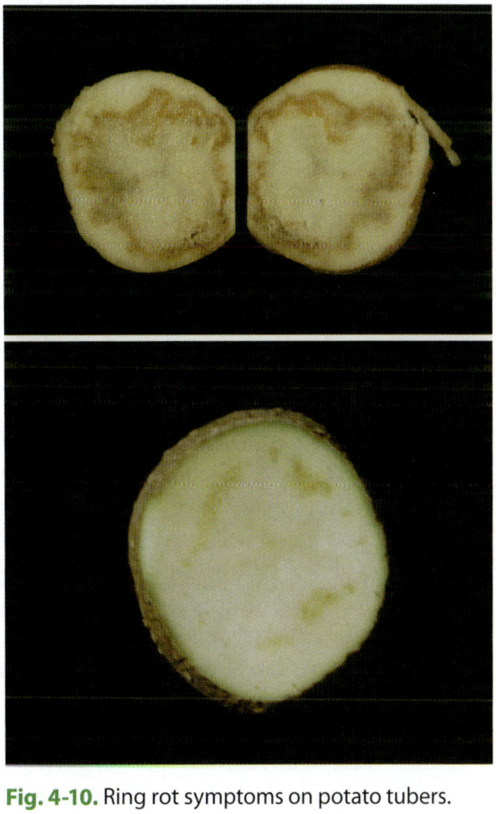

Fig. 4-10. Ring rot symptoms on potato tubers. **Top,** Advanced stage; **Bottom,** Early oozing stage with yellow bacterial exudates appearing around the vascular cylinder. (Courtesy R. Michael Davis)

produce rosettes of stunted terminal leaves when infected by *Clavibacter michiganensis* subsp. *sepedonicus*.

A second major category of vascular wilt diseases occurs as the result of colonization of the phloem rather than of the xylem. These diseases are caused by fastidious, phloem-limited prokaryotes called phytoplasmas, previously known as mycoplasma-like organisms or MLOs. Because the xylem is not colonized, the symptoms of these diseases are not those of water stress, except that they often cause stunting. The phytoplasmas are pleomorphic organisms, bound by a triple membrane rather than by a cell wall. Because of their fastidious nature, their presence in phloem is generally verified by electron microscopy or staining techniques with light microscopy, or polymerase chain reaction (PCR), rather than by culturing. The systemic symptoms include yellowing of the foliage and virescence (greening) of the flower petals. Flower parts may also revert back to leaflike structures within the flower heads of infected plants. The group of phytoplasmas collectively known as aster yellows (the '*Candidatus* Phytoplasma asteris' group) has a very broad host range, including marigold, coreopsis, cosmos, echinacea, and China aster. Aster yellows is vectored by the aster leafhopper. Aster yellows disease causes hosts to show chlorotic leaves and stems as well as flower abnormalities, often including virescence and phyllody.

Overgrowths and Fasciation

Within the families *Rhizobiaceae*, *Pseudomonadaceae*, *Enterobacteriaceae*, and *Nocardiaceae*, there are species that cause galls and fasciation. These species produce either auxin (indole-3-acetic acid), cytokinin, or both phytohormones, the result of which is hypertrophic growth of tissue surrounding the infection site. Infection is initiated through wounds that allow the bacteria to proliferate and begin elaborating the phytohormones. The ratio of the amount of auxin and cytokinin elaborated by the pathogen determines the type of overgrowth produced. Hairy root, a disease caused by *Agrobacterium rhizogenes*, is the result of excess cytokinin in relation to a lower amount of auxin produced. Likewise, fasciation, a disease of many flower crops caused by *Rhodococcus fascians*, is the result of high cytokinin production. Experimental inoculation of leaves by these bacteria causes the initial formation of galls at the site of inoculation. Inoculation of meristematic regions results in fasciation that is more typical of natural infections, which most often take the form of leafy galls.

Rhodococcus fascians naturally infects a number of hosts. This pathogen displays a relatively wide host range and produces an assortment of symptoms when experimentally inoculated at high dosages. Pathogenic *Rhodococcus fascians* isolates were associated with and isolated from eight species or hybrids not previously reported as hosts: *Acanthus mollis* L., *Campanula* × 'Sarastro', *Heliopsis helianthoides* (L.) Sweet 'Loraine Sunshine', *Nemesia* × 'Natalie', *Hosta* × 'Blue Umbrella', *Verbascum* × 'Sierra Sunset', *Veronica spicata* L. 'Minuet', and *Viola* × 'Purple Showers' (Putnam and Miller, 2006). Under natural field conditions, abnormal proliferation of leaves into a leafy gall (Fig. 4-11) is a primary symptom.

Rhodococcus fascians, a member of the family *Nocardiaceae*, contains a large conjugative linear plasmid. On this plasmid is a gene called *ipt* that encodes isopentenyl transferase. This enzyme catalyzes the formation of isopentenyl adenosine monophosphate, the precursor of one or more cytokinins.

Pantoea agglomerans pv. *gypsophilae* causes galls on baby's breath (*Gypsophila paniculata* L.). The pathogen produces both auxin and cytokinin, whose concentrations are equivalent and therefore result in gall formation instead of fasciation. Galls are usually formed in the crown region of the stem (see Fig. 8-6). Vegetative propagation spreads the pathogen, and galls are commonly observed at the cut ends of the vegetative piece. The pathogen may become systemic, moving through the xylem vessels.

Pseudomonas savastanoi pv. *savastanoi* causes olive knot, a gall disease of olive and oleander (Fig. 4-12). The pathogen gains entry into the host via wounds, such as leaf scars, and artificial breaks in the epidermis and periderm of susceptible

Fig. 4-11. Leafy gall at the base of a *Verbascum* sp. (Reprinted from Putnam and Miller, 2006)

Fig. 4-12. Olive knot of 'Mission' olive (left) and oleander knot (right) caused by *Pseudomonas savastanoi* pv. *savastanoi*.

parts of the host. Leaf scars remain susceptible for about a week after leaves fall. Bacterial cells move systemically through the lactifers (latex-bearing vessels), and galls (also called knots) are initiated at wound sites along the stems, branches, and flower buds. The pathogen is spread by wind-blown rain and bacteria-laden dust. The organism produces both auxin and cytokinin, which are elaborated during infection.

Among pathogens that cause overgrowth of plant tissue, the most information is available on members of the genus *Agrobacterium*. *Agrobacterium rhizogenes* causes hairy root on woody crops. *Agrobacterium vitis* causes galls on grapevine. *Agrobacterium tumefaciens* causes crown gall of hundreds of dicotylendonous plants and a few monocots. Experimental inoculations with *Agrobacterium tumefaciens* have demonstrated that plants from at least 93 families are susceptible to this gall-producing bacterium. Under natural field circumstances, crown gall is often found on woody hosts, such as *Prunus* species (almond, apricot, cherry, peach, plum), members of the family Juglandaceae (hickory, walnut), members of the family Rosaceae (apple, pear, rose), and grapevine. Ornamentals in the family Asteraceae, such as chrysanthemum and argyranthemum (marguerite daisy), also develop symptoms occasionally. Crown galls are most commonly found at the base of stems (or trunks) and underground on roots (Fig. 4-13).

Unlike other gall-forming pathogens, such as *Pantoea agglomerans* pv. *gypsophilae* and *Pseudomonas savastanoi* pv. *savastanoi*, *Agrobacterium rhizogenes*, *Agro-*

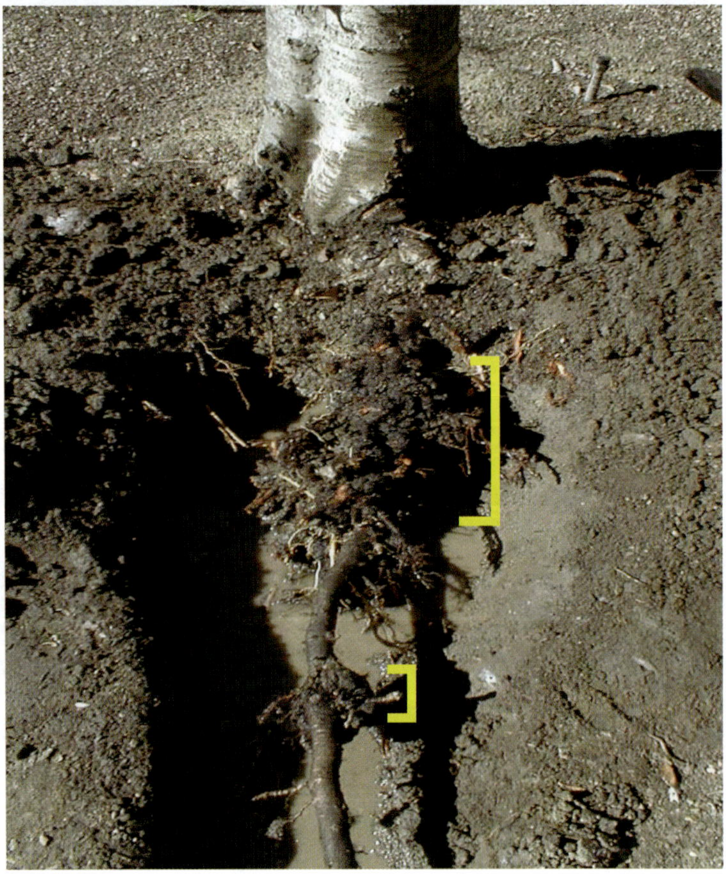

Fig. 4-13. Crown gall tumors (bracketed) formed underground on a major root of a cherry tree infected by *Agrobacterium tumefaciens*.

bacterium vitis, and *Agrobacterium tumefaciens* transfer phytohormone-encoding genes into their hosts by horizontal gene transfer events. The outcome is essentially the same in that both auxin and cytokinin are produced to initiate gall formation.

Diseases with Latent Symptoms

Most bacterial pathogens cause one or more of the four basic symptom types outlined above. There are, however, many diseases that do not always present symptoms. The damage caused by the pathogen may become apparent only some time after the crop is harvested, for example. In some cases, the pathogen has caused damage internally and the disease symptoms become visible when a fruit or nut is cut open. When pathogens are present in low numbers, cuttings of floral or nursery crops may be sold to other nurseries before symptoms of vascular wilt or galls have developed. Stock plants grown under cooler growing conditions may harbor inoculum of leaf spot-, soft rot-, or vascular wilt-causing bacteria that multiply and produce symptoms readily in the warmer environments used to propagate cuttings. In some instances, crops are susceptible to pathogens and have weed or wild relative reservoirs that show no symptoms. In other cases, damage comes from the conversion of natural products that become visible only when the crop is processed. An example is pink disease of pineapple, caused by *Tatumella citrea*. The internal fruit tissue of pineapple becomes infected with the bacterium via florets on the fruit. Within the fruit, *Tatumella citrea* oxidizes the glucose into gluconate, 2-ketogluconate, and 2,5-diketogluconate. The latter compound forms a red color upon heating. Hence, any canned diseased fruit turns red-brown and is unmarketable (Fig. 4-14).

An example of a bacterial pathogen that can remain latent in some hosts is *Xylella fastidiosa* subsp. *piercei*, which causes Pierce's disease of grapevine and leaf scorch of almond, oleander, and oak. More than 100 asymptomatic species are known to harbor *Xylella fastidiosa* subsp. *piercei*; its extensive host range includes both dicots and monocots. The xylem of wild plants, such as California blackberry, Sudan grass, ryegrass, brome grass, and Russian thistle, may harbor the pathogen in moderate to large numbers without symptom development. The bacterium is spread in the xylem after inoculation by a leafhopper that vectors the pathogen as it feeds.

Fig. 4-14. Pink disease of pineapple. Top row: uninfected fruit. Bottom row: fruit infected by *Tatumella citrea* and canned. (Reprinted from Kado, 2003)

Fig. 4-15. Asymmetric coloration of mandarin oranges caused by '*Candidatus* Liberibacter asiaticus'. (Reprinted from Gottwald et al., 2007)

The blue-green sharpshooter (*Graphocephala atropunctata* (Signoret)), the green sharpshooter (*Draeculacephala minerva* Ball), the glassy-winged sharpshooter (*Homalodisca coagulata* (Say)), and the red-headed sharpshooter (*Carneocephala fulgida* Nottingham) are known vectors. In some instances, disease symptoms comparable to Pierce's disease on grapevine may be induced in asymptomatic hosts by altering the environment, for example, by brief heat-shock treatment.

'*Candidatus* Liberibacter asiaticus', the causal organism of huanglongbing (HLB) of citrus, is transmitted by psyllids, such as the African citrus psyllid, *Trioza erytreae* (Del Guercio), and the Asian citrus psyllid, *Diaphorina citri* Kuwayama. HLB, also known as citrus greening, produces a wide range of symptoms in cultivated citrus. The symptoms can range from dieback, yellowing of shoot tips, and mottled chlorosis or corking of leaves to multiple coloration of fruits (Fig. 4-15) and asymmetric fruit maturation. HLB remains symptomless in wild members of the citrus family; *Vepris lanceolata* (Lam.) G. Don, for example, is an asymptomatic host of HLB (Gottwald et al., 2007).

'*Candidatus* Liberibacter africanus' and '*Candidatus* Liberibacter americanus' cause similar citrus greening diseases in Africa and South America, respectively.

Sequence comparisons at the 16S ribosomal DNA and 16S-23S interspacer regions appear identical among different '*Candidatus* Liberibacter asiaticus' Asian isolates (Tomimura et al., 2009). Further phylogenetic analyses of bacteriophage-type DNA polymerase sequences within the '*Candidatus* Liberibacter asiaticus' population revealed three clusters: one group consisted of isolates from Indonesia while the other clusters did not correlate well with any particular location (Tomimura et al., 2009). Although '*Candidatus* Liberibacter' species have not been cultured in vitro, a culture medium called "Liber A" containing citrus vein extract and a growth factor was recently reported to sustain the growth of '*Candidatus* Liberibacter' species (Sechler et al., 2009). The cultured strains of '*Candidatus* Liberibacter asiaticus' and a strain of '*Candidatus* Liberibacter americanus' were pathogenic on citrus and could be reisolated from noninoculated tissues of inoculated trees and seedlings.

REFERENCES

Argun, N., Momol, M. T., Maden, S., Momol, E. A., Reid, C. L., Çelek, H., and Burr, T. J. 2002. Characterization of *Agrobacterium vitis* strains isolated from Turkish grape cultivars in the Central Anatolia region. Plant Dis. 86:162-166.

Billing, E. 1987. Bacteria as Plant Pathogens. Van Nostrand Reinhold, New York.

Deng, W., Pu, X.-A., Goodman, R. N., Gordon, M. P., and Nester, E. W. 1995. T-DNA genes responsible for inducing a necrotic response on grape vines. Mol. Plant-Microbe Interact. 8:538-548.

Gardan, L., Shafik, H., Belouin, S., Brosch, R., Grimont, F., and Grimont, P. A. D. 1999. DNA relatedness among the pathovars of *Pseudomonas syringae* and description of *Pseudomonas tremae* sp. nov. and *Pseudomonas cannabina* sp. nov. (ex. Sutic and Dowson 1959). Int. J. Syst. Bacteriol. 49:469-478.

Gottwald, T. R., da Graça, J. V., and Bassanezi, R. B. 2007. Citrus huanglongbing: The pathogen and its impact. Online. Plant Health Progress doi:10.1094/PHP-2007-0906-01-RV.

Herlache, T. C., Zhang, H. S., Ried, C. L., Carle, S. A., Zheng, D., Basaran, P., Thaker, M., Burr, A. T., and Burr, T. J. 2001. Mutations that affect *Agrobacterium vitis*-induced grape necrosis also alter its ability to cause a hypersensitive response on tobacco. Phytopathology 91:966-972.

Hernandez-Anguiano, A. M., Suslow, T. V., Leloup, L., and Kado, C. I. 2004. Biosurfactants produced by *Pseudomonas fluorescens* and soft-rotting of harvested florets of broccoli and cauliflower. Plant Pathol. 53:596-601.

Hirano, S. S., and Upper, C. E. 1990. Population biology and epidemiology of *Pseudomonas syringae*. Annu. Rev. Phytopathol. 28:155-177.

Kado, C. I. 1994. Anti-host-defense systems are elaborated by plant pathogenic bacteria. Pages 581-591 in: Molecular Mechanisms of Bacterial Virulence. C. I. Kado and J. H. Crosa, eds. Kluwer Academic Publishers, Dordrecht, the Netherlands.

Kado, C. I. 2003. Pink disease of pineapple. APS*net* Feature. American Phytopathological Society, St. Paul, MN. www.APSnet.org/online/feature/pineapple/.

Latorre, B. A., and Jones, A. L. 1979. Evaluation of weeds and plant refuse as potential sources of inoculum of *Pseudomonas syringae* in bacterial canker of cherry. Phytopathology 69:1122-1125.

Mohan, S. K. 2008. Soft rots. Pages 59-62 in: Compendium of Onion and Garlic Diseases and Pests, 2nd ed. H. F. Schwartz and S. K. Mohan, eds. The American Phytopathological Society, St. Paul, MN.

Pennsylvania State University. Corn flea beetle and bacterial wilt. Entomological Notes, The Pennsylvania State University, University Park, PA. http://ento.psu.edu/extension/factsheets/corn-flea-beetle-and-bacterial-wilt.

Putnam, M. L., and Miller, M. 2006. Pathogenic isolates of *Rhodococcus fascians* from new hosts in the United States. Plant Dis. 90:526.

Schaad, N. W., Vidaver, A. K., Lacy, G. H., Rudolph, K., and Jones, J. B. 2000. Evaluation of proposed amended names of several pseudomonads and xanthomonads and recommendations. Phytopathology 90:208-213.

Sechler, A., Schuenzel, E. L., Cooke, P., Donnua, S., Thaveechai, N., Postnikova, E., Stone, A. L., Schneider, W. L., Damsteegt, V. D., and Schaad, N. W. 2009. Cultivation of '*Candidatus* Liberibacter asiaticus', '*Ca.* L. africanus', and '*Ca.* L. americanus' associated with huanglongbing. Phytopathology 99:480-486.

Stack, J. 2009. Stewart's wilt. CropWatch, Plant Disease Central, Corn. University of Nebraska-Lincoln Extension, Lincoln, NE. http://pdc.unl.edu/agriculturecrops/corn/stewartswilt.

Tomimura, K., Miyata, S.-I., Furuya, N., Kubota, K., Okuda, M., Subandiyah, S., Hung, T.-H., Su, H.-J., and Iwanami, T. 2009. Evaluation of genetic diversity among '*Candidatus* Liberibacter asiaticus' isolates collected in Southeast Asia. Phytopathology 99:1062-1069.

Zucker, M., and Hankin, L. 1970. Regulation of pectate lyase synthesis in *Pseudomonas fluorescens* and *Erwinia carotovora*. J. Bacteriol. 104:13-18.

Rapid Necrotizing Diseases

Necrogenic bacteria (those that cause necrotic symptoms) grow intercellularly in their hosts and rapidly kill host tissues at initial infection sites. They can either be compartmentalized to specific tissues or be highly invasive in a non-tissue-specific manner. Necrogenic bacteria produce toxic compounds that permeate and kill cells of tissues surrounding the site of infection. Many of the toxins produced are, either alone or in combination, metabolic inhibitors and deleteriously affect membrane potentials, resulting in loss of membrane integrity, efflux of electrolytes (primarily potassium ions), and influx of calcium ions. The killing reaction results from the initial leakage of electrolytes, which is seen as a water-soaking symptom, followed by the collapse of tissue and desiccation. The rapid killing potential of these necrogenic bacteria may be observed when these bacteria are infiltrated into tissues of nonhost plants. The killing reaction in nonhost plants is called the hypersensitive response (HR).

As described in Chapter 4, bacterial pathogens that rapidly kill host cells produce dramatic disease symptoms. Usually, rapid-necrosis-producing bacterial pathogens are members of the class *Gammaproteobacteria*. They include members of the families *Xanthomonadaceae*, *Pseudomonadaceae*, and *Enterobacteriaceae*. Necrotizing species of these families are equipped with type III secretion systems (see Chapter 10 for details) that inject various toxins and pathogenic effectors into the cells of the host tissue.

Fire Blight

Fire blight of pear and apple is the best example of a rapid necrogenic disease. *Erwinia amylovora* infects members of the Rosaceae family of plants, primarily members of the subfamily Pomoideae. Pears and apples are among the most economically important crops of this family, although many important ornamental species also are seriously affected. At least 32 genera within the family Rosaceae, including some *Prunus* species, are susceptible to this pathogen (Table 5-1). A recent survey showed that more than 200 species are susceptible to fire blight (Jock et al., 2000).

General Characteristics of *Erwinia amylovora*

Erwinia amylovora, a member of the family *Enterobacteriaceae*, possesses all the characteristics of the family with the exception that nitrate reductase is not produced. The bacteria are Gram-negative, non-spore-forming straight rods that are motile with peritrichous flagella. They are facultative anaerobes; produce organic acids from sugars, such as glucose and sucrose; and produce catalase but not cytochrome oxidase or urease. Growth is promoted by nicotinic acid, and cells accumulate glycogen as cellular reserve material. On rich medium containing sucrose,

Table 5-1. Host Range of *Erwinia amylovora*

Common Name	Scientific Name	Reference
Western serviceberry	*Amelanchier alnifolia* (Nutt.) Nutt. ex M. Roem.	O'Gara, 1910; Thomas and Ark, 1934
Downy serviceberry	*Amelanchier arborea* (F. Michx.) Fernald var. *austromontana* (Ashe) H. E. Ahles	Arthur, 1885
Allegheny serviceberry	*Amelanchier laevis* Weigand	Rosen, 1929
Red chokeberry	*Aronia arbutifolia* (L.) Pers.	Shaw, 1934
Goat's beard	*Aruncus sylvestris* Kostel.	Thomas and Ark, 1934
Dwarf Japanese flowering quince	*Chaenomeles japonica* (Thunb.) Lindl. ex Spach	Thomas and Ark, 1934
Chinese quince	*Chaenomeles sinensis* (Thouin) Koehne	Thomas and Thomas, 1931; Thomas and Ark, 1934
Cotoneaster	*Cotoneaster* spp.	Thomas and Ark, 1934
Cliff rose	*Cowania stansburiana* Torr.	Thomas and Ark, 1934
English hawthorn	*Crataegus oxyacantha* L. and other *Crataegus* spp.	Arthur, 1885; Shaw, 1934; Thomas and Ark, 1934; Billing et al., 1974
Quince	*Cydonia oblonga* Mill.	Burrill, 1880
Docynia	*Docynia delavayi* (Franch.) C. K. Schneid.	Thomas and Ark, 1934
Loquat	*Eriobotrya japonica* (Thunb.) Lindl.	Thomas and Ark, 1934
Pearlbush	*Exochorda* sp.	Parker et al., 1956
Strawberry (cultivated)	*Fragaria* × *ananassa* Duchesne	Munn, 1918; Pierstorff, 1931
Avens	*Geum* sp.	Parker et al., 1956
Creambush	*Holodiscus discolor* (Pursh) Maxim.	Thomas and Ark, 1934; Parker et al., 1956
Chilean hawthorn	*Kageneckia oblonga* Ruiz & Pav.	Thomas and Ark, 1934
Crabapple	*Malus* spp.	Crandall, 1898
Apple (cultivated)	*Malus* × *domestica* Borkh.	Burrill, 1880; Nelson, 1975; van der Zwet and Keil, 1979
Medlar	*Mespilus germanica* L.	Curtis, 1934; Hildebrand, 1954
Ulei	*Osteomeles anthyllidifolia* (Sm.) Lindl.	Thomas and Ark, 1934
Christmas berry	*Photinia arbutifolia* Lindl.	O'Gara, 1910; Thomas and Ark, 1934
Ninebark	*Physocarpus* sp.	Parker et al., 1956
Cinquefoil	*Potentilla* sp.	Parker et al., 1956
Prinsepia	*Prinsepia* sp.	Parker et al., 1956
Purple apricot	*Prunus dasycarpa* Ehrh.[a]	Thomas and Thomas, 1931
Garden plum	*Prunus domestica* L.[a]	U.S. Dep. Agric., 1960
Hollyleaf cherry	*Prunus ilicifolia* (Nutt. ex Hook. & Arn.) D. Dietr.[a]	Thomas and Ark, 1934
Oriental cherry	*Prunus serrulata* Lindl.[a]	U.S. Dep. Agric., 1960
Pyracantha	*Pyracantha coccinea* M. Roem. and other *Pyracantha* spp.	Thomas and Thomas, 1931; Thomas and Ark, 1934
Common pear	*Pyrus communis* L. and other *Pyrus* spp.	Burrill, 1878; Reimer, 1925; van der Zwet and Keil, 1970; Nelson, 1975
Indian hawthorn	*Rhaphiolepis umbellata* (Thunb.) Makino	Thomas and Ark, 1934
Rose	*Rosa* spp.	Pierstorff, 1931; Thomas and Ark, 1934
Red raspberry	*Rubus idaeus* L.	Lehman, 1933; Folsom, 1947; Starr et al., 1951
Mountain ash	*Sorbus americana* Marshall	Shaw, 1934; Billing et al., 1969; Wade, 1974
Vanhoutte spirea, mountain spirea	*Spiraea* × *vanhouttei* (Briot) Carriere, *Spiraea splendens* Baumann ex K. Koch	Groves, 1931; Thomas and Ark, 1934; Zeller, 1977
Chinese stranvaesia	*Stranvaesia davidiana* Decne.	Lelliott and Hayward, 1965

[a] Note: Although *Prunus* species (apricot, cherry, peach, prune) are listed as susceptible to *Erwinia amylovora* by artificial inoculation, natural infections of these crops are rare.

colonies of *Erwinia amylovora* are white and shiny because of the production of exopolysaccharide. Interestingly, sucrose metabolism by *Erwinia amylovora* is essential for colonization of the host (Bogs and Geider, 2000). In rich media, growth is relatively rapid with a generation time of 72 min at room temperature. (The generation time of *Escherichia coli* is about 48 min at room temperature and 20 min at 37°C.) Optimal growth of *Erwinia amylovora* occurs between 21 and 28°C.

History of Fire Blight

Fire blight was first observed affecting pear and quince in 1780 in the lower Hudson River Valley of New York, United States (Denning, 1794; Coxe, 1817). The etiological organism *Erwinia amylovora*, unknown at that time, is a natural resident of wild crabapple, hawthorn, serviceberry, and mountain ash in the northeastern United States. The disease became prevalent in commercial apples and pears and became destructive after 1826. Fire blight had spread from New York to Illinois by 1882 and westward to California by 1887, first appearing on the West Coast in Butte County, California. In Illinois, E. S. Hull successfully transmitted the disease by tissue grafts in 1870. By 1879, J. B. Turner reported transmission of fire blight using a knife. The following year, Thomas J. Burrill (Fig. 5-1) repeated the tissue graft transmission experiment and examined oozing material microscopically. The study led to the hypothesis that bacteria associated with dried exudates from infected twigs might be responsible for fire blight. By 1883, Burrill was convinced that a bacterium that he named *Micrococcus amylovorus* was the cause of fire blight (Burrill, 1883). (The species name "*amylovorus*" came from the observed absence of starch grains in infected tissue—although ironically the bacterium does not hydrolyze starch.) Burrill did not provide the necessary proof for the bacterial nature of this disease, although he has been credited for this in many textbooks (Baker, 1971).

Proof of the bacterial nature of fire blight came from the thesis work of J. C. Arthur, who isolated a bacterium from infected material into pure culture and demonstrated unequivocally that the bacterium and not the cell-free culture filtrate caused fire blight (Arthur, 1885). With this knowledge, M. B. Waite in 1891 observed that the fire blight organism overwintered in cankers and reappeared in plant exudates in the spring. Insects were attracted to exudates oozing from infected shoots and became infested with bacteria, which they subsequently transmitted to other pear trees. This was the first case in which an insect was actually demonstrated to be a vector of a plant pathogen, although Forbes (1884) had earlier hypothesized that the Lygus bug *Lygus pratensis* L. might be a vector of fire blight because it was often associated with the disease. By 1915, many insects were implicated as potential vectors of fire blight (Table 5-2), and the list continued to grow (reviewed in van der Zwet and Keil, 1979).

Fig. 5-1. Thomas J. Burrill.

During the decades from 1910 to 1930, fire blight epidemics caused the shift of apple and

Table 5-2. Insect Vectors of Fire Blight[a]

Insect	Common Name	Host Tested	Reference
Adelphocoris rapidus (Say)	Plant bug	Apple	Stewart and Leonard, 1915
Apis mellifera L.	Honey bee	Apple, pear	Waite, 1891; Reed, 1914; Gossard, 1916a,b; Gossard and Walton, 1922
Campylomma verbasci (Meyer-Dur)	Plant bug	Apple	Stewart and Leonard, 1915
Empoasca fabae (Harris)	Potato leafhopper	Apple	Burrill, 1915; Gossard, 1916a,b
Lygus pratensis (L.)	Tarnished plant bug	Apple	Forbes, 1884; Stewart, 1913
Macrosiphum avenae (Fabr.)	Aphid	Crabapple	Burrill, 1915
Orthotylus flavosparsus (Sahlb.)	Plant bug	Apple	Stewart and Leonard, 1915
Plagiognathus politus Uhler	Plant bug	Apple	Stewart, 1913
Polymerus basalis (Reut.)	Plant bug	Apple	Stewart and Leonard, 1915
Psylla pyricola Foerster	Pear psylla	Pear	Stewart and Leonard, 1915; Emmett and Baker, 1971

[a] For a complete list, see van der Zwet and Keil, 1979.

pear production westward from New York to the Midwest and then to California and Oregon. Although the disease had already been observed in 1887 in Butte County, California, it was not as serious in the drier climate of the West as it was in the East. Fire blight was observed in Ontario, Canada, in 1904, in British Columbia in 1911, and as far as New Zealand in 1919. By the 1960s, the disease was also observed in northern Europe, with outbreaks in England, 1957; Poland, 1966; the Netherlands, 1966; Denmark, 1968; Germany, 1971; France, 1972; and Belgium, 1973. Northern Africa and Mediterranean countries subsequently reported fire blight. Outbreaks of fire blight were reported in Mexico and South American countries, such as Chile and Guatemala. By 1997, fire blight became worldwide in distribution.

Symptomatology

Early symptoms of fire blight in pear easily can be confused with blossom blight caused by *Pseudomonas syringae* pv. *syringae*. However, small drops of creamy ooze appearing on the base of the receptacle cap or on the base of the peduncle or from the fruit itself (Fig. 5-2 bottom) are diagnostic signs for fire blight. The ooze appears on the surface as small white- or amber-colored globules (the color darkens as the ooze dries). Water-soaking occurs around surface wounds on young branches. This is followed by the collapse of tissues, which results in a scorched appearance at the site of infection (Fig. 5-2 top; see also Fig. 4-2). This symptom is commonly called blossom blight. Infected twigs and branches later develop a reddish brown tissue discoloration outside the cambial layer (see Fig. 4-1). As the killed tissue dehydrates, the dead area forms a depression, resulting in the formation of a canker. When the tree has high vigor, canker formation may be subtle and the affected area may show only tissue that is slightly darker than the adjacent healthy tissue. When blossoms are infected, they become blighted along with the petioles and display a blackened appearance on pear trees and a brownish red appearance on apple trees. The infected blossoms remain attached. At later stages of infection, infected immature pear fruits also show a blackened appearance. The infected fruits commonly remain attached to the blighted petiole, and blackened leaves tend to cling to infected shoots after normal leaf fall. The sight of these blighted leaves and fruits still hanging on trees in the winter is a good clue that there is fire blight in the orchard.

Disease Cycle

Erwinia amylovora is endemic in native or introduced rosaceous plants, where bacteria reside primarily on the stigmas of flowers. Infected native plants are rarely killed and serve as reservoirs of inoculum. In commercial orchard hosts, the inoculum occurs in the early spring when holdover cankers become active prior to the blossom period. Infection begins to extend into healthy bark tissues through the xylem simultaneously with the initial appearance of bacterial ooze on the surface. The ooze is composed, in part, of an arabinogalactan-peptide that insects find attractive. When ooze production is prolific together with a windless condition, fine aerial strands form from the slowly dripping ooze (Ivanoff and Keitt, 1937; Eden-Green and Billing, 1972). Bacteria account for about 20% of the dry weight of the ooze (Eden-Green and Knee, 1974) and contain live bacteria at a concentration of approximately 10^8 cells per milligram. The aerial strands can serve as a means of disseminating *Erwinia amylovora* as wind-blown particulates over relatively long distances. The dissolution of these aerial strand pieces by rain converts the strand to inoculum available to initiate disease. The pathogen can be efficiently and rapidly spread from blossom to blossom of healthy trees by wind-blown rain and especially by pollinating insects, such as honey bees and wasps.

Fig. 5-2. Typical scorched-looking twig blight of apple (top), ooze on a twig (middle), and ooze on a pear fruit (bottom). (Courtesy George W. Sundin; bottom photo by Mira Danilovich)

In the blossoms, infection occurs in the nectaries that secrete nectar rich in fructose (40%) and sucrose (60%). On leaves and petioles, most infections occur through injuries caused by hail or wind. Bacteria can enter through stomata and lenticels, but this mode of entry is uncommon. Infection is triggered by environmental changes causing dramatic shifts in the physiology of the tree so that bacteria can move intercellularly and systemically through the xylem vessels. The pathogen is favored by warm moist conditions with a relative humidity greater than 60%. *Erwinia amylovora* is favored by air temperatures ranging from 20 to 27°C for growth in planta, unlike other members of the family *Enterobacteriaceae*, whose growth is favored by temperatures near 37°C.

As the infection progresses, wilting of the infected stems and leaves takes place because of occlusion of the xylem vessels with extracellular polysaccharides produced by *Erwinia amylovora* (discussed below). Infection continues down the peduncles, petioles, and twigs into the main scaffold branches and finally into the trunk. *Erwinia amylovora* can reside in xylem elements of petioles and stems in the absence of symptoms (Hockenhull, 1979). Further invasion downward to the base of the trunk leads to collar blight or trunk blight. Progression of the infection into the rootstock, known as rootstock blight, is generally lethal to the tree.

Erwinia amylovora overwinters in the margins of the cankers that were formed from spring through fall in the branches and trunk. After the cankers have been formed and winter approaches, the disease cycle is completed.

Pathogenicity Tests

Young trees are desirable hosts for pathogenicity tests but immature fruits (usually pear, harvested when 2–3 cm in diameter) provide an alternative for testing putative isolates of *Erwinia amylovora*. Immature fruits or fruit slices are inoculated by stabbing the epidermis or cortex with a dissecting needle or pipet tip carrying inoculum (Fig. 5-3). Following incubation for 1–4 days, ooze is produced at the inoculation site, indicating that the bacterium is pathogenic.

Although not as desirable as the immature fruit assay, succulent shoots of pear, apple, cotoneaster, or pyracantha can also be used. The shoots are inoculated near the tip by injecting a suspension of bacteria with a small hypodermic needle and syringe. Necrosis surrounding the inoculated site indicates pathogenicity.

In addition to the immature fruit and young shoot assays, detached pear blossoms can also be used. Pear branches bearing dormant flower buds are cut from trees during the winter months and forced to bloom in a growth chamber set at 25°C. Freshly opened blossoms are removed by cutting at the peduncle base and are placed into a vial containing a 10% sucrose solution. The flowers are inoculated by placing a drop (20 µl) of inoculum (10^8 CFU/ml) on the stigma. Inoculated blossoms are incubated at 25°C in a moist chamber under standard day-length lighting (15–18 h of light, 7–9 h of dark). The inoculated blossoms quickly show necrosis and darkening of the blossom parts.

The results of the above assays can be complemented by testing for elicitation of the HR in tobacco leaves. Using a needleless syringe, panels of young tobacco leaves are infiltrated on the abaxial side (underside) with 10^8 CFU of the putative *Erwinia amylovora* cells. Within 24 h, the infiltrated site becomes flaccid, collapses, and becomes brown due to the death of the leaf tissue. A positive HR signifies a pathogenic isolate.

Fig. 5-3. Pathogenicity assay for *Erwinia amylovora* in an immature pear fruit. Creamy ooze is produced at the site inoculated with a pathogenic strain. (Courtesy George W. Sundin)

Pathogenicity and Virulence Factors

Several pathogenicity factors are elaborated by *Erwinia amylovora* using the type II and type III secretion systems (see Chapter 10). These have been reviewed (Oh and Beer, 2005). Some of the major factors that have effects on pathogenicity or virulence of the fire blight pathogen are described as follows.

Amylovoran

Extracellular polysaccharides are produced by many members of the family *Enterobacteriaceae*. When sugars, such as sucrose, are available, *Erwinia amylovora* produces levan, glucan, and a galactose-rich mucopolysaccharide called amylovoran. Amylovoran is an acidic heteropolymer with each polymeric unit containing one glucuronic acid residue and four residues of galactose as a repeating unit, on which acetyl and pyruvate groups are attached (Smith et al., 1990; Nimtz et al., 1996). The major heteropolysaccharide contained in ooze is amylovoran, which is produced when *Erwinia amylovora* is growing on the host or is cocultivated with host cells in in vitro culture (Hsu and Goodman, 1978). This suggests that the production of amylovoran is regulated by the pathogen–host interaction. Because a direct correlation between amylovoran production and virulence has been observed (Erskine, 1973; Goodman et al., 1974; Richie and Klos, 1977; Bennett and Billing, 1978; Ayers et al., 1979; Steinberg and Beer, 1988; Bellemann and Geider, 1992), amylovoran appears to be essential for virulence of *Erwinia amylovora*. However, levan and glucan, which are homopolymers of fructose and glucose, respectively, also play roles in virulence.

The copious production of amylovoran together with levan and glucan also may provide an effective means of evading host defenses. *Erwinia amylovora* mutants defective in their ability to produce amylovoran are nonpathogenic (Steinberger and Beer, 1988; Bellemann and Geider, 1992; Bernhard et al., 1993). Because of its long-chain polymeric structure, amylovoran causes the occlusion of xylem vessel elements, resulting in wilting of apple shoots (Hsu and Goodman, 1978; Sjulin and Beer, 1978). Amylovoran might also sequester important nutrients and cations, prevent cellular dehydration, and help protect against antimicrobial compounds produced by the host.

In *Erwinia amylovora*, extracellular polysaccharide synthesis is regulated by the *rcsA* (regulation of capsular synthesis) gene (Bernhard et al., 1990; Chatterjee et al., 1990; Roberts and Coleman, 1991). The Rcs product also enhances transcription of the *ams* operon shown in Figure 5-4. The genes of the *ams* operon are involved in amylovoran synthesis (Bugert and Geider, 1995). The products of *rcsA* and *rcsB* genes have been recognized as positive transcriptional regulators of the *cps* operon in *Escherichia coli* (Gottesman et al., 1985). Mutations in the *rcsA* gene of *Erwinia amylovora* reduced virulence and produced nonmucoid colonies (Bugert and Geider, 1995).

Secreted Proteins

Erwinia amylovora produces and secretes many proteins (Nissinen et al., 2007) that are involved in its interaction with plants (Oh and Beer, 2007). Effector proteins are virulence factors that are secreted by the bacterial cells and injected directly into host cells via an Hrp type III secretion system. This conserved secretion

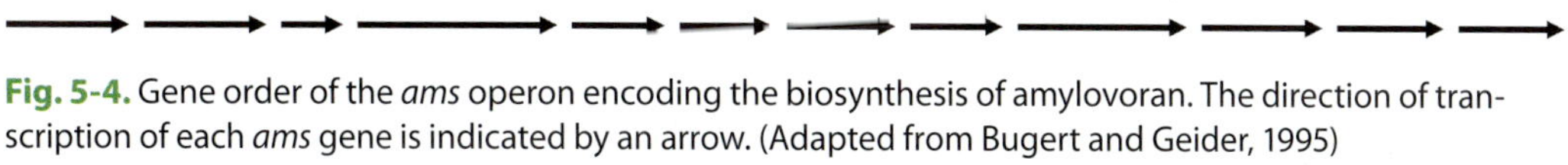

Fig. 5-4. Gene order of the *ams* operon encoding the biosynthesis of amylovoran. The direction of transcription of each *ams* gene is indicated by an arrow. (Adapted from Bugert and Geider, 1995)

system is encoded by a cluster of *hrc* and *hrp* genes designated as the *Erwinia* pathogenicity island 1 (PAI-1). These genes are primarily involved in encoding the highly conserved type III secretion apparatus; some regulate *hrc* and *hrp* genes and others encode proteins that are secreted by the apparatus, including the critically important disease-specific pathogenicity factor (*dsp*) and harpin (see Disease-Specific Protein A/E and Harpin below).

Other secreted and translocated proteins, either alone or in combination, are important as well. HrpA is a component of the type III secretion pilus that serves as a conduit for protein delivery (Jin and He, 2001; Jin et al., 2001; Li et al., 2002) (see Chapter 10). Several other putative virulence factors, including the hemagglutinin-like adhesin (HecA), a polygalacturonase (Peh), a membrane-bound lytic murein transglycosylase ($MltE_{EA}$), and the type III effector ortholog of HopPtoC from *Pseudomonas syringae* pv. *tomato* ($HopPtoC_{EA}$), have recently been identified using an in vivo expression technology (IVET) screen of genes expressed during infection of immature pear fruits by *Erwinia amylovora* (Zhao et al., 2005b).

Disease-Specific Protein A/E

DspA/E is the major disease-specific protein that is translocated into plant cells as an effector. It is encoded by a gene called *dspA/E*. DspA/E, a 198-kDa protein, displays homology (about 21%) to the avirulence (Avr) effector, AvrE, encoded by *Pseudomonas syringae* pv. *tomato* (Lorang and Keen, 1995; Bogdanove et al., 1998) (see below). DspA/E appears to be required for growth of *Erwinia amylovora* cells in the host plant and for incitement of disease (Barny et al., 1990; Gaudriault et al., 1997; Bogdanove et al., 1998), possibly by inducing oxidative stress at early stages of infection (Venisse et al., 2001) and by suppressing callose deposition (DebRoy et al., 2004). DspA/E appears on the surface of those *Erwinia amylovora* cells that have ingressed into the intercellular spaces of apple leaves (Boureau et al., 2006). HrpN is required for the translocation of DspA/E (Bocsanczy et al., 2008). The DspA/E protein interacts specifically with four apple proteins with leucine-rich repeat (LRR) receptor-like serine/threonine kinases (Meng et al., 2006). Host resistance genes encode serine/threonine kinases containing LRRs and a nucleotide-binding site. The binding of DspA/E to these receptor kinases could quench their activities in signal networks and cascades that lead to defense responses by the apple host. In addition, arabinogalactan proteins that predominate in the intercellular spaces of pear may be involved in plant defense (Chen et al., 1994).

Harpin

Harpin is a glycine-rich, 44-kDa protein that is secreted by *Erwinia amylovora* into plant intercellular spaces and elicits HR—but alone it does not cause fire blight (Wei et al., 1992; Wei and Beer, 1993). The initially described harpin is the product of the *hrpN* gene located within a cluster of genes encoding an Hrp type III secretion pathway (Bogdanove et al., 1996; Oh et al., 2005). HrpW, a second harpin, is encoded by another gene of the same cluster (Kim and Beer, 1998). HrpN and HrpW are members of the harpin family that are secreted and may be transported into host cells. They elicit the HR in nonhosts (Wei et al., 1992; Bogdanove et al., 1996; Gaudriault et al., 1998; Kim and Beer, 1998; Reboutier et al., 2007). Harpins, in addition, induce systemic acquired resistance in plants treated with the protein (Dong et al., 1999), and some have been registered by the

U.S. government for use in plant agriculture. In apple seedlings, HrpN moves into the intercellular spaces (Perino et al., 1999). HrpN protein interacts with HIPM, a 60-amino-acid plant protein of the host apple, and with orthologous proteins of several nonhosts, including Arabidopsis and rice (Oh and Beer, 2007). Interaction of the proteins may be involved in the development of disease, resistance, or both in plants.

Dihydrophenylalanine

Some virulent strains of *Erwinia amylovora* produce L-2,5-dihydrophenylalanine (DHP) (Feistner, 1988; Schwartz et al., 1991). DHP is an inhibitor of the shikimic acid pathway and inhibits the growth of pear cells (Schwartz et al., 1991). Growth inhibition of plant cells is caused by inhibiting hydroxylation of aromatic rings of phenylalanine and shikimic acid. Because many *Erwinia amylovora* isolates do not produce DHP, the compound may play a role in enhancing virulence but is not critical in causing the necrogenic reaction in host cells. DHP is therefore thought to provide certain strains of *Erwinia amylovora* a selective advantage in a particular environment.

Siderophores

Siderophores are natural iron-binding compounds that chelate ferric ions, forming colloidal hydroxides that are imported into bacteria via specific active transport systems. The major siderophore produced by *Erwinia amylovora* is proferrioxamine. Siderophores play an important role in pathogenesis. Pathogens need iron for growth in an iron-poor environment. Mutants defective in siderophore production are avirulent, supporting the view that *Erwinia amylovora* requires iron for invasive growth (Sneath et al., 1990; Vanneste and Expert, 1990). *Erwinia amylovora*, under the protective shield of amylovoran and growth-promoting siderophores, can then move freely intercellularly by continued elaboration of the proteinaceous effectors secreted by the type III secretion system.

Bacterial Canker

Bacterial canker of stone fruit trees is one of several diseases caused by *Pseudomonas syringae* pv. *syringae*. In addition, there are many strains of this pathogen that show specificity for plants besides *Prunus* species (Table 5-3). Certain strains of *Pseudomonas syringae* pv. *syringae* cause blossom blast of pear and apple. Such symptoms can be misidentified as fire blight, caused by *Erwinia amylovora* (Wilson, 1983). However, the extent of invasion in pear and apple trees is confined to short distances. *Pseudomonas syringae* pv. *syringae* strains from apple are pathogenic to Japanese plum and apricot but not to European plum, peach, and sweet cherry. There are strains that favor infecting citrus, while other strains are specific to lilac, common bean (*Phaseolus vulgaris* L.), spring wheat (*Triticum aestivum* L. cv. Bounty 309 and line L-1), or mulberry. In California, bacterial canker is also called gummosis because the disease causes the production of copious gummy exudates. In England, bacterial canker of cultivated *Prunus* species, such as cherry (*Prunus avium* L.) and plum (*Prunus domestica* L.), is caused by *Pseudomonas syringae* pv. *morsprunorum* rather than by gummosis-producing *Pseudomonas syringae* pv. *syringae*.

Table 5-3. Host Range of *Pseudomonas syringae* pv. *syringae*

Common Name	Scientific Name	Reference
Lime	*Citrus* × *aurantiifolia* (Christm.) Swingle	U.S. Dep. Agric., 1960
Lemon	*Citrus* × *limon* (L.) Burm. f.	U.S. Dep. Agric., 1960
King orange	*Citrus nobilis* Lour.	U.S. Dep. Agric., 1960
Common orange, sweet orange	*Citrus* × *sinensis* (L.) Osbeck	U.S. Dep. Agric., 1960
Tick trefoil	*Desmodium tortuosum* (Sw.) DC.	U.S. Dep. Agric., 1960
Cotton rosemallow	*Hibiscus mutabilis* L.	U.S. Dep. Agric., 1960
Saucer magnolia	*Magnolia soulangeana* Soul.-Bod.	U.S. Dep. Agric., 1960
Apple	*Malus* × *domestica* Borkh.	U.S. Dep. Agric., 1960
Mango	*Mangifera indica* L.	Cazorla et al., 1998
Velvet bean	*Mucuna pruriens* (L.) DC. var. *utilis* (Wall. ex Wight) Baker ex Burck	U.S. Dep. Agric., 1960
Witchgrass	*Panicum capillare* L.	U.S. Dep. Agric., 1960
Avocado	*Persea americana* Mill.	U.S. Dep. Agric., 1960
Plum	*Prunus americana* Marshall	Brzezinski, 1902
Almond	*Prunus amygdalus* Batsch	U.S. Dep. Agric., 1960
Apricot	*Prunus armeniaca* L.	Brzezinski, 1902; Barrett, 1918; U.S. Dep. Agric., 1960
Sweet cherry	*Prunus avium* L.	Brzezinski, 1902
Sour cherry	*Prunus cerasus* L.	Griffin, 1911
Garden plum	*Prunus domestica* L.	U.S. Dep. Agric., 1960
Peach	*Prunus persica* (L.) Batsch	Wilson, 1983
Pear	*Pyrus communis* L.	Wilson, 1934
Indian grass	*Sorghastrum elliottii* (C. Mohr) Nash	U.S. Dep. Agric., 1960
Johnsongrass	*Sorghum halepense* (L.) Pers.	U.S. Dep. Agric., 1960
Sudan grass	*Sorghum vulgare* Pers.	U.S. Dep. Agric., 1960
Persian lilac	*Syringa persica* L.	U.S. Dep. Agric., 1960
Carolina clover	*Trifolium carolinianum* Michx.	U.S. Dep. Agric., 1960
Clover	*Trifolium dubium* Sibth.	U.S. Dep. Agric., 1960
Asparagus bean	*Vigna sesquipedalis* (L.) Fruwirth	U.S. Dep. Agric., 1960
Cowpea	*Vigna sinensis* (Torner) Savi	U.S. Dep. Agric., 1960
Corn, maize	*Zea mays* L., *Zea mays* L. subsp. *parviglumis* Iltis & Doebley var. *huehuetenangensis* Iltis & Doebley	U.S. Dep. Agric., 1960

General Characteristics of *Pseudomonas syringae* pv. *syringae*

Pseudomonas syringae pv. *syringae* is a member of the class *Gammaproteobacteria* in the family *Pseudomonadaceae*. The bacteria are aerobic Gram-negative rods with polar flagellation. The flagellin subunits are glycosylated, a requirement for virulence and pathogenicity (Takeuchi et al., 2007). The bacteria can grow anaerobically when arginine is present, permitting generation of ATP via the *arc*-encoded pathway (Verhoogt et al., 1992). On iron-deficient media, most strains produce pigments that fluoresce under short-wavelength (254-nm) ultraviolet light and serve to sequester ferric iron from the environment. *Pseudomonas syringae* pv. *syringae* produces phenazines, which are tricyclic, six-membered, ringed compounds derived from the shikimate pathway via chorismic acid. Their affinity for ferric iron is usually higher than that produced by surrounding organisms and, therefore, phenazines are antimicrobial compounds. Many isolates of *Pseudomonas syringae* pv. *syringae* form ice nuclei between –5 and 5°C (Warren and Wolber, 1991; Lee et al., 1995; Warren, 1995). Interestingly, *Pseudomonas syringae* is a psychrotrophic bacterium that is capable of growing at 4°C because of the hydroxylated fatty acid content in its lipopolysaccharide (Kumar et al., 2002).

History of Bacterial Canker

Bacterial canker, also known as gummosis because of the exudation of gum from infected trees, primarily of the genus *Prunus*, was recognized in 1902 in Poland (Brzezinski, 1902) and in 1905 in Germany (Aderhold and Ruhland, 1905). In 1911, the bacterial nature of this disease was firmly established in Oregon (Griffin, 1911), and in 1918, the disease was recognized in California (Barrett, 1918). Currently, bacterial canker can be found worldwide in regions that grow apricot, plum, peach, and cherry. As shown in Table 5-3, a number of non-*Prunus* species also serve as hosts of *Pseudomonas syringae* pv. *syringae*.

Symptomatology

When trees are infected by *Pseudomonas syringae* pv. *syringae*, subepidermal tissues appear light to dark brown, become necrotic and moist, and emit a sour smell. The subepidermal tissue of young cankers display loosely connected, reddish brown streaks. Progressive killing of the subepidermal layers results in the girdling of the limb or trunk of the tree. As infected trees go into dormancy during the winter months, cankers of various sizes appear as depressed areas of the bark. Small cankers show brown to reddish brown spots and streaks in the outer bark that will grow in size and coalesce into brown and moist areas, extending mainly upward along branches and shoots. Depending on the tree variety, light-colored gum may be exuded from the affected tissues. Gum is more profusely produced in Japanese than in European varieties of peach and cherry. Plum usually does not produce gum.

Dormant buds can be sites of infection. Infected buds are no longer viable, characterizing the initial phase of bacterial canker. The buds appear dark and tissue just below the bud frequently appears water-soaked, leading to necrosis (Fig. 5-5).

Fig. 5-5. Dormant bud infection of peach by *Pseudomonas syringae* pv. *syringae*. This stage of infection is sometimes called bud failure blight because of the failure of the buds to start growth in the spring.

Disease Cycle

The highest period of susceptibility is during the winter months (Wilson, 1983). Cankers appear as small, brownish flecks in the cortex of the bark and are slow to develop external symptoms, such as profuse gum production. The cankers become depressed during the early spring and begin exuding gum. Cankers during early spring (before April) serve as the source of bacterial inoculum that is spread by splashing water during rain storms.

Dormant buds, broken twigs, and branches wounded from blowing sand serve as entry sites for *Pseudomonas syringae* pv. *syringae*. The pathogen may survive as a successful epiphyte on *Prunus* leaves as well as on plants neighboring the orchard. By April to June, cankers cease to progress. During the summer, the cankers are not very active and are quiescent. They resume their activity during October into November, extending further up the branches and trunk. Cankers are not found on the roots.

Pathogenicity Tests

There are several methods for validating the pathogenicity of *Pseudomonas syringae* pv. *syringae*. The most trustworthy involves inoculating intact stems of 'Lovell' peach seedlings 30–45 cm long. One or more drops of a heavy suspension of fresh *Pseudomonas syringae* pv. *syringae* cells are placed along the upper portion of each stem (bent horizontally), and a sterilized needle is then jabbed several times through the drop into and past the epidermis. The inoculated sites are wrapped with Parafilm (American National Can, Chicago, IL), and the seedlings are maintained for 1 week under greenhouse conditions. If the strain is pathogenic, necrosis is observed around the inoculated site and extends along the stem from the original site of inoculation.

A second method requires the use of green seed pods of bean that are surface sterilized with 70% ethanol. As with the stem assay, droplets of inoculum are deposited along the length of the pod and a sterilized needle is used to introduce the pathogen below the epidermis in essentially the same manner as described above. Inoculated pods are placed in sterile test tubes lined with a strip of wet filter paper and incubated for 2 days at room temperature. Pathogenic strains produce dark water-soaked spots that later turn brown (Fig. 5-6).

A third method utilizes immature cherry fruits. Inoculum of about 10^6 CFU/ml is applied as a droplet onto fresh puncture wounds of fresh cherries made with

Fig. 5-6. Common bean (*Phaseolus vulgaris*) pod assay for *Pseudomonas syringae* pv. *syringae*. Full-length green pods are inoculated by jabbing a needle through a drop of bacterial suspension placed on the pod surface.

a sterile needle. More than one wound can be made per fruit. The inoculated fruits show water-soaking followed by necrosis (darkening around the site of the puncture) after 5–6 h at room temperature.

Pathogenicity Factors

During infection, *Pseudomonas syringae* pv. *syringae* uses various secretion systems to inject a number of bacterial proteins into plant cells (see Chapter 10). Two clusters of genes called *hrp* and *hrc* clusters encode the structural components of the type III secretion system. The injected proteins, called effectors or Avr (avirulence) or Hop (Hrp-dependent out protein), are bacterial proteins that thwart innate defense systems of the plant, including the suppression of regulators of plant defense responses, and the activation of necrogenic responses (HR). Hop proteins are targeted to a type III Hrp secretion apparatus by N-terminal sequences. How Hop proteins are targeted to the Hrp secretion system remains to be elucidated but likely requires protein chaperones for escorting and protecting the effectors in the host cells. Requirements for translocation via the Hrp secretion system are that the first 12 residues of the N-terminal amino acids of most Hop proteins contain high serine content, proline, or an aliphatic amino acid, such as leucine, isoleucine, or valine, and no acidic amino acids (Schechter et al., 2004). Type III secretion systems are absent in naturally occurring, nonpathogenic isolates of *Pseudomonas syringae* (Mohr et al., 2008).

More than 150 effector genes have been identified in *Pseudomonas syringae* pathovars. These genes are located on the chromosome and on indigenous plasmids and are linked with an *hrp* pathogenicity island. *hrpZ*, *hrpL*, and *hrpS* of *Pseudomonas syringae* pv. *maculicola* are regulated by sigma factor RpoN and are required for virulence (Hendrickson et al., 2000a,b). A TetR-like regulator that is involved in virulence has been found in *Pseudomonas syringae* pv. *tomato* (Preiter et al., 2005). GacA/GacS plays a role as a master regulator of virulence of this pathovar (Chatterjee et al., 2003). Pathogenicity islands containing effector gene clusters have been found on resident plasmids of the pPT23A/pFKN family (Jackson et al., 1999; Vivian et al., 2001; Rohmer et al., 2003; Zhao et al., 2005a). Plasmids of this family also contain genes that encode a type IV secretion system of the subclasses IVA and IVB (Zhao et al., 2005a). These type IV genes are homologs of those comprising the *virB* operon and of the *virD4* gene of the Ti plasmid of *Agrobacterium tumefaciens.*

An adhesin, FhaB, is encoded by *Pseudomonas syringae* that interestingly contains an intein inserted into the β frame of the protein that can excise itself out by a protein-splicing reaction (Amital et al., 2003). This function may be needed for a pathogenic adaptation allowing *Pseudomonas syringae* to adhere to the host cell cytoplasmic membrane. FhaB is a counterpart of the translocated intimin receptor, Tir, which is inserted into the host cytoplasmic membrane where it functions as the receptor for intimin, an outer-membrane adhesin expressed by enteropathogenic *Escherichia coli* (Crawford and Kaper, 2002). Likewise, a nonfimbrial adhesin highly similar to those produced by animal pathogens is required for virulence of *Xanthomonas oryzae* pv. *oryzae* (Ray et al., 2002). Hence, the presence of FhaB in *Pseudomonas syringae* is not surprising.

In addition to protein effectors injected by the type II and type III secretion systems, secondary metabolites, such as iron scavenging compounds known as pyoverdines, are elaborated by members of the genus *Pseudomonas*, including *Pseudomonas syringae* pathovars (Budzikiewicz, 1993; Meyer, 2000). A specific achromobactin-like siderophore secreted by *Pseudomonas syringae* pv. *syringae* is likely to aid in virulence (Franza et al., 2005). The phytotoxin syringomycin, also thought to aid in virulence, is produced by most plant-pathogenic strains of *Pseudomonas syringae* (Mo and Gross, 1991). This compound is a member of a family of cyclic lipodepsinonapeptides consisting of a macrocyclic ring of nine amino acid residues with a fatty acid side chain that disrupts ion transport across the plasmalemma of affected host cells (Bidwai et al., 1987; Di Giorgio et al., 1994). Syringomycins and syringopeptins are produced by a nonribosomal peptide synthetase system (Bender et al., 1999). Likewise, mangotoxin is produced by a nonribosomal peptide synthesis mechanism (Arrebola et al., 2007). This toxin inhibits ornithine acetyl transferase. Syringomycin forms transmembrane pores in membrane bilayers (Hutchison et al., 1995). Likewise, syringopeptin, also produced by *Pseudomonas syringae*, is a lipodepsipeptide that causes pore formation in plant membranes (Hutchison and Gross, 1997). Syringomycin production is induced by flavonoid glycosides, such as arbutin (4-hydroxyphenyl-β-D-glucopyranoside), quercetin (2-(3,4-dihydroxyphenyl)-3,5,7-trihydroxy-4*H*-1-benzopyran-4-one), and D-fructose, compounds naturally present in cherry tissues (Mo and Gross, 1991) and perceived by the *syrB1* and *sypA* gene products (Quigley and Gross, 1994). Syringomycin and related toxins, such as syringotoxin, syringostatin, and pseudomycin, produced by *Pseudomonas syringae* lack host specificity and generally induce necrosis. These toxins aid in intercellular ingression as well as colonization of the host (reviewed in Bender et al., 1999). They exert a detergent-like action on the host plasma membrane and inhibit membrane-associated ATPase activity (Che et al., 1992). The type I secretion system delivers these toxins exocellularly (see Chapter 10 for a detailed description of these secretion systems).

Additional components are secreted by *Pseudomonas syringae* to enhance its virulence; they include extracellular polysaccharides, phytohormones, cell wall-degrading enzymes, and type IV pilin (Buell et al., 2003). Moreover, *Pseudomonas syringae* pv. *glycinea* produces alginate, which apparently increases fitness and virulence (Schenk et al., 2006). Hence, *Pseudomonas syringae* is well armed with pathogenicity factors of broad specificity that account for the wide host range of this organism.

Deep Bark Canker

Deep bark canker is a devastating disease of Persian walnuts (*Juglans regia* L. cultivars). *Brenneria rubrifaciens* (previously classified as *Erwinia rubrifaciens*), a member of the class *Gammaproteobacteria* in the family *Enterobacteriaceae*, is the causal organism of deep bark canker.

General Characteristics of *Brenneria rubrifaciens*

Brenneria rubrifaciens has all of the key features of the family *Enterobacteriaceae*. Cells are Gram-negative, small, and baciliiform, with at least six long peritrichous flagella. *Brenneria rubrifaciens* produces abundant acid from various simple sugars, such as glucose, mannose, sucrose, and arabinose, but not from lactose or xylose. The pathogen produces catalase and gelatinase (protease) but does not produce nitrate reductase, urease, or amylase. It does not ferment butanediol. *Brenneria rubrifaciens* produces a water-soluble and diffusible red pigment that is especially noticeable on yeast extract-dextrose-calcium carbonate (YDC) agar medium, which contains per liter 20 g of glucose, 10 g of yeast extract, 20 g of calcium carbonate, and 12 g of agar. The red pigment, identified as rubrifacine (Feistner et al., 1984), is abundantly produced in culture as well as in the walnut host. Rubrifacine is 3-(3-carboxymethyl-4,5-dihydro-4,5-dioxo-2-pyrrolyl)-4,5,6-trihydroxypyridine-2-acetic acid (Feistner and Budzikiewicz, 1985). *Brenneria rubrifaciens* can be stored and remains viable in the lyophilized state for 35 years at room temperature.

Limited Host Range

The host range of *Brenneria rubrifaciens* is limited to *Juglans regia*, *Juglans hindsii* (Jeps.) Jeps. ex R. E. Sm., and *Juglans nigra* L. Commercially, various horticulturally suitable cultivars from these *Juglans* species are usually grafted on *Juglans hindsii* and 'Paradox' (English walnut × black walnut hybrid) rootstocks to avoid infections by *Armillaria mellea* (Vahl:Fr.) P. Kumm., an insidious fungus that causes root rot. The degree of susceptibility of a large number of cultivars from these *Juglans* species to *Brenneria rubrifaciens* has been tested (Kado et al., 1977). The commercially desirable *Juglans regia* cv. Hartley is highly susceptible to *Brenneria rubrifaciens*. On 'Paradox' rootstock, 'Pioneer', 'Early Erhardt', 'Willson-Franquette', 'Sinensis-5', 'Conway-Mayette', 'Scharsch-Franquette', 'Meylan', and 'Sinensis-7' are the most susceptible. On *Juglans hindsii* rootstock, only 'Sinensis-5' and 'Sinensis-7' are more susceptible than 'Hartley'.

History of Deep Bark Canker

Deep bark canker, also called bacterial phloem canker, was first observed in 1955 in walnut orchards in the Sacramento Valley in California. *Juglans regia* 'Hartley' remains the preferred commercial cultivar in California. Unfortunately, the Hartley cultivar is also one of seven most susceptible to *Brenneria rubrifaciens* (Kado et al., 1977). Although deep bark canker was first observed in the Sacramento area, many infected trees were found in the walnut-growing regions of California. In Oregon, where the Hartley cultivar is rarely grown, deep bark canker has not been reported. The disease has continued to occur throughout the walnut-growing regions of California.

Symptomatology

Deep bark canker is characterized by progressive necrosis of the inner phloem of the walnut tree. As the disease progresses, black-colored pitting of the xylem occurs. During the hot summer months, deep longitudinal fissures are observed on the exterior of the walnut tree. Reddish brown liquid exudates are released from the inner phloem to the exterior, especially from fleshy scaffold branches (Fig. 5-7). Upon exposure to the air, the liquid hardens into dark red-brown dried exudate. The exudate harbors large quantities of *Brenneria rubrifaciens*. A milligram of hardened exudate contains at least 10 million viable cells of this pathogen. As few as five *Brenneria rubrifaciens* cells are sufficient for infection of 'Hartley' walnuts (Schaad et al., 1973). The bacteria move in the inner phloem and cause necrosis of the phloem tissues as they move systemically up into the remaining branches. The killing of phloem tissue around the branch girdles the branches, including the main trunk of the tree. Such extensive necrosis results in killing the tree.

Fig. 5-7. Deep bark canker of walnuts, *Juglans regia* cv. Hartley. **Left,** Three of the four scaffold limbs show cracking and oozing of *Brenneria rubrifaciens*-laden reddish sap that is characteristic of this disease. **Right,** Closer view of another infected tree.

Disease Cycle

Walnut trees infected by *Brenneria rubrifaciens* serve as the source of inoculum for deep bark canker. *Brenneria rubrifaciens* is spread readily by mechanical nut harvesters during the fall (Kado and Gardner, 1977). Wounds caused by the harvester are sites where the pathogen enters. Because as few as five bacterial cells can establish infection (Schaad et al., 1973), the coating of exudates from a diseased tree on the clamps or jaws of the harvester provides ample inoculum for transmission to many healthy trees. Infected trees become systemically infected (Gardner and Kado, 1973) and there is no cure.

Pathogenicity Tests

Pathogenicity of *Brenneria rubrifaciens* isolates is assessed by inoculating 10- to 12-year-old 'Hartley' walnut trees. Young seedlings are relatively resistant to infection. Small branches from mature trees may be inoculated by exposing the inner phloem tissue to bacterial inoculum. Necrosis of the phloem verifies that the isolate is pathogenic.

Pathogenicity Factors

Brenneria rubrifaciens elicits the HR in tobacco (*Nicotiana tabacum* L. cv. Havana) and, therefore, likely possesses a type III secretion system for delivering pathogenic factors. This pathogen secretes a pectate lyase that causes the HR in tobacco leaves (Gardner and Kado, 1976). The enzyme also causes necrosis of exposed inner phloem tissue of 'Hartley' walnuts. Mutational analysis of *Brenneria rubrifaciens* showed that there are additional factors that cause the tobacco HR (Azad and Kado, 1984). Nonetheless, a pectate lyase of *Brenneria rubrifaciens* is one of the components involved in the necrotizing reaction in walnuts. *Brenneria rubrifaciens* contains *hrp* genes that are essential for delivering or secreting pathogenic effectors. Further necrotizing factors remain to be characterized.

REFERENCES

Aderhold, R., and Ruhland, W. 1905. Über ein durch Bakterien hervorgerufenes Kirschensterben (Vorlaufige Mitteilung). Zentbl. Bakteriol. 15:376-377.

Amital, G., Belenkly, O., Dassa, B., Shainskaya, A., and Pietrokovski, S. 2003. Distribution and function of new bacterial intein-like protein domains. Mol. Microbiol. 47:61-73.

Arrebola, E., Cazorla, F. M., Romero, D., Pérez-García, A., and de Vicente, A. 2007. A non-ribosomal peptide synthetase gene (*mgoA*) of *Pseudomonas syringae* pv. *syringae* is involved in mangotoxin biosynthesis and is required for full virulence. Mol. Plant-Microbe Interact. 20:500-509.

Arthur, J. C. 1885. Proof that bacteria are the direct cause of the disease in trees known as pear blight. Am. Assoc. Adv. Sci. Proc. 34:295-298.

Ayers, A. R., Ayers, S. B., and Goodman, R. N. 1979. Extracellular polysaccharide of *Erwinia amylovora*: A correlation with virulence. Appl. Environ. Microbiol. 38:659-666.

Azad, H. R., and Kado, C. I. 1984. Relation of tobacco hypersensitivity to pathogenicity of *Erwinia rubrifaciens*. Phytopathology 74:61-64.

Baker, K. F. 1971. Fire blight of pome fruits: The genesis of the concept that bacteria can be pathogenic to plants. Hilgardia 40:603-633.

Barny, M. A., Guinebretière, M. H., Marcais, B., Coissac, E., Paulin, J. P., and Laurent, J. 1990. Cloning of a large gene cluster involved in *Erwinia amylovora* CFBP1430 virulence. Mol. Microbiol. 4:777-786.

Barrett, J. T. 1918. Bacterial gummosis of apricots—Prelimary report. Mo. Bull. Calif. State Comm. Hortic. 7:137-140.

Bellemann, P., and Geider, K. 1992. Localization of transposon insertions in pathogenicity mutants of *Erwinia amylovora* and their biochemical characterization. J. Gen. Microbiol. 138:931-940.

Bender, C. L., Alarcón-Chaidez, F., and Gross, D. C. 1999. *Pseudomonas syringae* phytotoxins: Mode of action, regulation, and biosynthesis by peptide and polyketide synthetases. Microbiol. Mol. Biol. Rev. 63:266-292.

Bennett, R. A., and Billing, E. 1978. Capsulation and virulence in *Erwinia amylovora*. Ann. Appl. Biol. 89:41-45.

Bernhard, F., Poetter, K., Geider, K., and Coplin, D. L. 1990. The *rcsA* gene from *Erwinia amylovora*: Identification, nucleotide sequence, and regulation of exopolysaccharide biosynthesis. Mol. Plant-Microbe Interact. 3:429-437.

Bernhard, F., Coplin, D. L., and Geider, K. 1993. A gene cluster for amylovoran synthesis in *Erwinia amylovora*: Characterization and relationship to *cps* genes in *Erwinia stewartii*. Mol. Gen. Genet. 239:158-168.

Bidwai, A. P., Zhang, L., Bachmann, R. C., and Takemoto, J. Y. 1987. Mechanism of action of *Pseudomonas syringae* phytotoxin, syringomycin: Stimulation of red beet plasma membrane ATPase activity. Plant Physiol. 83:39-43.

Billing, E., Fletcher, J. T., Glasscock, H. H., Jones, G. E., and Lelliott, R. A. 1969. Hosts of *Erwinia amylovora* (Burr.) Winslow. Plant Pathol. 8:152.

Billing, E., Bech-Andersen, J., and Lelliott, R. A. 1974. Fireblight in hawthorn in England and Denmark. Plant Pathol. 23:141-143.

Bocsanczy, A. M., Nissinen, R. M., Oh, C.-S., and Beer, S. V. 2008. HrpN of *Erwinia amylovora* functions in the translocation of DspA/E into plant cells. Mol. Plant Pathol. 9:425-434.

Bogdanove, A. J., Wei, Z.-M., Zhao, L., and Beer, S. V. 1996. *Erwinia amylovora* secretes harpin via a type III pathway and contains a homolog of *yopN* of *Yersinia*. J. Bacteriol. 178:1720-1730.

Bogdanove, A. J., Kim, J. F., Wei, Z., Kolchinsky, P., Charkowski, A. O., Conlin, A. K., Collmer, A., and Beer, S. V. 1998. Homology and functional similarity of an *hrp*-linked pathogenicity locus, *dspEF*, of *Erwinia amylovora* and the avirulence locus *avrE* of *Pseudomonas syringae* pathovar tomato. Proc. Natl. Acad. Sci. U.S.A. 95:1325-1330.

Bogs, J., and Geider, K. 2000. Molecular analysis of sucrose metabolism of *Erwinia amylovora* and influence on bacterial virulence. J. Bacteriol. 182:5351-5358.

Boureau, T., ElMaarouf-Bouteau, H., Garnier, A., Brisset, M.-N., Perino, C., Pucheu, I., and Barny, M.-A. 2006. DspA/E, a type III effector essential for *Erwinia amylovora* pathogenicity and growth in planta, induces cell death in host apple and nonhost tobacco plants. Mol. Plant-Microbe Interact. 19:16-24.

Brzezinski, J. 1902. Etiologie du chancre et de la gomme des arbres fruitiers. C. R. Acad. Sci. 134:1170-1173.

Budzikiewicz, H. 1993. Secondary metabolites from fluorescent pseudomonads. FEMS Microbiol. Rev. 10:209-228.

Buell, C. R., Joardar, V., Lindeberg, M., Selengut, J., Paulsen, I. T., Gwinn, M. L., Dodson, R. J., Deboy, R. T., Durkin, A. S., Kolonay, J. F., Madupu, R., Daugherty, S., Brinkac, L., Beanan, M., Haft, D. H., Nelson, W. C., Davidsen, T., Liu, J., Yuan, Q., Khouri, H., Fedorova, N., Tran, B., Russell, D., Berry, K., Utterback, T., Vanaken, S. E., Feldblyum, T. V., D'Ascenzo, M., Deng, W.-L., Ramos, A. R., Alfano, J. R., Cartinhour, S., Chatterjee, A. K., Delaney, T. P., Lazarowitz, S. G., Martin, G. B., Schneider, D. J., Tang, X., Bender, C. L., White, O., Fraser, C. M., and Collmer, A. 2003. The complete sequence of the Arabidopsis and tomato pathogen *Pseudomonas syringae* pv. *tomato* DC3000. Proc. Natl. Acad. Sci. U.S.A. 100:10181-10186.

Bugert, P., and Geider, K. 1995. Molecular analysis of the *ams* operon required for exopolysaccharide synthesis of *Erwinia amylovora*. Mol. Microbiol. 15:917-933.

Burrill, A. C. 1915. Insect control important in checking fire blight. Phytopathology 5:343-347.

Burrill, T. J. 1878. Pear-blight. Trans. Ill. State Hortic. Soc. 11:114-116.

Burrill, T. J. 1880. Pear and apple-tree blight. Trans. Ill. State Hortic. Soc. 14:157-167.

Burrill, T. J. 1883. New species of micrococcus (bacteria). Am. Nat. 17:319.

Cazorla, F. M., Torés, J. A., Olalla, L., Pérez-García, A., Farré, J. M., and de Vicente, A. 1998. Bacterial apical necrosis of mango in southern Spain: A disease caused by *Pseudomonas syringae* pv. syringae. Phytopathology 88:614-620.

Chatterjee, A., Chun, W., and Chatterjee, A. K. 1990. Isolation and characterization of an *rcsA*-like gene of *Erwinia amylovora* that activates extracellular polysaccharide production in *Erwinia* species, *Escherichia coli*, and *Salmonella typhimurium*. Mol. Plant-Microbe Interact. 3:144-148.

Chatterjee, A., Cui, Y., Yang, H., Collmer, A., Alfano, J. R., and Chatterjee, A. K. 2003. GacA, the response regulator of a two-component system, acts as a master regulator in *Pseudomonas syringae* pv. tomato DC3000 by controlling regulatory RNA, transcriptional activators, and alternate sigma factors. Mol. Plant-Microbe Interact. 16:1106-1117.

Che, F.-S., Kasamo, K., Fukuchi, N., Isogai, A., and Suzuki, A. 1992. Bacterial phytotoxins, syringomycin, syringostatin and syringotoxin, exert their effect on the plasma membrane H^+-ATPase partly by a detergent-like action and partly by inhibition of the enzyme. Physiol. Plant. 86:518-524.

Chen, C.-G., Pu, Z.-Y., Moritz, R. L., Simpson, R. J., Bacic, A., Clarke, A. E., and Mau, S.-L. 1994. Molecular cloning of a gene encoding an arabinogalactan-protein from pear (*Pyrus communis*) cell suspension culture. Proc. Natl. Acad. Sci. U.S.A. 91:10305-10309.

Coxe, W. 1817. Pears. Pages 175-176 in: A View of the Cultivation of Fruit Trees. M. Carey & Son, Philadelphia.

Crandall, C. S. 1898. Blight and other plant diseases: Blight of apple and pear trees. Colo. Agric. Exp. Stn. Bull. 41:1-14.

Crawford, J. A., and Kaper, J. B. 2002. The N-terminus of enteropathogenic *Escherichia coli* (EPEC) Tir mediates transport across bacterial and eukaryotic cell membranes. Mol. Microbiol. 46:855-866.

Curtis, K. M. 1934. Fireblight; A survey of current knowledge and recent advances. Cawthron Inst. Sci. Res., Mycol. Publ. 10.

DebRoy, S., Thilmony, R., Kwack, Y. B., Nomura, K., and He, S. Y. 2004. A family of conserved bacterial effectors inhibits salicylic acid-mediated basal immunity and promotes disease necrosis in plants. Proc. Natl. Acad. Sci. U.S.A. 101:9927-9932.

Denning, W. 1794. On the decay of apple trees. Trans. N. Y. Soc. Prom. Agric. Arts Manuf. 2:219-222.

Di Giorgio, D., Camoni, L., and Ballio, A. 1994. Toxins of *Pseudomonas syringae* pv. *syringae* affect H^+-transport across the plasma membrane of maize. Physiol. Plant. 91:741-746.

Dong, H., Delaney, T. P., Bauer, D. W., and Beer, S. V. 1999. Harpin induces disease resistance in Arabidopsis through the systemic acquired resistance pathway mediated by salicylic acid and NIM1 gene. Plant J. 20:207-215.

Eden-Green, S. J., and Billing, E. 1972. Fireblight: Occurrence of bacterial strands on various hosts under glasshouse conditions. Plant Pathol. 21:121-123.

Eden-Green, S. J., and Knee, M. 1974. Bacterial polysaccharide and sorbitol in fireblight exudates. J. Gen. Microbiol. 81:509-512.

Emmett, B. J., and Baker, L. 1971. Insect transmission of fire blight. Plant Pathol. 20:41-45.

Erskine, J. M. 1973. Characteristics of *Erwinia amylovora* bacteriophage and its possible role in the epidemiology of fire blight. Can. J. Microbiol. 19:837-845.

Feistner, G. J. 1988. (L)-2,5-Dihydrophenylalanine from the fireblight pathogen *Erwinia amylovora*. Phytochemistry 27:3417-3422.

Feistner, G., and Budzikiewicz, H. 1985. On the structure of rubrifacine. Can. J. Chem. 63:495-499.

Feistner, G. J., Korth, H., Budzikiewicz, H., and Pulverer, G. 1984. Rubrifacine from *Erwinia rubrifaciens*. Curr. Microbiol. 10:169-172.

Folsom, D. 1947. Bacterial twig and blossom blight of raspberry in Maine. Plant Dis. Rep. 31:324.

Forbes, S. A. 1884. The tarnished plant-bug. Farmers Rev. Feb. 28:150.

Franza, T., Mahé, B., and Expert, D. 2005. *Erwinia chrysanthemi* requires a second iron transport route dependent of the siderophore achromobactin for extracellular growth and plant infection. Mol. Microbiol. 55:261-275.

Gardner, J. M., and Kado, C. I. 1973. Evidence for systemic movement of *Erwinia rubrifaciens* in Persian walnuts by the use of double-antibiotic markers. Phytopathology 63:1085-1086.

Gardner, J. M., and Kado, C. I. 1976. Polygalacturonic acid *trans*-eliminase in the osmotic shock fluid of *Erwinia rubrifaciens*: Characterization of the purified enzyme and its effect on plant cells. J. Bacteriol. 127:451-460.

Gaudriault, S., Malandrin, L., Paulin, J. P., and Barny, M. A. 1997. DspA, an essential pathogenicity factor of *Erwinia amylovora* showing homology with AvrE of *Pseudomonas syringae*, is secreted via the Hrp secretion pathway in a DspB-dependent way. Mol. Microbiol. 26:1057-1069.

Gaudriault, S., Brisset, M. N., and Barny, M. A. 1998. HrpW of *Erwinia amylovora*, a new Hrp-secreted protein. FEBS Lett. 428:224-228.

Goodman, R. N., Huang, J. S., and Huang, P. Y. 1974. Host-specific phytotoxic polysaccharide from apple tissues infected by *Erwinia amylovora*. Science 183:1081-1082.

Gossard, H. A. 1916a. Is the hive a center for distributing fire blight? Is aphid honeydew a medium for spreading blight? J. Econ. Entomol. 9:59-62.

Gossard, H. A. 1916b. The role of insects as carriers of fire blight. Rep. Mont. State Hortic. Soc. 19:84-90.

Gossard, H. A., and Walton, R. C. 1922. Dissemination of fire blight. Ohio Agric. Exp. Stn. Bull. 357.

Gottesman, S., Trisller, P., and Torres-Cabassa, A. 1985. Regulation of capsular polysaccharide synthesis in *Escherichia coli* K-12: Characterization of three regulatory genes. J. Bacteriol. 162:1111-1119.

Griffin, F. L. 1911. A bacterial gummosis of cherries. Science 34:615-616.

Groves, A. B. 1931. Natural fire-blight infections on *Spiraea vanhouttei*. Phytopathology 21:89-91.

Hendrickson, E. L., Guevera, P., Peñaloza-Vàzquez, A., Shao, J., Bender, C., and Ausubel, F. M. 2000a. Virulence of the phytopathogen *Pseudomonas syringae* pv. *maculicola* is *rpoN* dependent. J. Bacteriol. 182:3498-3507.

Hendrickson, E. L., Guevera, P., and Ausubel, F. M. 2000b. The alternative sigma factor RpoN is required for *hrp* activity in *Pseudomonas syringae* pv. *maculicola* and acts at the level of *hrpL* transcription. J. Bacteriol. 182:3508-3516.

Hildebrand, E. M. 1954. Relative stability of fire blight bacteria. Phytopathology 44:192-197.

Hockenhull, J. 1979. In situ detection of *Erwinia amylovora* antigen in symptomless petiole and stem tissue by means of the fluorescent antibody technique. Pages 1-14 in: R. Vet. Agric. Univ. Yearb. 1979. Royal Veterinary and Agricultural University, Copenhagen.

Hsu, S.-T., and Goodman, R. N. 1978. Production of a host-specific, wilt-inducing toxin in apple cell suspension cultures inoculated with *Erwinia amylovora*. Phytopathology 68:351-354.

Hutchison, M. L., and Gross, D. C. 1997. Lipopeptide phytotoxins produced by *Pseudomonas syringae* pv. *syringae*: Comparison of the biosurfactant and ion channel-forming activities of syringopeptin and syringomycin. Mol. Plant-Microbe Interact. 10:347-354.

Hutchison, M. L., Tester, M. A., and Gross, D. C. 1995. Role of biosurfactant and ion channel-forming activities of syringomycin in transmembrane ion flux: A mode for the mechanism of action in the plant-pathogen interaction. Mol. Plant-Microbe Interact. 8:610-620.

Ivanoff, S. S., and Keitt, G. W. 1937. The occurrence of aerial bacterial strands on blossoms, fruits, and shoots blighted by *Erwinia amylovora*. Phytopathology 27:702-709.

Jackson, R. W., Athanassopoulos, E., Tsiamis, G., Mansfield, J. W., Sesma, A., Arnold, D. L., Bibbon, M. J., Murillow, J., Taylor, J. D., and Vivian, A. 1999. Identification of a pathogenicity island, which contains genes for virulence and avirulence, on a large native plasmid in the bean pathogen *Pseudomonas syringae* pathovar *phaseolicola*. Proc. Natl. Acad. Sci. U.S.A. 96:10875-10880.

Jin, Q., and He, S. Y. 2001. Role of the Hrp pilus in type III protein secretion in *Pseudomonas syringae*. Science 294:2556-2558.

Jin, Q., Hu, W., Brown, I., McGhee, G., Hart, P., Jones, A. L., and He, S. Y. 2001. Visualization of secreted Hrp and Avr proteins along the Hrp pilus during type III secretion in *Erwinia amylovora* and *Pseudomonas syringae*. Mol. Microbiol. 40:1129-1139.

Jock, S., Rodoni, B., Gillings, M., Kim., W.-S., Copes, C., Merriman, P., and Geider, K. 2000. Screening of ornamental plants from the botanic gardens of Melbourne and Adelaide for the occurrence of *Erwinia amylovora*. Australas. Plant Pathol. 29:251-258.

Kado, C. I., and Gardner, J. M. 1977. Transmission of deep bark canker of walnuts by the mechanical harvester. Plant Dis. Rep. 61:321-325.

Kado, C. I., Dutra, J. C., Moller, W. J., and Ramos, D. E. 1977. An assessment of the susceptibility of various walnut cultivars to deep bark canker. J. Am. Soc. Hortic. Sci. 102:698-702.

Kim, J. F., and Beer, S. V. 1998. HrpW of *Erwinia amylovora*, a new harpin that contains a domain homologous to pectate lyases of a distinct class. J. Bacteriol. 180:5203-5210.

Kumar, G. S., Jagannadham, M. V., and Ray, M. K. 2002. Low-temperature-induced changes in composition and fluidity of lipopolysaccharides in the Antarctic psychrotrophic bacterium *Pseudomonas syringae*. J. Bacteriol. 184:6746-6749.

Lee, R. E., Jr., Warren, G. J., and Gusta, L. V. 1995. Biological Ice Nucleation and Its Applications. American Phytopathological Society, St. Paul, MN.

Lehman, S. G. 1933. Bacterial twig and blossom blight of raspberry. (Abstr.) Phytopathology 23:21.

Lelliott, R. A., and Hayward, A. C. 1965. Further hosts of *Erwinia amylovora* (Burr.) Winslow et al., in England. Plant Pathol. 14:141-142.

Li, C. M., Brown, I., Mansfield, J., Stevens, C., Boureau, T., Romantschuk, M., and Taira, S. 2002. The Hrp pilus of *Pseudomonas syringae* elongates from its tip and acts as a conduit for translocation of the effector protein HrpZ. EMBO J. 21:1909-1915.

Lorang, J. M., and Keen, N. T. 1995. Characterization of *avrE* from *Pseudomonas syringae* pv. *tomato*: A *hrp*-linked avirulence locus consisting of at least two transcriptional units. Mol. Plant-Microbe Interact. 8:49-57.

Meng, X., Bonasera, J. M., Kim, J. F., Nissinen, R. M., and Beer, S. V. 2006. Apple proteins that interact with DspA/E, a pathogenicity effector of *Erwinia amylovora*, the fire blight pathogen. Mol. Plant-Microbe Interact. 19:53-61.

Meyer, J. M. 2000. Pyoverdines: Pigments, siderophores and potential taxonomic markers of fluorescent *Pseudomonas* species. Arch. Microbiol. 174:135-142.

Mo, Y. Y., and Gross, D. C. 1991. Plant signal molecules activate the *syrB* gene, which is required for syringomycin production by *Pseudomonas syringae* pv. *syringae*. J. Bacteriol. 173:5784-5792.

Mohr, T. J., Liu, H., Yan, S., Morris, C. E., Castillo, J. A., Jelenska, J., and Vinatzer, B. A. 2008. Naturally occurring nonpathogenic isolates of the plant pathogen *Pseudomonas syringae* lack a type III secretion system and effector gene orthologues. J. Bacteriol. 190:2858-2870.

Munn, M. T. 1918. Pathogenicity of *Bacillus amylovorus* (Burr.) Trev. for blossoms of the strawberry (*Fragaria* sp.). Phytopathology 8:33.

Nelson, S. H. 1975. Fireblight ratings on *Malus* and *Pyrus*. Rep. Proc. West. Can. Soc. Hortic. 31:73-75.

Nimtz, M., Mort, A., Domke, T., Wray, V., Zhang, Y., Qui, F., Coplin, D., and Geider, K. 1996. Structure of amylovoran, the capsular exopolysaccharide from the fireblight pathogen *Erwinia amylovora*. Carbohydr. Res. 287:59-76.

Nissinen, R. M., Ytterberg, A. J., Bogdanove, A. J., van Wijk, K., and Beer, S. V. 2007. Analyses of the secretomes of *Erwinia amylovora* and selected hrp mutants reveal novel type III secreted proteins and an effect of HrpJ on extracellular harpin levels. Mol. Plant Pathol. 8:55-67.

O'Gara, P. J. 1910. Control of pear blight on the Pacific Coast. Better Fruit 5:49-57.

Oh, C.-S., and Beer, S. V. 2005. Molecular genetics of *Erwinia amylovora* involved in the development of fire blight. FEMS Microbiol. Lett. 253:185-192.

Oh, C.-S., and Beer, S. V. 2007. AtHIPM, an ortholog of the apple HrpN-interacting protein, is a negative regulator of plant growth and mediates the growth-enhancing effect of HrpN in Arabidopsis. Plant Physiol. 145:426-436.

Oh, C.-S., Kim, J. F., and Beer, S. V. 2005. The Hrp pathogenicity island of *Erwinia amylovora* and identification of three novel genes required for systemic infection. Mol. Plant Pathol. 6:125-138.

Parker, K. G., Fisher, E. G., and Mills, W. D. 1956. Fire blight on pome fruits and its control. N. Y. State Coll. Agric. Ext. Bull. 96.

Perino, C., Gaudriault, S., Vian, B., and Barny, M. A. 1999. Visualization of harpin secretion *in planta* during infection of apple seedlings by *Erwinia amylovora*. Cell Microbiol. 1:131-141.

Pierstorff, A. L. 1931. Studies on the fire-blight organism, *Bacillus amylovorus*. N. Y. Agric. Exp. Stn. Mem. 136.

Preiter, K., Brooks, D. M., Penaloza-Vazquez, A., Sreedharan, A., Bender, C. L., and Kunkel, B. N. 2005. Novel virulence gene of *Pseudomonas syringae* pv. tomato strain DC3000. J. Bacteriol. 187:7805-7814.

Quigley, N. B., and Gross, D. C. 1994. Syringomycin production among strains of *Pseudomonas syringae* pv. *syringae*: Conservation of the *syrB* and *syrD* genes and activation of phytotoxin production by plant signal molecules. Mol. Plant-Microbe Interact. 7:78-90.

Ray, S. K., Rajeshwari, R., Sharma, Y., and Sonti, R. V. 2002. A high-molecular-weight outer membrane protein of *Xanthomonas oryzae* pv. *oryzae* exhibits similarity to non-fimbrial adhesins of animal pathogenic bacteria and is required for optimum virulence. Mol. Microbiol. 46:637-647.

Reboutier, D., Frankart, C., Briand, J., Biligui, B., Laroche, S., Rona, J.-P., Barny, M.-A., and Bouteau, F. 2007. The $HrpN_{ea}$ harpin from *Erwinia amylovora* triggers differential responses on the nonhost *Arabidopsis thaliana* cells and on the host apple cells. Mol. Plant-Microbe Interact. 20:94-100.

Reed, G. M. 1914. An unusual outbreak of apple blossom blight. Phytopathology 4:27-30.

Reimer, F. C. 1925. Blight resistance in pears and characteristics of pear species and stocks. Oreg. Agric. Exp. Stn. Bull. 214.

Richie, D. F., and Klos, E. J. 1977. Isolation of *Erwinia amylovora* bacteriophage from aerial parts of apple trees. Phytopathology 67:101-104.

Roberts, I. S., and Coleman, M. J. 1991. The virulence of *Erwinia amylovora*: Molecular genetic perspectives. J. Gen. Microbiol. 137:1453-1457.

Rohmer, L., Kjemtrup, S., Marchesini, P., and Dangl, J. L. 2003. Nucleotide sequence, functional characterization and evolution of pFKN, a virulence plasmid in *Pseudomonas syringae* pathovar *maculicola*. Mol. Microbiol. 47:1545-1562.

Rosen, H. R. 1929. The life history of the fire blight pathogen, *Bacillus amylovorus*, as related to the means of overwintering and dissemination. Ark. Agric. Exp. Stn. Bull. 244.

Schaad, N. W., Heskett, M. G., Gardner, J. M., and Kado, C. I. 1973. Influence of inoculum dosage, time after wounding, and season on infection of Persian walnut trees by *Erwinia rubrifaciens*. Phytopathology 63:327-329.

Schechter, L. M., Roberts, K. A., Jamir, Y., Alfano, J. R., and Collmer, A. 2004. *Pseudomonas syringae* type III secretion sytem targeting signals and novel effectors studied with a Cya translocation reporter. J. Bacteriol. 180:543-555.

Schenk, A., Berger, M., Keith, L. M., Bender, C. L., Muskhelishvili, G., and Ullrich, M. S. 2006. The *algT* gene of *Pseudomonas syringae* pv. *glycinea* and new insights into the transcriptional organization of the *algT-muc* gene cluster. J. Bacteriol. 188:8013-8021.

Schwartz, T., Bernhard, F., Theiler, R., and Geider, K. 1991. Diversity of the fire blight pathogen in production of dihydrophenylalanine, a virulence factor of some *Erwinia amylovora* strains. Phytopathology 81:873-878.

Shaw, L. 1934. Studies on resistance of apple and other rosaceous plants to fire blight. J. Agric. Res. 49:283-313.

Sjulin, T. M., and Beer, S. V. 1978. Mechanism of wilt induction by amylovoran in cotoneaster shoots and its relation to wilting of shoots infected by *Erwinia amylovora*. Phytopathology 68:89-94.

Smith, A. R. W., Rastall, R. A., Rees, N. H., and Hignett, R. C. 1990. Structure of the extracellular polysaccharide of *Erwinia amylovora*: A preliminary report. Acta Hortic. 273:211-219.

Sneath, B. J., Howson, J. M., and Beer, S. V. 1990. Siderophore genes of *Erwinia amylovora*: Cloning and putative role in virulence. Acta Hortic. 273:255-257.

Starr, M. P., Cardona, C., and Folsom, D. 1951. Bacterial fire blight of raspberry. Phytopathology 41:915-919.

Steinberger, E. M., and Beer, S. V. 1988. Creation and complementation of pathogenicity mutants of *Erwinia amylovora*. Mol. Plant-Microbe Interact. 1:135-144.

Stewart, V. B. 1913. The importance of the tarnished plant bug in the dissemination of fire blight in nursery stock. Phytopathology 3:273-276.

Stewart, V. B., and Leonard, M. D. 1915. The role of sucking insects in the dissemination of fire blight bacteria. Phytopathology 5:117-123.

Takeuchi, K., Ono, H., Yoshida, M., Ishii, T., Katoh, E., Taguchi, F., Miki, R., Murata, K., Kaku, H., and Ichinose, Y. 2007. Flagellin glycans from two pathovars of *Pseudomonas syringae* contain rhamnose in D and L configurations in different ratios and modified 4-amino-4,6-dideoxyglucose. J. Bacteriol. 189:6945-6956.

Thomas, H. E., and Ark, P. A. 1934. Fire blight of pears and related plants. Univ. Calif. Agric. Exp. Stn. Bull. 586.

Thomas, H. E., and Thomas, H. E. 1931. Plants affected by fire blight. Phytopathology 21:425-435.

United States Department of Agriculture. 1960. Index of plant diseases in the United States. U.S. Dep. Agric. Agric. Handb. 165.

van der Zwet, T., and Keil, H. L. 1970. Relative susceptibility of succulent and woody tissue of Magness pear to infection by *Erwinia amylovora*. Phytopathology 60:593-595.

van der Zwet, T., and Keil, H. L. 1979. Fire blight, a bacterial disease of rosaceous plants. U.S. Dep. Agric. Agric. Handb. 510.

Vanneste, J. L., and Expert, D. 1990. Detection and characterization of an iron uptake system in *E. amylovora*. Acta Hortic. 273:249-253.

Venisse, J. S., Gullner, G., and Brisset, M. N. 2001. Evidence for the involvement of an oxidative stress in the initiation of infection of pear by *Erwinia amylovora*. Plant Physiol. 125:2164-2172.

Verhoogt, H. J., Smit, H., Abee, T., Gamper, M., Driessen, A. J., Haas, D., and Konings, W. N. 1992. *arcD*, the first gene of the *arc* operon for anaerobic arginine catabolism in *Pseudomonas aeruginosa*, encodes an arginine-ornithine exchanger. J. Bacteriol. 174:1568-1573.

Vivian, A., Murillow, J., and Jackson, R. W. 2001. The roles of plasmids in phytopathogenic bacteria: Mobile arsenals? Microbiology 147:763-780.

Wade, E. K. 1974. Apple (*Malus* spp.), pear (*Pyrus* spp.), other trees including flowering crab, flowering quince, hawthorn, pyracantha, mountain ash disorder: Fireblight. Wis. Agric. Ext. Serv. Fact Sheet A1616.

Waite, M. B. 1891. Results from recent investigations in pear blight. (Abstr.) Proc. Am. Assoc. Adv. Sci. 40:315.

Warren, G. J. 1995. Identification and analysis of *ina* genes and proteins. Pages 85-99 in: Biological Ice Nucleation and Its Applications. R. E. Lee, Jr., G. J. Warren, and L. V. Gusta, eds. American Phytopathological Society, St. Paul, MN.

Warren, G., and Wolber, P. 1991. Molecular aspects of microbial ice nucleation. Mol. Microbiol. 5:239-243.

Wei, Z. M., and Beer, S. V. 1993. HarpI of *Erwinia amylovora* functions in secretion of harpin and is a member of a new protein family. J. Bacteriol. 175:7958-7967.

Wei, Z. M., Laby, R. J., Zumoff, C. H., Bauer, D. W., He, S. Y., Collmer, A., and Beer, S. V. 1992. Harpin, elicitor of the hypersensitive response produced by the plant pathogen *Erwinia amylovora*. Science 257:85-88.

Wilson, E. E. 1934. A bacterial canker of pear trees new to California. Phytopathology 24:534-537.

Wilson, E. E. 1983. Bacterial canker of stone-fruit trees in California. Hilgardia 8:83-123.

Zeller, W. 1977. Untersuchungen zur Feuerbrandkrankheit in der Bundes-republik Deutschland. 2. Die Anfalligkeit von Siergeholzen gegen *Erwinia amylovora*. Nachrbl. Dtsch. Pflanzenschutzd. 29:1-10.

Zhao, Y., Ma, A., and Sundin, G. W. 2005a. Comparative genomic analysis of the pPT23A plasmid family of *Pseudomonas syringae*. J. Bacteriol. 187:2113-2126.

Zhao, Y., Blumer, S. E., and Sundin, G. W. 2005b. Identification of *Erwinia amylovora* genes induced during infection of immature pear fruit. J. Bacteriol. 187:8088-8103.

6 Macerative Diseases and Their Pathogens

Macerative diseases are characterized by the dissolution of host tissues and the release of cellular fluids. These fluids attract secondary bacteria that proliferate and often release offensive odors, depending on the type of host infected: onions, for example, smell particularly unpleasant when affected by a macerative bacterial disease. This type of disease is commonly referred to as soft rot. Soft rot pathogens produce and secrete an arsenal of extracellular pectin- and pectate-cleaving enzymes, plus cellulases, nucleases, and proteases that attack cellulose, nucleic acids, and cellular proteins, respectively. Although mostly associated with postharvest diseases, soft rots also can be detected in the field. For example, soft rot of potato and root and bulb crops caused by *Pectobacterium carotovorum* subsp. *carotovorum* (formerly, *Erwinia carotovora* subsp. *carotovora* [Gardan et al., 2003]) is a devastating field disease. In subtropical regions, heart rot of pineapple and soft rot of banana plants occur in the field (Rohrbach, 1989; Kaneshiro et al., 2008). Other soft rot-causing pathogens, such as *Pseudomonas viridiflava*, which causes soft rot of harvested fruits and vegetables, are generally thought to be more important as postharvest pathogens although they also cause blights and leaf spots in the field. Typical examples of macerative diseases are described below.

Soft Rot and Blackleg of Potato

Pectobacterium carotovorum subsp. *carotovorum* infects a wide range of plants and is best known for causing soft rot of various types of produce delivered to the market (Ma et al., 2007). Blackleg of potato occurs under cool conditions and is attributed to the closely related *Pectobacterium atrosepticum*, a pathogen that appears to be restricted almost exclusively to potato in temperate regions (Perombelon and Kelman, 1980; Perombelon, 1992; Perombelon and Salmond, 1995). *Pectobacterium atrosepticum* gains entry into the potato plant through wounds of aboveground parts of the host and can ingress systemically, causing blackening of the invaded tissues and stem rot in the field. In contrast, *Pectobacterium carotovorum* subsp. *carotovorum* is not usually associated with stem symptoms, although it can cause stem rot of hydroponically grown plants (Schuerger and Batzer, 1993).

General Characteristics of *Pectobacterium carotovorum* subsp. *carotovorum* and *Pectobacterium atrosepticum*

Pectobacterium carotovorum subspecies and *Pectobacterium atrosepticum*, classified with the class *Gammaproteobacteria*, are members of the family *Enterobacteriaceae*. The species have all the general characteristics of the family: small, straight, Gram-negative rods with lateral peritrichous flagella; facultatively anaerobic; negative for cytochrome oxidase; positive for catalase; and acid producing when cultured on fructose, glucose, galactose, or sucrose. All soft rot-causing isolates produce pectinases and most isolates produce protease. Neither L-glutamic acid nor α-ketoglutaric acid supports growth as a sole carbon source (Grula et al., 1968). One point of differentiation for the two species is that *Pectobacterium carotovorum* subsp. *carotovorum* can grow at 37°C while *Pectobacterium atrosepticum* cannot. With *Pectobacterium atrosepticum*, growth on potato tuber tissue is substantially enhanced in the presence of citrate and the natural levels of citrate in the potato tuber are reduced (Urbany and Neuhaus, 2008).

When cultured on rich agar medium, *Pectobacterium carotovorum* subsp. *carotovorum* colonies are light grayish to creamy white, entire, smooth, and glistening. *Pectobacterium atrosepticum* contains a 5.064-Mb genome with a guanine-plus-cytosine (GC) content of 50.97% (Bell et al., 2003). Comparisons of the complete genome sequence of *Pectobacterium atrosepticum* (Bell et al., 2004) to that of *Pectobacterium carotovorum* subsp. *carotovorum* have helped elucidate the genetic and molecular similarities and differences between these pathogens. Molecular genetic analyses of various isolates from different hosts and environments (e.g., dicots vs. monocots) have revealed that *Pectobacterium carotovorum* subsp. *carotovorum* isolates have become specialized to particular hosts and environments because of early genomic divergence within this taxon (Yishay et al., 2008). It would, therefore, not be too surprising if *Pectobacterium atrosepticum* is determined to be one of these host-adapted *Pectobacterium carotovorum* subsp. *cartovorum* isolates. Comparative genome sequence analyses have revealed that the orthologous proteins encoded by *Pectobacterium carotovorum* subsp. *carotovorum* and *Pectobacterium atrosepticum* are 92% identical (Glasner et al., 2008). *Pectobacterium atrosepticum* possesses a *hrp–dsp* gene cluster that is similar in content and arrangement to that of *Erwinia amylovora*, but it is different to that in the more closely related soft rot potato pathogen, *Dickeya chrysanthemi* pv. *chrysanthemi* (Bell et al., 2002). The presence of *hrp* genes indicates that type III and type IV secretion systems are operating in *Pectobacterium carotovorum* subspecies (see below).

Host Range

Many different plant species are susceptible to *Pectobacterium carotovorum* subsp. *carotovorum*. Examples of host plants are listed in Table 6-1. Besides plants, *Pectobacterium carotovorum* subsp. *carotovorum* infects *Drosophila melanogaster* Meigen and induces an immune response that leads to the systemic expression of antimicrobial peptide genes in the fruit fly (Basset et al., 2000). *Drosophila melanogaster* is a natural vector for *Pectobacterium carotovorum* subsp. *carotovorum* and *Pectobacterium atrosepticum* (Perombelon and Kelman, 1980; Kloepper et al., 1981). The *Pectobacterium carotovorum* subsp. *carotovorum* gene designated *evf* specifies the interaction between the bacterial cell and the fruit fly (Basset et al., 2003).

History of Soft Rot and Blackleg

In 1901, L. R. Jones reported soft rot of stored cabbage, celery, and carrots and called the pathogen *Bacillus carotovorus*. The first published description of blackleg of potato appeared in 1903, given by Otto Appel (Appel, 1903). At that time, the pathogen was called *Bacillus phytophthorus* and the bacterial potato disease was called Schwarzbeinigkeit (blackleg). In 1910, Erwin F. Smith confirmed the findings of Jones and Appel (Smith, 1910; Smith, 1929). At that time, soft rot and blackleg were believed to be caused by very similar bacteria. In 1927, the soft rot

Table 6-1. Natural Host Range of *Pectobacterium carotovorum* subsp. *carotovorum*

Strain or Isolate	Plant Source	Test Host	Reference
765-1	Calla (*Zantedeschia aethiopica* (L.) Spreng.)	Potato	Pierce and McCain, 1992
665-2B	Carnation (*Dianthus caryophyllus* L.)	Potato	Pierce and McCain, 1992
325, 164	Christmas cactus (*Epiphyllum* sp.)	Potato	Pierce and McCain, 1992; Parent et al., 1996
277-3	Dieffenbachia (*Dieffenbachia* sp.)	Potato	Pierce and McCain, 1992
773	Garden balsam (*Impatiens balsamina* L.)	Potato	Pierce and McCain, 1992
576-1, 69III	Iris (*Iris* sp.)	Potato	Pierce and McCain, 1992; De Boer and Ward, 1995
976-2	English ivy (*Hedera helix* L.)	Potato	Pierce and McCain, 1992
412-2	Poinsettia (*Euphorbia pulcherrima* Willd. ex Klotzsch)	Potato	Pierce and McCain, 1992
538-3	Primrose (*Primula* sp.)	Potato	Pierce and McCain, 1992
425-2	Nephthytis (*Nephthytis afzelii* Schott)	Potato	Pierce and McCain, 1992
745-4	Arborvitae (*Thuja occidentalis* L.)	Potato	Pierce and McCain, 1992
Ecc26, 26V, 131	Potato (*Solanum tuberosum* L.)	nd[a]	De Boer and Ward, 1995; Parent et al., 1996
72VI	Chicory (*Cichorium intybus* L.)	nd	De Boer and Ward, 1995
372	Hibiscus (*Hibiscus* sp.)	nd	De Boer and Ward, 1995
376	Broccoli (*Brassica oleracea* L. var. *botrytis* L.)	nd	De Boer and Ward, 1995
441	Sunflower (*Helianthus annuus* L.)	nd	De Boer and Ward, 1995
462	Turnip (*Brassica rapa* L.)	nd	De Boer and Ward, 1995
539	Artichoke (*Cynara scolymus* L.)	nd	De Boer and Ward, 1995
172	Lily (*Lilium* sp.)	nd	Parent et al., 1996
182	Euphorbia (*Euphorbia* sp.)	nd	Parent et al., 1996
227, 93-824, PBJB	Cucumber (*Cucumis sativus* L.)	Chinese cabbage (pak choi or bak choi), broccoli, cauliflower, kale, kohlrabi, radish, cantaloupe	Parent et al., 1996; Schuerger and Batzer, 1993
375, 93-856	Lettuce (*Lactuca sativa* L.)	nd	Parent et al., 1996
93-823	Leek (*Allium porrum* L.)	nd	Parent et al., 1996
nd	Carrot (*Daucus carota* L.)	Carrot	Jones, 1901
nd	Cabbage (*Brassica oleracea* L. var. *capitata* L.)	Carrot	Jones, 1901
nd	Celery (*Apium graveolens* L. var. *dulce* (Mill.) DC.)	Celery	Jones, 1901

[a] nd = Not determined.

bacterium identified by Jones (1901) was renamed *Bacterium carotovorum*, and in 1945, it was renamed *Pectobacterium carotovorum*. In 1969, the organism was renamed *Erwinia carotovora* var. *carotovora* (Dye, 1969). Recent reclassification based on 16S rDNA comparisons has resurrected the genus *Pectobacterium* and shifted all soft rot-causing members of the genus *Erwinia* to the genera *Pectobacterium* and *Dickeya* of the class *Gammaproteobacteria* (Hauben et al., 1998; Gardan et al., 2003; Samson et al., 2005). Whole-genome sequence comparisons between these related soft-rotting bacteria showed an 8% difference between *Pectobacterium carotovorum* subsp. *carotovorum* and *Pectobacterium atrosepticum* (Bell et al., 2004; Glasner et al., 2008). Phylogenetic analyses have revealed multiple distinct subgroups of the genus *Pectobacterium*, some of which have been elevated to species status. Among these, the *Pectobacterium atrosepticum* clade is somewhat unusual with its narrow host range (primarily potato) and distinct pathogenesis potential (blackleg).

Symptomatology

Infection of the potato plant in the field by *Pectobacterium carotovorum* subsp. *carotovorum* results in a generalized collapse of the plant, which usually is enhanced by hot weather conditions. Soft rot of both immature and mature tubers is associated with the collapsed plant (see below).

On the other hand, localized infection by *Pectobacterium atrosepticum* is favored by cool weather, so that the plant remains relatively upright. Extensive invasion of the primary stalks and secondary stems by *Pectobacterium atrosepticum* results in the total collapse of the potato plant (see Fig. 4-6). During bacterial ingression, the leaves can undergo chlorosis and then become flagged, followed by wilting of the entire plant. The infected (rotted) stalks and stems are dark brown to black, a symptom characteristic of blackleg. All stem infections caused by *Pectobacterium atrosepticum* are considered stem rots, regardless of whether or not they produce blackening of the stem tissues (Sivasithamparam, 1982).

With *Pectobacterium carotovorum* subsp. *carotovorum*, tubers of various age and size are infected through wounds and lenticels. Infected tubers display a number of sites where *Pectobacterium carotovorum* subsp. *carotovorum* has gained entry and caused localized soft rot. Under semiarid conditions, the sites of infection on tubers are blackened, presumably due to oxidation of polyphenols in the tubers. Under conditions of high moisture levels and low oxygen concentrations (usually during storage of harvested tubers), the sites of infection coalesce into a generalized maceration of the entire tuber.

Unlike *Pectobacterium carotovorum* subsp. *carotovorum*, *Pectobacterium atrosepticum* colonizes tubers via a slower intercellular ingression of the parenchyma (Fox, 1972). The related species *Pectobacterium betavasculorum* causes substantial release of ethane and carbohydrates from inoculated potato tuber slices (Kuykendall and Hunter, 2008).

Disease Cycle

Pectobacterium carotovorum subsp. *carotovorum* and *Pectobacterium atrosepticum* overwinter in infected plant parts and tubers tilled into the field soil. Contaminated seed pieces can also be a source of inoculum. Planting of freshly cut seed pieces without sufficient time for suberization (corking over) predisposes the tuber to the soft rot bacterial cells overwintering on infected plant parts in the soil. During the early stages of growth of the potato plant, the proliferation (or growth) of *Pectobacterium carotovorum* subsp. *carotovorum* is slow because of temperatures below its optimum during the planting season. Soil moisture also plays an important role in infection. For both organisms, low moisture levels reduce the incidence and spread of disease. However, low oxygen levels promote infection. In fact, anaerobic conditions favor soft rot (Lund and Wyatt, 1972; De Boer and Kelman, 1978; Maher and Kelman, 1983).

Pectobacterium carotovorum subsp. *carotovorum*-caused soft rot problems occur following harvest, storage, and shipment. Root vegetables, such as carrot and potato, are washed in bulk and then crated or shipped. Any infected root or tuber serves as the primary source of inoculum and contaminates the wash water, which in turn serves as a disseminating agent. Fresh wounds made during harvest serve as excellent portals of entry for *Pectobacterium carotovorum* subsp. *carotovorum*. Washing with high-pressure water jets to clean roots or tubers increases disease incidence (Logan, 1964). High relative humidity and warm conditions facilitate soft rot incidence, particularly in those carrots beneath large piles of the vegetable, where the microclimate is highly conducive for disease progression.

Roots and immature tubers growing close to infected plant debris and tubers are sites of primary infection by *Pectobacterium atrosepticum*. As infection progresses systemically by this organism, stems at the base of the plant show dark brown to black symptoms characteristic of blackleg. As the infection girdles the main stem, upper foliage withers, followed by a generalized collapse of the entire plant (see Fig. 4-6). Left unchecked, these diseased plants are tilled back into the soil, where *Pectobacterium atrosepticum* overwinters in infected plant parts and debris. The pathogen also persists in surface water and on harvested vegetables, equipment, and boxes.

Pathogenicity Tests

Soft rot-causing bacteria can be isolated on various semiselective media, such as crystal violet pectate (CVP) agar (Cuppels and Kelman, 1974), which contains sodium polypectate as the substrate for pectinolytic enzymes and crystal violet, sodium nitrate, and manganese sulfate to inhibit various saprophytic bacteria.

The pathogenicity of *Pectobacterium carotovorum* subsp. *carotovorum* is verified by inoculating potato tubers using freshly grown bacterial cells suspended in water at a concentration of 10^7–10^8 CFU/ml. A 200-µl disposable plastic pipet tip holding the inoculum is jabbed into surface-sterilized potato tubers. Approximately 10–20 µl of cell suspension is deposited into the freshly made hole as the pipet tip is withdrawn from the tuber. The hole is sealed with sterile silicon grease or Vaseline. The tubers are wrapped with Mylar film or Saran wrap and stored in a moist chamber at 28°C. The presence of soft rot at the sites of inoculation is determined by using a round wooden toothpick to pierce into the inoculated sites. No resistance to the toothpick probing indicates that soft rotting is taking place.

When large numbers of bacterial samples are to be assessed, seedlings of plants such as cucumber, cabbage, or tobacco can be axenically grown in 24-well tissue culture plates filled with multiple shoot primordium (MSP) medium (Flow Laboratories, Rockville, MD) supplemented with 2% sucrose and solidified with 0.3% gellan gum (Pirhonen et al., 1991). The seedlings are surface inoculated by pipetting bacterial suspensions (10^8 CFU/ml) onto the plants without wounding. The inoculated seedlings are placed in a growth chamber and examined after 2 days.

Pectinolytic activity can be assayed by using the cantilevered potato slice method described by Dye (1960), modified as follows: fresh potato tissue slices (0.5 × 4 × 15 mm) are obtained using a sterilized scalpel blade or single-edged razor blade. A red potato is used for its low phenol oxidase content, in order to minimize darkening of the slices. A petri plate is filled with molten 1.5% agar, about 6 mm thick, which is allowed to solidify at room temperature. Two parallel trenches 11 mm wide are cut in the agar, 28 mm apart. Freshly cut potato slices are laid on the agar 4 mm apart, and about 6–7 mm of each slice is slid outward over the trench, unsupported. Each cantilevered slice is inoculated at the point where it leaves the support of the agar (Fig. 6-1). The plate is incubated at 28°C. The unsupported end of the slice will

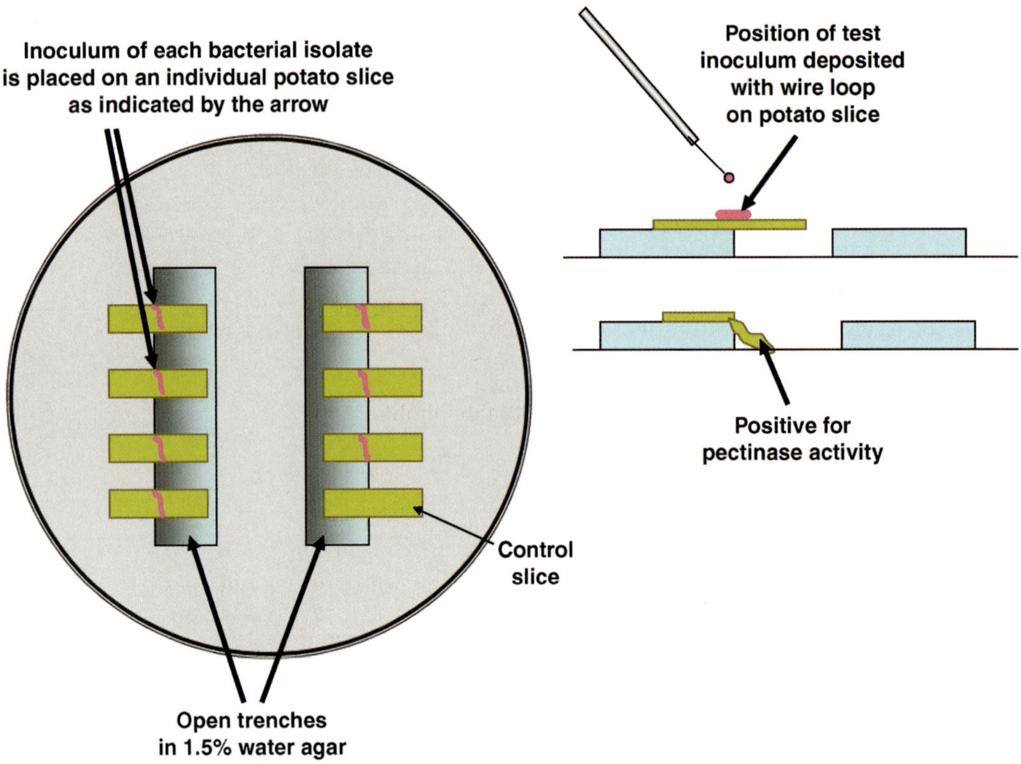

Fig. 6-1. Diagram of a simple assay for pectinase activity. **Left,** View from above, showing the arrangement of potato slices cantilevered over the trenches made in water agar. **Right,** Side view showing where the inoculum of each bacterial isolate should be placed.

droop and drop to the bottom of the petri plate if pectinolytic enzymes are produced by the inoculum. The results can be assessed within 4 h.

Pathogenicity Factors

There are a number of extracellular enzymes that are elaborated by *Pectobacterium carotovorum* subsp. *carotovorum*. The production of these enzymes is under the control of cell density-dependent signals called quorum-sensing *N*-acyl-homoserine lactones (AHLs) produced by this bacterium and controlled by various transcriptional regulators, such as *hrpL* (see Chapter 10 for details). For instance, *Pectobacterium carotovorum* subsp. *carotovorum* produces a small RNA-binding protein called RsmA (*r*epressor of *s*econdary *m*etabolites), which promotes the degradation of target mRNA, and a 259-nucleotide untranslated RNA called RsmB'. RsmA and RsmB' control the expression of several virulence or pathogenicity factors, including endopolygalacturonase, protease, and cellulolytic enzymes, as described below. RsmA binds to the target mRNA and induces its degradation via its ribonuclease activity. The degradation of the target mRNA is blocked by RsmB', which coats the targeted mRNA, protecting it from nuclease digestion (Johannsson and Cossart, 2003). The *rsmB* gene also encodes a regulatory RNA that binds to and neutralizes RsmA regulatory activity (Liu et al., 1998). ExpR encoded by *expR* is a LuxR homolog (viz., in *Vibrio fischeri* luciferase operon control) that activates *rsmA* transcription (Cui et al., 2005). Most strains of *Pectobacterium carotovorum* subsp. *carotovorum* possess two copies of *expR*, whereby ExpR1 activates transcription of *rsmA* in the presence of AHL and ExpR2 activates transcription of *rsmA* in the absence of AHL (Cui et al., 2006). *Pectobacterium carotovorum* subspecies elaborate AHLs of either 3-oxo-C8-*H*L (= *N*-3-oxooctanoyl-L-homoserine lactone) or 3-oxo-C6-*H*L (= 3-oxohexanoyl-L-homoserine lactone), depending on the strain (Chatterjee et al., 2005). At sufficient concentrations, the former quorum-sensing signal derepresses virulence gene expression that is normally repressed under the control of the regulator, VirR (Burr et al., 2006). Additional transcriptional factors and posttranscriptional regulators also play an important role in controlling the virulence of *Pectobacterium carotovorum* subspecies (Laasik et al., 2005) (see Chapter 10).

Pectin- and Cell Wall-Degrading Enzymes

Soft rot is the result of loss of skeletal components of the plant cell and the release of cellular fluids. *Pectobacterium carotovorum* subsp. *carotovorum* produces pectin methyl esterases, pectin lyase, pectin acetyl esterases, pectate lyases, and polygalacturonases. Collectively, these enzymes are called pectinases. Pectinase synthesis is induced by isolated plant cell wall components, pectin, and intermediates of the pectate degradation pathway. In planta studies showed that endopectate lyase, exopectate lyase, and endopolygalacturonase are induced sequentially, beginning 3 h after inoculation (Yang et al., 1992). The expression of genes encoding the major cellulase CelVI, the pectate lyases PelC and PelB, and polygalacturonase PehA is controlled globally by the *expI* gene (Pirhonen et al., 1993; Heikinheimo et al., 1995). Endopectate lyases are the primary enzymes that cause the soft rot symptoms. The polygalacturonate substrate is broken down to oligogalacturonides of various lengths, which are further degraded into small oligomers in the bacterial periplasm. Subsequently, the oligomers are converted into metabolizable monomers in the cytoplasm (Fig. 6-2). With incompatible (nonhost) plants, oligosaccharides resulting from pectin degradation elicit plant defense responses (Davis et al., 1986). However, the potential plant defense

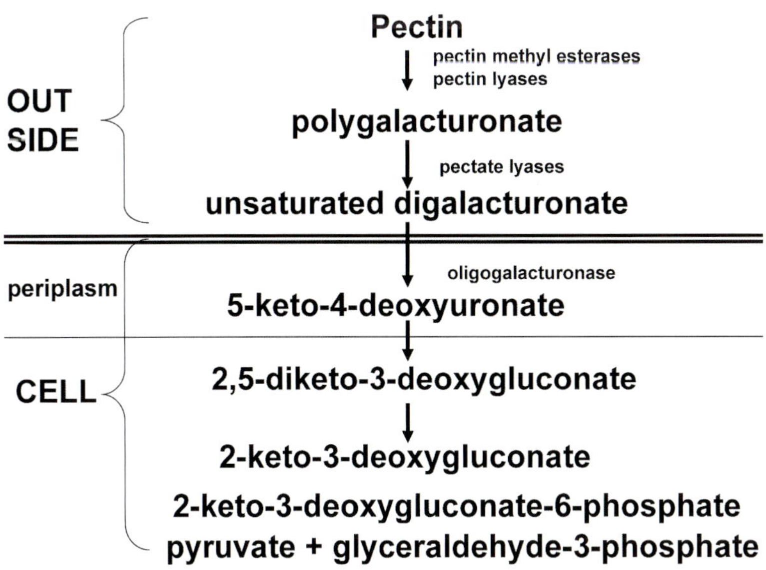

Fig. 6-2. Pathway of pectin assimilation by *Pectobacterium carotovorum* subspecies, *Pectobacterium atrosepticum*, and *Dickeya chrysanthemi*.

response elicitors are sequestered in the periplasm and quickly converted into utilizable monomers in the bacterial cell.

The expression of cell wall-degrading enzymes is controlled by quorum-sensing signal molecules, AHLs, that are produced in planta when the host is well colonized (Toth et al., 2004). Transgenic host plants expressing a lactonase derived from a *Bacillus* species displayed interesting resistance to soft rot caused by *Pectobacterium carotovorum* subsp. *carotovorum* (Dong et al., 2000; 2001). *N*-Acyl-homoserine lactonases may be useful in controlling soft rot through future generations of transgenic host plants expressing the lactonase. Additional chromosomal genes that modulate the expression of virulence factors have been found by random mutagenesis using mini-mu transposon insertions (Laasik et al., 2005).

Protease

For effective in planta growth of *Pectobacterium carotovorum* subsp. *carotovorum*, sources of nitrogen are essential. Small peptides and amino acids serve as a primary source of nitrogen as they are released during infection of the host. Very little is known about the characteristics of proteases elaborated by *Pectobacterium carotovorum* subsp. *carotovorum*. The gene *prt1* encodes an extracellular protease that appears to show a high degree of amino acid sequence identity to several bacterial zinc metalloproteases (Kyőstiő et al., 1991).

Cellulase

A β-1,4-endoglucanase encoded by the *celV* gene is a virulence factor since a ethylmethylsulfonate-induced mutation abolishing *celV* in *Pectobacterium carotovorum* subsp. *carotovorum* caused significant reduction in soft rot of potato (Walker et al., 1994). Interestingly, expression of *celV* in *Escherichia coli* failed to macerate potato, whereas pectate lyases expressed in *Escherichia coli* confer maceration activity on this heterologous enterobacterium (Zink and Chatterjee, 1985; Keen and Tamaki, 1986).

hrp and *hrc* Genes

Despite its ability to attack numerous plant species without causing a hypersensitive response, *Pectobacterium carotovorum* subsp. *carotovorum* also harbors *hrp* genes as listed in Table 6-2. The *hrp* and *hrc* genes as well as *out* genes encode proteins that make up the export pathway that is essential for pectinases, cellulose, and proteases to be released to the outer cellular surface. The *hrp* genes of *Pectobacterium carotovorum* subsp. *carotovorum* are clustered in three operons (Rantakari et al., 2001). Typically, *hrp* clusters are composed of about 20 genes organized in eight operons (Bell et al., 2002). A number of these genes are conserved in animal and plant pathogens. The conserved genes are called *hrc* (Bogdanove et al., 1996). Approximately nine *hrc* genes occur in animal and plant pathogens. These genes were likely distributed by lateral gene transfer among Gram-negative bacteria. Interestingly, eight of the nine *hrc* genes in *Pseudomonas* species show strong sequence similarity to genes involved in the assembly of the flagellar basal body (He, 1997). Hence, adaptation of the *fla* genes for aiding the survival of the pathogen could have transpired in the evolution of pathogenesis.

The genes identified for their role in virulence are listed in Tables 6-2 and 6-3.

Table 6-2. *hrp* and *hrc* Genes of *Pectobacterium carotovorum* subsp. *carotovorum* and *Pectobacterium atrosepticum*

Gene	Size of Gene Product (amino acid residues)	Gene Product	Reference
hrpA	80	Putative pilus subunit	Bell et al., 2002
hrpB	125	Putative regulator	Bell et al., 2002
hrcJ	264	Uncharacterized	Bell et al., 2002
hrpD	193	Uncharacterized	Bell et al., 2002
hrpE	200	Putative ATPase	Bell et al., 2002
hrpF	79	Unknown	Bell et al., 2002
hrpG	149	Uncharacterized	Bell et al., 2002
hrcC	676	Inner membrane structural protein	Bell et al., 2002
hrpT	69	Putative structural component	Bell et al., 2002
hrpV	120	Uncharacterized	Bell et al., 2002
hrpN	356; 403	Harpin, a hypersensitive response (HR) elicitor	Cui et al., 1996; Liu et al., 1999

Table 6-3. Virulence Genes of *Pectobacterium carotovorum* subsp. *carotovorum*

Gene	Size of Gene Product (kDa)	Gene Product	Reference
celS	27.7	Cellulase	Baumberger et al., 2003
celV	55	Cellulose	Cooper and Salmond, 1993; Walker et al., 1994
pehA	72	Polygalacturonase	Palomäki et al., 2002
pehB	50	Polygalacturonase	Hinton et al., 1989; Kluskens et al., 2005
pehC	34.2	Polygalacturonase	Hinton et al., 1989; Kluskens et al., 2005
pelA	42.0	Pectate lyase	Tamaki et al., 1988
pelB	37.4	Pectate lyase	Heikinheimo et al., 1995
pelC	38.9	Pectate lyase	Hinton et al., 1989

Secretion Systems of *Pectobacterium carotovorum* subsp. *carotovorum*

One or more mechanisms for excreting pectinases, cellulases, and proteases are essential for *Pectobacterium carotovorum* subsp. *carotovorum* to cause soft rot. The pathogen is equipped with a general secretion apparatus called the type II secretion system (Pugsley, 1993; Salmond and Reeves, 1993) (see Chapter 10 for details). Pectinases and cellulases synthesized in the bacterial cell are exported across the cytoplasmic membrane into the periplasm. During this process, the peptide leader (or signal sequence), which anchors proteins to the inner membrane, is cleaved away by a set of secretion helper proteins called Sec. Following protein processing, the pectinases and cellulases are exported by means of Out proteins (Reeves et al., 1993) to the cell surface. Metalloprotease is secreted by a one-step, Sec-independent pathway.

In addition to the type II secretion system, *Pectobacterium carotovorum* subsp. *carotovorum* possesses the type III secretion machinery that is encoded by *hrp* and *hrc* genes (Table 6-2). Type III protein secretion systems play an integral part in the pathogenesis of both animal- and plant-pathogenic Gram-negative bacteria (Mecsas and Strauss, 1996; Hueck, 1998; Galán and Collmer, 1999) (see Chapter 10 for details). At least one type IV secretion system is present in *Pectobacterium atrosepticum* and is proposed to secrete hemolysin-coregulated proteins (Mattinen et al., 2008).

Carotovoricin

Pectobacterium carotovorum subsp. *carotovorum* produces at least two bacteriocins, originally called carotovoricin (Hamon and Peron, 1961) and now called carotovoricin Era and Erb (Nguyen et al., 2001). These carotovoricins are a vestigial phage with antenna-like structure, sheath with a base plate, and tail fibers (Nguyen et al., 1999). The sheath is 22 nm in diameter and extends about 184 nm in length, including the base plate. These bacteriocins kill other *Pectobacterium carotovorum* subsp. *carotovorum* strains differentially (Nguyen et al., 2001). *Pectobacterium carotovorum* subsp. *carotovorum* strain 89-H-4 produces carocin S1, a low-molecular-weight bacteriocin (Chuang et al., 2007). The role of these bacteriocins in pathogenicity is unclear, but strains producing them might have competitive advantages over other strains, particularly during infection of the host and in the soil habitat.

Lipid A

Lipopolysaccharide is connected to the outer membrane of *Pectobacterium carotovorum* subsp. *carotovorum* by lipid A. The palmitoylation of lipid A requires the product (palmitoyltransferase) of the *pagP* gene, which may be required for virulence (Bishop, 2005). The expression of the *pagP* gene is induced by low magnesium concentrations and 21 to 37°C temperature shifts. Both *Pectobacterium carotovorum* subsp. *carotovorum* and *Dickeya chrysanthemi* pv. *chrysanthemi* have *pagP* homologs.

MopA and MopC Acquired Through Lateral Gene Transfer Events

MopA of *Pectobacterium carotovorum* subsp. *carotovorum* is similar in amino acid sequence to HrpQ of *Ralstonia solanacearum*, Spa33 of *Shigella flexneri*, and FliN of *Escherichia coli* and *Salmonella typhimurium* (Mulholland et al., 1993). Likewise, MopC of *Pectobacterium carotovorum* subsp. *carotovorum* shows similar amino acid sequence homology to HrpT of *Ralstonia solanacearum* and Spa24 of *Shigella flexneri* (Mulholland et al., 1993). These similarities may have resulted from convergent evolution, in which proteins with similar amino acid sequences evolved independently to perform similar functions. However, the high level of similarity between many of these genes suggests that they have a common evolutionary origin and were acquired by lateral gene transfer events.

Soft Rot Caused by *Dickeya chrysanthemi* pv. *chrysanthemi*

Soft rot in a wide range of plants is caused by *Dickeya chrysanthemi* pv. *chrysanthemi* (basonyms: *Pectobacterium chrysanthemi*, *Erwinia chrysanthemi* [Samson et al., 2005]). Like *Pectobacterium carotovorum* subsp. *carotovorum*, *Dickeya chrysanthemi* pv. *chrysanthemi* secretes a variety of cell wall-degrading enzymes that macerate plant tissues followed by plant cell lysis with the release of cellular fluids (liquefaction) into soft rot symptoms. Among these cell wall- and protopectin-degrading enzymes are pectinesterase, *endo*-D-galacturonosidase, pectate lyase, and pectin lyase, which play key roles in pathogenesis (Collmer and Keen, 1986). Strain variations in host preferences of *Erwinia chrysanthemi* prompted dividing strains into six pathovars: pv. *chrysanthemi*, pv. *dianthicola*, pv. *dieffenbachiae*, pv.

paradisiaca, pv. *parthenii*, and pv. *zeae* (Young et al., 1978; Lelliott and Dickey, 1984). *Erwinia chrysanthemi* pv. *paradisiaca* was renamed *Brenneria paradisiaca* (Hauben et al., 1998) based on 16S rDNA sequence comparative studies, while the other five pathovars of *Erwinia chrysanthemi* remained clustered with members of the genus *Pectobacterium*. Some members of the genus *Pectobacterium* (as well as *Brenneria paradisiaca*) were recently proposed to be changed to the genus *Dickeya* (Samson et al., 2005). As discussed in Chapter 3, names of genera and species will continue to change until comparative analyses of whole-genome sequences of all of the above species and strains are made.

General Characteristics of *Dickeya chrysanthemi* pv. *chrysanthemi*

Dickeya chrysanthemi pv. *chrysanthemi* is a member of the family *Enterobacteriaceae* in the class *Gammaproteobacteria*. It forms Gram-negative rods with peritrichous flagella. The bacterium is a facultative anaerobe that produces acid with simple sugars and sugar alcohols, such as arabinose, xylose, cellobiose, mannose, rhamnose, mannitol, sorbitol, and salicin. It secretes catalase and nitrate reductase but not cytochrome oxidase (it is oxidase negative). Analyses of the genome sequence of *Dickeya chrysanthemi* pv. *chrysanthemi* are certain to reveal additional novel genes involved in virulence.

In the presence of glycerol, a deep blue pigment called indigoidine (3,3'-bipyridyl) is produced by *Dickeya chrysanthemi* pv. *chrysanthemi* (Reverchon et al., 1994). Glucose and galactose repress indigoidine synthesis. Indigoidine apparently protects the organism against oxidative stress induced in the host. The crystal structure (at 1.9 Å resolution) of a 31.5-kDa indigoidine synthase isolated from *Thermotoga maritima* is similar to that in *Dickeya chrysanthemi* pv. *chrysanthemi* (Levin et al., 2005). Three structural genes called *indA*, *indB*, and *indC*, involved in indigoidine biosynthesis, have been characterized (Reverchon et al., 2002).

Host Range

The host range of *Dickeya chrysanthemi* pv. *chrysanthemi* is relatively broad. Although most host range studies on this organism revealed a wide range of dicotyledonous plants, monocotyledons are also susceptible to this pathogen (Table 6-4). Interestingly, *Dickeya chrysanthemi* pv. *chrysanthemi* strains that infect monocots, such as corn, produce substantially more xylanase than do strains from nongraminaceous hosts (Braun and Rodrigues, 1993).

Symptomatology

Initial infection occurs primarily at wound sites and under microaerophilic conditions. Early infections are recognized by water-soaking of the region surrounding the infection site or sites. The water-soaked tissues darken rapidly with progressive softening. As the infection continues and spreads, the original infected regions collapse and are further degraded into watery masses, accompanied by liquefaction. At this stage, the entire plant is no longer erect because of the loss of turgidity. Considerable withering is noticeable in the field and in the glasshouse where adequate sanitation is not practiced.

Table 6-4. Host Range of *Dickeya chrysanthemi* pv. *chrysanthemi*

Plant	Scientific Name	Reference
Chrysanthemum	*Chrysanthemum* × *morifolium* (Ramat.)	Burkholder et al., 1953; Chatterjee et al., 1983; Dahler et al., 1990; He and Collmer, 1990
Guayule	*Parthenium argentatum* A. Gray	Campbell, 1947
Tobacco	*Nicotiana tabacum* L. cv. Wisconsin 38 clone 19-3	Braun and Rodrigues, 1993; Bauer et al., 1995; Shevchik et al., 1998
Tobacco	*Nicotiana clevelandii* A. Gray	Rojas et al., 2002
Tomato	*Lycopersicon esculentum* Mill. var. *esculentum* Mill.	Boccara et al., 1991; Bauer et al., 1995
Pinto bean	*Phaseolus vulgaris* L.	Braun and Rodrigues, 1993
Chicory	*Cichorium intybus* L.	El Hassouni et al., 1999; Llama-Palacios et al., 2003; Shevchik and Hugovieux-Cotte-Pattat, 2003
Belgian endive	*Cichorium endivia* L.	Bauer et al., 1994; Barabote et al., 2003
Cotton	*Gossypium hirsutum* L.	Braun and Rodrigues, 1993
Cucumber	*Cucumis sativus* L.	Braun and Rodrigues, 1993
Corn, maize	*Zea mays* L., *Zea mays* L. subsp. *parviglumis* Iltis & Doebley var. *huehuetenangensis* Iltis & Doebley	Braun and Rodrigues, 1993
Oat	*Avena sativa* L.	Braun and Rodrigues, 1993
Pear	*Pyrus communis* L.	Llama-Palacios et al., 2003
Pepper	*Capsicum annuum* L.	Bauer et al., 1995
Rhubarb	*Rheum rhabarbarum* L.	Llama-Palacios et al., 2003
Cowpea	*Vigna sinensis* (Torner) Savi	Braun and Rodrigues, 1993
African violet	*Saintpaulia ionantha* H. Wendl.	Kotoujansky et al., 1982; Boccara et al., 1994; Bauer et al., 1995; El Hassouni et al., 1999; Franza et al., 2005; Yap et al., 2005
Petunia	*Petunia* sp.	Bauer et al., 1995
Geranium	*Pelargonium* sp.	Bauer et al., 1995
Squash	*Cucurbita* sp.	Bauer et al., 1995
Zinnia	*Zinnia* sp.	Bauer et al., 1995
Potato	*Solanum tuberosum* L. cv. Russet Burbank	Payne et al., 1987; Dahler et al., 1990; Boccara et al., 1991; Łojkowska et al., 1993
Lettuce	*Lactuca sativa* L.	Boccara et al., 1991
Aloe	*Aloe vera* (L.) Burm. f.	Laat et al., 1994
Banana	*Musa* spp.	Dickey and Victoria, 1980

Disease Cycle

Dickeya chrysanthemi pv. *chrysanthemi* survives primarily in diseased plant material and residues left behind after cropping. The bacterial cells also survival in soil and in compost. Replanting into infested soil is highly conducive to infection of seedlings or transplanted propagules. Infected plants tilled into the soil serve as reservoirs of inoculum for the next planting.

Pathogenicity Tests

Because *Dickeya chrysanthemi* pv. *chrysanthemi* has a broad host range, a number of plants can be used for pathogenicity testing (Table 6-4). African violets (*Saintpaulia ionantha* H. Wendl.) that can be grown under laboratory conditions with fluorescent lighting are convenient assay hosts that are popularly used by researchers. The plant also can be axenically cultured in test tubes (Boccara et al., 1988). The succulent leaves of African violets serve as the area of infection.

Inoculum is infiltrated by using a syringe without a needle on the abaxial side of the leaf. Infusion of the inoculum is visible as the liquid spreads between the epidermal layer and parenchymatous tissues. Inoculum can also be applied to an artificially wounded area on the leaf; pinching with a pair of tweezers, for example, inflicts wounds that can be used as the points of inoculation.

Potato tubers also serve as convenient indicator hosts. Freshly prepared inoculum (about 50 μl) is dispensed into holes made in the tuber with a plastic disposable pipet tip, to a depth of approximately 10–15 mm. Holes filled with bacterial suspension (10^9 cells/ml) are sealed with sterile Vaseline; water-filled holes similarly sealed serve as controls. Incubation at 30°C under high humidity facilitates invasion and pathogenesis.

Tomato seedlings also may be used. The inoculum is applied by stem puncture with a syringe needle. A suspension of about 10^8 CFU/ml is applied. Seedlings are placed at 29°C under 100% humidity for 15–20 h (Boccara et al., 1991). Collapse of the normally turgid seedling is an indication of maceration of tissues around the inoculation zone.

Pathogenicity Factors

Dickeya chrysanthemi pv. *chrysanthemi* secretes pectate lyases, pectin lyases, pectin esterase, exopolygalacturonase, cellulases, and proteases to aid in the infection and maceration of plant tissue. This pathogen utilizes the type II secretion system (see Chapter 10 for details) to deliver these hydrolytic enzymes to their surroundings and hosts (see also below). It was observed that *Dickeya chrysanthemi* pv. *chrysanthemi* strains isolated only from monocots, such as corn and grasses, possess xylanase activities (Braun and Kelman, 1987). The xylanase called XynA is a 42-kDa endoxylanase that is encoded by the *xynA* gene (Hurlbert and Preston, 2001). XynA is one of several enzymes involved in hemicellulose (= β-1,4-xylan) degradation. A β-glucosidase/xylosidase called BgxA is also present in *Dickeya chrysanthemi* pv. *chrysanthemi* (Vroemen et al., 1995). Although xylanase is produced by some strains of *Dickeya chrysanthemi*, it apparently does not play a role in pathogenesis (Keen et al., 1996). Pectate lyase isozymes with alkaline isoelectric points appear to play a key role in the pathogenesis of *Dickeya chrysanthemi* pv. *chrysanthemi* (Payne et al., 1987). The major pectinase genes, *pelA*, *pelB*, *pelC*, and *pem*, are moderately expressed while *pelD* and *pelE* are highly expressed and modulated by one or more unidentified plant compounds (Łojkowska et al., 1993). The expression of the *pelE* gene is complex because of the complexity of the promoter interaction with its regulatory proteins (Gold et al., 1992).

Dickeya chrysanthemi pv. *chrysanthemi* utilizes the Sec-dependent type I secretion pathway to secrete a metalloprotease and a type II secretion pathway (previously called the Out pathway) to deliver at least eight endopectate lyases called PelA, PelB, PelC, PelD, PelE, PelI, PelL, and PelZ; two exopectate lyases called PelX and PelW; two pectin methylesterases (PemA and PemB); and a pectin acetyl-esterase (PaeY). The Out secretion pathway composed of *outC*, *outD*, and *outE* operate as an operon while *outB*, *outT*, and *outS* are independent transcription units. The *outCDE* operon is regulated by the repressor KdgR (Condemine et al., 1992). The secreted pectate lyase PelI induces necrosis. For necrotizing activity, PelI must undergo posttranslational modification (Shevchik et al., 1997; Shevchik et al., 1998). Endoglucanases Z and Y apparently play a minor role in the virulence of *Dickeya chrysanthemi* pv. *chrysanthemi* (Boccara et al., 1994). These enzymes encoded by

Table 6-5. Genes Involved in the Virulence of *Dickeya chrysanthemi* pv. *chrysanthemi*

Gene	Encoded Function	Substrate	Reference
acsF-A operon	Achromobactin synthase, a siderophore	Iron	Franza et al., 2005
arbF	β-Glucoside-specific phosphotransferase system-dependent permease	Arbutin, salicin	El Hassouni et al., 1992
arbB	Phospho-β-glucosidase	β-Glucoside	El Hassouni et al., 1992
cbrA, *cbrB*, *cbrC*, *cbrD*	CbrA (34 kDa) and CbrB (55 kDa) negatively regulate chrysobactin biosynthesis via permease complex	Promoter region of siderophore chrysobactin; complex of periplasmic CbrA, integral membrane CbrB and CbrC, and ATP binder CbrD	Expert et al., 1992; Mahé et al., 1995
cbsA, *cbsB*, *cbsC*	Chrysobactin biosynthesis; CbsA, a 32-kDa 2,3-dihydro-2,3-dihydroxylbenzoate dehydrogenase; CbsC, a 38-kDa isochorismate-pyruvate hydrolase	Shikimate pathway precursors	Franza and Expert, 1991
celZ	EGZ cellulase secreted by Out general secretion pathway	Crystalline cellulose	Bortoll-German et al., 1994
expI	Producer of *N*-(hexanoyl)-homoserine lactone and *N*-(decanoyl)-homoserine lactone	*S*-Adenosyl methionine	Nasser et al., 1998
expR	Transcriptional activator of *expI–expR* locus	*N*-Acyl-homoserine lactone, a quorum-sensing signal	Reverchon et al., 1998
exuT	Galacturonic acid uptake	Galacturonic acid	San Francisco and Keenan, 1993
fct	A 80-kDa receptor of chrysobactin	Chrysobactin siderophore (*N*-α-(2,3-dihydroxybenzoyl)-D-lysyl-L-serine)	Enard et al., 1988; Franza and Expert, 1991
hecA	A 39-kDa adhesion with a signal peptide	Host epidermis	Rojas et al., 2002
hrpC	Presumed type III translocator component	Unknown	Yang et al., 2002
hrpG	Presumed type III translocator component	Unknown	Yang et al., 2002
hrpN$_{Ech}$	A 34.3-kDa, glycine-rich, hypersensitive response elicitor	Unknown	Bauer et al., 1994; Bauer et al., 1995
kdgA	A 22.2-kDa aldolase	2-Keto-3-deoxy-6-phosphogluconate	Hugouvieux-Cotte-Pattat and Robert-Baudouy, 1994
kdgK	A 36.4-kDa protein kinase involved in pectinolytic enzyme production	2-Keto-3-deoxygluconate	Hugouvieux-Cotte-Pattat et al., 1994
kdgR	A 30-kDa repressor of genes involved in pectinolysis and pectinase secretion	Upstream KdgR box AAAAAAGAAACATTGTTTCATTTGT and degenerate box (AAT(G/A) AAA(C/T)(G/A)N(G/A)(C/T)TTT(C/T) ATT)	Nasser et al., 1992; Nasser et al., 1994; Nomura et al., 1998
msrA	A 23.7-kDa methionine sulfoxide reductase	Oxidized proteins, promotes systemic invasion	El Hassouni et al., 1999
nucM	A 29.9-kDa deoxyribonuclease that appears ancillary to virulence effects	DNA	Moulard et al., 1993
paeY	Pectin acetylesterase	Pectin	Shevchik and Hugouvieux-Cotte-Pattat, 1997

(continued)

Table 6-5. Genes Involved in the Virulence of *Dickeya chrysanthemi* pv. *chrysanthemi* (continued)

Gene	Encoded Function	Substrate	Reference
paeX	Pectin acetylesterase (34.9 kDa); *paeX* gene induced by pectic catabolic products; regulated by KdgR and PecS; localized in periplasmic space	Pectin	Shevchik and Hugovieux-Cotte-Pattat, 2003
pecM	A 31.9-kDa regulator of pectinase, cellulose, and indigoidine production	Promoter region of *peh*, *cel*, and indigoidine biosynthesis genes	Reverchon et al., 1994
pecS	A 19.3-kDa transcriptional activator of *peh* genes; regulates indigoidine and cellulase production	Competitor for binding overlapping binding domains within and surrounding the KdgR box	Nasser et al., 1999
pehV	Polygalacturonase	Pectin-derived polygalacturonate	Nasser et al., 1999
pehW	Polygalacturonase	Pectin-derived polygalacturonate	Nasser et al., 1999
pehX	*exo*-Poly-α-D-galacturonidase (64.6 kDa)	Pectin-derived polygalacturonate	He and Collmer, 1990
pelA, *pelB*, *pelC*, *pelD*, *pelE*, *pelI*, *pelL*, *pelW*, *pelZ*	Pectic lyases: 42.1-kDa PelA, 44.8-kDa PelB, 37.7-kDa PelC, 47.1-kDa PelD, 46.2-kDa PelE, 41.3-kDa PelI, 51-kDa PelL, 66.4-kDa PelW, 50.4-kDa PelZ	Pectate	Keen et al., 1984; Thurn et al., 1987; Boccara et al., 1988; Tamaki et al., 1988; Barras et al., 1994; Łojkowska et al., 1995; Pissavin et al., 1996; Shevchik et al., 1997, 1999; Creze et al., 2008
pemA	Pectin methylesterase	Pectin	Shevchik et al., 1996
pemB	Pectin methylesterase	Outer membrane-linked lipoprotein	Shevchik et al., 1996
pir	A 30-kDa transcriptional regulator of *pelE*	A 35-bp binding site in the promoter of *pelE*	Nomura et al., 1998
plcA	A 39-kDa phospholipase that is ancillary to virulence effects	Phosphatidyl choline	Keen et al., 1992
phoQ	Sensory histidine kinase, two-component partner PhoP	Regulates *pel* gene expression via phosphorylation of transcriptional activator PhoP	Llama-Palacios et al., 2005
phoP	Transcriptional regulator of *vir* genes	Activated by phoQ kinase; regulates *pel* gene expression	Llama-Palacios et al., 2005
pme	A 39.3-kDa pectin methylesterase with a signal peptide	Pectin	Plastow, 1988
prtA, *prtB*, *prtC*	Metalloprotease export system	Intermembrane	Dahler et al., 1990; Létoffé et al., 1990; Faith et al., 1991
sapA, *sapB*, *sapC*, *sapD*, *sapE*, *sapF*	Resistance to host antimicrobial peptides	Host antimicrobial peptides	López-Solanilla et al., 1998
tolC	A 51-kDa outer membrane trimeric protein efflux pump	Antimicrobial plant chemicals	Barabote et al., 2003

the *celZ* and *celY* genes, respectively, are not essential for soft rot development and are limited to the saprophytic life of this pathogen.

Secretion of metalloproteases A, B, and C requires a type I secretion apparatus composed of three proteins, PrtD, PrtE, and PrtF, of which the first two are localized to the inner membrane and the third is localized to the outer membrane (Létoffé et al., 1990; Létoffé et al., 1991; Létoffé and Wandersman, 1992). PrtD is a member of a family of ATP-binding membrane proteins that are homologous to the mammalian multidrug resistance P-glycoproteins (Endicott and Ling, 1989). P-Glycoproteins are part of a larger family of efflux transporters that have developed a mechanism to protect the organism from harmful substances. ATP is used as an energy source to excrete hydrophobic substances.

Dickeya chrysanthemi pv. *chrysanthemi* is also equipped with a type III secretion system. One of its roles appears to be concerned with aggregate formation in the form of a pellicle that forms on the surface of liquid medium (Yap et al., 2005). Also, one of the *hrp* gene products, HrpN, is claimed to be involved in this cell aggregation (Yap et al., 2006). The type III apparatus itself is a long filament that could be involved in this phenomenon (see Chapter 10).

In response to defense reactions of the host, *Dickeya chrysanthemi* pv. *chrysanthemi* produces methionine sulfoxide reductase (MsrA) and Mn-superoxide dismutase (SodA) to repair and circumvent oxidative damage to its proteins (El Hassouni et al., 1999; Santos et al., 2001). The expression of several transport proteins is upregulated during infection of the host tissue (Okinaka et al., 2002). Interestingly, *Dickeya chrysanthemi* pv. *chrysanthemi* contains four *cyt* gene homologs of *Bacillus thuringiensis* that encode cytotoxins that are lethal to the pea aphid (*Acyrthosiphon pisum* (Harris)) (Grenier et al., 2006). Identified genes known to be involved in the virulence of *Dickeya chrysanthemi* pv. *chrysanthemi* are listed in Table 6-5. They include genes that encode products that enhance the ability of this pathogen to circumvent defense compounds produced by the host.

Soft Rot, Scale Rot, and Sour Skin Caused by *Burkholderia cepacia*

Burkholderia cepacia members are grouped into a species complex. Various members of this complex may, either alone or in combination, cause soft rot-type infections of plants or opportunistic infections in humans with cystic fibrosis, function as antagonists of soilborne plant pathogens, and provide plant growth-stimulating factors. *Burkholderia cepacia*, previously known as *Burkholderia cenocepacia* and *Pseudomonas cepacia*, causes soft rot of onion as well as postharvest diseases known as sour skin and scale rot. The bacterium can metabolize complex hydrocarbons, herbicides, and environmental pollutants, such as 2,4,5-trichlorophenoxyacetic acid and trichloroethylene, as well as antagonize soilborne pathogens (Kilbane et al., 1982; Nelson et al., 1987; Sangodkar et al., 1988). It is a common pathogen in opportunistic and nosocomial infections of man (Martone et al., 1987; Yamagishi et al., 1993; Lacy et al., 1994; Govan and Deretic, 1996). The unusually broad capacity, ranging from the pathogenic trait to its ability to catabolize complex halogenated hydrocarbons and kill soilborne pathogens, reflects a highly complex genome conferring numerous abilities to cope with diverse environments.

General Characteristics of *Burkholderia cepacia*

Burkholderia cepacia is a member of the class *Betaproteobacteria* and represents the type taxon of the genus. The bacterium is strictly aerobic and appears as short bacilliform rods with one to three polar flagella. On rich media, water-soluble, nonfluorescent pigments ranging from yellow to green are released into the agar. On iron-deficient media, a violet to dark purple pigment is produced. *Burkholderia cepacia* accumulates poly-β-hydroxybutyrate granules and produces nitrate reductase and protease. Nutritionally, the organism is nonfastidious and able to utilize complex hydrocarbons. The optimum growth temperature is 32°C, but some strains are capable of growth at 41°C.

Type strain *Burkholderia cepacia* ATCC 25416 harbors three circular chromosomes and a large circular cryptic plasmid. The chromosomes range from 3.4 to 3.7, 2.5 to 3.2, and 0.9 to 1.1 Mbp (Cheng and Lessie, 1994; Rodley et al., 1995; Lessie et al., 1996; Wigley and Burton, 2000). *Burkholderia cepacia* harbors at least nine distinct insertion sequence (IS) elements that have likely played a role in the plasticity of its genome via lateral transfer of environmental adaptation genes (Lessie et al., 1996). For pathogenicity, *Burkholderia cepacia* utilizes a type IV secretion system encoded by a resident plasmid to deliver various effectors involved in virulence (Engledow et al., 2004; Bernier and Sokol, 2005).

Host Range

Burkholderia cepacia appears to be an opportunistic pathogen for both plants and humans. It survives in soil and fresh water and infects onion in the field under warm, high-moisture-level conditions. Under high humidity and warm conditions, *Burkholderia cepacia* causes sour skin in harvested onion (Burkholder, 1950). Field infections are facilitated by wounds, high moisture levels, and high temperatures, such as 32°C (Cother, 1974; Kawamoto and Lorbeer, 1974). *Burkholderia cepacia* also causes respiratory tract infections in patients with cystic fibrosis (Isles et al., 1984; Goldman and Klinger, 1986) and with urinary tract infections (Li et al., 2003).

History of Soft Rot and Sour Skin

Although long attributed to fungi, the actual cause of soft rot of onion bulbs in the field and postharvest was determined to be *Pseudomonas cepacia* (renamed *Burkholderia cepacia*) (Burkholder, 1950). The postharvest disease was called sour skin (Burkholder, 1950), scale rot (Cother, 1974), and internal breakdown (Cother and Dowling, 1985).

Symptomatology

The disease is recognized in the field by the collapse of the leaves and the maceration of the inner bulb scales. Lifting the leaves off the bulb in the field is easy when the entire neck of the bulb where the leaves attach has been macerated. Downed leaves turn pale yellow and later brown as the leaves dry.

The postharvest symptom called sour skin is recognized by the maceration of the inner bulb scales, which turn yellow to light brown. The outer bulb scales, where the concentration of polyphenolic compounds is high, remain firm. The inner core of the bulb becomes fully soft rotted.

Disease Cycle

Field infections occur primarily through wounds (Kawamoto and Lorbeer, 1974). High moisture levels and temperatures provide *Burkholderia cepacia* the best opportunity to infect through open wounds in the leaves and subsequently to spread into the bulb. Recent studies have shown that multiple *Burkholderia cepacia* complex species colonize the onion rhizosphere and have the potential to cause sour skin of onion. In addition, the onion rhizosphere is a natural habitat and a potential environmental source of *Burkholderia cepacia* (Jacobs et al., 2008). Entry into the bulbs is promoted during harvest when the leaves are cut from the bulb tops. During storage and transportation to markets, infected bulb scales develop water-soaking symptoms that progressively develop into soft rot. As diseased onion bulbs are tilled into the field, *Burkholderia cepacia* overwinters and locally infests soil surrounding the original infected plant. As fields become flooded upon heavy rains, the bacterial pathogen is further disseminated. Infection takes place during the next season as the new crop of onions matures into bulbs.

Pathogenicity Tests

Because *Burkholderia cepacia* is an opportunistic pathogen, difficulties have been encountered in assessing its pathogenicity using intact plants (Kawamoto and Lorbeer, 1974). Experimental inoculation of onion leaves through needle pricks followed by incubation of the inoculated plants under high humidity and temperature (e.g., 32°C) conditions results in water-soaking symptoms that progress into soft rot of bulb scales. Onion bulbs are also used to evaluate pathogenicity (Gonzalez et al., 1997). Bulbs are transversely cut across the top and bottom and then sectioned into four parts, and the outer bulb scale is discarded. The bulb scale portions are placed with the abaxial part of the scale facing up in a sterile petri dish lined with filter paper (Whatman No. 1) that is premoistened with sterile water. The bulb scales are pierced at the center with a blunted 18-gauge sterile needle to just break the epidermal tissue, and 25 μl of inoculum (10^3–10^5 CFU per bulb scale) is introduced into the fresh wounded site. The bulb scales are incubated at 37°C in a plastic moist chamber. Soft rot and necrosis are evaluated 24 and 48 h after inoculation.

Trypan blue-tetracycline (TB-T) agar is a useful selective medium for isolating *Burkholderia cepacia*. The medium contains trypan blue, tetracycline (20 μg/ml), glucose, and L-asparagine (Hagedorn et al., 1987). The medium provides a useful means of isolating putative *Burkholderia cepacia* but does not help discriminate pathogenic strains from common soil- and water-borne strains. The use of species-specific 16S and 23S DNA probes aids in confirming the identity of *Burkholderia cepacia* (Leff et al., 1995). Primers with the sequence 5'CCTCTGTTCCGACCA3' (16S) and 5'CCCATCGCATCTAACAAT3' (23S) have been successfully used for identifying *Burkholderia cepacia*. An additional primer set designed against a particular pectate lyase encoded by a plasmid-borne *pehA* gene provides added confidence in the identification of plant-pathogenic strains of *Burkholderia cepacia* (Gonzalez et al., 1997). The production of pectate lyase is confirmed by streaking onto pectate agar medium D, pH 4.8–5.0 (Hildebrand, 1971).

Pathogenicity Factors

As discussed in this chapter, macerative diseases degrade host tissues through the action of pectinolytic enzymes. These important enzymes are also produced by *Burkholderia cepacia*. *Burkholderia cepacia* type strain ATCC 25416 isolated from onion does not appear to contain a type III secretion system, even though the type III secretion pathway is present in all members of genomovar III that are human pathogens (Parsons et al., 2001). For example, *Burkholderia pseudomallei* uses a type III pathway similar to that used by *Salmonella* and *Shigella* species to secrete uncharacterized Bip effector proteins (Stevens et al., 2002). In addition, *Burkholderia cepacia* can harbor a 92-kb plasmid that encodes a type IV secretion system (Engledow et al., 2004). Moreover, one of the chromosomes of this pathogen, which causes water-soaking symptoms on onion, also encodes a type IV secretion system. The plasmid-borne type IV secretion system appears to be responsible for the secretion of uncharacterized protein phytotoxins (Engledow et al., 2004). *Burkholderia cepacia* uses the general secretion type II pathway to deliver the pectinolytic enzymes to the cell surface. So far, pathogenicity factors homologous with HrpN of *Pectobacterium carotovorum* subsp. *carotovorum* have not been found in *Burkholderia cepacia*.

REFERENCES

Appel, O. 1903. Introduction to bacterial diseases of plants. Science 19:264.

Barabote, R. D., Johnson, O. L., Zetina, E., San Francisco, S. K., Fralick, J. A., and San Francisco, M. J. D. 2003. *Erwinia chrysanthemi tolC* is involved in resistance to antimicrobial plant chemicals and is essential for phytopathogenesis. J. Bacteriol. 185:5772-5778.

Barras, F., Van Gijsegem, F., and Chatterjee, A. K. 1994. Extracellular enzymes and pathogenesis of soft-rot *Erwinia*. Annu. Rev. Phytopathol. 32:201-234.

Basset, A., Khush, R. S., Braun, A., Gardan, L., Boccard, F., Hoffmann, J. A., and Lemaitre, B. 2000. The phytopathogenic bacteria *Erwinia carotovora* infects *Drosophila* and activates an immune response. Proc. Natl. Acad. Sci. U.S.A. 97:3376-3381.

Basset, A., Tzou, P., Lemaitre, B., and Boccard, F. 2003. A single gene that promotes interaction of a phytopathogenic bacterium with its insect vector, *Drosophila melanogaster*. EMBO Rep. 4:205-209.

Bauer, D. W., Bogdanove, A. J., Beer, S. V., and Collmer, A. 1994. *Erwinia chrysanthemi hrp* genes and their involvement in soft rot pathogenesis and elicitation of the hypersensitive response. Mol. Plant-Microbe Interact. 7:573-581.

Bauer, D. W., Wei, Z.-M., Beer, S. V., and Collmer, A. 1995. *Erwinia chrysanthemi* harpin$_{Ech}$: An elicitor of the hypersensitive response that contributes to soft-rot pathogenesis. Mol. Plant-Microbe Interact. 8:484-491.

Baumberger, I. C., Fraefel, N., Gőttfert, M., and Hennecke, H. 2003. New NodW- or NifA-regulated *Bradyrhizobium japonicum* genes. Mol. Plant-Microbe Interact. 16:342-351.

Bell, K. S., Avrova, A. O., Holeva, M. C., Cardle, L., Morris, W., DeJong, W., Toth, I. K., Waugh, R., Bryan, G. J., and Birch, P. R. J. 2002. Sample sequencing of a selected region of the genome of *Erwinia carotovora* subsp. *atroseptica* reveals candidate phytopathogenicity genes and allows comparison with *Escherichia coli*. Microbiology 148:1367-1378.

Bell, K. S., Parkhill, J., Bentley, S. D., Holden, M., Birch, P. R. J., and Toth, I. K. 2003. The complete genome sequence of *Erwinia carotovora* subsp. *atroseptica*. http://cmr.jcvi.org/tigr-scripts/CMR/GenomePage.cgi?database=ntec06.

Bell, K. S., Sebaihia, M., Pritchard, L., Holden, M. T. G., Hyman, L. J., Holeva, M. C., Thomson, N. R., Bentley, S. D., Churcher, L. J. C., Mungall, K., Atkin, R., Bason, N., Brooks, K., Chillingworth, T., Clark, K., Doggett, J., Fraser, A., Hance, Z., Hauser, H., Jagels, K., Moule, S., Norbertczak, H., Ormond, D., Price, C., Quail, M. A., Sanders, M., Walker, D., Whitehead, S., Salmond, G. P. C., Birch, P. R. J., Parkhill, J., and Toth, I. K. 2004. Genome sequence of the enterobacterial phytopathogen *Erwinia carotovora* subsp. *atroseptica* and characterization of virulence factors. Proc. Natl. Acad. Sci. U.S.A. 101:11105-11110.

Bernier, S. P., and Sokol, P. A. 2005. Use of suppression-subtractive hybridization to identify genes in the *Burkholderia cepacia* complex that are unique to *Burkholderia cenocepacia*. J. Bacteriol. 187:5278-5291.

Bishop, R. E. 2005. The lipid A palmitoyltransferase PagP: Molecular mechanisms and role in bacterial pathogenesis. Mol. Microbiol. 57:900-912.

Boccara, M., Diolez, A., Rouve, M., and Kotoujansky, A. 1988. The role of individual pectate lyases of *Erwinia chrysanthemi* strain 3937 in pathogenicity on *Saintpaulia* plants. Physiol. Mol. Plant Pathol. 33:95-104.

Boccara, M., Vedel, R., Lalo, D., Lebrun, M.-H., and Lafay, J. F. 1991. Genetic diversity and host range in strains of *Erwinia chrysanthemi*. Mol. Plant-Microbe Interact. 4:293-299.

Boccara, M., Aymeric, J.-L., and Camus, C. 1994. Role of endoglucanases in *Erwinia chrysanthemi* 3937 virulence on *Saintpaulia ionantha*. J. Bacteriol. 176:1524-1526.

Bogdanove, A. J., Beer, S. V., Bonas, U., Boucher, C. A., Collmer, A., Coplin, D., Cornelis, G. R., Huang, H. C., Hutcheson, S. W., Panopoulos, N. J., and Van Gijsegem, F. 1996. Unified nomenclature for broadly conserved *hrp* genes of phytopathogenic bacteria. Mol. Microbiol. 20:681-683.

Bortoll-German, I., Brun, E., Py, B., Chippaux, M., and Barras, F. 1994. Periplasmic disulphide bond formation is essential for cellulose secretion by the plant pathogen *Erwinia chrysanthemi*. Mol. Microbiol. 11:545-553.

Braun, E. J., and Kelman, A. 1987. Production of cell wall-degrading enzymes by corn stalk rot strains of *Erwinia chrysanthemi*. Pages 206-211 in: Plant Pathogenic Bacteria. E. Civerolo, A. Collmer, R. E. Davis, and A. G. Gillaspie, eds. M. Nidjhoff Publ., Dordrecht, the Netherlands.

Braun, E. J., and Rodrigues, C. A. 1993. Purification and properties of an endoxylanase from a corn stalk rot strain of *Erwinia chrysanthemi*. Phytopathology 83:332-338.

Burkholder, W. H. 1950. Sour skin, a bacterial rot of onion bulbs. Phytopathology 40:115-117.

Burkholder, W. H., McFadden, L. A., and Dimock, A. W. 1953. A bacterial blight of chrysanthemums. Phytopathology 43:522-526.

Burr, T., Barnard, A. M. L., Corbett, M. J., Pemerton, C. L., Simpson, N. J. L., and Salmond, G. P. C. 2006. Identification of the central quorum sensing regulator of virulence in the enteric phytopathogen, *Erwinia carotovora*: The VirR repressor. Mol. Microbiol. 59:113-125.

Campbell, W. A. 1947. A bacterial root and stem disease of guayule. Phytopathology 37:271-277.

Chatterjee, A. K., Thurn, K. K., and Feese, D. A. 1983. Tn5-induced mutations in the enterobacterial phytopathogen *Erwinia chrysanthemi*. Appl. Environ. Microbiol. 45:644-650.

Chatterjee, A., Cui, Y., Hasegawa, H., Leigh, N., Dixit, V., and Chatterjee, A. K. 2005. Comparative analysis of two classes of quorum-sensing signaling systems that control production of extracellular proteins and secondary metabolites in *Erwinia carotovora* subspecies. J. Bacteriol. 187:8026-8038.

Cheng, H.-P., and Lessie, T. G. 1994. Multiple replicons constituting the genome of *Pseudomonas cepacia* 17616. J. Bacteriol. 176:4034-4042.

Chuang, D.-Y., Chien, Y.-C., and Wu, H.-P. 2007. Cloning and expression of the *Erwinia carotovora* subsp. *carotovora* gene encoding the low-molecular-weight bacteriocin carocin S1. J. Bacteriol. 189:620-626.

Collmer, A., and Keen, N. T. 1986. The role of pectic enzymes in plant pathogenesis. Annu. Rev. Phytopathol. 24:383-409.

Condemine, G., Dorel, C., Hugouvieux-Cotte-Pattat, N., and Robert-Baudouy, J. 1992. Some of the *out* genes involved in the secretion of pectate lyases in *Erwinia chrysanthemi* are regulated by *kdgR*. Mol. Microbiol. 6:3199-3211.

Cooper, V. J. C., and Salmond, G. P. C. 1993. Molecular analysis of the major cellulase (CelV) of *Erwinia carotovora*: Evidence for an evolutionary "mix-and-match" of enzyme domains. Mol. Gen. Genet. 241:341-350.

Cother, E. J. 1974. Bacterial scale rot of onion. Aust. Plant Pathol. Soc. Newslett. 3:70-71.

Cother, E. J., and Dowing, V. 1985. Association of *Pseudomonas cepacia* with internal breakdown of onion—A new record for Australia. Aust. Plant Pathol. 14:10-11.

Creze, C., Castang, S., Derivery, E., Haser, R., Hugouvieux-Cotte-Pattat, N., Shevchik, V. E., and Gouet, P. 2008. The crystal structure of pectate lyase PelI from soft rot pathogen *Erwinia chrysanthemi* in complex with its substrate. J. Biol. Chem. 283:18260-18268.

Cui, Y., Madi, L., Mukherjee, A., Dumenyo, C. K., and Chatterjee, A. K. 1996. The RsmA⁻ mutants of *Erwinia carotovora* subsp. *carotovora* strain Ecc71 overexpress *hrpN*$_{Ecc}$ and elicit a hypersensitive reaction-like response in tobacco leaves. Mol. Plant-Microbe Interact. 9:565-573.

Cui, Y., Chatterjee, A., Hasegawa, H., Dixit, V., Leigh, N., and Chatterjee, A. K. 2005. ExpR, a LuxR homolog of *Erwinia carotovora* subsp. *carotovora*, activates transcription of *rsmA*, which specifies a global regulator RNA-binding protein. J. Bacteriol. 187:4792-4803.

Cui, Y., Chatterjee, A., Hasegawa, H., and Chatterjee, A. K. 2006. *Erwinia carotovora* subspecies produce duplicate variants of Expr, LuxR homologs that activate *rsmA* transcription but differ in their interactions with *N*-acylhomoserine lactone signals. J. Bacteriol. 188:4715-4726.

Cuppels, D., and Kelman, A. 1974. Evaluation of selective media for isolation of soft-rot bacteria from soil and plant tissue. Phytopathology 64:468-475.

Dahler, G. S., Barras, F., and Keen, N. T. 1990. Cloning of genes encoding extracellular metalloproteases from *Erwinia chrysanthemi* EC16. J. Bacteriol. 172:5803-5815.

Davis, K. R., Darvill, A. G., Albersheim, P., and Dell, A. 1986. Host pathogen interactions. XXX. Characterization of elicitors of phytoalexin accumulation in soybean released from cell walls by endopolygalacturonic acid lyase. Z. Naturforsch. Sect. C 41:39-48.

De Boer, S. H., and Kelman, A. 1978. Influence of oxygen concentration and storage factors on susceptibility of potato tubers to bacterial soft rot (*Erwinia carotovora*). Potato Res. 21:65-80.

De Boer, S. H., and Ward, L. J. 1995. PCR detection of *Erwinia carotovora* subsp. *atroseptica* associated with potato tissue. Phytopathology 85:854-858.

Dickey, R. S., and Victoria, J. I. 1980. Taxonomy and emended description of strains of *Erwinia* isolated from *Musa paradisiaca* Linnaeus. Int. J. Syst. Bacteriol. 30:129-134.

Dong, Y.-H., Xu, J.-L., Li, X.-Z., and Zhang, L.-H. 2000. AiiA, an enzyme that inactivates the acylhomoserine lactone quorum-sensing signal and attenuates the virulence of *Erwinia carotovora*. Proc. Natl. Acad. Sci. U.S.A. 97:3526-3531.

Dong, Y.-H., Wang, L.-H., Xu, J.-L., Zhang, H.-B., Zhang, X.-F., and Zhang, L.-H. 2001. Quenching quorum-sensing-dependent bacterial infection by an N-acyl homoserine lactonase. Nature 411:813-817.

Dye, D. W. 1960. Pectolytic activity in *Xanthomonas*. N. Z. J. Sci. 3:61-69.

Dye, D. W. 1969. A taxonomic study of the genus *Erwinia*. II. The "Carotovora" group. N. Z. J. Sci. 12:81-97.

El Hassouni, M., Henrissat, B., Chippaux, M., and Barras, F. 1992. Nucleotide sequences of the *arb* genes, which control β-glucoside utilization in *Erwinia chrysanthemi*: Comparison with the *Escherichia coli bgl* operon and evidence for a new β-glycohydrolase family including enzymes from eubacteria, archeabacteria, and humans. J. Bacteriol. 174:765-777.

El Hassouni, M. E., Chambost, J. P., Expert, D., Van Gijsegem, F., and Barras, F. 1999. The minimal gene set member *msrA*, encoding peptide methionine sulfoxide reductase, is a virulence determinant of the plant pathogen *Erwinia chrysanthemi*. Proc. Natl. Acad. Sci. U.S.A. 96:887-892.

Enard, C., Diolez, A., and Expert, D. 1988. Systemic virulence of *Erwinia chrysanthemi* 3937 requires a functional iron assimilation system. J. Bacteriol. 170:2419-2426.

Endicott, J. A., and Ling, V. 1989. The biochemistry of P-glycoprotein-mediated multidrug resistance. Annu. Rev. Biochem. 58:137-171.

Engledow, A. S., Medrano, E. G., Mahenthiralingam, E., LiPuma, J. J., and Gonzalez, C. F. 2004. Involvement of a plasmid-encoded type IV secretion system in the plant tissue watersoaking phenotype of *Burkholderia cenocepacia*. J. Bacteriol. 186:6015-6024.

Expert, D., Sauvage, C., and Neilands, J. B. 1992. Negative transcriptional control of iron transport in *Erwinia chrysanthemi* involves an iron-responsive two-factor system. Mol. Microbiol. 6:2009-2017.

Faith, M. J., Skvirsky, R. C., and Kolter, R. 1991. Functional complementation between bacterial MDR-like export systems: Colicin V, alpha-hemolysin, and *Erwinia* protease. J. Bacteriol. 173:7549-7556.

Fox, R. T. V. 1972. The ultrastructure of potato tubers infected by the soft rot organism *Erwinia carotovora* var. *atroseptica*. Pages 95-120 in: Proc. Int. Conf. Plant Pathogenic Bacteria, 3rd. H. P. Maas Geesteranus, ed. Center for Agricultural Publications and Documents, Wageningen, the Netherlands.

Franza, T., and Expert, E. 1991. The virulence-associated chrysobactin iron uptake system of *Erwinia chrysanthemi* 3937 involves an operon encoding transport and biosynthetic functions. J. Bacteriol. 173:6874-6881.

Franza, T., Mahé, B., and Expert, D. 2005. *Erwinia chrysanthemi* requires a second iron transport route dependent of the siderophore achromobactin for extracellular growth and plant infection. Mol. Microbiol. 55:261-275.

Galán, J. E., and Collmer, A. 1999. Type III secretion machines: Bacterial devices for protein delivery into host cells. Science 284:1322-1328.

Gardan, L., Gouy, C., Christen, R., and Samson, R. 2003. Elevation of three subspecies of *Pectobacterium carotovorum* to species level: *Pectobacterium atrosepticum* sp. nov., *Pectobacterium betavasculorum* sp. nov. and *Pectobacterium wasabiae* sp. nov. Int. J. Syst. Evol. Microbiol. 53:381-391.

Glasner, J. D., Marquez-Villavicencio, M., Kim, H.-S., Jahn, C. E., Ma, B., Biehl, B. S., Rissman, A. I., Mole, B., Yi, X., Yang, C.-H., Dangl, J. L., Grant, S. R., Perna, N. T., and Charkowski, A. O. 2008. Niche-specificity and the variable fraction of the *Pectobacterium* pan-genome. Mol. Plant-Microbe Interact. 21:1549-1560.

Gold, S., Nishio, S., Tsuyumu, S., and Keen, N. T. 1992. Analysis of the *pelE* promoter in *Erwinia chrysanthemi* EC16. Mol. Plant-Microbe Interact. 5:170-178.

Goldman, D. A., and Klinger, J. D. 1986. *Pseudomonas cepacia*: Biology, mechanisms of virulence, epidemiology. J. Pediatr. 108:806-812.

Gonzalez, C. F., Pettit, E. A., Valadez, V. A., and Provin, E. M. 1997. Mobilization, cloning, and sequence determination of a plasmid-encoded polygalacturonase from a phytopathogenic *Burkholderia* (*Pseudomonas*) *cepacia*. Mol. Plant-Microbe Interact. 10:840-851.

Govan, J. R. W., amd Deretic, V. 1996. Microbial pathogenesis in cystic fibrosis mucoid *Pseudomonas aeruginosa* and *Burkhholderia cepacia*. Microbiol. Rev. 60:539-574.

Grenier, A.-M., Dupont, G., Pagès, S., Condemine, G., and Rahbé, Y. 2006. The phytopathogen *Dickeya dadantii* (*Erwinia chrysanthemi* 3937) is a pathogen of the pea aphid. Appl. Environ. Microbiol. 72:1956-1965.

Grula, M. M., Smith, R. W., Parham, C. F., and Grula, E. A. 1968. Cell division in a species of *Erwinia*. XI. Some aspects of the carbon and nitrogen nutrition of *Erwinia* species. Can. J. Microbiol. 14:1217-1224.

Hagedorn, C., Gould, W. D., Bardinelli, T. R., and Gustavson, D. R. 1987. A selective medium for enumeration and recovery of *Pseudomonas cepacia* biotypes from soil. Appl. Environ. Microbiol. 53:2265-2268.

Hamon, Y., and Peron, Y. 1961. Les propriete antagonists reciproques parmi les *Erwinia*. Discussion de la position taxonomique de ce genre. C. R. Acad. Sci. 253:913-915.

Hauben, L., Moore, E. R. B., Vauterin, L., Steeackers, M., Mergaert, J., Verdonck, L., and Swings, J. 1998. Phylogenetic position of phytopathogens within the *Enterobacteriaceae*. Syst. Appl. Microbiol. 21:384-397.

He, S. Y. 1997. Hrp-controlled interkingdom protein transport: Learning from flagellar assembly? Trends Microbiol. 5:489-495.

He, S. Y., and Collmer, A. 1990. Molecular cloning, nucleotide sequence, and marker exchange mutagenesis of the exo-poly-α-D-galacturonosidase-encoding *pehX* gene of *Erwinia chrysanthemi* EC16. J. Bacteriol. 172:4988-4995.

Heikinheimo, R., Flego, D., Pirhonen, M., Karlsson, M.-B., Eriksson, A., Mäe, A., Kõiv, V., and Palva, E. T. 1995. Characterization of a novel pectate lyase from *Erwinia carotovora* subsp. *carotovora*. Mol. Plant-Microbe Interact. 8:207-217.

Hildebrand, D. C. 1971. Pectate and pectin gels for differentiation of *Pseudomonas* sp. and other bacterial plant pathogens. Phytopathology 61:1430-1436.

Hinton, J. C., Sidebotham, J. M., Gill, D. R., and Salmond, G. P. 1989. Extracellular and periplasmic isoenzymes of pectate lyase from *Erwinia carotovora* subspecies *carotovora* belong to different gene families. Mol. Microbiol. 3:1785-1795.

Hueck, C. J. 1998. Type III protein secretion systems in bacterial pathogens of animals and plants. Microbiol. Mol. Biol. Rev. 62:379-433.

Hugouvieux-Cotte-Pattat, N., and Robert-Baudouy, J. 1994. Molecular analysis of the *Erwinia chrysanthemi* region containing the *kdgA* and *zwf* genes. Mol. Microbiol. 11:67-75.

Hugouvieux-Cotte-Pattat, N., Nasser, W., and Robert-Baudouy, J. 1994. Molecular characterization of the *Erwinia chrysanthemic kdgK* gene involved in pectin degradation. J. Bacteriol. 176:2386-2392.

Hurlbert, J. C., and Preston, J. F., III. 2001. Functional characterization of a novel xylanase from a corn strain of *Erwinia chrysanthemi*. J. Bacteriol. 183:2093-2100.

Isles, A., Maclusky, I., Corey, M., Gold, R., Prober, C., Fleming, P., and Levison, H. 1984. *Pseudomonas cepacia* infection in cystic fibrosis: An emerging problem. J. Pediatr. 104:206-210.

Jacobs, J. L., Fasi, A. C., Ramette, A., Smith, J. J., Hammerschmidt, R., and Sundin, G. W. 2008. Identification and onion pathogenicity of *Burkholderia cepacia* complex isolates from the onion rhizosphere and onion field soil. Appl. Environ. Microbiol. 74:3121-3129.

Johansson, J., and Cossart, P. 2003. RNA-mediated control of virulence gene expression in bacterial pathogens. Trends Microbiol. 11:280-285.

Jones, L. R. 1901. *Bacillus carotovorus* n. sp. die Ursache einer weichen Fäulnis der Mőhre. Zentbl. Bakteriol. II 7:12-21.

Kaneshiro, W. S., Burger, M., Vine, B. G., de Silva, A. S., and Alvarez, A. M. 2008. Characterization of *Erwinia chrysanthemi* from a bacterial heart rot of pineapple outbreak in Hawaii. Plant Dis. 92:1444-1450.

Kawamoto, S. O., and Lorbeer, J. W. 1974. Infection of onion leaves by *Pseudomonas cepacia*. Phytopathology 64:1440-1445.

Keen, N. T., and Tamaki, S. 1986. Structure of two pectate lyase genes from *Erwinia chrysanthemi* EC16 and their high level expression in *E. coli*. J. Bacteriol. 168:595-606.

Keen, N. T., Dahlbeck, D., Staskawicz, B., and Belser, W. 1984. Molecular cloning of pectate lyase genes from *Erwinia chrysanthemi* and their expression in *Escherichia coli*. J. Bacteriol. 159:825-831.

Keen, N. T., Ridgway, D., and Boyd, C. 1992. Cloning and characterization of a phospholipase gene from *Erwinia chrysanthemi* EC16. Mol. Microbiol. 6:179-187.

Keen, N. T., Boyd, C., and Henrissat, B. 1996. Cloning and characterization of a xylanase gene from corn strains of *Erwinia chrysanthemi*. Mol. Plant-Microbe Interact. 9:651-657.

Kilbane, J. J., Chatterjee, D. K., Karns, J. S., Kellogg, S. T., and Chakrabarty, A. M. 1982. Biodegradation of 2,4,5-trichlorophenoxyacetic acid by a pure culture of *Pseudomonas cepacia*. Appl. Environ. Microbiol. 44:72-78.

Kloepper, J. W., Brewer, J. W., and Harrison, M. D. 1981. Insect transmission of *Erwinia carotovora* var. *carotovora* and *Erwinia carotovora* var. *atroseptica* to potato plants in the field. Am. Potato J. 58:165-175.

Kluskens, L. D., Van Alebeek, G.-J. W. M., Walther, J., Voragen, A. G. J., De Vos, W. M., and Van der Oost, J. 2005. Characterization and mode of action of an exopolygalacturonase from the hyperthermophilic bacterium *Thermotoga maritima*. FEBS J. 272:5464-5473.

Kotoujansky, A., Lematre, M., and Boistard, P. 1982. Utilization of a thermosensitive episome bearing transposon Tn*10* to isolate Hfr donor strains of *Erwinia carotovora* subsp. *chrysanthemi*. J. Bacteriol. 150:122-131.

Kuykendall, L. D., and Hunter, W. J. 2008. Carbohydrate and ethane release with *Erwinia carotovora* subspecies *betavasculorum*-induced necrosis. Curr. Microbiol. 56:128-133.

Kyőstiő, S. R. M., Cramer, C. L., and Lacy, G. H. 1991. *Erwinia carotovora* subsp. *carotovora* extracellular protease: Characterization and nucleotide sequence of the gene. J. Bacteriol. 173:6537-6546.

Laasik, E., Ojarand, M., Pajunen, M., Savilahti, H., and Mäe, A. 2005. Novel mutants of *Erwinia carotovora* subsp. *carotovora* defective in the production of plant cell wall degrading enzymes generated by Mu transposome-mediated insertion mutagenesis. FEMS Microbiol. Lett. 243:93-99.

Laat, P. C. A., Verhoeven, J. T. W., and Janse, J. D. 1994. Bacterial leaf rot of *Aloe vera* L., caused by *Erwinia chrysanthemi* biovar 3. Eur. J. Plant Pathol. 100:81-84.

Lacy, D. E., Spencer, D. A., Goldstein, A., Weller, P. H., and Darbyshire, P. 1994. Chronic granulomatous disease presenting in childhood with *Pseudomonas cepacia* septicaemia. J. Infect. 27:301-304.

Leff, L. G., Kernan, R. M., McArthur, J. V., and Shimkets, L. J. 1995. Identification of aquatic *Burkholderia* (*Pseudomonas*) *cepacia* by hybridization with species-specific rRNA gene probes. Appl. Environ. Microbiol. 61:1634-1636.

Lelliott, R. A., and Dickey, R. S. 1984. Genus VII. *Erwinia* Winslow, Broadhurst, Buchanan, Krumwiede, Rogers and Smith 1920, 209AL. Pages 469–476 in: Bergey's Manual of Systematic Bacteriology, Vol. 1. N. R. Krieg and J. G. Holt, eds. Williams & Wilkins, Baltimore.

Lessie, T. G., Hendrickson, W., Manning, B., and Devereux, R. 1996. Genomic complexity and plasticity of *Burkholderia cepacia*. FEMS Microbiol. Lett. 144:117-128.

Létoffé, S., and Wandersman, C. 1992. Secretion of CyaA-PrtB and Hlya-PrtB fusion proteins in *Escherichia coli*: Involvement of the glycine-rich repeat domain of *Erwinia chrysanthemi* protease B. J. Bacteriol. 174:4920-4927.

Létoffé, S., Delepelaire, P., and Wandersman, C. 1990. Protease secretion by *Erwinia chrysanthemi*: The specific secretion functions are analogous to those of *Escherichia coli* α-haemolysin. EMBO J. 9:1375-1382.

Létoffé, S., Delepelaire, P., and Wandersman, C. 1991. Cloning and expression in *Escherichia coli* of the *Serratia marcescens* metalloprotease gene: Secretion of the protease from *E. coli* in the presence of the *Erwinia chrysanthemi* protease secretion functions. J. Bacteriol. 173:2160-2166.

Levin, I., Miller, M. D., Schwarzenbacher, R., McMullan, D., Abdubek, P., Ambing, E., Biorac, T., Cambell, J., Canaves, J. M., Chiu, H.-J., Ceacon, A. M., DiDonato, M., Elsliger, M.-A., Godzik, A., Grittini, C., Grzechnik, S. K., Hale, J., Hampton, E., Han, G. W., Haugen, J., Hornsby, M., Jaroszewski, L., Karlak, C., Klock, H. E., Koesema, E., Kreusch, A., Kuhn, P., Lesley, S. A., Morse, A., Moy, K., Nigoghossian, E., Ouyang, J., Page, R., Quijano, K., Reyes, R., Robb, A., Sims, E., Spaggon, G., Stevens, R. C., van den Bedem, H., Velasquez, J., Vincent, J., Wang, X., West, B., Wolf, G., Xu, Q., Zagnitko, O., Hodgson, K. O., Wooley, J., and Wilson, I. A. 2005. Crystal structure of an indigoidine synthase A (IndA)-like protein (TM1464) from *Thermotoga maritima* at 1.90 Å resolution reveals a new fold. Proteins Struct. Funct. Bioinf. 59:864-868.

Li, F. K., Chan, K. W., Chan, T. M., and Lai, K. N. 2003. *Burkholderia* urinary tract infection after renal transplantation. Transplant. Infect. Dis. 5:59-61.

Liu, Y., Cui, Y., Mukherjee, A., and Chatterjee, A. K. 1998. Characterization of a novel RNA regulator of *Erwinia carotovora* ssp. *carotovora* that controls production of extracellular enzymes and secondary metabolites. Mol. Microbiol. 29:219-234.

Liu, Y., Jiang, G., Cui, Y., Mukherjee, A., Ma, L. E., and Chatterjee, A. 1999. $kdgR_{Ecc}$ negatively regulates genes for pectinases, cellulose, protease, $harpin_{Ecc}$ and a global RNA regulator in *Erwinia carotovora* subsp. *carotovora*. J. Bacteriol. 181:2411-2422.

Llama-Palacios, A., López-Solanilla, E., Poza-Carrion, C., Garcia-Olmedo, F., and Rodriguez-Palenzuela, P. 2003. The *Erwinia chrysanthemi phoP-phoQ* operon plays an important role in growth at low pH, virulence and bacterial survival in plant tissue. Mol. Microbiol. 49:347-357.

Llama-Palacios, A., López-Solanilla, E., and Rodriguez-Palezuela, P. 2005. Role of the PhoP-PhoQ system in the virulence of *Erwinia chrysanthemi* strain 3937: Involvement in sensitivity to plant antimicrobial peptides, survival at acid pH, and regulation of pectolytic enzymes. J. Bacteriol. 187:2157-2162.

Logan, C. 1964. Bacterial hard rot of potato. Eur. Potato J. 7:45-56.

Łojkowska, E., Dorel, C., Reignault, P., Hugouvieux-Cotte-Pattat, N., and Robert-Baudouy, J. 1993. Use of GUS fusion to study the expression of *Erwinia chrysanthemi* pectinase genes during infection of potato tubers. Mol. Plant-Microbe Interact. 6:488-494.

Łojkowska, E., Masclaux, C., Boccara, M., Robert-Baudouy, J., and Hugouvieux-Cotte-Pattat, N. 1995. Characterization of the *pelL* gene encoding a novel pectate lyase of *Erwinia chrysanthemi* 3937. Mol. Microbiol. 16:1183-1195.

López-Solanilla, E., Garcia-Olmedo, F., and Rodriguez-Palenzuela, P. 1998. Inactivation of the *sapA* to *sapF* locus of *Erwinia chrysanthemi* reveals common features in plant and animal bacterial pathogenesis. Plant Cell 10:917-924.

Lund, B. M., and Wyatt, G. M. 1972. The effect of oxygen and carbon dioxide concentrations on bacterial soft rot. I. King Edward potatoes inoculated with *Erwinia cartovora* var. *atroseptica*. Potato Res. 15:174-179.

Ma, B., Hibbing, M. E., Kim, H.-S., Reedy, R. M., Yedidia, I., Breuer, J., Breuer, J., Glasner, J. D., Perna, N. T., Kelman, A., and Charkowski, A. O. 2007. Host range and molecular phylogenies of the soft rot enterobacterial genera *Pectobacterium* and *Dickeya*. Phytopathology 97:1150-1163.

Mahé, B., Masclaux, C., Rauscher, L., Enard, C., and Expert, D. 1995. Differential expression of two siderophore-dependent iron-acquisition pathways in *Erwinia chrysanthemi* 3937: Characterization of a novel ferrisiderophore permease of the ABC transporter family. Mol. Microbiol. 18:33-43.

Maher, E. A., and Kelman, A. 1983. Oxygen status of potato tuber tissue in relation to maceration by pectic enzymes of *Erwinia carotovora*. Phytopathology 73:536-539.

Martone, W. J., Tablan, O. C., and Jarvis, W. R. 1987. The epidemiology of nosocomial epidemic *Pseudomonas cepacia* infections. Eur. J. Epidemiol. 3:222-232.

Mattinen, L., Somervuo, P., Kykyri, J., Nissinen, R., Kouvonen, P., Corthais, G., Auvinen, P., Aittamaa, M., Valkonen, J. P., and Perhonen, M. 2008. Microarray profiling of host-extract-induced genes and characterization of the type VI secretion cluster in the potato pathogen *Pectobacterium atrosepticum*. Microbiology 154:2387-2396.

Mecsas, J., and Strauss, E. J. 1996. Molecular mechanisms of bacterial virulence: Type III secretion and pathogenicity islands. Emerg. Infect. Dis. 2:1-16.

Moulard, M., Condemine, G., and Robert-Baudouy, J. 1993. Characterization of the *nucM* gene coding for a nuclease of the phytopathogenic bacteria *Erwinia chrysanthemi*. Mol. Microbiol. 8:685-695.

Mulholland, V., Hinton, J. C., Sidebotham, J., Toth, I. K., Hymany, L. J., Perombelon, M. C., Reeves, P. J., and Salmond, G. P. 1993. A pleiotropic reduced virulence (Rvi$^-$) mutant of *Erwinia carotovora* subspecies *atroseptica* is defective in flagella assembly proteins that are conserved in plant and animal bacterial pathogens. Mol. Microbiol. 9:343-356.

Nasser, W., Reverchon, S., and Robert-Baudouy, J. 1992. Purification and functional characterization of the KdgR protein, a major repressor of pectinolysis genes of *Erwinia chrysanthemi*. Mol. Microbiol. 6:257-265.

Nasser, W., Reverchon, S., Condemine, G., and Robert-Baudouy, J. 1994. Specific interactions of *Erwinia chrysanthemi* KdgR repressor with different operators of genes involved in pectinolysis. J. Mol. Biol. 236:427-440.

Nasser, W., Bouillant, M. L., Salmond, G., and Reverchon, S. 1998. Characterization of the *Erwinia chrysanthemi expI-expR* locus directing the synthesis of two N-acy-homoserine lactone signal molecules. Mol. Microbiol. 29:1391-1405.

Nasser, W., Shevchik, V. E., and Hugovieux-Cotte-Pattat, N. 1999. Analysis of three clustered polygalacturonase genes in *Erwinia chrysanthemi* 3937 revealed an anti-repressor function for the PecS regulator. Mol. Microbiol. 34:641-650.

Nelson, M. J. K., Montgomery, S. O., Mahaffey, W. R., and Pritchard, P. H. 1987. Biodegradation of trichloroethylene and involvement of an aromatic biodegradative pathway. Appl. Environ. Microbiol. 53:949-954.

Nguyen, A. H., Tomita, T., Hirota, M., Sato, T., and Kamio, Y. 1999. A simple purification method and morphology and component analyses for carotovoricin from the plant pathogen *Erwinia carotovora* Er. Biosci. Biotechnol. Biochem. 63:1360-1369.

Nguyen, A. H., Tomita, T., Hirota, M., Kaneko, J., Hayashi, T., and Kamio, Y. 2001. DNA inversion in the tail fiber gene alters the host range specificity of carotovoricin Er, a phage-tail-like bacteriocin of phytopathogenic *Erwinia carotovora* subsp. *carotovora* Er. J. Bacteriol. 183:6274-6281.

Nomura, K., Nasser, W., Kawagishi, H., and Tsuyumu, S. 1998. The *pir* gene of *Erwinia chrysanthemi* EC16 regulates hyperinduction of pectate lyase virulence genes in response to plant signals. Proc. Natl. Acad. Sci. U.S.A. 95:14034-14039.

Okinaka, Y., Yang, C.-H., Perna, N. T., and Keen, N. T. 2002. Microarray profiling of *Erwinia chrysanthemi* 3937 genes that are regulated during plant infection. Mol. Plant-Microbe Interact. 15:619-629.

Palomäki, T., Pickersgill, R., Riekki, R., Romantschuk, M., and Saarilahti, H. T. 2002. A putative three-dimensional targeting motif of polygalacturonase (PehA), a protein secreted through the type II (GSP) pathway in *Erwinia carotovora*. Mol. Microbiol. 43:585-596.

Parent, J.-G., Lacroix, M., Pagé, D., Vézina, L., and Végiard, S. 1996. Identification of *Erwinia carotovora* from soft rot diseased plants by random amplified polymorphic DNA (RAPD) analysis. Plant Dis. 80:494-499.

Parsons, Y. N., Glendinning, K. J., Thornton, V., Hales, B. A., Hart, C. A., and Winstanley, C. 2001. A putative type III secretion gene cluster is widely distributed in the *Burkholderia cepacia* complex but absent from genomovar I. FEMS Microbiol. Lett. 203:103-108.

Payne, J. H., Schoedel, C., Keen, N. T., and Collmer, A. 1987. Multiplication and virulence in plant tissues of *Escherichia coli* clones producing pectate lyase isozymes Plb and Ple at high levels and of an *Erwinia chrysanthemi* mutant deficient in Ple. Appl. Environ. Microbiol. 53:2315-2320.

Perombelon, M. C. M. 1992. Potato blackleg: Epidemiology, host-pathogen interaction and control. Neth. J. Plant Pathol. 92:135-146.

Perombelon, M. C., and Kelman, A. 1980. Ecology of the soft rot Erwinias. Annu. Rev. Phytopathol. 18:361-387.

Perombelon, M. C. M., and Salmond, G. P. C. 1995. Bacterial soft rots. Pages 1-20 in: Pathogenesis and Host Specificity in Plant Diseases, Vol. 1. U. S. Singh, R. P. Singh, and K. Kohmoto, eds. Pergamon Press, Oxford.

Pierce, L., and McCain, A. H. 1992. Selective medium for isolation of pectolytic *Erwinia* sp. Plant Dis. 76:382-384.

Pirhonen, M., Saarilahti, H., Karlsson, M.-B., and Palva, E. T. 1991. Identification of pathogenicity determinants of *Erwinia carotovora* subsp. *carotovora* by transposon mutagenesis. Mol. Plant-Microbe Interact. 4:276-283.

Pirhonen, M., Flego, D., Heikinheimo, R., and Palva, E. T. 1993. A small diffusible signal molecule is responsible for the global control of virulence and exoenzyme production in the plant pathogen *Erwinia carotovora*. EMBO J. 12:2467-2476.

Pissavin, C., Robert-Baudouy, J., and Hugouvieux-Cotte-Pattat, N. 1996. Regulation of *pelZ*, a gene of the *pelBC* cluster encoding a new pectate lyase in *Erwinia chrysanthemi* 3937. J. Bacteriol. 178:7187-7196.

Plastow, G. S. 1988. Molecular cloning and nucleotide sequence of the pectin methyl esterase gene of *Erwinia chrysanthemi* B347. Mol. Microbiol. 2:247-254.

Pugsley, A. P. 1993. The complete general secretory pathway in Gram-negative bacteria. Microbiol. Rev. 57:50-108.

Rantakari, A., Virtaharju, O., Vähämiko, S., Taira, S., Palva, E. T., Saarilahti, H. T., and Romantschuk, M. 2001. Type III secretion contributes to the pathogenesis of the soft-rot pathogen *Erwinia carotovora*: Partial characterization of the *hrp* gene cluster. Mol. Plant-Microbe Interact. 14:962-968.

Reeves, P. J., Whitcombe, D., Wharam, S., Gibson, M., Allison, G., Bunce, N., Barallon, R., Douglas, P., Mulholland, V., Stevens, S., Walker, D., and Salmond, G. P. C. 1993. Molecular cloning and characterization of 13 *out* genes from *Erwinia carotovora* subspecies *carotovora*: Genes encoding members of a general secretion pathway (GSP) widespread in Gram-negative bacteria. Mol. Microbiol. 8:443-456.

Reverchon, S., Nasser, W., and Robert-Baudouy, J. 1994. *pecS*: A locus controlling pectinase, cellulose and blue pigment production in *Erwinia chrysanthemi*. Mol. Microbiol. 11:1127-1139.

Reverchon, S., Bouillant, M. L., Salmond, G., and Nasser, W. 1998. Integration of the quorum-sensing system in the regulatory networks controlling virulence factor synthesis in *Erwinia chrysanthemi*. Mol. Microbiol. 29:1407-1418.

Reverchon, S., Rouanet, C., Expert, D., and Nasser, W. 2002. Characterization of indigoidine biosynthetic genes in *Erwinia chrysanthemi* and role of this blue pigment in pathogenicity. J. Bacteriol. 184:654-665.

Rodley, P. D., Rőling, U., and Tűmmler, B. 1995. A physical genome map of the *Burkholderia cepacia* type strain. Mol. Microbiol. 17:57-67.

Rohrbach, K. G. 1989. Unusual tropical fruit diseases with extended latent periods. Plant Dis. 73:607-609.

Rojas, C. M., Ham, J. H., Deng, W.-L., Doyle, J. J., and Collmer, A. 2002. HecA, a member of a class of adhesins produced by diverse pathogenic bacteria, contributes to the attachment, aggregation, epidermal cell killing, and virulence phenotypes of *Erwinia chrysanthemi* EC16 on *Nicotiana clevelandii* seedlings. Proc. Natl. Acad. Sci. U.S.A. 99:13142-13147.

Salmond, G. P. C., and Reeves, P. J. 1993. Membrane traffic wardens and protein secretion in Gram-negative bacteria. Trends Biochem. Sci. 18:7-12.

Samson, R., Legendre, J. B., Christen, R., Fischer-Le Saux, M., Achouak, W., and Gardan, L. 2005. Transfer of *Pectobacterium chrysanthemi* (Burkholder *et al.* 1953) Brenner *et al.* 1973 and *Brenneria paradisiaca* to the genus *Dickeya* gen. nov. as *Dickeya chrysanthemi* comb. nov. and *Dickeya paradisiaca* comb. nov. and delineation of four novel species, *Dickeya dadantii* sp. nov., *Dickeya dianthicola* sp. nov., *Dickeya dieffenbachiae* sp. nov. and *Dickeya zeae* sp. nov. Int. J. Syst. Evol. Microbiol. 55:1415-1427.

San Francisco, M. J. D., and Keenan, R. W. 1993. Uptake of galacturonic acid in *Erwinia chrysanthemi* EC16. J. Bacteriol. 175:4263-4265.

Sangodkar, U. M. X., Chapman, P. J., and Chakrabarty, A. M. 1988. Cloning, physical mapping and expression of chromosomal genes specifying degradation of the herbicide 2,4,5-T in *Pseudomonas cepacia* AC1100. Gene 71:267-277.

Santos, R., Franza, T., Laporte, M. L., Sauvage, C., Touati, D., and Expert, D. 2001. Essential role of superoxide dismutase on the pathogenicity of *Erwinia chrysanthemi* strain 3937. Mol. Plant-Microbe Interact. 14:758-767.

Schuerger, A. C., and Batzer, J. C. 1993. Identification and host range of an *Erwinia* pathogen causing stem rots on hydroponically grown plants. Plant Dis. 77:472-477.

Shevchik, V. E., and Hugovieux Cotte Pattat, N. 1997. Identification of a bacterial pectin acetyl esterase in *Erwinia chrysanthemi*. Mol. Microbiol. 24:1285-1301.

Shevchik, V. E., and Hugovieux-Cotte-Pattat, N. 2003. PaeX, a second pectin acetylesterase of *Erwinia chrysanthemi* 3937. J. Bacteriol. 185:3091-3100.

Shevchik, V. E., Condemine, G., Hugovieux-Cotte-Pattat, N., and Robert-Baudouy, J. 1996. Characterization of pectin methylesterase B, an outer membrane lipoprotein of *Erwinia chrysanthemi* 3937. Mol. Microbiol. 19:455-466.

Shevchik, V. E., Robert-Baudouy, J., and Hugouvieux-Cotte-Pattat, N. 1997. The pectate lyase PelI of *Erwinia chrysanthemi* belongs to a new family. J. Bacteriol. 179:7321-7330.

Shevchik, V. E., Boccara, M., Vedel, R., and Hugouvieux-Cotte-Pattat, N. 1998. Processing of the pectate lyase PelI by extacellular proteases of *Erwinia chrysanthemi* 3937. Mol. Microbiol. 29:1459-1469.

Shevchik, V. E., Condemine, G., Robert-Baudouy, J., and Hugouvieux-Cotte-Pattat, N. 1999. The exopolygalacturonate lyase PelW and the oligogalacturonate lyase Ogl, two cytoplasmic enzymes of pectin catabolism in *Erwinia chrysanthemi* 3937. J. Bacteriol. 181:3912-3919.

Sivasithamparam, K. 1982. Black leg—A confusing potato disease. J. Agric. West. Aust. 11:17-18.

Smith, E. F. 1910. *Bacillus phytophthorus* Appel. Science 31:748-749.

Smith, E. F. 1929. Fifty years of pathology. Proc. Int. Congr. Plant Sci. 1:31.

Stevens, M. P., Wood, M. W., Taylor, L. A., Monaghan, P., Hawes, P., Jones, P. W., Wallis, T. S., and Galyov, E. E. 2002. An Inv/Mxi-Spa-like type III protein secretion system in *Burkholderia pseudomallei* modulates intracellular behaviour of the pathogen. Mol. Microbiol. 46:649-659.

Tamaki, S. J., Gold, S., Robeson, M., Manulis, S., and Keen, N. T. 1988. Structure and organization of the *pel* genes from *Erwinia chrysanthemi* EC16. J. Bacteriol. 170:3468-3478.

Thurn, K. K., Barras, F., Kegoya-Yoshino, Y., and Chatterjee, A. K. 1987. Pectate lyases of *Erwinia chrysanthemi*: PelE-like polypeptides and *pelE* homologous sequences in strains isolated from different plants. Physiol. Mol. Plant Pathol. 31:429-439.

Toth, I. K., Newton, J. A., Hyman, L. J., Lees, A. K., Daykin, M., Ortori, C., Williams, P., and Fray, R. G. 2004. Potato plants genetically modified to produce *N*-acylhomoserine lactones increase susceptibility to soft rot Erwiniae. Mol. Plant-Microbe Interact. 17:880-887.

Urbany, C., and Neuhaus, H. E. 2008. Citrate uptake into *Pectobacterium atrosepticum* is critical for bacterial virulence. Mol. Plant-Microbe Interact. 21:547-554.

Vroemen, S., Heldens, J., Boyd, C., Henrissat, B., and Keen, N. T. 1995. Cloning and characterization of the *bgxA* gene from *Erwinia chrysanthemi* D1 which encodes a β-glucosidase/xylosidase enzyme. Mol. Gen. Genet. 246:465-477.

Walker, D. S., Reeves, P. J., and Salmond, G. P. C. 1994. The major secreted cellulose, CelV, of *Erwinia carotovora* subsp. *carotovora* is an important soft rot virulence factor. Mol. Plant-Microbe Interact. 7:425-431.

Wigley, P., and Burton, N. F. 2000. Multiple chromosomes in *Burkholderia cepacia* and *B. gladioli* and their distribution in clinical and environmental strains of *B. cepacia*. J. Appl. Microbiol. 88:914-918.

Yamagishi, Y., Fujita, J., Tagigawa, K., Negayama, K., Nakazawa, T., and Takahara, J. 1993. Clinical features of *Pseudomonas cepacia* pneumonia in an epidemic among immuncompromised patients. Chest 103:1706-1709.

Yang, C.-H., Gavilanes-Ruiz, M., Okinaka, Y., Vedel, R., Berthuy, I., Boccara, M., Chen, J. W.-T., Perna, N. T., and Keen, N. T. 2002. *hrp* Genes of *Erwinia chrysanthemi* 3937 are important virulence factors. Mol. Plant-Microbe Interact. 15:472-480.

Yang, Z., Cramer, C. L., and Lacy, G. H. 1992. *Erwinia carotovora* subsp. *carotovora* pectic enzymes: *In planta* gene activation and roles in soft-rot pathogenesis. Mol. Plant-Microbe Interact. 5:104-112.

Yap, M.-N., Yang, C.-H., Barak, J. D., Jahn, C. E., and Charkowski, A. O. 2005. The *Erwinia chrysanthemi* type III secretion system is required for multicellular behavior. J. Bacteriol. 187:639-648.

Yap, M.-N., Rojas, C. M., Yang, C.-H., and Charkowski, A. O. 2006. Harpin mediates cell aggregation in *Erwinia chrysanthemi* 3937. J. Bacteriol. 189:2280-2284.

Yishay, M., Burdman, S., Valverde, A., Luzzatto, T., Ophir, R., and Yedidia, I. 2008. Differential pathogenicity and genetic diversity among *Pectobacterium carotovorum* ssp. *carotovorum* isolates from monocot and dicot hosts support early genomic divergence within this taxon. Environ. Microbiol. 10:2746-2759.

Young, J. M., Dye, D. W., Bradbury, J. F., Panagopoulos, C. G., and Robbs, C. F. 1978. A proposed nomenclature and classification for plant-pathogenic bacteria. N. Z. J. Agric. Res. 21:153-177.

Zink, R. T., and Chatterjee, A. K. 1985. Cloning and expression in *E. coli* of pectinase genes of *Erwinia carotovora* ssp. *carotovora*. Appl. Environ. Microbiol. 49:714-717.

Vascular Wilt Diseases and Their Pathogens

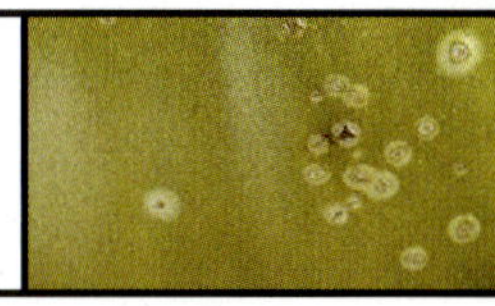

7

Plant-pathogenic bacteria that invade, proliferate, and colonize mainly the vascular tissues and cause wilting of the plant host are categorized as vascular wilt pathogens. Generally, these wilt-inducing pathogens gain entry into their hosts via the root system. Wilt-inducing pathogens infect intercellular spaces of the cortical and vascular parenchyma tissues and penetrate into the xylem vessels followed by systemic infection. Systemically infected plants wilt, especially under water-stress conditions. Completely flagged plants in the field suggest the presence of a vascular wilt pathogen. Some of the important wilt diseases induced by aggressive bacterial pathogens are described here.

Bacterial Wilt and Canker of Tomato

Bacterial canker of tomato is also known as bird's eye canker (English), chancre bactérien (French), Bakterienwelke (German), marchitez bacteriana (Spanish), and cancro batterico (Italian). Bacterial canker can cause major economic losses in commercial tomato production worldwide. The disease is caused by *Clavibacter michiganensis* subsp. *michiganensis.*

General Characteristics of *Clavibacter michiganensis* subsp. *michiganensis*

Clavibacter michiganensis subsp. *michiganensis* is in the family *Microbacteriaceae* of the class *Actinobacteria.* Its cells are Gram-positive, nonmotile, curved as well as straight rods that are occasionally elongated and bent. They grow aerobically at 25°C, but relatively slowly on nutrient agar and faster on more complex media. The colonies (2–3 mm in diameter) on agar medium are usually convex, smooth, pale yellow, opaque, and semifluidal in appearance and glisten when young. With time, the intensity of the yellow colony pigmentation increases. The pigments are composed of carotenoids (Saperstein et al., 1954), and their formation is dependent on the concentration of thiamine in the growth medium (Braun, 1949). The organism produces acid from mannose; utilizes acetate, citrate, fumarate, malate, and succinate; and weakly produces hydrogen sulfide from peptone (Dye and Kemp, 1977). It does not produce nitrate reductase, urease, or high amounts of acid from sugars. The chromosome of *Clavibacter michiganensis* subsp. *michiganensis* is a circular DNA molecule of 3,298 Mb and 72.6% guanine-plus-cytosine (GC) content (Gartemann et al., 2008). The pathogen contains two circular plasmids, pCM1 (27.4 kb; 67.6% GC) and pCM2 (69.9 kb; 66.5% GC) (Gartemann et al., 2008). Plasmid pCM1 contains a *celA* gene and plasmid pCM2 contains a *pat-1* gene (see Pathogenicity Factors below).

Clavibacter michiganensis subsp. *michiganensis* can be grown on rich media, such as medium 523 (Kado and Heskett, 1970) and TBY medium (Meletzus and Eichenlaub, 1991). The semiselective medium CNS (Gross and Vidaver, 1979) has also been used with the omission of the ingredient lithium chloride (Gitaitis et al., 1991). Semiselective media D2anx (Chun, 1982) and Cmm1 (Alvarez et al., 2005), based on medium D2 (Kado and Heskett, 1970), and SCM (Fatmi and Schaad, 1988) are commonly used for isolating *Clavibacter michiganensis* subsp. *michiganensis* from infested seeds. Another semiselective medium has been described that contains a number of antibiotics for selection (Dhanvantari, 1987). Strain variations and differences require titration of each antibiotic. King's B and nutrient agar are not recommended since they were designed for growth of fluorescent *Pseudomonas* species and enteric bacteria (*Escherichia coli*), respectively, and growth of *Clavibacter michiganensis* subsp. *michiganensis* is particularly slow on these media.

Host Range

The host range of *Clavibacter michiganensis* subsp. *michiganensis* appears to be restricted to members of the Solanaceae family of plants. Plants in the Gramineae family have been reported as susceptible when artificially inoculated (Stamova and Sotirova, 1987). However, natural infections of grasses by *Clavibacter michiganensis* subsp. *michiganensis* have not been reported. Documented hosts of *Clavibacter michiganensis* subsp. *michiganensis* are listed in Table 7-1.

History of Bacterial Wilt and Canker

Because it was first discovered near Grand Rapids, Michigan, bacterial wilt and canker of tomato was first called the Grand Rapids disease (Smith, 1910). The cause was identified as a yellow schizomycete called *Aplanobacter michiganense*. The culturable causal organism was next called *Bacterium michiganense* (Smith, 1910) and changed to *Phytomonas michiganensis* (Bergey et al., 1923). *Corynebacterium michiganense*, proposed by Jensen (1934), was the preferred name, even when it was briefly changed to *Mycobacterium flavum* subsp. *michiganense* (Krasil'nikov, 1941). A pathovar-specific name was then attached, *Corynebacterium michiganense* pv. *michiganense* (Dye and Kemp, 1977), and later changed to a subspecies, *Corynebacterium michiganense* subsp. *michiganense*, because the traits extended beyond host range differences (Carlson and Vidaver, 1982). Plant pathogens in the genus *Corynebacterium* were placed in the genus *Clavibacter*, based on a combination of chemical composition of the cell wall, including peptidoglycan differences (diaminobutyrate), one-dimensional protein electrophoretogram profiles, and physiological and nutritional traits (Davis et al., 1984). The *Clavibacter* generic name will likely again change based on forthcoming whole-genome classifications of coryneform bacteria (Konstantinidis and Tiedje, 2005).

Table 7-1. Host Range of *Clavibacter michiganensis* subsp. *michiganensis*

Common Name	Scientific Name	Reference
Jimson weed	*Datura stramonium* L.	U.S. Dep. Agric., 1960
Pepper	*Capsicum frutescens* L.	U.S. Dep. Agric., 1960
Tomato	*Lycopersicon esculentum* Mill. var. *esculentum* Mill.	U.S. Dep. Agric., 1960
Horsenettle	*Solanum carolinense* L.	U.S. Dep. Agric., 1960
Greenspot nightshade	*Solanum douglasii* Dunal	Thyr et al., 1975
Black nightshade	*Solanum nigrum* L.	Thyr et al., 1975
Cutleaf nightshade	*Solanum triflorum* Nutt.	Thyr et al., 1975
Tree tomato	*Cyphomandra betacea* (Cav.) Sendtn.	U.S. Dep. Agric., 1960

Symptomatology

Symptom development is highly dependent on environmental conditions, age of the host plant, cultivar susceptibility, and initial means of entry into the plant (Gleason et al., 1993). Systemic infections cause blockage of the xylem vessels, which often results in wilting, particularly under water-stress conditions. Unilateral wilting, a withering on only one side of a leaf, may occur. Leaves on young plants can curl and later show necrosis or growth distortion. With severe vascular infections, the epidermis and outer cortical tissue are easily removed from the inner stem when stems are squeezed between thumb and finger. Squeezing cut stems produces yellowish exudates from the xylem vessels. Developing fruits can bear spots or appear marbled.

Clavibacter michiganensis subsp. *michiganensis* can multiply on the plant surface or within surface wounds and stomata. Surface colonization can result in leaves appearing crusted on the surface because of raised yellowish white blisters. Infected leaves bear yellow to dark brown regions of marginal necrosis that occur through infection of hydathodes (Carlton et al., 1998). Superficial infections of tomato fruit result in raised, pale green or white pustules with a light brown center and a chlorotic halo that give the appearance of a bird's eye (Fig. 7-1). Interestingly, canker symptoms are rarely observed on fruits, although cankers are commonly observed on stems.

Disease Cycle

Clavibacter michiganensis subsp. *michiganensis* overwinters on diseased plant debris mainly on the surface of harvested fields. New plantings of susceptible crops in fields bearing diseased plant materials, such as tomatoes, invariably lead to increased bacterial wilt, including fruit infections. The pathogen is introduced into new fields via latently infected seedling transplants and into seedling-rearing glasshouses mainly via infested seeds (Gitaitis et al., 1991). *Clavibacter michiganensis* subsp. *michiganensis* remains viably recoverable from seed coats even after several years of storage. Foliar surface populations of *Clavibacter michiganensis* subsp.

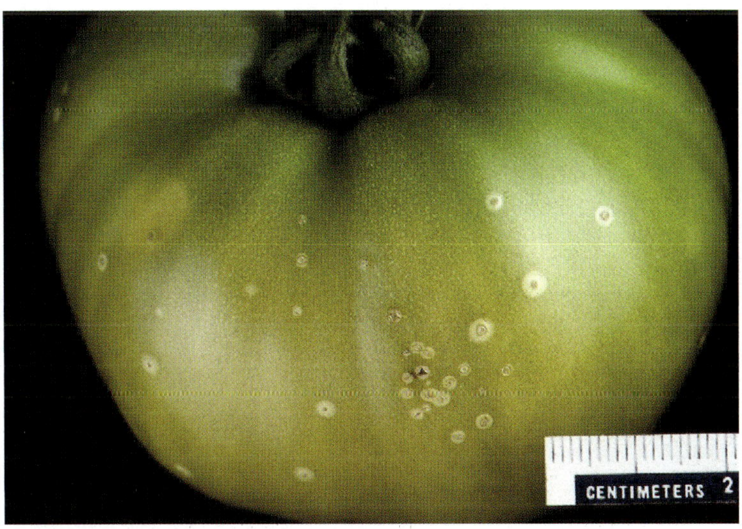

Fig. 7-1. Brownish pustules with white halo bird's eye fruit symptoms. (Courtesy Diane A. Cuppels)

michiganensis serve as local sources of inocula and the foci for secondary spread of the pathogen. Leaf surface populations of this bacterium have been reported as high as 10^9 CFU/g fresh weight at least 2.7 m from the focus of infection (Chang et al., 1992). Analysis of the genome sequence of *Clavibacter michiganensis* subsp. *michiganensis* revealed the apparent lack of a sulfate reduction pathway, which makes *Clavibacter michiganensis* subsp. *michiganensis* dependent on reduced sulfur compounds for growth. This deficiency is probably the reason for the poor survival of this pathogen in soil (Gartemann et al., 2008).

Pathogenicity Tests

Plant-isolated and pure-cultured bacteria are suspended in sterile distilled water (10^9 cells/ml) to serve as the test inoculum. A reference strain of *Clavibacter michiganensis* subsp. *michiganensis* should be used as a positive control. Tomato seedlings at the cotyledon or two-leaf stage are used as indicator plants. Seedlings are inoculated by injection with a syringe equipped with a 27-gauge needle into the stem at the cotyledons. The plants are incubated at 28°C under relatively high humidity (ca. 70% relative humidity). After 6–8 days of incubation, the inoculated seedlings are observed for wilt symptoms. Systemic infection is confirmed by sampling stem sections distal to the site of inoculation. Thin slices of stem section are positioned on one of the selective agar media (Fatmi and Schaad, 1988; Gitatis et al., 1991; Alvarez et al., 2005) and incubated for 3 days at 28°C.

Seedlings with expanding cotyledons can also be used. A drop (50 µl) of inoculum is deposited on the surface of each cotyledon previously lightly dusted with 400-mesh sterile Carborundum. The inoculation is performed by using a cotton swab to gently rub the inoculum on the surface of the cotyledons. The young seedlings are covered with a polyethylene bag and incubated at 28°C for 2–3 days. White blisters on the inoculated surface of the cotyledons are presumptive indications of *Clavibacter michiganensis* subsp. *michiganensis*. Confirmatory identification of the pathogen may be determined by polymerase chain reaction (PCR) methods, such as repetitive-sequence–based (rep)-PCR (Louws et al., 1998; Kaneshiro et al., 2006).

Pathogenicity Factors

Although exopolysaccharides (EPS; also called extracellular polysaccharides) are copiously produced by *Clavibacter michiganensis* subsp. *michiganensis*, their role in directly affecting water transport by plugging the xylem vessels is suspect (Bermpohl et al., 1996). Their role may actually be to protect the bacterium from harmful agents in the environment and from defense reactants in the plant host. EPS may also play a role in recognition of the pathogen by the host during the initial phase of infection (Van Alfen, 1989; Bermpohl et al., 1996). The identity of genes in *Clavibacter michiganensis* subsp. *michiganensis* that are directly involved in virulence remains unclear. Strain NCPPB382 harbors two plasmids, pCM1 of 27.5 kb (66.5% GC) and pCM2 of 70–72 kb (67.6% GC), that are involved in conferring virulence (Meletzus and Eichenlaub, 1991; Meletzus et al., 1993; Gartemann et al., 2008). Both plasmids are required for full virulence of strain NCPPB382. A derivative of NCPPB382 lacking the plasmids no longer causes disease; however, it still can colonize the tomato plant (Gartemann et al., 2003). The genes conferring bacterial establishment and growth in the plant are therefore located on the chromosome of *Clavibacter michiganensis* subsp. *michiganensis*, whereas pCM1 encodes CelA, a 78-kDa *endo*-β-1,4-glucanase, and pCM2 encodes a putative serine protease, Pat-1 (Jahr et al., 2000; Gartemann et al., 2003). CelA is a major wilt-inducing factor. How this enzyme contributes to wilting remains unknown. The enzyme may catalyze the breakdown of cell wall materials of the vascular system, with the breakdown products interfering with water transport. Interestingly, other pathogenic strains of *Clavibacter michiganensis* subsp. *michiganensis* apparently do not harbor either plasmid (Meletzus et al., 1993). However, a homologous virulence locus called *pat-1* is located on the chromosome of these strains. The serine protease encoded by *pat-1* possesses high substrate specificity that apparently contributes to virulence (Dreier et al., 1997). There are at least seven serine proteases (called ChpA, ChpB, ChpC, ChpD, ChpE, ChpF, and ChpG) that belong to a trypsin subfamily with homologies to *pat-1*. ChpC appears to be required for effective interaction of *Clavibacter michiganensis* subsp. *michiganensis* with its host (Stork et al., 2008).

The pathogen also produces a heat-stable proteinaceous substance that elicits hypersensitive responses (HRs) in four o'clock (*Mirabilis jalapa* L.) and tobacco (*Nicotiana tabacum* L.) (Alarcón et al., 1998). The identity of the proteinaceous substance remains unknown but is thought to be similar to harpin and PopA of Gram-negative pathogens. A list of virulence genes and their products are shown in Table 7-2.

Host response to *Clavibacter michiganensis* subsp. *michiganensis* is reflected by transcriptional activation of numerous genes in tomato. Microarray analyses revealed that 122 of 9,254 tomato genes are differentially expressed in *Clavibacter michiganensis* subsp. *michiganensis*-infected plants, compared with mock-inoculated plants (Balaji et al., 2008). The genome of *Clavibacter michiganensis* subsp. *michiganensis* contains a 130-kb pathogenicity island (also known as *chp/tom* region) (Bentley et al., 2008). The *tom* region contains the *tomA* gene that encodes *endo*-1,4-β-glycosidase. The bacterial interaction with tomato activates defense-related protein production and turnover, the scavenging of free oxygen radicals, and hormone synthesis, including a subset of tomato genes involved in ethylene biosynthesis and response, as well as sapronin (α-tomatine) production. This latter compound is degraded by the pathogen's β-glycosidase (Kaup et al., 2005).

Table 7-2. Virulence Genes and Encoded Pathogenicity Factors of *Clavibacter michiganensis* subsp. *michiganensis*

Gene	Encoded Function	Reference
celA	A 78-kDa *endo*-β-1,4-glucanase CelA encoded by 27.5-kb plasmid pCM1	Meletzus et al., 1993; Dreier et al., 1997; Jahr et al., 2000; Gartemann et al., 2003
pat-1 (or *sep*)	A 29.7-kDa serine protease Pat-1 encoded by 72-kb plasmid pCM2	Dreier et al., 1997; Gartemann et al., 2003
hrpN	A putative necrogenic effector present in the culture supernatant of *Clavibacter michiganensis* subsp. *michiganensis*	Alarcón et al., 1998
furA	A catalase-encoding gene expressed in planta	Yang et al., 2005
tomA	An *endo*-1,4-β-glycosidase implicated in degrading α-tomatine	Kaup et al., 2005

Ring Rot of Potatoes

Ring rot occurs in potato-growing regions worldwide. The name ring rot is derived from the characteristic creamy necrosis of the vascular cylinder of the potato tuber. The disease is caused by *Clavibacter michiganensis* subsp. *sepedonicus*, a highly contagious pathogen that is quarantined in many countries because of an implemented policy of zero tolerance for its detectable presence in a field, storage bin, or seed stock. If *Clavibacter michiganensis* subsp. *sepedonicus* is detected, entire potato fields and harvested seed lots are either totally destroyed or directly sold as table stock and not certified for seed use. Potato seed is the primary source of the spread of this pathogen. Vegetative propagation of seed potatoes ensures local as well as interregional and international spread of the bacteria. Heavy losses in yield and quality of potato tubers result from ring rot.

General Characteristics of *Clavibacter michiganensis* subsp. *sepedonicus*

The bacterium was originally named *Bacterium sepedonicum* by Spieckermann and Kotthoff in 1914 and was changed to *Aplanobacter sepedonicum* by Smith in 1920, to *Phytomonas sepedonica* by Magrou in 1937, to *Corynebacterium sepedonicum* by Skaptason and Burkholder in 1942, to *Mycobacterium sepedonicum* or *Pseudobacterium sepedonicum* by Krasil'nikov in 1949, to *Corynebacterium michiganense* pv. *sepedonicum* by Dye and Kemp in 1977. The pathovar was elevated to subspecies by Carlson and Vidaver in 1982 and the genus was changed to *Clavibacter michiganensis* subsp. *sepedonicus* by Davis et al. in 1984.

The Gram-positive cells of this pathogen are predominantly wedge shaped with occasional rods that are curved or straight. On rich agar medium, colonies are small, smooth, glistening, and convex and appear pale yellow or whitish cream depending on the medium used. The bacterium is strictly aerobic and requires biotin, thiamine, nicotinic acid, histidine, purines, and pyrimidines for proficient growth. The organism grows best at room temperature (23°C). It utilizes acetate and succinate as carbon sources, produces acid from mannitol and sorbitol, and hydrolyzes potato starch. The 3.2-Mb genome of *Clavibacter michiganensis* subsp. *sepedonicus* has been completely sequenced (Bentley et al., 2008). Inspection of the genome sequence revealed 106 insertion sequence elements, which suggests that considerable rearrangements of the chromosome have taken place when compared with the genome sequence of *Clavibacter michiganensis* subsp. *michiganensis* (Bentley et al., 2008). *Clavibacter michiganensis* subsp. *sepedonicus* strains harbor linear plasmids pCSL1 of 94 kb and pCSL2 of 140 kb, and a circular plasmid, pCS1 of 50 kb (Brown et al., 2002; Bentley et al., 2008). The plasmids are apparently related because they contain regions of homology as detected by Southern hybridization. The biological roles of these plasmids have not been elucidated.

Host Range

Clavibacter michiganensis subsp. *sepedonicus* has a narrow host range, primarily restricted to solanaceous species. Its natural host is potato (*Solanum tuberosum* L.), whereas eggplant and tomato are susceptible by artificial inoculation. *Clavibacter michiganensis* subsp. *sepedonicus* persists in non-solanaceous species; it survives at least 8 weeks in artificially inoculated maize, wheat, barley, oat, bush bean, broad bean, rape, and pea (van der Wolf et al., 2005). While some solanaceous species, such as watermelon nightshade (*Solanum citrullifolium* A. Braun) and South American potato vine (*Solanum commersonii* Dunal), produce symptoms when artificially inoculated (Knorr, 1948; Slack, 1987), other members of the family Solanaceae can be symptomless carriers of *Clavibacter michiganensis* subsp. *sepedonicus* (Bonde, 1942; Knorr, 1948; Zizz and Harrison, 1991; van der Wolf et al., 2005).

History of Ring Rot

Ring rot of potato was first described by Spieckermann in Germany in 1913 (Spieckermann, 1913). Ring rot was discovered in the United States in 1932 in a field of potatoes in Aroostook County, Maine (Bonde, 1937). By 1939, the disease was reported to occur in 31 states of the United States and in Canada (Howard, 1940; Bodine and Durrell, 1941; Eddins, 1941; Richards, 1941; Thornberry, 1941; Weiss, 1941; Skaptason, 1943). Because potato seed pieces are distributed widely, the disease became a worldwide problem. The imperative institution of zero-tolerance seed piece certification programs worldwide has helped immensely in controlling the spread of ring rot.

Symptomatology

Ring rot of potato is recognized by the wilting of leaves, severe withering of leaflets, and chlorosis. The lower leaves show rolled margins and appear pale green with interveinal yellowing and necrosis (Fig. 7-2). Leaf death occurs as the disease progresses. The leaf margins become necrotic, followed by browning and drying to a brittle consistency. Stems become yellow and wilted, unable to hold up foliage. Symptoms can be delayed during cold growing seasons.

Ring rot derives its name from the characteristic yellow to brown necrotic discoloration of the vascular cylinder revealed by a cross section of the tuber (Fig. 7-3). If the tuber is squeezed, a creamy to light yellow bacterial ooze can be pressed out

Fig. 7-2. Foliar leaf curling symptoms of ring rot of potato. (Courtesy R. Michael Davis)

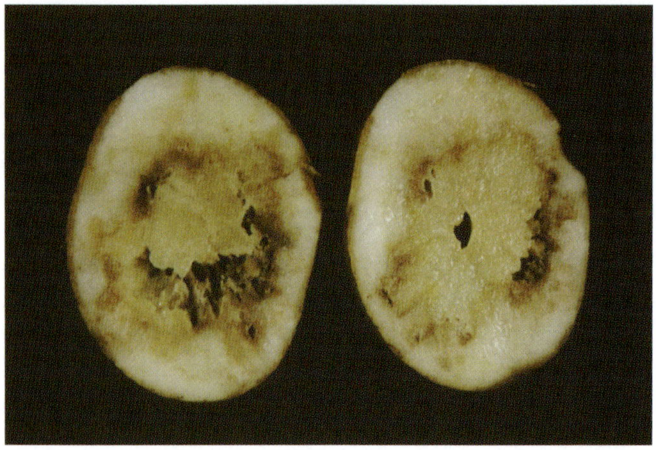

Fig. 7-3. Tuber symptoms of ring rot of the vascular ring. (Courtesy R. Michael Davis)

from the infected vascular ring. Externally, severely infected tubers display sunken areas. Cracks in the tubers provide access to opportunistic secondary invaders that quickly rot the tuber completely.

Diagnosis of ring rot in seed lots is highly dependent on external disease symptoms, which may not be obvious (see Disease Cycle below). The disease can often be asymptomatic in potato tubers (Knorr, 1948). Therefore, screening of tubers in the laboratory using serological and PCR methods may be needed.

Disease Cycle

Clavibacter michiganensis subsp. *sepedonicus* can persist as dried ooze for as long as 2 years in infected plant debris and tubers. Partially infected volunteer plants originating from buried infected potato tubers can serve as reservoirs of primary inoculum. Usually, the introduction of ring rot occurs when noncertified potato seed pieces are planted in the field. Poor sanitation measures in tuber storage facilities or for handling, grading, and planting machinery can also allow transmission of the pathogen onto seed pieces (Bonde, 1942). Because wounds are required for establishing infection, *Clavibacter michiganensis* subsp. *sepedonicus* cells are easily spread to new sites when using contaminated equipment to cut and plant potato seed pieces. Contamination can also occur when using pathogen-infested carry-all bins to transport seed pieces. About 300 CFU of *Clavibacter michiganensis* subsp. *sepedonicus* cells are required for effective infection of the potato seed pieces (Nelson, 1982). Plants originating from infected seed pieces are stunted and show mild chlorosis of the lower leaves as *Clavibacter michiganensis* subsp. *sepedonicus* moves up the vascular system of the lower parts of the plant. The bacterial cells eventually move through the stolons to infect growing tubers. The bacteria persist in the tubers in the field and rarely persist in soil in the absence of plant debris or tubers. Disease development is favored by temperatures ranging from 18 to 23°C.

Pathogenicity Tests

Pathogenicity tests are performed on tubers that are inoculated, planted, and assessed 8 weeks later. Potato tubers are infected by injecting three to four eyes with a suspension containing 10^8 cells/ml prior to planting (van der Wolf et al., 2005). The inoculated tubers are subsequently examined for symptoms of infection of the vascular ring. Jimson weed (*Datura stramonium* L.) and eggplant (*Solanum melongena* L. cv. Black Beauty) are also used as sensitive test hosts (van der Wolf et al., 2005). These plants are grown in the glasshouse to a four- to eight-leaf stage. Plants are inoculated by using a syringe equipped with a 23-gauge needle to inject a solution of 10^8 CFU/ml into the stem about 1 cm below the petiole of the first leaf of each indicator plant. After 2 months of growing in the glasshouse, stem sections 2 cm above the inoculation point are surface sterilized with acetone, dried in a hood, and triturated in sterile 1 mM sodium phosphate buffer, pH 7.0. The extract is then tested serologically, such as by enzyme-linked immunosorbent assay (ELISA) or by direct immunofluorescence of pelleted bacteria, for the presence of *Clavibacter michiganensis* subsp. *sepedonicus* (De Boer et al., 1994). Moreover, enzyme-linked oligonucleotide-sorbent assay (ELOSA) can be performed on the extract (Baer et al., 2001). ELOSA is a solid-phase sandwich hybridization assay that allows a very specific and sensitive detection of a target nucleic acid. ELOSA is similar to ELISA and involves simple washing steps and a 4-h incubation step.

Various PCR amplification methods for *Clavibacter michiganensis* subsp. *sepedonicus* genome sequences have been used in an attempt to see strain differences and for positive identification (Hu et al., 1995; Mills et al., 1997; Lee et al., 2001). Recent studies have shown, however, that various primer sets employed have failed to reveal significant differences (Fousek and Mraz, 2003). An insertion sequence called IS*1121*, present in multiple copies in the genome of *Clavibacter michiganensis* subsp. *sepedonicus*, has been targeted as a specific PCR marker (Mogen and Oleson, 1987). IS*1121* is also present in 33 copies in the genome of *Clavibacter michiganensis* subsp. *insidiosus* (Samac et al., 1998). The lack of knowledge and identification of pathogenicity genes of *Clavibacter michiganensis* subsp. *sepedonicus* has impeded the use of PCR methods to accurately identify pathogenic versus nonpathogenic strains of *Clavibacter michiganensis* subsp. *sepedonicus*. The transposon Tn*1409C* used in identifying virulence genes in *Clavibacter michiganensis* subsp. *michiganensis* might be considered for this task (Kirchner et al., 2001).

The stem extract described above can also be plated on semiselective agar media, such as NCP-88 (de la Cruz et al., 1992). Alternatively, the stem sections are vegetatively propagated and foliar wilt symptoms are assessed 18–20 days after inoculation. Micropropagation of potato stem sections (such as *Solanum tuberosum* L. cv. Red Pontiac) on tissue culture agar medium are exposed to test inoculum by immersing the trimmed roots of the plantlets for 30 min in inoculum of 10^8 CFU/ml. The plantlets are planted in soil and foliar wilt symptoms are assessed 12 weeks after inoculation (Nissinen et al., 1997).

The tobacco leaf HR test can also be used to distinguish pathogenic from nonpathogenic strains of *Clavibacter michiganensis* subsp. *sepedonicus* (Nissinen et al., 1997). Cells of this organism are infiltrated using a needleless syringe in the same manner as carried out for HR assays of necrogenic pathogens (see Chapter 5). Generally, a concentration of 10^8–10^9 CFU/ml is used as the inoculum to infiltrate tobacco (*Nicotiana tabacum* L. cv. Samsun). Elicitation of the HR occurs within 48 h following infiltration. A strong and rapid elicitation of an HR indicates a highly virulent strain of *Clavibacter michiganensis* subsp. *sepedonicus*, while a weak HR is associated with mildly virulent strains (Nissinen et al., 1997).

Pathogenicity Factors

EPS and glycopeptides produced by *Clavibacter michiganensis* subsp. *sepedonicus* were originally thought to play a direct role in virulence. It has become clear, however, that EPS contribute toward providing habitat enhancement benefits, such as protection against environmental extremes (e.g., dehydration) on the organism, rather than directly to pathogenicity and virulence. There is an obvious need to use molecular genetic approaches to identify novel virulence genes in *Clavibacter michiganensis* subsp. *sepedonicus*. The paucity of molecular biological tools specific for *Clavibacter* species has impeded the identification of specific virulence genes. Like *Clavibacter michiganensis* subsp. *michiganensis*, *Clavibacter michiganensis* subsp. *sepedonicus* elaborates *endo*-β-1,4-glucanase as well as other cellulose-degrading enzymes (Baer and Gudmestad, 1995). The endoglucanase has been implicated as a contributing virulence factor of *Clavibacter michiganensis* subsp. *michiganensis* (Table 7-2). The protein or proteins produced by *Clavibacter michiganensis* subsp. *sepedonicus* that cause HR have not been fully characterized (Nissinen et al., 1997). Polyacrylamide gel electrophoresis of a semipurified culture extract revealed a unique protein staining band of ca. 25 kDa. The specific role of this protein in virulence was not determined.

Recent studies have identified at least six genes thought to be involved in virulence. They are a plasmid-encoded cellulose gene called *celA*, a second cellulose gene called *celB*, a gene encoding a pectate lyase, a gene encoding a xylanase, and two genes, named *pat-1* and *pat-2*, encoding a putative serine protease (Holtsmark et al., 2008). Interestingly, there are eight *pat-1* genes scattered throughout the chromosome of *Clavibacter michiganensis* subsp. *sepedonicus* (Bentley et al., 2008). This finding suggests the relatively recent acquisition of host-adaptive and unidentified virulence/pathogenicity genes. In order to identify genes responsible for virulence, a quick assay system for virulence is needed so that it can be used to screen for avirulent mutants that will lead to the identification of additional virulence genes for this pathogen.

Alfalfa Wilt

Alfalfa wilt, also known as bacterial blight and root rot (English), jaunissement bactérien or flétrissement bactérien (French), Bakterielle Luzernewelke (German), and podredumbre de la alfalfa (Spanish), is found in alfalfa-growing regions worldwide. *Clavibacter michiganensis* subsp. *insidiosus* is the causal organism of alfalfa wilt. The pathogen is spread via the mechanical harvester and infested seeds. The introduction and use of resistant cultivars of alfalfa have reduced the incidence of this disease.

General Characteristics of *Clavibacter michiganensis* subsp. *insidiosus*

The bacterium was originally named *Aplanobacter insidiosum* by McCulloch in 1925, *Corynebacterium insidiosum* by Jensen in 1934, *Corynebacterium michiganense* pv. *insidiosum* by Dye and Kemp in 1977, and *Corynebacterium michiganense* subsp. *insidiosum* by Carlson and Vidaver in 1982. The current name is *Clavibacter michiganensis* subsp. *insidiosus*.

McCulloch (1925) described the bacterium as a short, nonmotile, Gram-positive rod, measuring 0.4–0.5 × 0.7–1.0 μm for single rods and 0.4–0.5 × 1.8–2.0 μm for paired rods. The bacterium is aerobic and grows slowly, with an optimal growth temperature of 22°C (Jones and McCulloch, 1926). Biotin, nicotinic acid, and thiamine appear to be required for optimal growth (Starr, 1949). YDC medium (Wilson et al., 1967) is a suitable medium. Young colonies are white and turn pale yellow with age. They are circular with a smooth shining surface, slightly convex, and slightly viscous. After 7 days at 18°C on YDC agar, the colonies form a blue insoluble extracellular pigment that appears as granules composed of indigoidine (A_{606} = max) (Starr, 1958). Temperatures higher than 18°C can result in low blue pigment production. Also, pH 7.0 is optimum for pigment production. The pigment is soluble in dimethylsulfoxide but not in acetone, benzene, chloroform, or diethyl ether. Acid is weakly produced from glucose, lactose, sucrose, and mannose. The bacterium does not produce nitrate reductase, cytochrome oxidase, urease, lipase, β-galactosidase, or tyrosinase. Hydrogen sulfide is produced from cysteine hydrochloride (Dye and Kemp, 1977).

Clavibacter michiganensis subsp. *insidiosus* is grown on a modified semiselective D2 agar medium (Kado and Heskett, 1970). Cultures are incubated at 28°C for 4 days. A differential tetrazolium chloride-containing medium described by Kelman (Kelman, 1954) (see Southern Bacterial Wilt below) has been used to discriminate between virulent and weakly virulent or avirulent colony forms (Carroll and Lukezic, 1971). Three colony types appear on this medium as follows: 1) a large, butyrous, deep red-centered, smooth colony with a white border; 2) a smaller, light pink to white colony, sometimes showing a small, dark pink center with an irregular border; and 3) a small, green colony with a round to oblong shape, one-fourth as large as the pink type (Carroll and Lukezic, 1971). The pink type of colony is the virulent type of *Clavibacter michiganensis* subsp. *insidiosus*. The red and green colony types are avirulent derivatives.

Infected alfalfa roots can be used to maintain *Clavibacter michiganensis* subsp. *insidiosus*. Infected roots are washed, shredded, packed into polyethylene resealable-zipper storage bags, and frozen at –20°C (Kernkamp and Hemerick, 1952). The frozen roots are thawed in sterile tap water to allow the bacteria to move out from the vascular cylinder. A portion of the water containing *Clavibacter michiganensis* subsp. *insidiosus* is streaked on D2 and YDC agar medium.

Host Range

The host range of *Clavibacter michiganensis* subsp. *insidiosus* is apparently confined to *Medicago* and *Melilotus* species (*Medicago falcata* L. [yellow-flowered alfalfa], *Melilotus alba* Desr. [white sweet clover], and *Medicago sativa* L. [alfalfa, lucerne]). *Medicago sativa* appears to be its primary host.

History of Alfalfa Wilt

Alfalfa wilt was first noted in 1924 in Freeport, Illinois, by L. F. Garber and was soon recognized as caused by a bacterium (Jones, 1925). The isolated bacterium was called *Aplanobacter insidiosum* and it was found to produce blue granules (McCulloch, 1925) later shown to be composed of indigoidine (Starr, 1958). The species name, *insidiosum*, was selected for this bacterium because of the insidious infection it caused, with the slow outward appearance of symptoms long after it had seriously infected the host plant (McCulloch, 1925).

Symptomatology

Field sightings of stunted, chlorotic alfalfa plants with a proliferation of short stem leaflets that show various degrees of chlorosis (light green to yellow) are characteristic symptoms of bacterial wilt (Fig. 7-4). Leaves of infected plants develop rounded, curved tips. Infected plants wilt to different degrees depending on the extent of water-stress conditions. Chronically infected plants may stay wilted and death ensues. Partially infected plants become turgid at night but wilt again during the heat of the day. Stunted plants are recognized during the regrowth of mowed (harvested) fields. The vascular tissue of the stems and taproots of infected plants display dark brown to yellow streaks where *Clavibacter michiganensis* subsp. *insidiosus* has invaded and colonized.

Disease Cycle

Clavibacter michiganensis subsp. *insidiosus* requires wounds to establish infection (Peltier, 1934). Debris from infected plants serves as the primary source of inoculum. *Clavibacter michiganensis* subsp. *insidiosus* remains viable in dry infected tissues and seeds for extended periods, from 3 to 10 years (Cormack, 1961; Erwin, 1990). Seeds from severely infected plants also aid in the dissemination of this pathogen. Transmission by the nematode *Ditylenchus dipsaci* (Kühn) Filipjev has been reported (Hawn, 1963; Hunt et al., 1971). The bacteria are also spread by the mechanical harvester and by tillage (Malvick, 1988; Davis and Frate, 2006). Fresh wounds, which are made by the blades of the mower of the harvester, provide new sites of infection. The mower blades also serve to spread the pathogen from plant to plant. The blades become contaminated from diseased plants and spread inoculum to healthy plants. *Clavibacter michiganensis* subsp. *insidiosus* enters primarily through wounds and colonizes the vascular tissues (xylem vessels) and, in mature plants, can occupy the intercellular spaces of the seed parenchyma (endosperm aleurone layer). The pathogen remains resident in most parts of the alfalfa plant and overwinters well under temperate climatic conditions.

Fig. 7-4. Stunted, chlorotic symptoms of alfalfa wilt on alfalfa (*Medicago sativa* L.). (Courtesy R. Michael Davis)

Pathogenicity Tests

Inoculation of cotyledons and roots of alfalfa seedlings has been used in assessing pathogenicity.

Cotyledon Assay

After light scarification between sheets of 200-grit sand paper, alfalfa seeds of a susceptible cultivar of alfalfa are sown under a 5- to 8-mm layer of moist sand layered on top of potting soil in plastic pots. The seedlings are grown at 20–23°C in a glasshouse or growth facility. The seedlings emerge after 7–8 days. They are thinned to about 20 per pot. The tip or one-third of each cotyledon is clipped with a pair of sharp dissecting scissors. The test bacterium growing on agar is used as inoculum. With the aid of a cotton swab, bacteria are swabbed onto the freshly made cut tip of each cotyledon. Bacteria can also be applied with an atomizer. Each pot of seedlings is enclosed in a clear plastic bag for 24 h, is removed after that time, and receives regular watering without splashing. After 28–32 days, the upper taproot is examined for discoloration (yellow to reddish brown streaks in the central cylinder) and assayed on D2 agar medium (Kado and Heskett, 1970). Some seedlings should show wilt symptoms. Cotyledon assays using direct inoculation have been shown to be highly efficient (95–97%) (Kreitlow, 1963; Frosheiser, 1966).

Root Assay

Seedlings prepared as described above are gently washed free of soil. They are bundled in groups of 10–15. The roots are trimmed with sterile dissecting scissors to expose freshly wounded tissue. The trimmed roots are dipped into fresh inoculum and then transplanted in potting mix. Assessment of pathogenicity is made 4 weeks after inoculation.

Pathogenicity Factors

Unfortunately, very little work has been carried out to identify novel genes that confer virulence on *Clavibacter michiganensis* subsp. *insidiosus*. The Gram-positive feature of this organism precludes the use of many useful molecular tools that have been used to elucidate virulence mechanisms in Gram-negative plant pathogens.

Southern Bacterial Wilt

Southern bacterial wilt, also referred to as bacterial wilt, Moko disease, and potato brown rot, is caused by *Ralstonia solanacearum* (previously called *Pseudomonas solanacearum* and, briefly, *Burkholderia solanacearum* [Yabuuchi et al., 1995]). Bacterial wilt is one of the most destructive plant diseases worldwide, with a high impact on growers in the tropics and warm temperate zones. It is a progressive disease that is characterized by complete wilting and frequently death of the infected plant. Originally, bacterial wilt was called brown rot disease and, in tobacco, the Granville wilt disease (see History of Bacterial Wilt below). Although endemic in subtropical and tropical regions, some strains of *Ralstonia solanacearum* have also become established in temperate regions, as in the Northern Tablelands of New South Wales, Australia, and in northern Europe (Lloyd, 1976; Olsson, 1976;

Janse, 1996). The pathogen has a highly promiscuous host range that covers more than 50 plant families (Hayward, 2000) with an increasing list of hosts (Table 7-3). However, no single strain has this large a host range, and host range varies among *Ralstonia solanacearum* isolates (Allen et al., 2005). Historically, *Ralstonia solanacearum* strains were grouped into races based loosely on host range (Buddenhagen et al., 1962). Race 1 infects many hosts, race 2 infects triploid bananas (Moko disease), *Heliconia* spp., or both. Race 3 infects potato, other members of the family Solanaceae, and several species in other families. Later, race 4, which infects ginger (Pegg and Moffett, 1971), and race 5, which infects mulberry (He et al., 1983), were added. A biovar subclassification system was also proposed, based on differences in catabolic properties (sugar utilization, acid formation), presence of nitrate reductase, and phage, serological, and bacteriocin typing (Hayward, 1964; Hayward,

Table 7-3. Host Range of *Ralstonia solanacearum* (Incomplete)

Common Name	Scientific Name	Reference
Banana	*Musa* spp.	Rorer, 1909
Belladonna	*Atropa belladonna* L.	U.S. Dep. Agric., 1960
Bell pepper	*Capsicum annuum* L.	Kelman, 1954; Hsu and Chen, 1977; He et al., 1983; Swanepoel and Young, 1988; Hsu, 1991
Bird-of-paradise	*Strelitzia reginae* Aiton	Hsu, 1991
Bittersweet nightshade	*Solanum dulcamara* L.	He et al., 1983
Castor bean	*Ricinus communis* L.	Hsu, 1991
Casuarina	*Casuarina* sp.	He et al., 1983
Comfrey	*Symphytum officinale* L.	Hsu, 1991
Cutleaf nightshade	*Solanum triflorum* Nutt.	U.S. Dep. Agric., 1960
Eggplant	*Solanum melogena* L.	Smith, 1896
Eucalyptus, gum	*Eucalyptus* sp.	Dristig and Dianese, 1990
Henbane	*Hyoscyamus niger* L.	U.S. Dep. Agric., 1960
Geranium	*Pelargonium × hortorum* L. H. Bailey	Williamson et al., 2002
Ginger	*Zingiber officinale* Roscoe	Pegg and Moffett, 1971; He et al., 1983
Jimson weed	*Datura stramonium* L.	U.S. Dep. Agric., 1960
Large thorn apple	*Datura ferox* L.	Swanepoel and Young, 1988
Mulberry	*Morus* sp.	He et al., 1983
Nettle	*Urtica nivea* L.	He et al., 1983
Olive	*Olea europaea* L.	He et al., 1983
Peanut	*Arachis hypogaea* L.	Winstead and Kelman, 1952; He et al., 1983; Swanepoel and Young, 1988; Hsu, 1991
Perilla	*Perilla crispa* (Thunb.) Tanaka	Hsu et al., 1993
Plantain	*Musa* sp.	Rorer, 1909
Potato	*Solanum tuberosum* L.	Smith, 1896
Radish	*Raphanus sativus* L.	Yang and Ho, 1998
Sesame	*Sesamum indicum* L.	He et al., 1983
Stinging nettle	*Urtica dioica* L.	Wenneker et al., 1999
Strawberry	*Fragaria chiloensis* (L.) Mill.	Hsu, 1991
Strong-spined medick	*Medicago truncatula* Gaertn.	Vailleau et al., 2007
Sugar apple	*Annona squamosa* L.	Hsu, 1991
Sunflower	*Helianthus annuus* L.	Swanepoel and Young, 1988
Sweet potato	*Ipomoea batatas* (L.) Lam.	He et al., 1983
Thale cress	*Arabidopsis thaliana* (L.) Heynh.	Deslandes et al., 1998; Yang and Ho, 1998
Tobacco	*Nicotiana tabacum* L.	Winstead and Kelman, 1952; Hsu and Chen, 1977; He et al., 1983; Xu et al., 1988; Negishi et al., 1993
Tomato	*Lycopersicon esculentum* Mill. var. *esculentum* Mill.	Smith, 1896
Townsville stylo	*Stylosanthes humilis* (Kunth) Hester	Lloyd, 1976

1994). Race 3 strains fall into biovar 2; otherwise, race and biovar do not correlate well with one another. Neither subclassification sytem was satisfactory since it was not predictive (e.g., race 1 was a very large and heterologous group), and accumulating DNA evidence showed that they did not reflect phylogenetic relationships (Cook and Sequeira, 1994). A new subclassification system based on sequences of the *egl*, *hrpB*, and *mutS* genes and the internal transcribed spacer (ITS) regions divides the *Ralstonia solanacearum* species complex into four phylotypes that correspond well with phylogenetic and geographic origin (Fegan and Prior, 2005). A single multiplex PCR reaction distinguishes among phylotypes. Phylotype I strains originate in Asia, phylotype II strains originate in the Americas, phylotype III strains originate in Africa, and phylotype IV strains (which include the as-yet-unnamed banana blood disease bacterium) originate in Indonesia. These phylotypes are further subdivided into 23 sequevars based on a 750-bp region of the *egl* (endoglucanase) gene.

Recent genomic studies suggest that the genes conferring virulence and pathogenicity are still undergoing adaptive evolutionary change. For example, numerous virulence-related genes are located in apparent genomic islands whose characteristics indicate that they were acquired via lateral gene transfer from other species (Guidot et al., 200) (see General Characteristics of *Ralstonia solanacearum* below). A comparison of the genome sequences of an Asian phylotype I strain with an American phylotype II strain showed that around one-tenth of the genes in the American strain were not present in the Asian strain, underlining the very high diversity among strains in this species complex (Gabriel et al., 2006).

General Characteristics of *Ralstonia solanacearum*

Ralstonia solanacearum is a member of the class *Betaproteobacteria* in the family *Ralstoniaceae*. It has Gram-negative, aerobic, rod-shaped cells that are motile with one to four polar flagella. Cells can generate polar fimbriae, which confer twitching motility. They form inclusion bodies of poly-β-hydroxybutyric acid that are stainable with Sudan Black B (Hayward, 1960). Colonies on complex agar medium are irregularly round, fluidal, mucoid, and cream colored. As colonies age, a brown diffusible melanin pigment is produced by some strains. Green fluorescent pigment is not produced. *Ralstonia solanacearum* produces catalase, nitrate reductase, and cytochrome oxidase. It is usually nonfastidious for growth requirements. Some strains produce a 65-kDa bacteriocin (Cuppels et al., 1978). *Ralstonia solanacearum* grows poorly in culture media containing more than 0.5% sodium chloride. No growth occurs in nutrient broth containing 2% sodium chloride. Optimal growth temperatures range between 27 and 35°C.

Because there is no direct correlation between physiological characteristics and pathogenicity, the biotype designation in the laboratory is not useful in determining the potential host range of an isolate (He et al., 1983). As stated above, intraspecific groupings (biovars) of *Ralstonia solanacearum* based on physiological criteria and host range were proposed in 2001 (Table 7-3). Genetic variability reflects the adaptiveness of *Ralstonia solanacearum* to the wide range of soil and host conditions to which this pathogen has been exposed. Indeed, *Ralstonia solanacearum* becomes highly competent to take up DNA during infection of its host (Bertolla et al., 1999). This state of competence may permit this pathogen to acquire additional genetic material through natural transformation (Bertolla et al., 1997; Coupat et al., 2008). Both plasmid and genomic DNA can be acquired through natural transformation; such acquisitions may have allowed this bacterium to take on numerous genes from other microorganisms (Fall et al., 2007). This capacity may explain *Ralstonia solanacearum*'s broad host range and adaptability.

Host Range

At least 50 different plant families representing more than 200 different plant species are susceptible to *Ralstonia solanacearum* (Hayward, 2000). A representative list of plants susceptible to *Ralstonia solanacearum* is presented in Table 7-3.

History of Bacterial Wilt

Bacterial wilt was first noted by Thomas Burrill in 1890 (Burrill, 1890). In 1896, Erwin F. Smith first described the causal organism of bacterial wilt as *Bacillus solanacearum* (Smith, 1896). Bacterial wilt was originally called brown rot disease on castor bean (Smith and Godfrey, 1918) and Granville wilt on tobacco (Smith, 1908). James Rorer described Moko disease on banana and plantain in 1909 and noted the similarity of this pathogen to *Bacillus solanacearum* (Rorer, 1909). The ensuing years revealed the presence of bacterial wilt worldwide in semitropical and tropical climatic regions.

Symptomatology

As indicated by its name, bacterial wilt (or Southern wilt) is recognized by the appearance of wilted foliage of the infected host. Wilting can occur overnight or gradually, following wilting of leaves on one branch or one side of a plant. Compared with uninfected plants, diseased plants are often stunted. Tubers, corms, and enlarged roots of infected plants are often found decayed when unearthed. Most diagnostic for bacterial wilt is the formation of brownish vascular tissues in infected stems. Milky bacterial ooze can be forced out from a cut end of a diseased stem and streams from cut stems placed in water. A small portion of the ooze can be streaked over a semiselective agar medium, such as SMSA, with certain modifications for isolation of *Ralstonia solanacearum*; the pathogen grows relatively slowly, with colonies appearing in 36–48 h at 28°C (Denny and Hayward, 2001).

Diagnosis

Wilting symptoms, vascular browning, and bacterial streaming from stems are enough for a preliminary diagnosis of bacterial wilt. *Ralstonia solanacearum* can be detected and identified using traditional culturing methods followed by biochemical and virulence tests (Denny and Hayward, 2001). Several DNA-based PCR primer pairs have been developed that identify the pathogen to species or, in some cases, to subgroup (Weller et al., 2000; Fegan and Prior, 2005). A commercially available rapid immunostrip test based on an anti-EPS monoclonal antibody can identify infected plants in the field in 10 min without equipment.

Management

Control of bacterial wilt is difficult. Planting resistant hosts is the best method, but breeders have had only moderate success at developing agronomically desirable resistant cultivars of most important hosts (Boshou, 2005). Because *Ralstonia solanacearum* forms latent infections, it is imperative to plant pathogen-free cuttings of vegetatively propagated hosts, such as potato, ginger, banana, and plantain. Keeping uninfested fields free of the pathogen is also important, and cultural practices, such as tool sanitation and control of irrigation waters, can reduce incidence. Soil fumigation is neither environmentally desirable nor effective. Rotations of less than 5 years have limited impact under most conditions. Biological control and alternative disease management practices, such as fumigation with plant essential oils, have had some success on experimental scales (Saddler, 2005).

Disease Cycle

Under warm, moist conditions, *Ralstonia solanacearum* can persist in loamy and sandy soils for extended periods. High populations of *Ralstonia solanacearum* can persist as long as 4 years (McCarter, 1976). This pathogen survives well in the rhizosphere of weeds (Devi et al., 1981). *Ralstonia solanacearum* persists in soils when corn, cotton, and peanut are used for rotational crops in fields where Southern wilt (Granville wilt) is a problem. Soybean or small grains are good alternative rotational crops.

Motility and chemotaxis are required for efficiently locating and entering host roots (Tans-Kersten et al., 2001; Yao and Allen, 2006). Wounds and natural openings serve as portals for penetrating beyond the root surface into the cortex. Bacterial wilt is synergistically increased by the presence of a root-knot nematode, probably because it creates wounds facilitating pathogen entry (Hayward, 1994). Infection of the vascular parenchyma and invasion of the xylem takes place soon after. When bacterial cell density exceeds 10^7 cells/ml, *Ralstonia solanacearum* begins producing acidic EPS and β-1,4-endoglucanase (Clough et al., 1997), materials that are not required for bacterial invasion but appear to promote colonization of the plant (Saile et al., 1997). A quorum-sensing system mediated by the volatile extracellular compound 3-hydroxy-palmitic acid methyl ester is produced at high cell density and regulates several virulence factors in *Ralstonia solanacearum*, including EPS, endoglucanase, motility, and a type III secretion apparatus (Clough et al., 1994; Flavier et al., 1997). *Ralstonia solanacearum* readily forms latent infections, in which plants are symptomless but contain bacteria at 10^5–10^8 CFU/g; both latently infected and symptomatic plants with active infections shed abundant numbers of bacteria from their roots, increasing the likelihood of further infections (Kelman and Sequeira, 1965; Swanson et al., 2005).

Pathogenicity and Virulence Tests

Eggplant (*Solanum melongena* L. cv. Black Beauty) can be used for testing the pathogenicity of *Ralstonia solanacearum* (Xu et al., 1988). Eggplant seedings are propagated from surface-sterilized seeds on agar medium containing Hoagland mineral salts in test tubes. Seedlings are grown for 2 weeks at 28°C under fluorescent lights. The primary leaves of the seeding are inoculated at freshly made wounds inflicted with a scalpel, and the cut edge is rubbed with a cotton swab saturated with a heavy suspension (10^9 cells/ml) of *Ralstonia solanacearum*. After 4–5 days of incubation in a growth chamber, completely withered or wilted seedlings signify a pathogenic strain.

Tobacco and tomato seedlings can also be used for pathogenicity tests. Here, a sharp needle is forced into the stem through a drop of inoculum placed in the axil of the second or third expanded leaf below the apex of 4-week-old plants (Winstead and Kelman, 1952). Care must be taken to use susceptible cultivars of these test plants. Inoculated test plants are maintained under warm glasshouse conditions for 6–10 days and scored for infection.

A more naturalistic quantitative virulence assay uses inoculation of unwounded roots (Winstead and Kelman, 1952; Trigalet and Demery, 1986; Tans-Kersten et al., 2001). Seedlings about 10–15 cm tall are inoculated by adding 30 ml of inoculum (10^9 cells/ml) to each pot. Inoculated plants are grown at 28°C in a glasshouse or growth chamber and scored after 6–12 days for flagging and wilt symptoms.

Serial culturing of *Ralstonia solanacearum* in broth culture results in loss of virulence (Kelman and Hruschka, 1973). Smith noted in 1896 that maintaining virulence in the laboratory is troublesome. Pathogenic bacteria can be grown on tetrazolium chloride indicator agar medium at 28°C (Kelman, 1954). Virulent isolates form irregular, fluidal, cream-colored colonies with pink or red centers that can be readily distinguished from the small, round, dark red, butyrous colonies of avirulent or weakly virulent isolates that do not produce EPS (Kelman, 1954).

Serological tests can be employed in the detection of Hrp-secreted proteins such as PopA and PopC. A modified *hrp*-derepressing minimal culture medium (Huynh et al., 1989) with 10 mM L-glutamic acid and 10 mM sucrose added is used in obtaining Hrp-secreted proteins from *Ralstonia solanacearum* (Tamura et al., 2005).

Pathogenicity Factors

Ralstonia solanacearum has two chromosomes of 3.7 and 2.1 Mb (Coenye and Vandamme, 2003). Both chromosomes carry genes required for virulence and for ingression and survival in the host plant, but it has been suggested that the smaller replicon (known as the megaplasmid) carries a disproportionate share of the genes encoding for interactions with plant hosts (Salanoubat et al., 2002). For example, a cluster of genes encoding a type III secretion system, designated the *hrp* cluster, are located on the megaplasmid (Boucher et al., 1986).

A megaplasmid of about 1,000 kb is harbored in most *Ralstonia solanacearum* strains regardless of race or geographical origin (Rosenberg et al., 1982). Transposon Tn*5* insertional mutagenesis to screen for virulence genes of *Ralstonia solanacearum* revealed the fact that most of the virulence genes, designated the *hrp* cluster, are located on the megaplasmid (Boucher et al., 1986; Boucher et al., 1987). Plasmids that are much smaller than the megaplasmid can also be harbored in *Ralstonia solanacearum* strains. Apparently, the smaller plasmids do not play a direct role in virulence (Currier and Morgan, 1981; Morales and Sequeira, 1985). However, a 6.6-kb plasmid, pJTPS1, confers loss of virulence in pathogenic strain U-7, converting it into a nonfluidal colony-forming type that also has reduced levels of endoglucanase and EPS production (Negishi et al., 1993). The mechanism of this plasmid-mediated virulence suppression is unknown. Cross talk between plasmid genes and chromosomal genes appear to be operating in this *Ralstonia solanacearum* strain. Plasmid and chromosomal virulence gene products are known to communicate with each ether (Kado, 1992).

The broad host range of *Ralstonia solanacearum* suggests that pathogenicity effectors have broad substrate specificities. The following components produced by *Ralstonia solanacearum* play a role in virulence.

EPS

All virulent strains of *Ralstonia solanacearum* produce a high-molecular-weight slime composed of an unbranched polymer of *N*-acetylgalactosamine, *N*-acetylgalactosaminuronic acid, and *N*-acetylbacillosamine decorated with 3-hydroxybutyric acid (Orgambide et al., 1991). A portion of the unusual acidic EPS identified as EPS I is not required for invasive growth in planta but is apparently involved in blocking vascular transport of water (Denny et al., 1990). The biosynthesis of EPS I is encoded by the 16-kb cluster of *eps* genes, and *eps*$^{-}$ mutants almost never wilt or kill plants, although they can weakly colonize roots and lower stems (Saile et al., 1997; McGarvey et al., 1999). Strains with mutations in the EPS I gene cluster reduce EPS production but retain virulence that is lower than that of the wild type (Kao et al., 1992). The biosynthesis of EPS I requires the presence of the 16-kb cluster of *eps* genes. The EPS was shown to reduce considerably the fimbriae-mediated agglutination of *Ralstonia solanacearum* with tobacco cell wall preparations (Young and Sequeira, 1986).

There is an interesting observation that a mutation in a chromosomal gene called *ros* in *Agrobacterium tumefaciens* causes the loss of acidic EPS without the loss of virulence. However, the *ros* mutation causes an unusual sensitivity of the mutant to low temperatures (Close et al., 1987; Cooley et al., 1991). It is quite possible that besides its ancillary role in virulence, EPS I is required for the protection of *Ralstonia solanacearum* against low temperatures and environmental stresses, such as free radicals, drought, and high temperatures. In other words, EPS I may be a capsular material required for suitable habitat survival.

Pectinases and Cellulases

To facilitate the degradation of cell wall-cementing materials and the cell wall itself, *Ralstonia solanacearum* strains produce β-1,4-endoglucanase (Egl), endopolygalacturonases (PehA, also called PglA), exopolygalacturonases (PehB and PehC), pectin methylesterase (PmeA), and cellulases (CelA). Egl, PehA, PehB, and PehC each contribute quantitively to the wilting phenotype of *Ralstonia solanacearum* (Denny et al., 1990; Huang and Allen, 2000; González and Allen, 2003). Apparently, Pme has no direct effect on virulence (Tans-Kersten et al., 1998). Interestingly, a mutant defective in PehA, PehB, and PehC enzyme production had increased virulence, suggesting that pectin oligomers may trigger plant defense responses (González and Allen, 2003).

Plant Growth Substances

Although plant-pathogenic bacteria that cause abnormal growths require auxin and cytokinin synthesis in order to be virulent (see Chapter 8), the role of auxin, cytokinin, and ethylene in virulence of *Ralstonia solanacearum* remains unclear. *Ralstonia solanacearum* produces these plant growth compounds (Freebairn and Buddenhagen, 1964; Phelps and Sequeira, 1968; Akiyoshi et al., 1987). They may be important effectors for establishing initial infection or in preparing plant tissue substrates for efficient invasion, but at present, the role of plant growth factors in bacterial wilt remains unclear.

Type II-Secreted Virulence Effectors

Ralstonia solanacearum is equipped with a type II secretion apparatus that secretes EPS and plant cell wall-degrading enzymes and is absolutely required for virulence (Kang et al., 1994); this is similar to the situation in many members of the family *Enterobacteriaceae* (see Chapter 10). Recent studies implicate a variety of additional, unidentified type II-secreted effectors that contribute to virulence and colonization of the pathogen (Liu et al., 2005).

Type III Secretion

Transposon mutagenesis followed by screening for complete loss of virulence identified a large cluster of genes absolutely required for pathogenicity on host plants and induction of the HR on nonhost plants (known as the *hrp* cluster) (Boucher et al., 1987). Sequence analyses and genetic experiments determined that this cluster encodes synthesis and function of a type III secretion system that enables the pathogen to directly inject virulence effectors into the host cell (Gough et al., 1992). The precise functions of most of the individual effectors remain unknown, but results with paralogous proteins from other systems suggest that most type III effectors suppress host defense responses (Jones and Dangl, 2006). These effectors appear to be functionally redundant since virulence is unaffected by mutation of individual candidate effector genes and mutation of three at once also had no discernable effect (Genin and Boucher, 2004). Expression of the *Ralstonia solanacearum* type III secretion system is controlled by a complex environmentally responsive regulatory network that responds to direct contact with plant cells, to low-nutrient environments, and to population density via the Phc regulatory network (Schell, 2000; Genin et al., 2005; Denny, 2006).

Virulence Genes and Their Products

Using molecular genetic approaches, genes encoding many virulence factors of *Ralstonia solanacearum* have been identified (Table 7-4). In addition to genes for production of EPS I, motility, and the cell wall-degrading enzymes described above, they include genes involved in recognition of the host, type III secretion, suppression of host defenses, suspected effectors of unknown function, and a broad array of regulators that control expression of virulence factors.

The genes expressing the virulence and pathogenicity factors are regulated by an elaborate network of distinct, interacting sensory-signal transduction pathways that include the above *phc* gene products that regulate the production of cell wall-degrading enzymes and EPS (Schell, 2000). PopP2 is an *Ralstonia solanacearum* type III effector protein that belongs to the YopJ/AvrRxv protein family. This effector interacts with the *Arabidopsis thaliana* (L.) Heynh. RRS1-R (resistant to *Ralstonia solanacearum* 1-R) resistance protein. This resistance protein contains the Toll/Interleukin1 receptor-nucleotide binding site that is rich in leucine repeats and contains a C-terminal WRKY DNA-binding site. PopP2 interacts with a cysteine protease (RD19) that is responsive to dehydration. RD19 is required for RRS1-R-mediated resistance and localizes in mobile vacuole-associated compartments but, upon coexpression with PopP2, RD19 is specifically relocalized to the plant nucleus, where the two proteins physically interact. RD19 apparently associates with PopP2

Table 7-4. Virulence Genes and the Functions of Their Products in *Ralstonia solanacearum*

Gene	Function of Encoded Protein	Substrate or Target	Reference
acrAB	Multidrug efflux pump	Antimicrobial compounds generated by the host	Brown et al., 2007
dinF	Multidrug efflux pump	Antimicrobial compounds generated by the host	Brown et al., 2007
egl	A 43-kDa β-1,4-endoglucanase Egl	Soluble cellulose	Schell, 1987; Roberts et al., 1988
eps	Extracellular polysaccharide biosynthesis, EPS I cluster	Excreted surface component involved in colonization	Denny and Baek, 1991; Kao et al., 1992; Flavier et al., 1997
fliC	Swimming motility	A 28.3-kDa FliC flagellar subunit	Tans-Kersten et al., 2001
hrpA	A 60.4-kDa protein of type III secretion apparatus	Encodes Hrp pilin	Gough et al., 1992
hrpB	A 53.3-kDa AraC family member, similar to XylS of *Pseudomonas putida*	AraC-type transcriptional activator; activates the synthesis of 3-hydroxy-oxindole; regulates expression of the type III secretion system; secretes effectors and other genes of unknown function	Genin et al., 1992; Mukaihara et al., 2004; Tamura et al., 2005; Delaspre et al., 2007
hrpG	OmpG subclass of two-component response regulators, resembles *hrpG* of *Xanthomonas vesicatoria*	Regulates expression of the type III secretion system in response to nutrient status and cell contact	Brito et al., 1999; Brito et al., 2002
hrpH (now *hrcC*)	Part of the type III secretion system	Membrane	van Gijsegem et al., 1995
hrpI	A 28.9-kDa lipoprotein of the type III secretion apparatus	Membrane	Gough et al., 1992
hrpO	A 74-kDa protein of the type III secretion apparatus	Membrane	Gough et al., 1992
opsA-D	Lipopolysaccharide biosynthesis	Rhamnose, glucose, heptose, 2-keto-deoxy-octonate	Kao and Sequeira, 1991
pehA or *pglA*	A 52.4-kDa endopolygalacturonase	Pectate	Schell et al., 1994; Yoder and Schell, 1995; Liu et al., 2005
pehB	Exopolygalacturonase	Polygalacturonate	Huang and Allen, 1997
phcA	Member of the LysR family of transcriptional regulators that regulates *eps, egl, pilA* expression	Quorum-responsive regulation of *eps, egl, pilA*	Brumbley and Denny, 1990; Brumbley et al., 1993; Genin et al., 2005
phcB	S-Adenosyl methionine-dependent methyltransferase	3-OH-PAME synthesis	Flavier et al., 1997
pilA	Type 4 pilin	Twitching motility	T. Denny cited by Genin and Boucher, 2002
pme	A 41-kDa pectin methylesterase	Pectin, provides oligosaccharides for nutrition	Spök et al., 1991; Tans-Kersten et al., 1998
popA	Hypersensitive response elicitor secreted by type III Hrp secretion pathway	Lipid target through a calcium-mediated oligomerization leading to membrane pore formation	Arlat et al., 1994; Racapé et al., 2005; Kanda et al., 2008
popB	A 18-kDa protein with a functional bipartite nuclear localization signal	Apparently modulates host gene expression	Guéneron et al., 2000
popC	A 111-kDa protein with 22 tandem leucine-rich repeats	Unknown function and substrate	Guéneron et al., 2000
popP1, popP2, popP3	Cysteine proteases	Suppression of host defense	Lavie et al., 2004; Bernoux et al., 2008

(continued)

Table 7-4. Virulence Genes and the Functions of Their Products in *Ralstonia solanacearum* (continued)

Gene	Function of Encoded Protein	Substrate or Target	Reference
rpoS	Sigma factor involved in the acyl homoserine lactone-dependent autoinduction system	RNA polymerase alternate sigma factor	Flavier et al., 1998
vsrA	A 53-kDa membrane-associated histidine kinase that regulates *eps* expression	Two-component virulence regulator	Schell et al., 1993
vsrB	A 63-kDa membrane-associated histidine kinase regulates *pglA* and *eps* expression	Two-component virulence regulator	Huang et al., 1993; Schell et al., 1994
vsrC	A 24-kDa response regulator	Regulates *eps* and other *vir* genes, including colonization	Huang et al., 1995
vsrD	A 24-kDa response regulator	Regulates *eps*, colonization	Huang et al., 1995
xpsR	Very basic (pI = 10.6) 35-kDa regulatory protein	Indirectly controls the *eps* promoter via the VsrA/VsrD and PhcA signal transduction systems	Huang et al., 1995

to form a nuclear complex that is required for activation of the RRS1-R-mediated resistance response (Bernoux et al., 2008).

Stewart's Wilt of Corn

Stewart's wilt is a destructive disease principally of sweet corn and to a lesser degree of popcorn or dent corn. The disease occurs in the eastern and southern corn belt regions of the United States. Introductions of resistant cultivars of corn have reduced the incidence of Stewart's wilt. *Pantoea stewartii* subsp. *stewartii* is the causal organism of the disease. The pathogen is spread by the corn flea beetle and is seedborne.

General Characteristics of *Pantoea stewartii* subsp. *stewartii*

Pantoea stewartii subsp. *stewartii* was originally named *Pseudomonas stewarti* after Fred Carlton Stewart. Difficulty with the taxonomic placement of this bacterium is reflected in numerous name changes in the genus from *Pseudomonas* to *Bacterium* to *Aplanobacter* to *Phytomonas* to *Xanthomonas* to *Pseudobacterium* to *Erwinia* and finally to *Pantoea* in 1993.

Pantoea stewartii subsp. *stewartii* is a member of the class *Gammaproteobacteria* in the family *Enterobacteriaceae*. Its Gram-negative cells are motile by means of peritrichous flagella. It is a facultative anaerobe that produces a capsular polysaccharide called stewartan, whose structure has been defined (Yang et al., 1996). The bacterium produces acid from arabinose, lactose, sucrose, galactose, glucose, fructose, mannose, ribose, melibiose, raffinose, xylose, trehalose, mannitol, and sorbitol. It utilizes organic acids, such as acetate, fumarate, gluconate, malate, and succinate. It grows best at 28°C. It produces catalase and nitrate reductase but not cytochrome oxidase.

Pantoea stewartii subsp. *stewartii* produces a water-insoluble, yellow, carotenoid pigment. Various strains of this bacterium harbor a large number of cryptic plasmids (Coplin et al., 1981). Most virulent strains analyzed contained 11–13 cryptic plasmids of 2.8–210 MDa. Avirulent strains appear to harbor fewer plasmids than do virulent strains. Plasmids are needed to confer habitat enhancement on the bacterial cell (Kado, 1998).

Host Range

Although popcorn, flint, dent, and flour corn cultivars are susceptible to *Pantoea stewartii* subsp. *stewartii* infection, sweet corn cultivars are the primary hosts. The pathogen can also infect eastern gamagrass (*Tripsacum dactyloides* (L.) L.) and Mexican teosinte (*Zea mexicana* (Schrad.) Kuntze). Weed members of the family Poaceae can be symptomless hosts of the pathogen and can serve as a reservoir of inoculum.

History of Stewart's Wilt

Stewart's wilt was first described in the late 1800s on Long Island, New York, by Fred Carlton Stewart, a student of W. G. Farlow. In February 1897, Stewart isolated the causal organism from sweet corn and described the disease in a New York Agricultural Experiment Station bulletin (Bulletin 130) in December 1897. In March 1898, the pathogen was sent to E. F. Smith, who named the organism *Pseudomonas stewarti* (Smith, 1898). On November 7, 1923, Smith indicated in a letter to Paul Moore of the National Research Council that Frederick V. Rand had shown that Stewart's wilt is transmitted by a beetle (Smith, *unpublished*). Stewart's wilt has appeared worldwide in sweet corn-producing regions.

Symptomatology

Vector-borne transmission of *Pantoea stewartii* subsp. *stewartii* results in the formation of lesions at and around the feeding scars left by the flea beetle. The lesions appear as water-soaked tissues that turn light green to yellow and extend along the vascular bundle. As the lesions grow in size, they coalesce with one another, forming large necrotic streaks. As the bacterial cells colonize and proliferate systemically in the vascular elements, portions of the plant begin to wilt, particularly under water-stress conditions. Plants unable to recover from water deficiency eventually die. Infected plants are dwarfed and produce bleached tassels. Inspection of the cross section of stalks from diseased plants reveals yellowish bacterial ooze appearing at the vascular bundles. Necrotic tissues along the vasculature show various degrees of browning. Interestingly, fertilization with high ammoniacal nitrogen and phosphate can increase susceptibility of the corn plant to Stewart's wilt.

Disease Cycle

Pantoea stewartii subsp. *stewartii* is transmitted by the corn flea beetle (*Chaetocnema pulicaria* Melsheimer), a coleopteran member of the family Chrysomelidae. The beetle introduces the pathogen into the xylem and intercellular spaces of corn seedlings. The bacterial pathogen colonizes the vascular elements, causing vascular occlusion of the pit membranes, apparently due to the accumulation of EPS (Braun, 1982), including the EPS called stewartan (Yang et al., 1996). *Pantoea stewartii* subsp. *stewartii* overwinters in the hibernating flea beetle. An infested beetle harbors and transmits the bacterial pathogen during its entire life. The toothed flea beetle (*Chaetocnema denticulata* (Illiger)), the southern corn rootworm (*Diabrotica undecempunctata howardi* Barber), the larvae of the seed corn maggot (*Delia platura* (Meigen)), the wheat wireworm (*Agriotes mancus* (Say)), corn rootworm (*Diabrotica longicornis* (Say)), and white grubs (*Phyllophaga* spp.) can also transmit Stewart's wilt but less efficiently.

Pantoea stewartii subsp. *stewartii* is also transmitted by seed. Seedlings from infested seed are colonized and the infected plants become a primary source of inoculum. Inbred seed parents with high levels of resistance to Stewart's wilt are unlikely to transmit *Pantoea stewartii* subsp. *stewartii* through seed because systemic movement of the pathogen in the host is restricted (Khan et al., 1996). Besides residence in flea beetles, the bacteria can overwinter in corn stalks. The degree of disease incidence can be predicted for the forthcoming season by the severity of the average winter temperature. When the sum of the average temperatures of December, January, and February is less than 20–24°C, the incidence of Stewart's wilt disease is low. If the sum of the average temperatures for the winter months is high, e.g., 32–38°C, the disease incidence is likely to be severe. Larger populations of hibernating flea beetles are killed at the lower temperatures, reducing the inoculum potential for the forthcoming season.

Pathogenicity Tests

Pathogenicity tests are made on sweet corn seedlings. Seeds of a susceptible cultivar are sown in potting mix in 15- to 20-cm pots. Seedlings are grown at 28°C and inoculated after 8–10 days of growth. The tops of the seedlings are cut with dissecting scissors and a suspension of freshly grown bacteria (10^7–10^8 CFU/ml) is pipetted onto the fresh cut stalk. After 10–14 days, the plants are assessed. Alternatively, inoculum of the same concentration is prepared in 10 mM potassium phosphate buffer, pH 7.0, containing 0.2% Tween 40 and deposited in the whorls of the seedlings (Coplin et al., 2002). After 10–14 days, the inoculated plants are examined for lesions and bacterial ooze formation.

Tobacco plants (*Nicotiana tabacum* L. cv. Wisconsin-38) also can be used to assay HR elicitation. A fresh suspension of *Pantoea stewartii* subsp. *stewartii* cells, adjusted to a concentration of 10^8 CFU/ml, is infiltrated into the underside of an attached tobacco leaf. Infiltration is performed with a needleless syringe pressed lightly to the epidermis, allowing the injection of the bacterial suspension into the interveinal regions of the leaf. A water-soaked appearance is readily observed between the leaf epidermis and parenchyma. HR development is assessed after 24 h of incubation in the glasshouse.

Pathogenicity Factors

Type III Secretion and Water-Soaking

Pantoea stewartii subsp. *stewartii* harbors a *hrp/wts* gene cluster that is required for lesion formation and wilting on corn (Coplin et al., 1992). A number of genes encoding pathogenicity factors are located in the *hrp/wts* cluster (Fig. 7-5). The genes of this cluster have sequence homology with *hrp* genes of *Erwinia amylovora* (see Chapter 5) and *Pantoea agglomerans* pv. *gypsophilae* (see Chapter 8). The genetic organization and regulation also are similar among these two organisms. Transposon mutations in the *hrp* operon abolished pathogenicity and HR elicita-

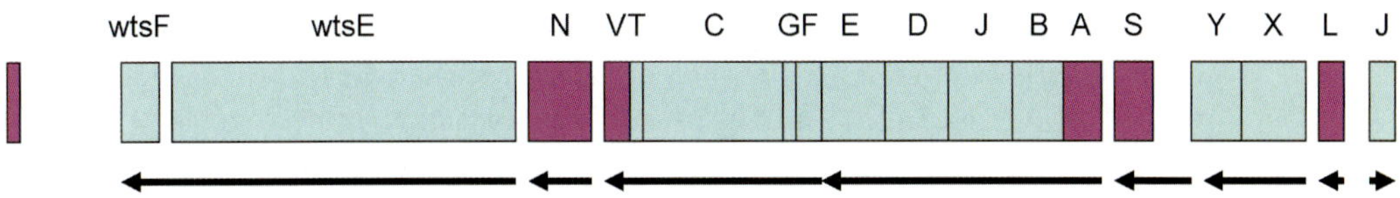

Fig. 7-5. Arrangement of open reading frames of the 28-kb *hrp/wts* cluster that is required for pathogenicity and hypersensitive response elicitation by *Pantoea stewartii* subsp. *stewartii.* The *hrp* genes are designated in capital letters. The direction of transcription is indicated by the arrows. Open reading frames that encode pathogenicity traits are shaded in blue. (Adapted from Frederick et al., 2001)

tion, except for *hrpN* and *hrpV* mutants, which remained pathogenic (Frederick et al., 2001). *wtsE* encodes a 201-kDa protein that is similar to DspA/E in *Erwinia amylovora* and to AvrE in *Pseudomonas syringae* pv. *tomato.* This protein appears to play a major role in water-soaking activity. WtsE induces cell death in *Nicotiana benthamiana*, tobacco, beet, and *Arabidopsis thaliana* when it is transiently produced in plant cells following agroinfiltration or is translocated into plant cells from *Pantoea stewartii* subsp. *stewartii*, *Escherichia coli*, or *Pseudomonas syringae* pv. *phaseolicola*. WtsE-induced cell death in *Nicotiana benthamiana*, tobacco, and beet resembled an HR. The HR in *Nicotiana benthamiana* was delayed by cycloheximide (Ham et al., 2008). The mechanism of action by DspA/E to trigger water-soaking is unknown. The 198-kDa DspA/E of *Erwinia amylovora* interacts with host receptor kinases bearing leucine-rich repeats of apple (Meng et al., 2006).

The *hrpN* gene encodes a 39.5-kDa protein that appears as a 43-kDa band in acrylamide gel electrophoretograms (Ahmad et al., 2001). Interestingly, purified HrpN has no visible adverse effects on cells of the susceptible host plant and yet is able to induce HR in tobacco and Jimson weed. HrpN from *Pantoea stewartii* subsp. *stewartii* and from *Pantoea agglomerans* pv. *gypsophilae* is not essential for pathogenicity (Ahmad et al., 2001; Mor et al., 2001). The remaining *hrp* genes encode the type III secretion system that is required for virulence (Frederick et al., 2001). The *hrp* secretion and *wts* genes are controlled by an HrpX/HrpY two-component signal transduction system that maps to the right of the *hrp* cluster (Fig. 7-5). HrpY activates *hrpS*, which in turn activates *hrpL*, and HrpL controls expression of the remaining *hrp* and *wts* genes.

EPS (Stewartan)

In addition to WtsE involved in water-soaking, EPS seem to play a role in virulence of *Pantoea stewartii* subsp. *stewartii.* In particular, the EPS called stewartan appears to be directly involved, since mutations affecting stewartan formation cause loss of wilting and systemic ingression (Coplin and Majerczak, 1990). Stewartan may assist in the mobilization of the pathogen in the vascular system of the host by preventing attachment to the cell walls of the vessel elements (von Bodman et al., 2003). In addition, stewartan has homologies to polysaccharide biosynthesis motifs in *Sinorhizobium meliloti*, *Ralstonia solanacearum*, *Erwinia amylovora*, and *Xanthomonas campestris*, whose function may be to provide protection against natural oxidants (Yildiz and Schoolnik, 1999). Stewartan is synthesized by *cps* (capsular polysaccharide synthesis) genes arranged in the order *cpsA*, *cpsB*, *cpsI*, *cpsC*, *cpsD*, *cpsE*, *cpsF*, *cpsG*, *cpsH*, *cpsJ*, *cpsK*, *cpsL*, and *cpsM* (Coplin et al., 1996). Interestingly, both *Erwinia amylovora* and *Pantoea stewartii* subsp. *stewartii* form related EPS capsules, which are the products of the *ams* and *cps* gene clusters, respectively. Sequencing of the 18.5-kb *cps* cluster revealed CpsA, CpsB, CpsI, CpsC, CpsH, CpsJ, CpsK, CpsL, and CpsM, proteins that have more than 59% identity (>80% similarity) to the products of *ams* genes of *Erwinia amylovora*. The sequence data suggest that CpsA is a galactose transferase that joins this moiety to a lipid carrier, while CpsB and CpsC appear to be integral membrane proteins that may be involved in EPS export. CpsD, CpsE, CpsG, and CpsK look like glycosyl transferases. CpsI has homologies to acid phosphatases.

The *cps* genes are regulated by a two-component signal transducing (phosphorelay) system with a sensory kinase called RcsC that detects environmental signals and transduces the signal by phosphorylating a response regulator called RcsB. The nature of the environmental signals is unknown. Phosphorylation converts RcsB into a transcriptional activator of the *cps* genes. As with many two-component regulatory systems, e.g., ChvE in the *Agrobacterium tumefaciens* two-component regulation of virulence genes (see Chapter 8), signal perception and transduction are enhanced by a second protein identified in *Pantoea stewartii* subsp. *stewartii* as RcsA.

Superimposed on the two-component regulation, EPS synthesis is also regulated by a quorum-sensing cascade (von Bodman and Farrand, 1995; von Bodman et al., 2003). An autoinducer, acyl homoserine lactone (AHL) in the form of *N*-(3-oxo-hexanoyl)-L-homoserine lactone (3-oxoC6), is constitutively synthesized by EsaI encoded by a chromosomal *esaI* gene. AHL is required for EPS synthesis and virulence as indicated by the loss of virulence (wilting and water-soaking) by an *esaI* mutation (von Bodman et al., 1998). RcsA expression is negatively regulated by a DNA-binding repressor, EsaR, that binds within the *rcsA* promoter and autoregulates its own gene (Minogue et al., 2005). EsaR does not bind to the *cspA* promoter region.

Given the interrelationship of two-component signal transduction with quorum sensing in *Pantoea stewartii* subsp. *stewartii*, it would not be at all surprising to find that this pathogen harbors multiple quorum-sensing systems that act in parallel, one of which activates the expression of small RNAs called quorum regulatory RNA (Qrr) that regulate quorum sensing by binding to a quorum-sensing master regulator (Lenz et al., 2005). The level of pathogenic complexity of *Pantoea stewartii* subsp. *stewartii* appears very high since it targets a specific type of host, colonizes a specialized niche (vascular system) of the host, and persists and is vectored by the flea beetle.

REFERENCES

Ahmad, M., Majerczak, D. R., Pike, S., Hoyos, M. E., Novacky, A., and Coplin, D. L. 2001. Biological activity of harpin produced by *Pantoea stewartii* subsp. *stewartii*. Mol. Plant-Microbe Interact. 14:1223-1234.

Akiyoshi, D. E., Regier, D. A., and Gordon, M. P. 1987. Cytokinin production by *Agrobacterium* and *Pseudomonas* spp. J. Bacteriol. 169:4242-4248.

Alarcón, C., Castro, J., Muñoz, F., Arce-Johnson, P., and Delgado, J. 1998. Protein(s) from the gram-positive bacterium *Clavibacter michiganensis* subsp. *michiganensis* induces a hypersensitive response in plants. Phytopathology 88:306-310.

Allen, C., Prior, P., and Hayward, A. C. 2005. Bacterial Wilt Disease and the *Ralstonia solanacearum* Species Complex. American Phytopathological Society, St. Paul, MN.

Alvarez, A. M., Kaneshiro, W. S., and Vine, B. G. 2005. Diversity of *Clavibacter michiganensis* subsp. *michiganensis* populations in tomato seed: What is the significance? Acta Hortic. 695:205-213.

Anonymous. 1941. Known distribution, by states, of bacterial ring rot (*Phytomonas sepedonica*) on potato in the United States in 1940. Plant Dis. Rep. 25:130.

Arlat, M., van Gijsegem, F., Huet, J. C., Pernollet, J. C., and Boucher, C. A. 1994. PopA1, a protein which induces a hypersensitivity-like response on specific *Petunia* genotypes, is secreted via the Hrp pathway of *Pseudomonas solanacearum*. EMBO J. 13:543-553.

Baer, D., and Gudmestad, N. C. 1995. In vitro cellulolytic activity of the plant pathogen *Clavibacter michiganensis* subsp. *sepedonicus*. Can. J. Microbiol. 41:877-888.

Baer, D., Mitzel, E., Pasche, J., and Gudmestad, N. C. 2001. PCR detection of *Clavibacter michiganensis* subsp. *sepedonicus*-infected tuber samples in a plate capture assay. Am. J. Potato Res. 78:269-277.

Balaji, V., Mayrose, M., Sherf, O., Jacob-Hirsch, J., Eichenlaub, B., Iraki, N., Manulis-Sasson, S., Rechavi, G., Barash, I., and Sessa, G. 2008. Tomato transcriptional changes in response to *Clavibacter michiganensis* subsp. *michiganensis* reveal a role for ethylene in disease development. Plant Physiol. 146:1797-1809.

Bentley, S. D., Corton, C., Brown, S. E., Barron, A., Clark, L., Doggett, J., Harris, B., Ormond, D., Quail, M. A., May, G., Francis, D., Knudson, D., Parkhill, J., and Ishimaru, C. A. 2008. Genome of the actinomycete plant pathogen *Clavibacter michiganensis* subsp. *sepedonicus* suggests recent niche adaptation. J. Bacteriol. 190:2150-2160.

Bergey, D. H., Harrison, F. C., Breed, R. S., Hammer, B. W., and Huntoon, F. M. 1923. Bergey's Manual of Determinative Bacteriology, 2nd ed. Williams and Wilkins Co., Baltimore, MD.

Bermpohl, A., Dreier, J., Bahro, R., and Eichenlaub, R. 1996. Exopolysaccharides in the pathogenic interaction of *Clavibacter michiganensis* subsp. *michiganensis* with tomato plants. Microbiol. Res. 151:391-399.

Bernoux, M., Timmers, T., Jauneau, A., Brière, C., de Wit, P. J., Marco, Y., and Deslandes, L. 2008. RD19, an Arabidopsis cysteine protease required for RRS1-R-mediated resistance, is relocalized to the nucleus by the *Ralstonia solanacearum* PopP2 effector. Plant Cell 20:2252-2264.

Bertolla, F., van Gijsegem, R., Nesme, X., and Simonet, P. 1997. Conditions for natural transformation of *Ralstonia solanacearum*. Appl. Environ. Microbiol. 63:4965-4968.

Bertolla, F., Frostegård, A., Brito, B., Nesme, X., and Simonet, P. 1999. During infection of its host, the plant pathogen *Ralstonia solanacearum* naturally develops a state of competence and exchanges genetic material. Mol. Plant-Microbe Interact. 12:467-472.

Bodine, E. W., and Durrell, L. W. 1941. Plant diseases in Colorado in 1941. Plant Dis. Rep. 25:484-485.

Bonde, R. 1937. A bacterial wilt and soft rot of the potato in Maine. Phytopathology 27:106-108.

Bonde, R. 1942. Ring rot in volunteer plants. Am. Potato J. 19:131-133.

Boshou, L. 2005. A broad review and perspective on breeding for resistance to bacterial wilt. Pages 225-238 in: Bacterial Wilt Disease and the *Ralstonia solanacearum* Species Complex. C. Allen, P. Prior, and A. C. Hayward, eds. American Phytopathological Society, St. Paul, MN.

Boucher, C., Martinel, A., Barberis, P., Alloing, G., and Zischek, C. 1986. Virulence genes are carried by a megaplasmid of the plant pathogen *Pseudomonas solanacearum*. Mol. Gen. Genet. 205:270-275.

Boucher, C. A., van Gijsegem, F., Barberis, P. A., Arlat, M., and Zischek, C. 1987. *Pseudomonas solanacearum* genes controlling both pathogenicity on tomato and hypersensitivity on tobacco are clustered. J. Bacteriol. 169:5626-5632.

Braun, A. C. 1949. The role of thiamin and certain related compounds in pigment production by *Corynebacterium michiganense* (Erw. Smith) Jensen. Phytopathology 39:171-172.

Braun, E. J. 1982. Ultrastructural investigation of resistant and susceptible maize inbreds infected with *Erwinia stewartii*. Phytopathology 72:159-166.

Brito, B., Marenda, M., Barberis, P., Boucher, C., and Genin, S. 1999. *prhJ* and *hrpG*, two new components of the plant signal-dependent regulatory cascade controlled by PrhA in *Ralstonia solanacearum*. Mol. Microbiol. 31:237-251.

Brito, B., Aldon, D., Barberis, P., Boucher, C., and Genin, S. 2002. A signal transfer system through three compartments transduces the plant cell contact-dependent signal controlling *Ralstonia solanacearum hrp* genes. Mol. Plant-Microbe Interact. 15:109-119.

Brown, D. G., Swanson, J. K., and Allen, C. 2007. Two host-induced *Ralsonia solanacearum* genes, *acrA* and *dinF*, encode multidrug efflux pumps and contribute to bacterial wilt virulence. Appl. Environ. Microbiol. 73:2777-2786.

Brown, S. E., Knudson, D. L., and Ishimaru, C. A. 2002. Linear plasmid in the genome of *Clavibacter michiganensis* subsp. *sepedonicus*. J. Bacteriol. 184:2841-2844.

Brumbley, S. M., and Denny, T. P. 1990. Cloning of wild-type *Pseudomonas solanacearum phcA*, a gene that when mutated alters expression of multiple traits that contribute to virulence. J. Bacteriol. 172:5677-5685.

Brumbley, S. M., Carney, B. F., and Denny, T. P. 1993. Phenotype conversion in *Pseudomonas solanacearum* due to spontaneous inactivation of PhcA, a putative LysR transcriptional regulator. J. Bacteriol. 175:5477-5487.

Buddenhagen, I., Sequeira, L., and Kelman, A. 1962. Designation of races in *Pseudomonas solanacearum*. (Abstr.) Phytopathology 52:726.

Burrill, T. J. 1890. Preliminary notes upon the rotting of potatoes. Proc. Annu. Meet. Soc. Promot. Agric. Sci. 11:21-22.

Carlson, R. R., and Vidaver, A. K. 1982. Taxonomy of *Corynebacterium* plant pathogens, including a new pathogen of wheat, based on polyacrylamide gel electrophoresis of cellular proteins. Int. J. Syst. Bacteriol. 32:315-326.

Carlton, W. M., Braun, E. J., and Gleason, M. L. 1998. Ingress of *Clavibacter michiganensis* subsp. *michiganensis* into tomato leaves through hydathodes. Phytopathology 88:525-529.

Carroll, R. B., and Lukezic, F. L. 1971. Methods of preservation of *Corynebacterium insidiosum* isolates in relation to virulence and colony appearance on a tetrazolium chloride medium. Phytopathology 61:1423-1425.

Chang, R. J., Ries, S. M., and Pataky, J. K. 1992. Local sources of *Clavibacter michiganensis* ssp. *michiganensis* in the development of bacterial canker on tomatoes. Phytopathology 82:553-560.

Chun, W. C. C. 1982. Identification and detection of *Corynebacterium michiganense* in tomato seed using the indirect enzyme-linked immunosorbent assay. M.S. thesis. University of Hawaii, Honolulu, HI.

Close, T. J., Rogowsky, P. M., Kado, C. I., Winans, S. C., Yanofsky, M. F., and Nester, E. W. 1987. Dual control of *Agrobacterium tumefaciens* Ti plasmid virulence genes. J. Bacteriol. 169:5113-5118.

Clough, S. J., Schell, M. A., and Denny, T. P. 1994. Evidence for involvement of a volatile extracellular factor in *Pseudomonas solanacearum* virulence gene expression. Mol. Plant-Microbe Interact. 7:621-630.

Clough, S. J., Flavier, A. B., Schell, M. A., and Denny, T. P. 1997. Differential expression of virulence genes and motility in *Ralstonia* (*Pseudomonas*) *solanacearum* during exponential growth. Appl. Environ. Microbiol. 63:844-850.

Coenye, T., and Vandamme, P. 2003. Simple sequence repeats and compositional bias in the bipartite *Ralstonia solanacearum* GMI1000 genome. BMC Genomics 4:10.

Cook, D., and Sequeira, L. 1994. Strain differentiation of *Pseudomonas solanacearum* by molecular genetic methods. Pages 77-93 in: Bacterial Wilt: The Disease and Its Causative Agent *Pseudomonas solanacearum*. A. C. Hayward and G. L. Hartman, eds. CAB International, Wallingford, United Kingdom.

Cooley, M. B., D'Souza, M. R., and Kado, C. I. 1991. The *virC* and *virD* operons of the *Agrobacterium* Ti plasmid are regulated by the *ros* chromosomal gene: Analysis of the cloned *ros* gene. J. Bacteriol. 173:2608-2616.

Coplin, D. L., and Majerczak, D. R. 1990. Extracellular polysaccharide genes in *Erwinia stewartii*: Directed mutagenesis and complementation analysis. Mol. Plant-Microbe Interact. 3:286-292.

Coplin, D. L., Rowan, R. G., Chisholm, D. A., and Whitmoyer, R. E. 1981. Characterization of plasmids in *Erwinia stewartii*. Appl. Environ. Microbiol. 42:599-604.

Coplin, D. L., Federick, R. D., Majerczak, D. R., and Tuttle, L. D. 1992. Characterization of a gene cluster that specifies pathogenicity in *Erwinia stewartii*. Mol. Plant-Microbe Interact. 5:81-88.

Coplin, D. L., Majerczak, D. R., Bugert, P., and Geider, K. 1996. Nucleotide sequence analysis of the *Erwinia stewartii cps* gene cluster for synthesis of stewartan and comparison to the *Erwinia amylovora ams* cluster for synthesis of amylovoran. Acta Hortic. 411:251-257.

Coplin, D. L., Majerczak, D. R., Zhang, Y., Kim, W.-S., Jock, S., and Geider, K. 2002. Identification of *Pantoea stewartii* subsp. *stewartii* by PCR and strain differentiation by PFGE. Plant Dis. 86:304-311.

Cormack, M. W. 1961. Longevity of the bacterial wilt organism in alfalfa hay, pod debris, and seed. Phytopathology 51:260-261.

Coupat, B., Chaumelle-Dole, F., Fall, S., Prior, P., Simonet, P., Nesme, X., and Bertolla, F. 2008. Natural transformation in the *Ralstonia solanacearum* species complex: Number and size of DNA that can be transferred. FEMS Microbiol. Ecol. 66:14-24.

Cuppels, D. A., Hanson, R. S., and Kelman, A. 1978. Isolation and characterization of a bacteriocin produced by *Pseudomonas solanacearum*. J. Gen. Microbiol. 109:295-303.

Currier, T. C., and Morgan, M. K. 1981. Plasmids are not associated with formation of non-capsulated variants of *Pseudomonas solanacearum*. Pages 420-425 in: Proc. Int. Conf. Plant Pathogenic Bacteria, 5th. CIAT, Cali, Colombia.

Davis, M. J., Gillaspie, A. G., Vidaver, A. K., and Harris, R. W. 1984. *Clavibacter*, a new genus containing some phytopathogenic coryneform bacteria, including *Clavibacter xyli* subsp. *xyli* sp. nov., subsp. nov. and *Clavibacter xyli* subsp. *cynodontis* subsp. nov., pathogens that cause ratoon stunting disease of sugarcane and Bermudagrass stunting disease. Int. J. Syst. Bacteriol. 34:107-117.

Davis, R. M., and Frate, C. R. 2006. Alfalfa bacterial wilt. Univ. Calif. Agric. Nat. Resourc. Publ. 3430.

De Boer, S. H., Stead, D. E., Alivizatos, A. S., Janse, J. D., Van Vaerenbergh, J., De Haan, T. L., and Mawhinney, J. 1994. Evaluation of serological tests for detection of *Clavibacter michiganensis* subsp. *sepedonicus* in composite potato stem and tuber samples. Plant Dis. 78:725-729.

de la Cruz, A. R., Wiese, M. V., and Schaad, N. W. 1992. A semiselective agar medium for isolation of *Clavibacter michiganensis* subsp. *sepedonicus* from potato tissues. Plant Dis. 76:830-834.

Delaspre, F., Nieto-Peñalver, C. G., Saurel, O., Kiefer, P., Gras, E., Milon, A., Boucher, C., Genin, S., and Vorholt, J. A. 2007. The *Ralsonia solanacearum* pathogenicity regulator HrpB induces 3-hydroxy-oxindole synthesis. Proc. Natl. Acad. Sci. U.S.A. 104:15870-15875.

Denny, T. P. 2006. Plant pathogenic *Ralstonia* species. Pages 573-644 in: Plant-Associated Bacteria. S. S. Gnanamanickam, ed. Springer Publishing, Dordrecht, the Netherlands.

Denny, T. P., and Baek, S.-R. 1991. Genetic evidence that extracellular polysaccharide is a virulence factor of *Pseudomonas solanacearum*. Mol. Plant-Microbe Interact. 4:198-206.

Denny, T. P., and Hayward, A. C. 2001. *Ralstonia*. Pages 151-174 in: Laboratory Guide for Identification of Plant Pathogenic Bacteria, 3rd ed. N. W. Schaad, J. B. Jones, and W. Chun, eds. American Phytopathological Society, St. Paul, MN.

Denny, T. P., Carney, B. F., and Schell, M. A. 1990. Inactivation of multiple virulence genes reduces the ability of *Pseudomonas solanacearum* to cause wilt symptoms. Mol. Plant-Microbe Interact. 3:293-300.

Deslandes, L., Pileur, F., Liaubet, L., Camut, S., Can, C., Williams, K., Holub, E., Beynon, J., Arlat, M., and Marco, Y. 1998. Genetic characterization of *RRS1*, a recessive locus in *Arabidopsis thaliana* that confers resistance to the bacterial soilborne pathogen *Ralstonia solanacearum*. Mol. Plant-Microbe Interact. 11:659-667.

Devi, L. R., Menon, M. R., and Aiyer, R. S. 1981. Survival of *Pseudomonas solanacearum* in soil. Plant Soil 62:169-182.

Dhanvantari, B. N. 1987. Comparison of selective media for isolation of *Clavibacter michiganense* pv. *michiganense*. (Abstr.) Phytopathology 77:1694.

Dreier, J., Meletzus, D., and Eichenlaub, R. 1997. Characterization of the plasmid encoded virulence region *pat*-1 of phytopathogenic *Clavibacter michiganensis* subsp. *michiganensis*. Mol. Plant-Microbe Interact. 10:195-206.

Dristig, M. C. G., and Dianese, J. C. 1990. Characterization of *Pseudomonas solanacearum* biovars based on membrane protein patterns. Phytopathology 80:641-646.

Dye, D. W., and Kemp, W. J. 1977. A taxonomic study of plant pathogenic *Corynebacterium* species. N.Z. J. Agric. Res. 20:563-582.

Eddins, A. H. 1941. Losses from potato disease in the Hastings section, Florida, in 1941. Plant Dis. Rep. 25:353-354.

Erwin, D. C. 1990. Bacterial wilt. Pages 5-6 in: Compendium of Alfalfa Diseases, 2nd ed. D. L. Stuteville and D. C. Erwin, eds. American Phytopathological Society, St. Paul, MN.

Fall, S., Mercier, A., Bertolla, F., Calteau, A., Gueguen, L., Perrière, G., Vogel, T. M., and Simonet, P. 2007. Horizontal gene transfer regulation in bacteria as a "spandrel" of DNA repair mechanisms. PLos One 2(10):e1055. doi:10.1371/journal.pone.0001055.

Fatmi, M., and Schaad, N. W. 1988. Semiselective agar medium for isolation of *Clavibacter michiganense* subsp. *michiganense* from tomato seed. Phytopathology 78:121-126.

Fegan, M., and Prior, P. 2005. How complex is the "*Ralstonia solanacearum* species complex"? Pages 449-461 in: Bacterial Wilt Disease and the *Ralstonia solanacearum* Species Complex. C. Allen, P. Prior, and A. C. Hayward, eds. American Phytopathological Society, St. Paul, MN.

Flavier, A. B., Clough, S. J., Schell, M. A., and Denny, T. P. 1997. Identification of 3-hydroxypalmitic acid methyl ester as a novel autoregulator controlling virulence in *Ralstonia solanacearum*. Mol. Microbiol. 26:251-259.

Flavier, A. B., Schell, M. A., and Denny, T. P. 1998. An RpoS (sigmaS) homologue regulates acylhomoserine lactone-dependent autoinduction in *Ralstonia solanacearum*. Mol. Microbiol. 28:475-486.

Fousek, J., and Mraz, I. 2003. Determination of genetic differences between fluid and nonfluid variants of *Clavibacter michiganensis* subsp. *sepedonicus* using rep-PCR technique. Folia Microbiol. 48:682-686.

Frederick, R. D., Ahmad, M., Majerczak, D. R., Arroyo-Rodriguez, A. S., Manulis, S., and Coplin, D. L. 2001. Genetic organization of the *Pantoea stewartii* subsp. *stewartii hrp* gene cluster and sequence analysis of the *hrpA*, *hrpC*, *hrpN*, and *wtsE* operons. Mol. Plant-Microbe Interact. 14:1213-1222.

Freebairn, H. T., and Buddenhagen, I. W. 1964. Ethylene production by *Pseudomonas solanacearum*. Nature 202:313-314.

Frosheiser, F. I. 1966. Alfalfa cotyledon inoculation with bacterial wilt inoculum prepared from infected alfalfa roots. Phytopathology 56:566-567.

Gabriel, D. W., Allen, C., Schell, M., Denny, T. P., Greenberg, J. T., Duan, Y. P., Flores-Cruz, Z., Huang, Q., Clifford, J. M., Presting, G., González, E. T., Reddy, J., Elphinstone, J., Swanson, J., Yao, J., Mulholland, V., Liu, L., Farmerle, W., Patnaikuni, M., Balogh, B., Norman, D., Alvarez, A., Castillo, J. A., Jones, J., Saddler, G., Walunas, T., Zhukov, A., and Mikhailova, N. 2006. Identification of open reading frames unique to a select agent: *Ralstonia solanacearum* race 3 biovar 2. Mol. Plant-Microbe Interact. 19:69-79.

Gartemann, K. H., Kirchner, O., Engemann, J., Gräfen, I., Eichenlaub, R., and Burger, A. 2003. *Clavibacter michiganensis* subsp. *michiganensis*: First steps in the understanding of virulence of a Gram-positve phytopathogenic bacterium. J. Biotechnol. 106:179-191.

Gartemann, K. H., Abt, B., Bekel, T., Burger, A., Engemann, J., Flügel, M., Gaigalat, L., Goesmann, A., Gräfen, I., Kalinowski, J., Kaup, O., Kirchner, O., Krause, L., Linke, B., McHardy, A., Meyer, F., Pohle, S., Rückert, C., Schneiker, S., Zellermann, E. M., Pühler, A., Eichenlaub, R., Kaiser, O., and Bartels, D. 2008. The genome sequence of the tomato-pathogenic actinomycete *Clavibacter michiganensis* subsp. *michiganensis* NCPPB382 reveals a large island involved in pathogenicity. J. Bacteriol. 190:2138-2149.

Genin, S., and Boucher, C. 2002. *Ralstonia solanacearum*: Secrets of a major pathogen unveiled by analysis of its genome. Mol. Plant Pathol. 3:111-118.

Genin, S., and Boucher, C. 2004. Lessons learned from the genome analysis of *Ralstonia solanacearum*. Annu. Rev. Phytopathol. 42:107-134.

Genin, S., Gough, C. L., Zischek, C., and Boucher, C. A. 1992. Evidence that the *hrpB* gene encodes a positive regulator of pathogenicity genes from *Pseudomonas solanacearum*. Mol. Microbiol. 6:3065-3076.

Genin, S., Brito, B., Denny, T. P., and Boucher, C. 2005. Control of the *Ralstonia solanacearum* type III secretion system (Hrp) genes by the global virulence regulator PhcA. FEBS Lett. 579:2077-2081.

Gitaitis, R. D., Beaver, R. W., and Voloudakis, A. E. 1991. Detection of *Clavibacter michiganensis* subsp. *michiganensis* in symptomless tomato transplants. Plant Dis. 75:834-838.

Gleason, M. L., Gitaitis, R. D., and Ricker, M. D. 1993. Recent progress in understanding and controlling bacterial canker of tomato in eastern North America. Plant Dis. 77:1069-1076.

González, E. T., and Allen, C. 2003. Characterization of a *Ralstonia solanacearum* operon required for polygalacturonate degradation and uptake of galacturonic acid. Mol. Plant-Microbe Interact. 16:536-544.

Gough, C. L., Genin, S., Zischek, C., and Boucher, C. A. 1992. *hrp* Genes of *Pseudomonas solanacearum* are homologous to pathogenicity determinants of animal pathogenic bacteria and are conserved among plant pathogenic bacteria. Mol. Plant-Microbe Interact. 5:384-389.

Gross, D. C., and Vidaver, A. K. 1979. A selective medium for isolation of *Corynebacterium nebraskense* from soil and plant parts. Phytopathology 69:82-87.

Guéneron, M., Timmers, A. C. J., Boucher, C., and Arlat, M. 2000. Two novel proteins, PopB, which has functional nuclear localization signals, and PopC, which has a large leucine-rich repeat domain, are secreted through the Hrp-secretion apparatus of *Ralstonia solanacearum*. Mol. Microbiol. 36:261-277.

Guidot, A., Prior, P., Schoenfeld, J., Carrère, S., Genin, S., and Boucher, C. 2007. Genomic structure and phylogeny of the plant pathogen *Ralstonia solanacearum* inferred from gene distribution analysis. J. Bacteriol. 189:377-387.

Ham, J. H., Majerczak, D., Ewert, S., Sreerekha, M. V., Mackey, D., and Coplin, D. 2008. WtsE, an AvrE-family type III effector protein of *Pantoea stewartii* subsp. *stewartii* causes cell death in non-host plants. Mol. Plant Pathol. 9:633-643.

Hawn, E. J. 1963. Transmission of bacterial wilt of alfalfa by *Ditylenchus dipsaci* (Kühn). Naematologica 9:65-68.

Hayward, A. C. 1960. A method for characterizing *Pseudomonas solanacearum*. Nature 186:405-406.

Hayward, A. C. 1964. Characteristics of *Pseudomonas solanacearum*. J. Appl. Bacteriol. 27:265-277.

Hayward, A. C. 1994. Systematics and phylogeny of *Pseudomonas solanacearum* and related bacteria. Pages 123-135 in: Bacterial Wilt: The Disease and Its Causative Agent, *Pseudomonas solanacearum*. A. C. Hayward and G. L. Hartman, eds. CAB International, Wallingford, United Kingdom.

Hayward, A. C. 2000. Biology and epidemiology of bacterial wilt caused by *Pseudomonas solanacearum*. Annu. Rev. Phytopathol. 29:65-87.

He, L. Y., Sequeira, L., and Kelman, A. 1983. Characteristics of strains of *Pseudomonas solanacearum* from China. Plant Dis. 67:1357-1361.

Holtsmark, I., Takle, G. W., and Brurberg, M. B. 2008. Expression of putative virulence factors in the potato pathogen *Clavibacter michiganensis* subsp. *sepedonicus* during infection. Arch. Microbiol. 189:131-139.

Howard, F. L. 1940. Bacterial ring rot in Rhode Island. Plant Dis. Rep. 25:32.

Hsu, S.-T. 1991. Ecology and control of *Pseudomonas solanacearum* in Taiwan. Plant Prot. Bull. (Taiwan) 33:72-79.

Hsu, S.-T., and Chen, J.-Y. 1977. Physiological variation among isolates of *Pseudomonas solanacearum* from Taiwan. Plant Prot. Bull. (Taiwan) 19:124-132.

Hsu, S.-T., Hong, W.-F., Tzeng, K.-C., and Chen, C.-C. 1993. Bacterial wilt of perilla caused by *Pseudomonas solanacearum* and its transmission. Plant Dis. 77:674-677.

Hu, X., Lai, F.-M., Reddy, A. S. N., and Ishimaru, C. A. 1995. Quantitative detection of *Clavibacter michiganensis* subsp. *sepedonicus* by competitive polymerase chain reaction. Phytopathology 85:1468-1473.

Huang, J., Denny, T. P., and Schell, M. A. 1993. VsrB, a regulator of virulence genes of *Pseudomonas solanacearum*, is homologous to sensors of the two-component regulatory family. J. Bacteriol. 175:6169-6178.

Huang, J., Carney, B. F., Denny, T. P., Weissinger, A. K., and Schell, M. A. 1995. A complex network regulates expression of *eps* and other virulence genes of *Pseudomonas solanacearum*. J. Bacteriol. 177:1259-1267.

Huang, Q., and Allen, C. 1997. An exo-poly-alpha-D-galacturonosidase, PehB, is required for wild-type virulence of *Ralstonia solanacearum*. J. Bacteriol. 179:7369-7378.

Huang, Q., and Allen, C. 2000. Polygalacturonases are required for rapid colonization and full virulence of *Ralstonia solanacearum* on tomato plants. Physiol. Mol. Plant Pathol. 57:77-83.

Hunt, O. J., Griffin, G. D., Murray, J. J., Pedersen, M. W., and Peaden, R. N. 1971. The effects of root knot nematodes on bacterial wilt in alfalfa. Phytopathology 61:256-259.

Huynh, T., Dahlbeck, D. V., and Staskawicz, B. J. 1989. Bacterial blight of soybean: Regulation of a pathogen gene determining host cultivar specificity. Science 245:1374-1377.

Jahr, H., Dreier, J., Meletzus, D., Bahro, R., and Eichenlaub, R. 2000. The endo-β-1,4-glucanase CelA of *Clavibacter michiganensis* subsp. *michiganensis* is a pathogenicity determinant required for induction of bacterial wilt of tomato. Mol. Plant-Microbe Interact. 13:703-714.

Janse, J. D. 1996. Potato brown rot in western Europe—History, present occurrence and some remarks on possible origin, epidemiology and control strategies. Bull. OEPP 26:679-695.

Jensen, H. L. 1934. Studies on saprophytic Mycobacteria and Corynebacteria. Proc. Linn. Soc. N.S.W. 59:19-61.

Jones, F. R. 1925. A new bacterial disease of alfalfa. Phytopathology 15:243-244.

Jones, F. R., and McCulloch, L. 1926. A bacterial wilt and root rot of alfalfa caused by *Aplanobacter insidiosum* McCulloch. J. Agric. Res. 33:493-521.

Jones, J. D., and Dangl, J. L. 2006. The plant immune system. Nature 444:323-332.

Kado, C. I. 1992. Chromosomal- and Ti plasmid-mediated regulation of *Agrobacterium* virulence genes: Molecular intercommunication. Pages 201-208 in: Molecular Signals in Plant-Microbe Communications. D. P. S. Verma, ed. CRC Press, Boca Raton, FL.

Kado, C. I. 1998. Origin and evolution of plasmids. Antonie Leewuwenhoek 73:117-126.

Kado, C. I., and Heskett, M. G. 1970. Selective media for isolation of *Agrobacterium*, *Corynebacterium*, *Erwinia*, *Pseudomonas*, and *Xanthomonas*. Phytopathology 60:969-976.

Kanda, A., Ohnishi, K., Kiba, A., and Hikichi, Y. 2008. Implication of C-terminal mutation of PopA of *Ralstonia solanacearum* strain OE1-1 in suppression of bacterial wilt. Plant Pathol. 58:159-169.

Kaneshiro, W. S., Mizumoto, C. Y., and Alvarez, A. M. 2006. Differentiation of *Clavibacter michiganensis* subsp. *michiganensis* from seed-borne saprophytes using ELISA, Biolog and 16S rDNA sequencing. Eur. J. Plant Pathol. 116:45-56.

Kang, Y., Huang, J., Mao, G., He, L.-Y., and Schell, M. A. 1994. Dramatically reduced virulence of mutants of *Pseudomonas solanacearum* defective in export of extracellular proteins across the outer membrane. Mol. Plant-Microbe Interact. 7:370-377.

Kao, C. C., and Sequeira, L. 1991. A gene cluster required for coordinated biosynthesis of lipopolysaccharide and extracellular polysaccharide also affects virulence of *Pseudomonas solanacearum*. J. Bacteriol. 173:7841-7847.

Kao, C. C., Barlow, E., and Sequeira, L. 1992. Extracellular polysaccharide is required for wild-type virulence of *Pseudomonas solanacearum*. J. Bacteriol. 174:1068-1071.

Kaup, O., Gräfen, I., Zellermann, E.-M., Eichenlaub, R., and Gartemann, K.-H. 2005. Identification of a tomatinase in the tomato-pathogenic actinomycete *Clavibacter michiganensis* subsp. *michiganensis* NCPPB382. Mol. Plant-Microbe Interact. 18:1090-1098.

Kelman, A. 1954. The relationship of pathogenicity in *Pseudomonas solanacearum* to colony appearance on a tetrazolium medium. Phytopathology 44:693-695.

Kelman, A., and Hruschka, J. 1973. The role of motility and aerotaxis in the selective increase of avirulent bacteria in still broth cultures of *Pseudomonas solanacearum*. J. Gen. Microbiol. 76:177-188.

Kelman, A., and Sequeira, L. 1965. Root-to-root spread of *Pseudomonas solanacearum*. Phytopathology 55:304-309.

Kernkamp, M. F., and Hemerick, G. 1952. A "deep-freeze" method of maintaining virulent inoculum of the alfalfa wilt bacterium, *Corynebacterium insidiosum*. (Abstr.) Phytopathology 42:13.

Khan, A., Ries, S. M., and Pataky, J. K. 1996. Transmission of *Erwinia stewartii* through seed of resistant and susceptible field and sweet corn. Plant Dis. 80:398-403.

Kirchner, O., Gartemann, K.-H., Zellermann, E.-M., Eichenlaub, R., and Burger, A. 2001. A highly efficient transposon mutagenesis system for the tomato pathogen *Clavibacter michiganensis* subsp. *michiganensis*. Mol. Plant-Microbe Interact. 14:1312-1318.

Knorr, L. C. 1948. Suspect range of the potato ring rot bacterium. Am. Potato J. 25:361-371.

Konstantinidis, K. T., and Tiedje, J. M. 2005. Towards a genome-based taxonomy for prokaryotes. J. Bacteriol. 187:6258-6264.

Krasil'nikov, N. A. 1941. Page 119 in: Guide to the Bacteria and Actinomycetes. Akad. Nauk SSSR, Moscow.

Kreitlow, K. W. 1963. Infecting seven-day-old alfalfa seedlings with wilt bacteria through wounded cotyledons. Phytopathology 53:800-803.

Lavie, M., Seunes, B., Prior, P., and Boucher, C. 2004. Distribution and sequence analysis of a family of type III-dependent effectors correlate with the phylogeny of *Ralstonia solanacearum* strains. Mol. Plant-Microbe Interact. 17:931-940.

Lee, I.-M., Lukaesko, L. A., and Maroon, C. J. M. 2001. Comparison of dig-labeled PCR, nested PCR, and ELISA for the detection of *Clavibacter michiganensis* subsp. *sepedonicus* in field-grown potatoes. Plant Dis. 85:261-266.

Lenz, D. H., Miller, M. B., Zhu, J., Kulkarni, R. V., and Bassler, B. L. 2005. CsrA and three redundant small RNAs regulate quorum sensing in *Vibrio cholerae*. Mol. Microbiol. 58:1186-1202.

Liu, H., Zhang, S., Schell, M. A., and Denny, T. P. 2005. Pyramiding unmarked deletions in *Ralstonia solanacearum* shows that secreted proteins in addition to plant cell-wall-degrading enzymes contribute to virulence. Mol. Plant-Microbe Interact. 18:1296-1305.

Lloyd, A. B. 1976. Bacterial wilt in a cold-temperature climate of Australia. Pages 134-135 in: Proc. Int. Plan. Conf. Workshop Ecology and Control of Bacterial Wilt Caused by *Pseudomonas solanacearum*, 1st. L. Sequeira and A. Kelman, eds. North Carolina State University, Raleigh, NC.

Louws, F. J., Bell, J., Medina-Mora, C. M., Smart, C. D., Opgenorth, D., Ishimaru, C. A., Hausbeck, M. K., de Bruijn, F. J., and Fulbright, D. W. 1998. rep PCR mediated genomic fingerprinting: A rapid and effective method to identify *Clavibacter michiganensis*. Phytopathology 88:862-868.

Malvick, D. K. 1988. Bacterial wilt of alfalfa. Univ. Ill. Ext. Rep. Plant Dis. (RPD) No. 300.

McCarter, S. M. 1976. Persistence of *Pseudomonas solanacearum* in artificially infested soils. Phytopathology 66:998-1000.

McCulloch, L. 1925. *Aplanobacter insidiosum* n. sp. Phytopathology 15:497.

McGarvey, J. A., Denny, T. P., and Schell, M. A. 1999. Spatial-temporal and quantitative analysis of growth and EPS I production by *Ralstonia solanacearum* in resistant and susceptible tomato cultivars. Phytopathology 89:1233-1239.

Meletzus, D., and Eichenlaub, R. 1991. Transformation of the phytopathogenic bacterium *Clavibacter michiganense* subsp. *michiganense* by electroporation and development of a cloning vector. J. Bacteriol. 173:184-190.

Meletzus, D., Bermpohl, A., Dreier, J., and Eichenlaub, R. 1993. Evidence for plasmid-encoded virulence factors in the phytopathogenic bacterium *Clavibacter michiganensis* subsp. *michiganensis* NCPPB382. J. Bacteriol. 175:2131-2136.

Meng, X., Bonasera, J. M., Kim, J. F., Nissinen, R. M., and Beer, S. V. 2006. Apple proteins that interact with DspA/E, a pathogenicity effector of *Erwinia amylovora*, the fire blight pathogen. Mol. Plant-Microbe Interact. 19:53-61.

Mills, D., Russell, B. W., and Hanus, J. W. 1997. Specific detection of *Clavibacter michiganensis* subsp. *sepedonicus* by amplification of three unique DNA sequences isolated by subtraction hybridization. Phytopathology 87:853-861.

Minogue, T. D., Carlier, A. L., Koutsoudis, M. D., and von Bodman, S. B. 2005. The cell density-dependent expression of stewartan exopolysaccharide in *Pantoea stewartii* ssp. *stewartii* is a function of EsaR-mediated repression of the *rcsA* gene. Mol. Microbiol. 56:189-203.

Mogen, B. D., and Oleson, A. E. 1987. Homology of pCS1 plasmid sequences with chromosomal DNA in *Clavibacter michiganense* subsp. *sepedonicum*: Evidence for the presence of a repeated sequence and plasmid integration. Appl. Environ. Microbiol. 53:2476-2481.

Mor, H., Manulis, S., Zuck, M., Nizan, R., Coplin, D. L., and Barash, I. 2001. Genetic organization of the *hrp* gene cluster and *dspAE/BF* operon in *Erwinia herbicola* pv. *gypsophilae*. Mol. Plant-Microbe Interact. 14:431-436.

Morales, V. M., and Sequeira, L. 1985. Indigenous plasmids in *Pseudomonas solanacearum*. Phytopathology 75:767-771.

Mukaihara, T., Tamura, N., Murata, Y., and Iwabuchi, M. 2004. Genetic screening of Hrp type III-related pathogenicity gene controlled by the HrpB transcriptional activator in *Ralstonia solanacearum*. Mol. Microbiol. 54:863-875.

Negishi, H., Yamada, T., Shiraishi, T., Oku, H., and Tanaka, H. 1993. *Pseudomonas solanacearum*: Plasmid pJTPS1 mediates a shift from the pathogenic to nonpathogenic phenotype. Mol. Plant-Microbe Interact. 6:203-209.

Nelson, G. A. 1982. *Corynebacterium sepedonicum* in potato: Effect of inoculum concentration on ring rot symptoms and latent infection. Can. J. Plant Pathol. 4:129-133.

Nissinen, R., Lai, F.-M., Laine, M. J., Bauer, P. J., Reilley, A. A., Li, X., De Boer, S. H., Ishimaru, C. A., and Metzler, M. C. 1997. *Clavibacter michiganensis* subsp. *sepedonicus* elicits a hypersensitive response in tobacco and secretes hypersensitive response-inducing protein(s). Phytopathology 87:678-684.

Olsson, K. 1976. Overwintering of *Pseudomonas solanacearum* in Sweden. Pages 105-109 in: Proc. Int. Plan. Conf. Workshop Ecology and Control of Bacterial Wilt Caused by *Pseudomonas solanacearum*, 1st. L. Sequeira and A. Kelman, eds. North Carolina State University, Raleigh, NC.

Orgambide, G., Montrozier, H., Servin, P., Roussel, J., Trigalet-Demery, D., and Trigalet, A. 1991. High heterogeneity of the exopolysaccharides of *Pseudomonas solanacearum* strain GM1000 and the complete structure of the major polysaccharide. J. Biol. Chem. 266:8312-8321.

Pegg, K. G., and Moffett, M. 1971. Host range of the ginger strain of *Pseudomonas solanacearum* in Queensland. Aust. J. Exp. Agric. Anim. Husb. 11:696-698.

Peltier, G. L. 1934. The inability of *Aplanobacter insidiosum* to enter alfalfa seedlings in the absence of wounds. Phytopathology 24:1044-1045.

Phelps, R. H., and Sequeira, L. 1968. Auxin biosynthesis in a host-parasite complex. Pages 197-212 in: Biochemistry and Physiology of Plant Growth Substances. F. Wightman and G. D. Setterfield, eds. Runge Press, Ottawa, Canada.

Racapé, J., Belbahri, L., Engelhardt, S., Lacome, B., Lee, J., Lochman, J., Marais, A., Nicole, M., Nürnberger, T., Parlange, F., Puverel, S., and Keller, H. 2005. Ca^{2+}-dependent lipid binding and membrane integration of PopA, a harpin-like elicitor of the hypersensitive response in tobacco. Mol. Microbiol. 58:1406-1420.

Richards, M. C. 1941. Bacterial ring rot of potato in New Hampshire. Plant Dis. Rep. 25:419.

Roberts, D. P., Denny, T. P., and Schell, M. A. 1988. Cloning of the *egl* gene of *Pseudomonas solanacearum* and analysis of its role in phytopathogenicity. J. Bacteriol. 170:1445-1451.

Rorer, J. B. 1911. A bacterial disease of bananas and plantains. Phytopathology 1:45-49.

Rosenberg, C., Casse-Delbart, F., Dusha, I., David, M., and Boucher, C. 1982. Megaplasmids in the plant associated *Rhizobium meliloti* and *Pseudomonas solanacearum*. J. Bacteriol. 150:402-406.

Saddler, G. S. 2005. Management of bacterial wilt disease. Pages 121-132 in: Bacterial Wilt Disease and the *Ralstonia solanacearum* Species Complex. C. Allen, P. Prior, and A. C. Hayward, eds. American Phytopathological Society, St. Paul, MN.

Saile, E., McGarvey, J. A., Schell, M. A., and Denny, T. P. 1997. Role of extracellular polysaccharide and endoglucanase in root invasion and colonization of tomato plants by *Ralstonia solanacearum*. Phytopathology 87:1264-1271.

Salanoubat, M., Genin, S., Artiguenave, F., Gouzy, J., Mangenot, S., Arlat, M., Billault, A., Brottier, P., Camus, J. C., Cattolico, L., Chandler, M., Choisne, N., Claudel-Renard, C., Cunnac, S., Demange, N., Gaspin, C., Lavie, M., Mosian, A., Robert, C., Saurin, W., Schiex, T., Siguier, P., Thebault, P., Whalen, M., Wincker, P., Levy, M., Seissenbach, J., and Boucher, C. A. 2002. Genome sequence of the plant pathogen *Ralstonia solanacearum*. Nature 415:497-502.

Samac, D. A., Nix, R. J., and Oleson, A. E. 1998. Transmission frequency of *Clavibacter michiganensis* subsp. *insidiosus* to alfalfa seed and identification of the bacterium by PCR. Plant Dis. 82:1362-1367.

Saperstein, S., Starr, M. P., and Filfus, J. A. 1954. Alterations in carotenoid synthesis accompanying mutation in *Corynebacterium michiganense*. J. Gen. Microbiol. 10:85-92.

Schell, M. A. 1987. Purification and characterization of an endoglucanase from *Pseudomonas solanacearum*. Appl. Environ. Microbiol. 53:2237-2241.

Schell, M. A. 2000. Control of virulence and pathogenicity genes of *Ralstonia solanacearum* by an elaborate sensory network. Annu. Rev. Phytopathol. 38:263-292.

Schell, M. A., Denny, T. P., and Huang, J. 1993. VsrA, a second two-component sensor regulating virulence genes of *Pseudomonas solanacearum*. Mol. Microbiol. 11:489-500.

Schell, M. A., Denny, T. P., Clough, S. J., and Huang, J. 1994. Extracellular virulence factors of *Pseudomonas solanacearum*: Role in disease and their regulation. Pages 311-324 in: Molecular Mechanism of Bacterial Virulence. C. I. Kado and J. H. Crosa, eds. Kluwer Academic Publishers, Dordrecht, the Netherlands.

Skaptason, J. B. 1943. Studies on the bacterial ring-rot disease of potatoes. Cornell Univ. Agric. Exp. Stn. Mem. 250.

Slack, S. A. 1987. Biology and ecology of *Corynebacterium sepedonicum*. Am. Potato J. 64:665-670.

Smith, E. F. 1896. Bacterial disease of tomato, eggplant, and Irish potato (*Bacillus solanacearum* n. sp.). U.S. Dep. Agric. Div. Veg. Physiol. Pathol. Bull. 12:1-28.

Smith, E. F. 1898. Notes on Stewart's sweet-corn germ, *Pseudomonas stewarti*, new species. Proc. Am. Assoc. Adv. Sci. 47:422-426.

Smith, E. F. 1908. Granville wilt of tobacco. Bur. Plant Ind. Bull. 141, Part II.

Smith, E. F. 1910. A new tomato disease of economic importance. Science 31:794-796.

Smith, E. F. 1911. Pages 209-299 in: Bacteria in Relation to Plant Diseases. Carnegie Inst. Wash. Publ. 2.

Smith, E. F., and Godfrey, G. H. 1918. Brown rot of Solanaceae on *Ricinus*. Science 48:42-43.

Spieckermann, A. 1913. Zur Kenntuis der in Deutschland auf trenten den Gefasskrankheiten der Kartoffelplanze. Landwirtsch. Ztg. 33:680-682.

Spök, A., Stubenrauch, G., Schorgendorfer, K., and Schwab, H. 1991. Molecular cloning of a pectin esterase gene from *Pseudomonas solanacearum*. J. Gen. Microbiol. 137:131-140.

Stamova, L., and Sotirova, V. 1987. Reaction of different crops to artificial inoculation with *Corynebacterium michiganense*. Arch. Phytopathol. Pflanzenschutz 23:211-216.

Starr, M. P. 1949. The nutrition of phytopathogenic bacteria. III. The Gram-positive phytopathogenic *Corynebacterium* species. J. Bacteriol. 57:253-258.

Starr, M. P. 1958. The blue pigment of *Corynebacterium insidiosum*. Arch. Mikrobiol. 30:325-334.

Stork, I., Gartemann, K. H., Burger, A., and Eichenlaub, R. 2008. A family of serine proteases of *Clavibacter michiganensis* subsp. *michiganensis*: *chpC* Plays a role in colonization of the host plant tomato. Mol. Plant Pathol. 9:599-608.

Swanepoel, A. E., and Young, B. W. 1988. Characterisitics of South African strains of *Pseudomonas solanacearum*. Plant Dis. 72:403-405.

Swanson, J. K., Yao, J., Tans-Kersten, J., and Allen, C. 2005. Behavior of *Ralstonia solanacearum* race 3 biovar 2 during latent and active infection of geranium. Phytopathology 95:136-143.

Tamura, N., Murata, Y., and Mukaihara, T. 2005. Isolation of *Ralstonia solanacearum hrpB* constitutive mutants and secretion analysis of *hrpB*-regulated gene products that share homology with known type III effectors and enzymes. Microbiology 151:2873-2884.

Tans-Kersten, J., Guan, Y. F., and Allen, C. 1998. *Ralstonia solanacearum* pectin methylesterase is required for growth on methylated pectin but not for bacterial wilt virulence. Appl. Environ. Microbiol. 64:4918-4923.

Tans-Kersten, J., Huang, H., and Allen, C. 2001. *Ralstonia solanacearum* needs motility for invasive virulence on tomato. J. Bacteriol. 183:3597-3605.

Thornberry, H. H. 1941. Bacterial ring rot of potatoes in Illinois. Plant Dis. Rep. 25:509-510.

Thyr, B. D., Samuel, M. J., and Brown, P. G. 1975. New solanaceous host records for *Corynebacterium michiganense*. Plant Dis. Rep. 59:595-598.

Trigalet, A., and Demery, D. 1986. Invasiveness in tomato plants of Tn5-induced avirulent mutants of *Pseudomonas solanacearum*. Physiol. Plant Pathol. 28:423-430.

United States Department of Agriculture. 1960. Index of plant diseases in the United States. U.S. Dep. Agric. Agric. Handb. 165.

Vailleau, F., Sartorel, E., Jardinaud, M.-F., Chardon, F., Genin, S., Huguet, T., Gentzbittel, L., and Petitprez, M. 2007. Characterization of the interaction between the bacterial wilt pathogen *Ralstonia solanacearum* and the model legume plant *Medicago truncatula*. Mol. Plant-Microbe Interact. 20:159-167.

Van Alfen, N. K. 1989. Reassessment of plant wilt toxins. Annu. Rev. Phytopathol. 27:533-550.

van der Wolf, J. M., van Beckhoven, J. R. C. M., Hukkanen, A., Karjalainen, R., and Müller, P. 2005. Fate of *Clavibacter michiganensis* ssp. *sepedonicus*, the causal organism of bacterial ring rot of potato, in weeds and field crops. J. Phytopathol. 153:358-365.

van Gijsegem, F., Gough, C., Zischek, C., Nigueux, E., Arlat, M., Genin, S., Barberis, P., German, S., Castello, P., and Boucher, C. 1995. The *hrp* gene locus of *Pseudomonas solanacearum*, which controls the production of a type III secretion system, encodes eight proteins related to components of the bacterial flagellar biogenesis complex. Mol. Microbiol. 15:1095-1114.

von Bodman, S. B., and Farrand, S. K. 1995. Capsular polysaccharide biosynthesis and pathogenicity in *Erwinia stewartii* require induction by an *N*-acylhomoserine lactone autoinducer. J. Bacteriol. 177:5000-5008.

von Bodman, S. B., Majerczak, D. R., and Coplin, D. L. 1998. A negative regulator mediates quorum-sensing control of exopolysaccharide production in *Pantoea stewartii* subsp. *stewartii*. Proc. Natl. Acad. Sci. U.S.A. 95:7687-7692.

von Bodman, S. B., Bauer, W. D., and Coplin, D. L. 2003. Quorum sensing in plant-pathogenic bacteria. Annu. Rev. Phytopathol. 41:455-482.

Weller, S. A., Elphinstone, J. G., Smith, N. C., Boonham, N., and Stead, D. E. 2000. Detection of *Ralstonia solanacearum* strains with a quantitative, multiplex, real-time fluorogenic PCR (TaqMan) assay. Appl. Environ. Microbiol. 66:2853-2858.

Wenneker, M., Verdel, M. S. W., Groeneveld, R. M. W., Kempenaar, C., Van Beuningen, A. R., and Janse, J. D. 1999. *Ralstonia* (*Pseudomonas*) *solanacearum* race 3 (biovar 2) in surface water and natural weed hosts: First report on stinging nettle (*Urtica dioica*). Eur. J. Plant Pathol. 105:307-315.

Williamson, L., Nakaho, K., Hudelson, B., and Allen, C. 2002. *Ralsonia solanacearum* race 3, biovar 2 strains isolated from geranium are pathogenic on potato. Plant Dis. 86:987-991.

Wilson, E. E., Zeitoun, F. M., and Fredrickson, D. L. 1967. Bacterial phloem canker, a new disease of Persian walnut trees. Phytopathology 57:618-621.

Winstead, N. N., and Kelman, A. 1952. Inoculation techniques for evaluating resistance to *Pseudomonas solanacearum*. Phytopathology 42:628-634.

Xu, P., Leong, S., and Sequeira, L. 1988. Molecular cloning of genes that specify virulence in *Pseudomonas solanacearum*. J. Bacteriol. 170:617-622.

Yabuuchi, E., Kosako, Y., Yano, I., Hotta, H., and Nishiuchi, Y. 1995. Transfer of two *Burkholderia* and an *Alcaligenes* species to *Ralstonia* gen. nov.; proposal of *Ralstonia pickettii* (Ralston,

Palleroni and Doudoroff 1973) comb. nov., *Ralstonia solanacearum* (Smith 1896) com. nov. and *Ralstonia eutropha* (Davis 1969) comb. nov. Microbiol. Immunol. 39:897-904.
Yang, B. Y., Gray, J. S. S., and Montgomery, R. 1996. The structure of stewartan, a capsular polysaccharide produced by *Erwinia stewartii* strain DC283. Carbohydr. Res. 296:183-201.
Yang, C.-H., and Ho, G.-D. 1998. Resistance and susceptibility of *Arabidopsis thaliana* to bacterial wilt caused by *Ralstonia solanacearum*. Phytopathology 88:330-334.
Yang, W., Coaker, G. L., Kinter, M. T., and Francis, D. M. 2005. Proteomic and gene expression analysis of the tomato-*Clavibacter* interaction: A role for oxidative defense and counter-defense? Page 482 in: Proc. Plant Anim. Genomes XIII Conf.
Yao, J., and Allen, C. 2006. Chemotaxis is required for virulence and competitive fitness in the bacterial wilt pathogen *Ralstonia solanacearum*. J. Bacteriol. 188:3697-3708.
Yildiz, F. H., and Schoolnik, G. K. 1999. *Vibrio cholerae* 01 El Tor: Identification of a gene cluster required for the rugose colony type, exopolysaccharide production, chlorine resistance, and biofilm formation. Proc. Natl. Acad. Sci. U.S.A. 96:4028-4033.
Yoder, M. D., and Schell, M. A. 1995. X-ray analysis of crystals of polygalacturonase A from *Pseudomonas solanacearum*. Acta Cryst. D 51:1097-1098.
Young, D. H., and Sequeira, L. 1986. Binding of *Pseudomonas solanacearum* fimbriae to tobacco leaf cell walls and its inhibition by bacterial extracellular polysaccharides. Physiol. Plant Pathol. 28:393-402.
Zizz, J. D., and Harrison, M. D. 1991. Detection of *Clavibacter michiganense* subsp. *sepedonicum* (Spieck & Kotth.) (Carlson & Vidaver) in common weed species found in Colorado potato fields. (Abstr.) Phytopathology 81:1348.

Hyperplastic Overgrowth Diseases

8

Some of the most interesting symptoms caused by plant-pathogenic bacteria are nodules, tumors, and excessive root, leaf, and floral proliferations. Basically, overgrowths of plant parts are caused by plant hormonal dysfunctions orchestrated by the infecting pathogen. Excessive production of indoleacetic acid (IAA) and of cytokinin is the primary cause of overgrowth diseases. Why bacteria cause overgrowth of the host has been an intriguing question. Presumably, the nodules, tumors, and other tissue proliferations serve as specialized, protective niches for best survival fitness against microbial competitors. Overgrowth diseases limit the growth of the infected host plant and can severely decrease crop yields.

Crown Gall

One of the most thoroughly studied diseases in plant pathology is crown gall, caused by *Agrobacterium tumefaciens*. Although many plants have been shown to produce crown gall upon experimental infection (reviewed in De Cleene and De Ley, 1976), crown gall naturally occurs primarily in woody crops, including members of the family Rosaceae. Cane gall is caused by *Agrobacterium rubi*, whereas hairy root is caused by *Agrobacterium rhizogenes*.

Crown gall tumors are observed most often at the base of the trunk and belowground on major roots. Once initiated, crown gall tumors continue to grow into relatively large, friable masses that dissociate into smaller pieces as the tumor ages and dries. Disseminated tumor fragments serve as sources of inoculum. Where there are sandy soils, tree crops, such as cherry, become infected in their scaffold branches and produce crown galls that are visible in low aerial portions of the tree. Aerial galls are presumably caused by windblown sand bearing inoculum to open wound sites, such as twig and leaf scars.

General Characteristics of *Agrobacterium tumefaciens*

Agrobacterium tumefaciens is a non-spore-forming, Gram-negative member of the class *Alphaproteobacteria* in the family *Rhizobiaceae*. The bacterium proliferates as small rods that are highly motile by the use of circumthecally arranged flagella. The flagella are not arranged peritrichously. Both virulent and avirulent strains produce common pili of 3-nm diameter. Only pathogenic strains produce an inducible T pilus of 10-nm diameter (Lai and Kado, 2000). On rich medium, colonies are nonpigmented and glossy when young and lose their glossy appearance with age. *Agrobacterium tumefaciens* is not fastidious; nutrient requirements are broad and simple. The organism grows aerobically (although *Agrobacterium tumefaciens* is capable of anaerobic growth via denitrification [Baek et al., 2008]). Certain strains produce 3-ketosugars, such as 3-ketolactose from lactose. All pathogenic strains contain a large circular plasmid known as the Ti (tumor-inducing) plasmid that is more than 200 kb long. A majority of the genes (at least 24) required for tumorigenicity are located on the Ti plasmid. Furthermore, genes conferring the biosynthesis and assimilation of auxin, cytokinin, sugar uptake gene (6b), and opines are located on the T-DNA of Ti and Ri (root-inducing) plasmids. The Ti and Ri plasmids bear conserved inducible transfer determinants (Tra genes) utilized in conjugative transfer of these plasmids between *Agrobacterium* strains.

Additional virulence genes are located on the chromosomes of *Agrobacterium tumefaciens* (reviewed in Charles and Nester, 1994; Suzuki et al., 2001). Interestingly, *Agrobacterium tumefaciens* contains two chromosomes, one of which is circular while the other is linear. The chromosomal virulence genes *chvA*, *chvB*, *chvD*, *chvE*, *chvG*, *chvH*, and *chvI* and those not directly affecting virulence, such as *acvB*, *miaA*, and *ros*, are located on the circular chromosome. Two additional genes, *pgm* (*exoC*) and *glgP*, which affect virulence, are located on the linear chromosome (Suzuki et al., 2001) (see Pathogenicity Factors below for details). The linear chromosome contains a relaxase and an associated *oriT* core (Goodner et al., 2001; Wood et al., 2001), which suggests that this chromosome could have originated from a large plasmid and might be functionally active.

Host Range

Agrobacterium tumefaciens is capable of causing tumors on many plants. More than 90 families of plants are susceptible to this pathogen (De Cleene and De Ley, 1976). Many of these plants were shown to be susceptible by experimental inoculations; symptoms were not observed to occur naturally in the field. Under natural field conditions, plants reported with crown gall are mainly woody crops and some field and glasshouse crops. Plants naturally infected by *Agrobacterium tumefaciens* in their native environments are listed in Table 8-1.

History of Crown Gall

Crown gall was described in biblical times on trees and grapevines as galls and nodules that were thought to be caused by foul air and vermin. In 1853, E. Fabre and F. Dunal published a detailed description of galls on grapevines (Fabre and Dunal, 1853). Although many investigators have credited Erwin F. Smith and C. O. Townsend for first isolating and identifying the crown gall bacterium (Smith and Townsend, 1907), it was actually Fridiano Cavara (Cavara, 1897) who should be credited with being the first investigator to successfully isolate the causal organism of crown gall. As director of the Royal Botanical Gardens at the University of Naples, Cavara noticed galls growing at the base of grapevines in the botanical garden and subsequently isolated on agar medium bacteria with dimensions of 0.5 × 1.5–2.0 μm. Inoculation of the pure culture into healthy grapevines gave rise to galls identical to those from which he first isolated the bacterium. In 1904, George Hedgcock isolated bacteria that produced white colonies on agar medium and caused the same type of galls as those from which he isolated the bacterium (Hedgcock, 1905). In 1907, Erwin F. Smith and C. O. Townsend isolated a bacterium from crown gall tumors and showed that the purified organism caused tumors on chrysanthemum, marguerite daisy, tobacco, tomato, potato, and sugar beet and on peach roots (Smith and Townsend, 1907). Host range studies were continued by Smith. By 1920, numerous reports were published on crown gall on stone fruit trees.

Table 8-1. Plants Naturally Found with Crown Gall Bearing *Agrobacterium tumefaciens*

Host	Host Family	Reference
Clockvine (*Thunbergia grandiflora* Roxb.)	Acanthaceae	U.S. Dep. Agric., 1960
Silver maple (*Acer saccharinum* L.)	Aceraceae	U.S. Dep. Agric., 1960
Bunya-bunya (*Araucaria bidwillii* Hook.)	Araucariaceae	U.S. Dep. Agric., 1960
Aster (*Aster* spp.)	Asteraceae	U.S. Dep. Agric., 1960
Marguerite daisy (*Chrysanthemum frutescens* L.)	Asteraceae	U.S. Dep. Agric., 1960
Mum (*Chrysanthemum morifolium* (Ramat.) Hemsl.)	Asteraceae	U.S. Dep. Agric., 1960
Dahlia (*Dahlia variabilis* (Willd.) Desf.)	Asteraceae	U.S. Dep. Agric., 1960
Begonia (*Begonia* sp.)	Begoniaceae	U.S. Dep. Agric., 1960
European hazel nut (*Corylus avellana* L.)	Betulaceae	U.S. Dep. Agric., 1960
Saguaro (*Carnegiea gigantea* (Engelm.) Britton & Rose)	Cactaceae	U.S. Dep. Agric., 1960
Common hop (*Humulus lupulus* L.)	Cannabaceae	U.S. Dep. Agric., 1960
Swampfly honeysuckle (*Lonicera oblongifolia* (Goldie) Hook.)	Caprifoliaceae	U.S. Dep. Agric., 1960
Common snowberry (*Symphoricarpos albus* (L.) S. F. Blake)	Caprifoliaceae	U.S. Dep. Agric., 1960
European cranberrybush (*Viburnum opulus* L.)	Caprifoliaceae	U.S. Dep. Agric., 1960
Weigela (*Weigela* spp.)	Caprifoliaceae	U.S. Dep. Agric., 1960
Carnation (*Dianthus caryophyllus* L.)	Caryophyllaceae	U.S. Dep. Agric., 1960
Oriental bittersweet (*Celastrus orbiculata* Thunb.)	Celastraceae	U.S. Dep. Agric., 1960
Beet (*Beta vulgaris* L.)	Chenopodiaceae	U.S. Dep. Agric., 1960
Dogwood (*Cornus florida* L.)	Cornaceae	U.S. Dep. Agric., 1960
Horseradish (*Armoracia rusticana* G. Gaertn., B. Mey. & Scherb.)	Cruciferae	U.S. Dep. Agric., 1960
Rutabaga (*Brassica campestris* L. var. *napobrassica* (L.) DC.)	Cruciferae	U.S. Dep. Agric., 1960
Turnip (*Brassica rapa* L.)	Cruciferae	U.S. Dep. Agric., 1960
White cedar (*Chamaecyparis lawsoniana* (A. Murray) Parl.)	Cupressaceae	U.S. Dep. Agric., 1960
Cypress (*Cupressus arizonica* Greene)	Cupressaceae	U.S. Dep. Agric., 1960
Chinese juniper (*Juniperus chinensis* L.)	Cupressaceae	U.S. Dep. Agric., 1960
Creeping juniper (*Juniperus horizontalis* Moench)	Cupressaceae	U.S. Dep. Agric., 1960
Persimmon (*Diospyros virginiana* L.)	Ebenaceae	U.S. Dep. Agric., 1960
Russian olive (*Elaeagnus angustifolia* L.)	Elaeagnaceae	U.S. Dep. Agric., 1960
Madrone (*Arbutus menziesii* Pursh)	Ericaceae	U.S. Dep. Agric., 1960
Blueberry (*Vaccinium corymbosum* L.)	Ericaceae	U.S. Dep. Agric., 1960
Castor bean (*Ricinus communis* L.)	Euphorbiaceae	U.S. Dep. Agric., 1960
Oak (*Quercus* spp.)	Fagaceae	U.S. Dep. Agric., 1960
Witch hazel (*Hamamelis* spp.)	Hamamelidaceae	U.S. Dep. Agric., 1960
Hippocratea (*Hippocratea obtusifolia* Roxb.)	Hippocrateaceae	U.S. Dep. Agric., 1960
Pecan (*Carya illinoinensis* (Wangenh.) K. Koch)	Juglandaceae	Smith, 1912
English walnut (*Juglans regia* L.)	Juglandaceae	U.S. Dep. Agric., 1960
Bird of paradise shrub (*Caesalpinia gilliesii* (Hook.) Wall. ex D. Dietr.)	Leguminosae	U.S. Dep. Agric., 1960
Pigeonpea (*Cajanus cajan* (L.) Millsp.)	Leguminosae	U.S. Dep. Agric., 1960
Sweet pea (*Lathyrus odoratus* L.)	Leguminosae	U.S. Dep. Agric., 1960
Alfalfa (*Medicago sativa* L.)	Leguminosae	U.S. Dep. Agric., 1960; Palumbo et al., 1998
Bean (*Phaseolus vulgaris* L.)	Leguminosae	U.S. Dep. Agric., 1960
Mesquite (*Prosopis* spp.)	Leguminosae	U.S. Dep. Agric., 1960
Cotton (*Gossypium hirsutum* L.)	Malvaceae	U.S. Dep. Agric., 1960
Cotton rosemallow (*Hibiscus mutabilis* L.)	Malvaceae	U.S. Dep. Agric., 1960
Rosemallow (*Hibiscus palustris* L.)	Malvaceae	U.S. Dep. Agric., 1960
India rubber tree (*Ficus elastica* Roxb. ex Hornem.)	Moraceae	U.S. Dep. Agric., 1960
Ironbark gum (*Eucalyptus leucoxylon* F. Muell.)	Myrtaceae	U.S. Dep. Agric., 1960
Goldenbells (*Forsythia suspensa* (Thunb.) Vahl)	Oleaceae	U.S. Dep. Agric., 1960
Jasmine (*Jasminum* spp.)	Oleaceae	U.S. Dep. Agric., 1960
Privet (*Ligustrum* spp.)	Oleaceae	U.S. Dep. Agric., 1960

(continued)

Table 8-1. Plants Naturally Found with Crown Gall Bearing *Agrobacterium tumefaciens (continued)*

Host	Host Family	Reference
Lilac (*Syringa vulgaris* L.)	Oleaceae	U.S. Dep. Agric., 1960
Peony (*Paeonia officinalis* L.)	Paeoniaceae	U.S. Dep. Agric., 1960
Summer phlox (*Phlox paniculata* L.)	Polemoniaceae	U.S. Dep. Agric., 1960
Rhubarb (*Rheum rhaponticum* L.)	Polygonaceae	U.S. Dep. Agric., 1960
Virgin's bower (*Clematis* spp.)	Ranunculaceae	Smith, 1912; U.S. Dep. Agric., 1960
Delphinium (*Delphinium × cultorum* Voss)	Ranunculaceae	U.S. Dep. Agric., 1960
Ceanothus (*Ceanothus* spp.)	Rhamnaceae	U.S. Dep. Agric., 1960
Japanese flowering quince (*Chaenomeles japonica* (Thunb.) Lindl. ex Spach)	Rosaceae	Smith, 1912; U.S. Dep. Agric., 1960
Quince (*Cydonia oblonga* Mill.)	Rosaceae	U.S. Dep. Agric., 1960
Apple (*Malus sylvestris* (L.) Mill.)	Rosaceae	U.S. Dep. Agric., 1960
Almond (*Prunus amygdalus* Batsch)	Rosaceae	U.S. Dep. Agric., 1960
Apricot (*Prunus armeniaca* L.)	Rosaceae	U.S. Dep. Agric., 1960
Cherry (*Prunus avium* L.)	Rosaceae	U.S. Dep. Agric., 1960
David peach (*Prunus davidiana* (Carrière) Franch.)	Rosaceae	U.S. Dep. Agric., 1960
Garden plum (*Prunus domestica* L.)	Rosaceae	U.S. Dep. Agric., 1960
Pear (*Pyrus communis* L.)	Rosaceae	U.S. Dep. Agric., 1960
Rose (*Rosa* spp.)	Rosaceae	U.S. Dep. Agric., 1960
Blackberry/dewberry/raspberry (*Rubus* spp.); black raspberry (*Rubus occidentalis* L.)	Rosaceae	Banfield, 1935; U.S. Dep. Agric., 1960
American mountain ash (*Sorbus americana* Marshall)	Rosaceae	U.S. Dep. Agric., 1960
Lemon (*Citrus × limon* (L.) Burm. f.)	Rutaceae	Smith, 1912
Grapefruit (*Citrus × paradisi* Macfad.)	Rutaceae	U.S. Dep. Agric., 1960
Cottonwood (*Populus* spp.)	Salicaceae	U.S. Dep. Agric., 1960
White poplar (*Populus alba* L.)	Salicaceae	U.S. Dep. Agric., 1960
Willow (*Salix alba* L.)	Salicaceae	U.S. Dep. Agric., 1960
Snapdragon (*Antirrhinum majus* L.)	Scrophulariaceae	U.S. Dep. Agric., 1960
Tobacco (*Nicotiana tabacum* L.)	Solanaceae	U.S. Dep. Agric., 1960
Nightshade (*Solanum dulcamara* L.)	Solanaceae	U.S. Dep. Agric., 1960
Camellia (*Camellia japonica* L.)	Theaceae	U.S. Dep. Agric., 1960
Parsnip (*Pastinaca sativa* L.)	Umbelliferae	U.S. Dep. Agric., 1960
Verbena (*Lippia* spp.)	Verbenaceae	Unger et al., 1985; L. W. Moore, personal commununication
Grape (*Vitis vinifera* L.)	Vitaceae	U.S. Dep. Agric., 1960

The crown gall bacterium was first named *Bacterium tumefaciens* and then changed to *Phytomonas tumefaciens* and eventually to *Agrobacterium tumefaciens*.

In 1930, an era of searching for the tumor-inducing factor began. There were three schools of thought on the source of this hypothetical factor. The favored idea was that *Agrobacterium tumefaciens* itself produced an endotoxin, plus ammonia and IAA, which caused the crown gall tumor. The second idea was that tumor formation was the result of the bacterium being a constant irritant that perturbed the plant to form tumors. A third idea was that the bacterium simply awakened a latent schizomycete that then caused the tumor—and the plant had to be "conditioned" before tumor development occurred. A major breakthrough on the crown gall problem was evidenced unequivocally through the demonstration that normal plant cells may be irreversibly converted into tumor cells in short periods of time by *Agrobacterium tumefaciens* and that the tumor cells can be axenically cultured in the absence of phytohormones (White and Braun, 1942; Braun and White, 1943; de

Ropp, 1947). The plant tissue culture studies provided strong evidence that crown gall tissue remains in a transformed state in the absence of bacteria. Subsequently, the so-called tumor-inducing principle was sought through studies on differences in the physiology and biochemistry between crown gall and healthy tissues and between *Agrobacterium tumefaciens* and other plant tumor-inducing bacteria, such as *Pseudomonas savastanoi* (now called *Pseudomonas savastanoi* pv. *savastanoi*). The direct study of *Agrobacterium tumefaciens* paid dividends when the bacterium became avirulent when cultured at 37°C or when treated with a DNA-intercalating dye, ethidium bromide. Based on results of acridine dye-mediated curing of the F plasmid in *Escherichia coli* (Hirota and Lijima, 1957; Hirota, 1960) and on the effect of elevated temperatures on curing plasmids in *Escherichia coli*, it was shown that *Agrobacterium tumefaciens* lost virulence when incubated at 37°C (Hamilton and Fall, 1971) or when in the presence of ethidium bromide (Lin and Kado, 1977). These results suggested that an extrachromosomal element is involved and plays a key role in virulence in *Agrobacterium tumefaciens*.

This idea was pursued by several laboratories by comparing *Agrobacterium tumefaciens* with *Agrobacterium radiobacter*, a naturally occurring avirulent relative. When these species were inoculated as a mixture and bacteria were subsequently recovered from tumors, the avirulent *Agrobacterium radiobacter* isolate was shown to have been converted in planta to the virulent form (Kerr, 1969; Kerr, 1971). Again, based on the F plasmid analogy of conjugative transfer between *Escherichia coli* strains, the nicely demonstrated transfer of virulence in planta strongly suggested that a plasmid carries the virulence factor or factors. Direct analysis of *Agrobacterium radiobacter* isolates and *Agrobacterium tumefaciens* for plasmids revealed a large 200-kb plasmid harbored in the virulent *Agrobacterium tumefaciens* strain and not in the avirulent *Agrobacterium radiobacter* isolate (Zaenen et al., 1974). Curiously, careful examination of a number of other *Agrobacterium radiobacter* isolates revealed the presence of plasmids similar in size to those found in virulent *Agrobacterium tumefaciens* isolates (Sheikholeslam et al., 1978; Sheikholeslam et al., 1979). It was, therefore, only by luck that the *Agrobacterium radiobacter* isolates used in the study by Zaenen et al. (1974) appeared plasmid free (Jeff Schell, personal communication). The correlation between the presence of the large plasmid (the Ti plasmid) and the tumor-forming ability of *Agrobacterium tumefaciens* was substantiated by genetic evidence that the Ti plasmid was conjugatively transferred in vitro from a virulent strain to an avirulent strain and that the avirulent recipient was converted into a virulent one (Van Larebeke et al., 1975; Watson et al., 1975).

Studies on crown gall tumors in axenic culture had revealed some clues as to the genetic nature of the tumor-inducing factor. Chemical analyses of extracts from crown gall tumors showed the presence of unique arginine derivatives (or guanidines) (Lioret, 1956; Morel, 1956), one of which was identified as *N*-carboxyethyl arginine (Ménagé and Morel, 1964). This interesting compound was first identified by Henze (Henze, 1908; Henze, 1911) and Morizawa (1927) as octopine in the muscle of invertebrates, such as *Octopus* spp., king scallop (*Pecten maximus* L.), and peanut worm (*Sipunculus nudus* L.), where octopine serves as a functional analog of lactic acid, i.e., NADH produced by glycolysis is oxidized to NAD during the synthesis of octopine, and NAD can be reduced to NADH by the reversal of the same process. In plants, octopine was unique to crown gall tissue and suggested that a novel biochemical pathway had been either activated or inserted into the crown gall cells.

Further chemical studies revealed the presence of additional unique guanidines, such as nopaline from a crown gall tumor derived from cactus (*Opuntia vulgaris* Mill.) (Goldmann et al., 1969). The particular guanidine synthesized in the crown gall tumor was determined by the particular strain of *Agrobacterium tumefaciens* that had induced the tumor and not by the genotype of the transformed plant cell (Goldmann et al., 1968; Petit et al., 1970). The unique guanidine compounds synthesized by crown gall tumor cells were collectively called opines (Tempé and Schell, 1977). The transfer of the Ti plasmid between isogenic *Agrobacterium* strains revealed that the Ti plasmids specified the type of opine produced by the tumor cell. A strain of *Agrobacterium tumefaciens* that specifies a particular type of opine that will be synthesized by the tumor cell also specifically utilizes the same opine for carbon and nitrogen nutrition. The pathogen has acquired the capacity to catabolize a given opine that is not readily utilized by competing free-living bacteria. Thus, *Agrobacterium tumefaciens* has developed a sophisticated means to genetically engineer the plant cell to produce a specific opine for its sole use.

Mutational studies of the Ti plasmid demonstrated that only a certain region of the circular DNA molecule is needed for tumorigenesis. Thus, only a certain portion of the Ti plasmid must be introduced into the plant cell. The original thinking regarding the Ti plasmid was that the entire plasmid was transferred into the plant during transformation. DNA:DNA reassociation kinetic studies using the Ti plasmid DNA as a probe failed to reveal its presence in axenically grown crown gall cells (reviewed by Chilton, 1982). However, methodical testing of individual fragments of the Ti plasmid cleaved by restriction enzymes demonstrated the presence of a signal in the DNA isolated from a tobacco crown gall tumor tissue culture line E9 (Chilton et al., 1977). *Sma*I restriction fragments 3b and 10c reassociated more rapidly with tumor DNA than with normal tobacco tissue DNA. These two fragments were found to be contiguous on the Ti plasmid physical map. The sector of the Ti plasmid containing *Sma*I fragments 3b and 10c overlapped with *Hin*dIII restriction fragment I (Chilton et al., 1978; Depicker et al., 1978). The transferred 25-kb DNA (T-DNA) was later found to have been integrated into the chromosome of the plant host and appeared to insert in a wide variety of chromosomal sites.

The T-DNA contains genes responsible for the synthesis of opines, auxin, and cytokinin, as well as for sugar uptake. These genes are more strongly expressed in the plant tumor than in the free-living bacterium. The auxin and cytokinin produced in the plant cell are critical compounds leading to abnormal cellular proliferation that culminates in tumor formation. Interestingly, deletion of the auxin- and cytokinin-encoding genes does not prevent insertion of the truncated or "disarmed" T-DNA into plant chromosomes because the borders of the T-DNA are the sites processed for T-DNA delivery and integration into the plant cell and its chromosomes. *Agrobacterium tumefaciens* transforms its hosts by transferring single-stranded copies of the T-DNA from the Ti plasmid into the plant cell and integrating it into the host genome. The VirD2 protein is covalently linked to the transferred T-DNA strand at its 5' end. The VirE2 protein is translocated into the same host cell independently of the T-DNA strand and associates with the T-DNA strand in the plant cytosol. The complex is then translocated into the nucleus, mediated by karyopherin-α, which binds to the VirD2 terminal protein. Nuclear import is further facilitated by the host VirE2-interacting protein VIP1, which serves as an adaptor (Citovsky et al., 2007). A second VirE2-interacting protein, VIP2, is required for *Agrobacterium tumefaciens*-mediated stable transformation but not for transient transformation (Anand et al., 2007).

Although it was not completely understood how the T-DNA was transferred into the plant cell and incorporated into the nuclear genome, work focused on those Ti plasmid genes involved in virulence and plant host-associated proteins opened new horizons and opportunities on the utility of the T-DNA delivery system (reviewed in Tinland and Hohn, 1995; Kado, 1998; Hansen and Chilton, 1999; Zhu et al., 2000; Zupan et al., 2000; Gelvin, 2003; Citovsky et al., 2007). The virulence regulon comprising all of the virulence genes in six operons on the Ti plasmid was fully sequenced in 1990 (Rogowsky et al., 1990). By 2001, the circular and linear chromosomes of *Agrobacterium tumefaciens* C58 were completely sequenced (Goodner et al., 2001; Wood et al., 2001). The updated, revised, annotated sequence of *Agrobacterium tumefaciens* C58 is available on GenBank and can be accessed at http://agro.vbi.vt.edu/public. The information on the sequence of the virulence regulon helped further characterize the nature and function of each virulence gene, and these were found to encode the T-DNA transfer machinery and apparatus. Genome sequence comparisons between *Agrobacterium tumefaciens* strains and their evolutionary significance have been recently reported (Slater et al., 2009).

The development of plant biotechnology can be credited to all the workers who contributed to our fundamental understanding of *Agrobacterium tumefaciens* and the crown gall disease. Crop plants transformed with novel genes conferring a number of commercially interesting properties through the use of Ti plasmid-mediated transfer of modified and disarmed T-DNAs have promoted the field of genetic engineering of crop plants. Genes conferring resistance to herbicides, insects, and various diseases caused by fungi, bacteria, and viruses have prevented huge crop damages and losses and have increased food and fiber production throughout the world. Moreover, genes conferring nutritional and health-benefiting traits in transgenic plants also have been brought to the market by *Agrobacterium tumefaciens*-mediated transformations. In addition, a complex of *Agrobacterium tumefaciens* virulence proteins VirD2 and VirE2, together with an in vitro synthesized T-DNA, can be used to transform mammalian cells. This has led to a means of developing novel transgenic cell lines (Pelczar et al., 2004). All of these benefits resulted from the application of molecular biology to a single plant disease and its pathogen.

Symptomatology

Crown gall tumors are the culmination of *Agrobacterium tumefaciens*-mediated transformation of cells surrounding the infection court. Infection courts where bacteria proliferate are established mainly through injuries caused by mechanical trauma, grafting, freezing temperatures, root-invading insects, and root-feeding animals. Conditions with high moisture levels favor infections leading to transformation. Transformation takes place in less than 3 h. Transformed cells proliferate into visible abnormal outgrowths of the woody tissues and vary in size according to age and conditions of growth. As the tumors grow, *Agrobacterium tumefaciens* is shed into the surrounding soil, particularly when rain or sprinkler irrigation occurs. Old tumors and their friable tissues no longer harbor *Agrobacterium tumefaciens*. It is therefore difficult to isolate the pathogen from old crown gall tumors. Woody galls formed on trees grow to very large sizes. Crown galls on canes and vines are formed along the vascular tissues, owing to the systemic invasiveness of *Agrobacterium tumefaciens* cells in these crops. During the winter months, *Agrobacterium tumefaciens* cells remain in the stationary phase of growth and begin to proliferate during the warmer spring months. Any winter injury, such as

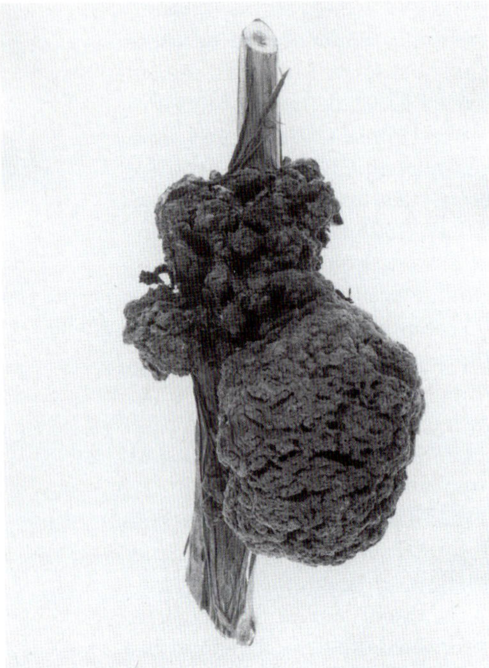

Fig. 8-1. Crown gall on a grapevine (*Vitis vinifera* L. cv. Sauvignon Blanc).

frost damage of tissues harboring the bacterial cells, initiates numerous tumors along the vascular regions of the vine or cane (Fig. 8-1).

Disease Cycle

Agrobacterium tumefaciens cells colonize the surface of freshly growing crown gall tumors and are shed onto other plant surfaces and into the soil surrounding the tumors. Subsequent plantings in the same soil expose the newly planted crop to the pathogen in the soil. *Agrobacterium tumefaciens* can persist on roots of crops such as cherry for at least 2 years (Stockwell et al., 1993). The pathogen can reside on the roots of native plants, including monocots, such as members of the family Poaceae (Bouzar and Moore, 1987). In fact, bindweed (*Convolvulus sepium* L.), commonly found in orchards, can harbor *Agrobacterium tumefaciens* in small root galls (C. I. Kado, *unpublished observations*). The bindweed can be used as an indicator plant to test for the presence of *Agrobacterium tumefaciens* in soil. Cultivating tools and plowing disks pulled behind tractors help disseminate the infested soil to other parts of the orchard. In nurseries, the practice of replanting in the infested soil results in *Agrobacterium tumefaciens*-contaminated rootstocks and scions being used in grafting. Crown gall appears at graft unions when infected scions are used. The cycle is repeated when *Agrobacterium tumefaciens* from freshly growing crown gall tumors is shed back into the soil and onto plant surfaces. Experiments with pinto bean leaves have shown that the minimum time required to convert a normal cell to a tumor cell is less than 2 h and may be as short as a few minutes (Lippincott and Lippincott, 1967). Hence, infections and transformations are likely to be established relatively quickly.

Pathogenicity Tests

Virulence can be assessed by inoculating various herbaceous plants, usually on fleshy stems, with a freshly grown culture of *Agrobacterium tumefaciens* delivered by several jabs with a sharp toothpick or needle. Test plants can vary in their susceptibility to *Agrobacterium tumefaciens*. For example, although succulent plants, such as *Kalanchoë* (previously called *Bryophyllum*) species, were first used by Levine (1919) and Smith (1921), they are not the most sensitive test plants for assessing the virulence of *Agrobacterium tumefaciens*. Moreover, *Kalanchoë* species have a tendency to form leafy tissues during gall formation (Levine, 1919; Smith, 1921). Nonetheless, the leaves of *Kalanchoë* species are used because of convenience in making pairwise comparisons of strains or mutants on a single leaf (Fig. 8-2). Some of the more sensitive test plants are Jimson weed (*Datura stramonium* L.), tomato (*Lycopersicon esculentum* Mill. var. *esculentum* Mill.), and cosmos (*Cosmos bipinnatus* Cav.) (Fig. 8-3).

Fig. 8-2. *Kalanchoë* leaf experimentally inoculated with *Agrobacterium tumefaciens*, bearing crown gall tumors. (Photo by Jeff Hall)

Fig. 8-3. Crown gall tumor formation as assayed on *Cosmos* and *Datura* plants.

In the absence of plant growth facilities, virulence tests can be performed directly in the laboratory. Inoculations with a cotton swab dipped in inoculum are made on freshly cut surfaces of surface-sterilized tubers of Jerusalem artichoke (*Helianthus tuberosus* L.) (Ark and Thompson, 1960), disks of carrot roots (Ark and Schroth, 1958), and potato tubers (Anand and Heberlein, 1977). Inoculated tubers or roots are stored in a moist chamber or in large petri dishes lined on the bottom with moistened filter paper. Tumors are formed in about 2–3 weeks.

Bacterial isolates from tumors can be recognized as *Agrobacterium tumefaciens* by the absorption of aniline blue dye. A mannitol agar medium containing (per liter) 5.0 g of mannitol, 1.0 g of yeast extract, 0.2 g of magnesium sulfate, 0.2 g of dipotassium phosphate, 0.2 g of sodium chloride, 0.1 g of calcium sulfate, 0.1 g of aniline blue, and 15 g of agar, adjusted to pH 7.0, is used. *Agrobacterium tumefaciens* absorbs aniline blue, while *Agrobacterium rhizogenes* and *Agrobacterium radiobacter* do not (Hendrickson et al., 1934). *Agrobacterium tumefaciens* also selectively absorbs malachite green in medium D1M (Perry and Kado, 1982). Medium D1M is a semiselective medium used in the isolation of *Agrobacterium tumefaciens* from soil extracts, as well as from suspected tumors.

Pathogenicity Factors

At least 38 genes confer virulence functions on *Agrobacterium tumefaciens*, and another four genes encode oncogenic compounds in the plant. The Ti plasmid contains 24 *vir* genes in the *vir* regulon (as defined by Rogowsky et al., 1990) that encode regulatory proteins responsible for upregulating the genes involved in the assembly of the type IV secretion apparatus (Christie and Vogel, 2000) (see Chapter 10) and for the processing and delivering of the T-DNA to host cells via the T pilus (Lai and Kado, 1998; Lai and Kado, 2000). The conjugative F pilus of *Escherichia coli* serves as the conduit for its DNA (Babic et al., 2008). Table 8-2 lists all of the

Table 8-2. Virulence Genes and Oncogenes and Their Function in *Agrobacterium tumefaciens*

Gene	Location	Function	References
virA	Ti plasmid	A 91.2-kDa sensory, trans-membrane histidine kinase homodimer that phosphorylates aspartic acid residue 52 of VirG, perceives methoxy-phenolic compounds (lignin precursors) with specificity, and is promoted by monosaccharides, low pH, and low phosphate; negatively regulates motility; phosphatase activity may be present; low pH sensing of VirA is dependent on VirA-ChvE interaction	Leroux et al., 1987; Melchers et al., 1989; Morel et al., 1989; Y. Huang et al., 1990; Jin et al., 1990a; Lee et al., 1998; Lai et al., 2000; Brencic et al., 2004a; Gao and Lynn, 2005; Wise et al., 2005; Wise et al., 2010
virB1	Ti plasmid	A 26.0-kDa bipartite protein with amino-terminal lytic transglycosylase (lysozyme) and a C-terminal VirB1*, associates with the T pilus; complemented by homologues; interacts with VirB8 and VirB10; a chaperone of VirB2 pilin subunits for T pilus biogenesis; nopaline type VirB1* is essential for T pilus formation	Bayer et al., 1995; Mushegian et al., 1996; Koraimann, 2003; Höppner et al., 2004; Zupan et al., 2007
virB2	Ti plasmid	A 12.3-kDa pro-pilin processed into a cyclic 7.2-kDa pilin subunit in T pilus biogenesis; interacts with T-strand in situ; secreted by the type IV secretion system	Shirasu and Kado, 1993a,b; Jones et al., 1996; Lai and Kado, 1998; Eisenbrandt et al., 1999; Lai et al., 2000; Lai et al., 2002; Cascales and Christie, 2004
virB3	Ti plasmid	A 11.6-kDa outer membrane protein stabilized by VirB4; interacts with VirB5; a component of the type IV secretion system; temperature sensitive above 20°C	Jones et al., 1994; Baron et al., 2001; Shamaei-Tousi et al., 2004
virB4	Ti plasmid	A 87.4-kDa cytoplasmic membrane ATPase, whose peptide segments interact with itself, VirB8, VirB10, and VirB11 and is required for configuring the trans-envelope channel; dedicated substrate exporter in the type IV secretion system	Shirasu et al., 1994; Dang and Christie, 1997; Dang et al., 1999; Ward et al., 2002
virB5	Ti plasmid	A 23.3-kDa single domain protein with three helices and a globular appendage; interacts with VirB3; a component of the type IV secretion system associated with T pilus biogenesis	Schmidt-Eisenlohr et al., 1999; Yeo et al., 2003; Shamaei-Tousi et al., 2004
virB6	Ti plasmid	A 31.8-kDa trans-membrane protein that interacts with T-strand in situ; a VirB7-VirB11 dependent localization to a cell pole; a component of the type IV secretion system; interacts with VirB5 for stabilization	Hapfelmeier et al., 2000; Jakubowski et al., 2003; Cascales and Christie, 2004; Judd et al., 2005
virB7	Ti plasmid	A 5.9-kDa outer membrane lipoprotein; stabilizes VirB9 by anchoring it to the outer membrane; interacts with VirB8 and VirB10; a component of the type IV secretion system	Fernandez et al., 1996; Spudich et al., 1996; Das and Xie, 2000; V. Sagulenko et al., 2001
virB8	Ti plasmid	A 26.4-kDa protein that interacts with T-strand in situ and initiates assembly of the type IV secretion apparatus by interacting with VirB9 and VirB10; self interacts into dimers	Das and Xie, 2000; Kumar et al., 2000; Kumar and Das, 2001; Cascales and Christie, 2004; Bailey et al., 2006

(continued)

Table 8-2. Virulence Genes and Oncogenes and Their Function in *Agrobacterium tumefaciens* (continued)

Gene	Location	Function	References
virB9	Ti plasmid	A 32.1-kDa protein interacts with T-strand in situ; forms disulfide linkage with VirB7 (outer membrane complex); a component of the type IV secretion system: contributes to substrate selection and translocation, including T pilus biogenesis and aids channel subunit contacts	Anderson et al., 1996; Spudich et al., 1996; Das and Xie, 2000; Cascales and Christie, 2004; Jakubowski et al., 2005
virB10	Ti plasmid	A 40.6-kDa protein; interacts with VirB9; associated with the T pilus; a component of the type IV secretion system	Berger and Christie, 1994; Das and Xie, 2000
virB11	Ti plasmid	A 38.1-kDa hexameric self-assembling ATPase; interacts with T-strand in situ; a component of the type IV secretion system; TrwD/PulE homolog	Rashkova et al., 2000; E. Sagulenko et al., 2001; Machón et al., 2002; Cascales and Christie, 2004
virC1	Ti plasmid	A 25.5-kDa protein whose expression is repressed by Ros; binds to overdrive sequence near the right T-DNA border; essential for infection of *Zea mays* L.	Close et al., 1987a; Grimsley et al., 1989; Toro et al., 1989
virC2	Ti plasmid	A 22.1-kDa protein whose expression is repressed by Ros; essential for infection of *Zea mays* L.	Close et al., 1987a; Grimsley et al., 1989
virD1	Ti plasmid	A 16.1-kDa DNA-relaxing enzyme	Ghai and Das, 1989
virD2	Ti plasmid	A 49.5-kDa protein with endonuclease activity in the N-terminal half that covalently links to 5' end of T-strands, contains two nuclear localization sequences for efficient nuclear localization; a MobA equivalent, binds to *Arabidopsis* cyclophilins	Yanofsky et al., 1986; Ward and Barnes, 1988; Young and Nester, 1988; Dűrrenberger et al., 1989; Steck et al., 1990; Howard et al., 1992; Vogel and Das, 1992a; Filichkin and Gelvin, 1993; Pansegrau et al., 1993; Jasper et al., 1994; Deng et al., 1998; Relić et al., 1998; Bravo-Angel et al., 1999
virD3	Ti plasmid	A 72.6-kDa protein that is not required for virulence and is not involved in T-DNA integration; however, it is involved in chemotaxis	Vogel and Das, 1992b; Koukolíková-Nicola et al., 1993; Lin and Kado, 1993; O. Chesnokova and C. I. Kado, *unpublished observations*
virD4	Ti plasmid	A 73.5-kDa ATPase required for T-strand transport to plant cells and their nuclei; interacts with VirB4 and VirB11; temperature sensitive above 20°C; required for transport of VirE2	Lin and Kado, 1993; Baron et al., 2001; Escudero et al., 2003; Atmakuri et al., 2004
virD5 (*orf5*)	Ti plasmid	A 97.4-kDa protein not required for virulence, independent of acetosyringone induction	Lin and Kado, 1993
virE0	Ti plasmid	Unknown function, acid inducible	Yuan et al., 2008
virE1	Ti plasmid	A 9.0-kDa protein associates with VirE2 and serves as a secretion chaperone of VirE2 protein in the type IV secretory pathway, including formation of channels; confers resistance for VirE2-mediated transformation	Deng et al., 1999; Sundberg and Ream, 1999; Zhao et al., 2001; Duckely et al., 2005; Humann et al., 2006

(continued)

Table 8-2. Virulence Genes and Oncogenes and Their Function in *Agrobacterium tumefaciens* *(continued)*

Gene	Location	Function	References
virE2	Ti plasmid	A 7.1-kDa T-strand chaperone in planta aiding delivery of the T-strand into the plant nucleus; the C-terminal region is required for its transport into plant cells; interacts with VirD4 coupling protein at the cell poles; complexes with VirE1 to form membrane channels	Christie et al., 1988; Citovsky et al., 1989; Sen et al., 1989; Yusibov et al., 1994; Rossi et al., 1996; Zupan et al., 1996; Sundberg and Ream, 1999; Zhou and Christie, 1999; Simone et al., 2001; Atmakuri et al., 2003; Citovsky et al., 2004; Duckely et al., 2005
virE3	Ti plasmid	A 63.4-kDa hydrophilic protein involved in host-range efficiency; binds to plant karyopherin alpha; facilitates VirE2 nuclear import function	Hirooka and Kado, 1986; Hirooka et al., 1987; Lacroix et al., 2005
virF (*pinF*)	Ti plasmid	A 22-kDa hydrophilic protein restricting host range; transferred to plant cells during tumor induction	Hooykaas et al., 1984; Otten et al., 1985; Melchers et al., 1990; Jarchow et al., 1991; Regensburg-Tuïnk and Hooykaas, 1993
virG	Ti plasmid	A 27.1-kDa transcriptional regulator of the *vir* regulon; the C-terminal half binds to *vir* boxes; VirG together with VirA is a member of the two-component signal transducing family of proteins	Jin et al., 1990b; Pazour and Das, 1990; Powell and Kado, 1990
virH1	Ti plasmid	A 46-kDa P450 monooxygenase; not required for efficient tumorigenesis	Kalogeraki and Winans, 1998
virH2	Ti plasmid	A 45.7-kDa *O*-demethylase inactivating *vir*-inducing phenolic compounds, rendering them less toxic	Kalogeraki et al., 1999; Brencic et al., 2004b
virI	Ti plasmid	Response regulatory protein; not required for efficient tumorigenesis	Kalogeraki and Winans, 1998
virJ	Located between *virA* and *virB* genes of octopine-type Ti plasmids	A 28.0-kDa periplasmic protein encoded by octopine-type Ti plasmids, a plant inducible gene of unknown function; homolog of chromosomal *acvB* gene; mediates protein-protein interactions between type IV substrates and apparatus	Wirawan et al., 1993; Kalogeraki and Winans, 1995; Pan et al., 1995; Pantoja et al., 2002
virK	Ti plasmid	A 15.9-kDa protein; not required for efficient tumorigenesis	Kalogeraki and Winans, 1998
virL	Ti plasmid	Not required for efficient tumorigenesis	Kalogeraki and Winans, 1998
virM	Ti plasmid	Not required for efficient tumorigenesis	Kalogeraki and Winans, 1998
chvA	Circular chromosome	A 135.2-kDa inner membrane glucan exporter with ATP binding activity; required for attachment and root nodulation; homolog of *Sinorhizobium ndvA* and *Brucella cgt* genes	Van Veen et al., 1987; Iñon de Iannino and Ugalde, 1989; Roset et al., 2004
chvB	Circular chromosome	A 235-kDa calcium-binding protein (rhicadhesin) for attachment to plant roots and nodulation; complemented by the *cviB* locus of *Azospirillum basilense*	Douglas et al., 1985; Van Veen et al., 1987; Zorreguieta et al., 1988; Swart et al., 1993; Raina et al., 1995

(continued)

Table 8-2. Virulence Genes and Oncogenes and Their Function in *Agrobacterium tumefaciens* *(continued)*

Gene	Location	Function	References
chvD	Circular chromosome	An ATP-binding transporter that potentiates *virG* expression at low pH and during phosphate starvation	Liu et al., 2001
chvE	Circular chromosome	Periplasmic monosaccharide-binding protein regulated by gbpR; interacts with the periplasmic domain of VirA	Melchers et al., 1989; Cangelosi et al., 1990; M. W. Huang et al., 1990; Shimoda et al., 1990; Doty et al., 1993
chvG	Circular chromosome	A 76-kDa protein with amino acid similarity to the family of bacterial histidine kinases	Charles and Nester, 1993; Mantis and Winans, 1993
chvH	Circular chromosome	A 51-kDa homolog of *Escherichia coli* elongation factor P	Peng et al., 2001
chvI	Circular chromosome	A 25.3-kDa response regulatory protein with 35% identity to PhoB of *Escherichia coli*; detoxifies tryptone, peptone, casamino acid containing media	Mantis and Winans, 1993
CS	Circular chromosome	A citrate synthase of 430 amino acids required for optimal virulence	Suksomtip et al., 2005
att	On pAtC58 between *avhB2* and *traA*	Host selective attachment to roots and root hair surfaces	Matthysse et al., 2000; Matthysse and McMahan, 2001; Chen et al., 2002; Matthysse et al., 2008
acvB	Chromosome	A 34.8-kDa protein with some similarity to the open reading frame downstream of virA	Wirawan et al., 1993; Pan et al., 1995
avhB	pAtC58	Type IV secretion system required for conjugal transfer of pAtC58 between agrobacteria	Chen et al., 2002
aviR	Chromosome of *Agrobacterium vitis* strain F2/5	A *luxR* homolog required for hypersensitive response (HR) on tobacco and necrosis on grapevine	Hao et al., 2005
glgP	Linear chromosome	Glycogen phosphorylase involved in exopolysaccharide synthesis	Suzuki et al., 2001
Hcp	Circular chromosome	A hemolysin-coregulated protein whose secretion is dependent on type VI secretion	Wu et al., 2008
Integrated inverted T DNA genes	Ti plasmid	Induces siRNAs that are recognized and suppressed	Dunoyer et al., 2006
Ipt, aux, 6a	T-DNA	Oncogenes synthesizing auxin, cytokinin, an opine such as octopine, and conferring sugar uptake by roots	Goodner et al., 2001; Suzuki et al., 2001; Wood et al., 2001; Clément et al., 2007
katA	Linear chromosome	A 78.7-kDa catalase that detoxifies hydrogen peroxide; regulated by OxyR, induced by plant tissue section acidic pH and low phosphate	Xu and Pan, 2000; Xu et al., 2001; Panek and O'Brien, 2004; Jamet et al., 2005; Yuan et al., 2005
miaA	Circular chromosome	A 33-kDa tRNA:isopentenyl transferase involved in the expression of *virB, virD, virE, pinF,* and *virG*; also present and functions in *Methylobacterium* spp. to form *trans*-zeatin	Gray et al., 1992; Koenig et al., 2002
ohrR	Circular chromosome	Transcriptional regulator of *ohr* and *ohrR*, induced by hydroperoxides to produce detoxification enzyme that degrades hydroperoxide	Chuchue et al., 2006

(continued)

Table 8-2. Virulence Genes and Oncogenes and Their Function in *Agrobacterium tumefaciens* (continued)

Gene	Location	Function	References
picA	Circular chromosome	A 25.5-kDa protein induced by pectin-derived acidic and neutral polysaccharides, mainly rhamnogalacturonan, arabinose, and galacturonic acid; these sugars also induce the *pgl* locus	Rong et al., 1991; Rong et al., 1994
pgmA	Linear chromosome	Phosphoglucose mutase involved in exopolysaccharide synthesis	Suzuki et al., 2001
pksA homologs	Linear chromosome (chromosome 2)	Polyketide and fatty acid synthases involved in grapevine necrosis by *Agrobacterium vitis*	Carle et al., 2008
pscA (*exoC*)	Circular chromosome	Succinoglycan synthetase; related to *exoC* gene of *Sinorhizobium meliloti*	Marks et al., 1987
psdA	Circular chromosome	Depresses succinoglycan production	Kamoun et al., 1989
ros	Circular chromosome	A 15.5-kDa repressor of *virC* and *virD* operons, and the T-DNA *ipt* gene; also regulates acidic succinoglycan synthesis	Close et al., 1985; Close et al., 1987b; Cooley et al., 1991; Kado, 2002b
sodBI, sodBII, sodBIII	Circular chromosome	Superoxide dismutase ferroisozymes	Saenkham et al., 2007
soxR	Circular chromosome	A member of the MerR family of transcriptional regulators; senses and regulates *atu5152* gene expression for superoxide stress protection	Eiamphungporn et al., 2006
Vp1	Located on pAtC58	VirD2 binding protein with putative nucleotidyltransferase motif	Guo et al., 2007
Vp2	Linear chromosome	VirD2 binding protein with putative nucleotidyltransferase motif	Guo et al., 2007
Vp3	Linear chromosome	VirD2 binding protein with putative nucleotidyltransferase motif	Guo et al., 2007

known genes involved in virulence, oncogenicity, defense, and niche adaptation. Additional genes involved in virulence may be uncovered by further mining the genomic and Ti plasmid sequences that have become available (Goodner et al., 2001; Suzuki et al., 2001; Wood et al., 2001).

Genetic Engineering of Crop Plants

As described in the section above, our knowledge gained from studies on *Agrobacterium tumefaciens* and its T-DNA delivery system that incorporates a novel gene or gene sets into the chromosome of various plants opened new opportunities in plant biotechnology. A number of binary vector systems, i.e., two plasmids, of which one contains the *vir* regulon and the other contains the disarmed T-DNA, were developed and the *Agrobacterium tumefaciens*-mediated transformation was optimized (these vector systems are now commercially available). Multiple copies of the *vir* regulon (Zyprian and Kado, 1990) and the addition of polyamines enhanced the transformation efficiency (Kumar and Rajam, 2005). Insertion of a novel gene into specific genes in a crop plant has been achieved (Offringa et al., 1992). At least 13 different genetically engineered crop plants are currently commercially propagated. They include cotton, canola, corn, sweet corn, flax, papaya, chicory, soybean, squash, sugar beet, tomato, potato, and sugarcane. In the near future, many more genetically engineered crops will be introduced commercially to enhance agricultural production and food and fiber quality. Interestingly, public objections to the use of genetically engineered crops remain. Unknown to most objectors is the fact that the human population has been consuming *Agrobacterium rhizogenes* T-DNA genes for centuries. These genes are naturally present in several crops, for example, carrot (Fründt et al., 1998; Kado, 2002a).

Hairy Root

Hairy root is caused by *Agrobacterium rhizogenes*, a member of the class *Alphaproteobacteria* in the family *Rhizobiaceae*. Like crown gall, hairy root is initiated at wound sites on mainly woody crops, such as apple, although row crops can become infected when planted in soil that has been cropped previously with hairy root-diseased plants. Hairy root is typified by the mass of numerous fibrous roots growing from a gall on roots and at graft unions. The roots are fleshy when their outgrowths are recent, and they become fibrous and intertwined with age. Moisture promotes the growth of fleshy roots. Because hairy roots grow from a basal gall, the disease was often referred to by nurserymen as woolly knot.

General Characteristics of *Agrobacterium rhizogenes*

Agrobacterium rhizogenes is a Gram-negative, rod-shaped, non-spore-forming bacterium that is highly motile. Cells are propelled by circumthecally arranged flagella. The organism is more fastidious than *Agrobacterium tumefaciens* in that it is unable to utilize ammonium ions and amino acids, except glutamic acid, as sources of nitrogen. Biotin is a critical growth factor. The organism grows well in glycerol phosphate-mannitol (GPM) agar medium composed of (per liter) 20 g of mannitol, 0.8 g of calcium glycerol phosphate, 5 g of potassium nitrate, 3.8 g of sodium chloride, 0.1 g of potassium chloride, 1.0 g of magnesium chloride, 0.6 g of magnesium sulfate, 100 µg of biotin, and 0.1 g of glutamic acid. Unlike *Agrobacterium tumefaciens*, colonies do not absorb Congo red or aniline blue on GPM agar containing either dye (Hendrickson et al., 1934). This is a handy and useful distinguishing phenotype for differentiating between *Agrobacterium rhizogenes* and *Agrobacterium tumefaciens*.

Agrobacterium rhizogenes contains a 3.7-Mbp circular chromosome and a second 2.7-Mbp chromosome that can be either linear or circular (Urbanczyk et al., 2003). Pathogenic strains harbor a large, circular, 217-kb plasmid called the Ri plasmid. In Ri plasmids of agropine-producing strains of *Agrobacterium rhizogenes*, the T-DNA is split into two regions, a 21.1-kb TL-DNA and a 28-kb TR-DNA, both of which are integrated into the genome of the host plant cells. Both regions contribute to the hairy root phenotype. The genes of the TR-DNA code for the biosynthesis of auxin and cytokinin, whereas the genes of the TL-DNA are needed for hairy root production. These genes may play a negative regulatory role during pathogenesis (Lemcke and Schmulling, 1998). The TR-DNA contains several root-inducing genes designated as *rol* genes, such as *rolA*, *rolB*, *rolC*, and *rolD* (White et al., 1985). Interestingly, the *rolA* gene contains a spliceosomal intron located in the untranslated leader region of its mRNA (Magrelli et al., 1994). Among the *rol* genes, *rolB* is considered the most important control gene of the hairy root phenotype (Tanaka et al., 2001). The exact role of *rolB* in hairy root production is unclear. Recently, the ca. 30-kDa RolB protein was shown to localize in the nucleus in spite of the lack of a typical nuclear localization signal sequence (Moriuchi et al., 2004). RolB interacts with host 14-3-3 proteins that contain a nuclear export signal.

Table 8-3. Host Range of *Agrobacterium rhizogenes*

Host	Host Family	Reference
Tatarian honeysuckle (*Lonicera tatarica* L.)	Caprifoliaceae	U.S. Dep. Agric., 1960
Common snowberry (*Symphoricarpos albus* (L.) S. F. Blake)	Caprifoliaceae	U.S. Dep. Agric., 1960
Cucumber (*Cucumis sativus* L.)	Cucurbitaceae	Weller et al., 2000
Russian olive (*Elaeagnus angustifolia* L.)	Elaeagnaceae	U.S. Dep. Agric., 1960
Silverberry (*Elaeagnus commutata* Bernh. ex Rydb.)	Elaeagnaceae	U.S. Dep. Agric., 1960
English walnut (*Juglans regia* L.)	Juglandaceae	Riker et al., 1930
Pea tree (*Caragana arborescens* Lam.)	Leguminosae	U.S. Dep. Agric., 1960
Honey locust (*Gleditsia triacanthos* L.)	Leguminosae	U.S. Dep. Agric., 1960
Soybean (*Glycine max* (L.) Merr.)	Leguminosae	Tolin and Lacy, 2004
Hollyhock (*Althaea rosea* (L.) Cav.)	Malvaceae	U.S. Dep. Agric., 1960
White mulberry (*Morus alba* L.)	Moraceae	U.S. Dep. Agric., 1960
White ash (*Fraxinus americana* L.)	Oleaceae	U.S. Dep. Agric., 1960
Cotoneaster (*Cotoneaster* spp.)	Rosaceae	U.S. Dep. Agric., 1960
Quince (*Cydonia oblonga* Mill.)	Rosaceae	U.S. Dep. Agric., 1960
Apple (*Malus* × *domestica* L. Borkh.)	Rosaceae	Riker et al., 1930
Almond (*Prunus amygdalus* Batsch)	Rosaceae	Riker et al., 1930
Rose (*Rosa* spp.)	Rosaceae	Riker et al., 1930
Blackberry (*Rubus* spp.)	Rosaceae	Riker et al., 1930
Red raspberry (*Rubus idaeus* L.)	Rosaceae	U.S. Dep. Agric., 1960
Western blackberry (*Rubus ursinus* Cham. & Schltdl.)	Rosaceae	U.S. Dep. Agric., 1960
Spirea (*Spiraea* spp.)	Rosaceae	U.S. Dep. Agric., 1960

Host Range

The host range of *Agrobacterium rhizogenes* is relatively broad when the organism is tested experimentally. Natural infections of hairy root are most commonly associated with woody crops, such as apple, and some row crops planted in infested soil. Listed in Table 8-3 is the host range of *Agrobacterium rhizogenes* naturally occurring in the field.

It should be noted that there are several herbaceous hosts that can be used for experimentally assaying *Agrobacterium rhizogenes*. For example, tomato and tobacco have been commonly used. *Agrobacterium rhizogenes* has also been used to initiate roots in the vegetative propagation of pine, larch, jujube, hazelnut, almond, walnut, pear, pine, etc. (viz., Damiano and Monticelli, 1998).

History of Hairy Root

Prior to 1929, the terms hairy root and crown gall were used interchangeably because tumors or galls always formed prior to abnormal proliferation of roots. There was much confusion over whether the then-called *Bacterium tumefaciens* was the sole causal organism or whether there were different strains, such as the apple strain that seemed to cause hairy root sometimes and not at other times. Workers had argued that hairy root or woolly knot might be due to excessive graft injury. In 1929, E. A. Siegler (Siegler, 1929) reported that M. B. Waite was able to control crown gall and woolly gall (hairy root) by using a solution of formaldehyde applied to the scions and stock previous to grafting. The experiments were repeated by Siegler using a solution of hydroxymercurichlorophenol as a disinfectant. Siegler concluded that the woolly knot type of malformation is pathogenic in its nature and not caused by wounding alone at graft unions. By 1930, W. H. Wright and colleagues separated the crown gall organism from the hairy root organism by a tedious single-bacterial-cell isolation technique (Wright et al., 1930). The workers showed conclusively that one of the single-celled strains consistently showed the typical characteristics of the crown gall organism, while the other three consistently showed those of the hairy root organism. Hence, these organisms were distinct, and the hairy root organism was named *Phytomonas rhizogenes* n. sp. (Riker et al., 1930).

Symptomatology

From its name, hairy root or woolly knot is characterized by abnormal hyper-proliferation of roots from the infection site (Fig. 8-4). Hairy root primordia are initiated from the neoformed cambial layers derived from the stem cambium (Bercetche et al., 1987).

Disease Cycle

The cycle of *Agrobacterium rhizogenes* infection begins when plants with hairy root are pulled out of the soil, leaving behind the roots and rooted galls of the infected plant. The bacterial cells remain resident in the soil and infect new seedlings as they are planted in the vicinity of *Agrobacterium rhizogenes* cells. Nursery stocks that were planted in infested soil serve to spread infected seedlings into new, often distant, areas that were originally free of *Agrobacterium rhizogenes.* Row crops planted in *Agrobacterium rhizogenes*-infested soil–straw beds become infected and produce massive root hair balls. The planting beds serve as the source of primary inoculum in subsequent plantings.

Pathogenicity Tests

The presence of *Agrobacterium rhizogenes* can easily be determined by inoculating sterilized carrot disks (Fig. 8-4), which are placed in petri plates lined on the bottom with moistened filter paper. Alternatively, surface-sterilized cucumber seeds may be germinated on water agar, and after a few days, the hypocotyls and cotyledons of the germinated seedlings are inoculated. Hairy roots appear from the small galls formed at the inoculation sites.

Pathogenicity Factors

The T-DNA located on the Ri plasmid of *Agrobacterium rhizogenes* and the TL-DNA of agropine strains of *Agrobacterium rhizogenes* contain the root-inducing genes. At least four genes, called *rolA*, *rolB*, *rolC*, and *rolD*, are involved in root induction. Two genes of the T-DNA encode tryptophan monooxygenase and indoleacetamide hydratase, which catalyze the formation of auxin. The product of the *rolB* gene alone is sufficient to induce adventitious roots (Cardarelli et al., 1987; Spena et al., 1987). To a lesser extent, the products of *rolA*, *rolC*, and *rolD* can independently induce root formation (Spena et al., 1987).

Fig. 8-4. Carrot root disk inoculated with *Agrobacterium rhizogenes*. Roots proliferating from the pericyclic cambium can be grown in tissue culture.

The T-DNA on the Ri plasmid is transferred by the same type IV secretion machinery as that encoded by the *vir* regulon on the Ti plasmid of *Agrobacterium tumefaciens*. The characteristics of the *vir* genes contained within the *vir* regulon are like those that are listed in Table 8-2. However, *Agrobacterium rhizogenes* encodes and secretes a protein called GALLS that can functionally substitute for VirE2 encoded by the *virE* operon of *Agrobacterium tumefaciens* (Hodges et al., 2004). Yet, GALLS does not resemble VirE2; instead, it shares amino acid sequences with TraA, which is essential for Ti plasmid conjugal transfer via the type IV secretion system. Its sequence also contains domains for ATP binding and nuclear localization (Hodges et al., 2006). Like VirE2, GALLS is probably localized to the plant nuclei. The built-in functional redundancies in pathogenic bacteria that require highly sophisticated adaptive mechanisms are reflected by GALLS. For example, the type III effector protein, HopPtoV, plays a functionally redundant role in *Pseudomonas syringae* pathogenesis (Wehling et al., 2004) (see Chapter 10).

Olive and Oleander Knot

This tumor disease is caused by *Pseudomonas savastanoi* pv. *savastanoi*, a pathogen whose host range is confined to members of the family Oleaceae, principally olive and oleander. The pathogen enters through natural or artificial breaks in the epidermis or periderm of the host. Fresh leaf scars created by strong winds are sites of infection since the scars remain susceptible to infection for about a week (Hewitt, 1938). Once infection is established, *Pseudomonas savastanoi* pv. *savastanoi* cells move systemically in the laticifers (latex vessel elements) to distal regions of the tree or shrub (Wilson and Magie, 1964). Subsequent wounding by windblown rain and sand results in leaf scars and abrasive injuries, sites that are foci for gall formation. Also, wounds created by pruning serve as infection foci.

The early stages of infection are recognized by water-soaking and dissolution of tissues surrounding the infection site. There is disorganization of various tissues followed by proliferation of cells into tumors or knots. The development of these knots occurs in the spring when the host is most active in growth. Knots persist for many years and no longer harbor the pathogen. Fresh tumors or knots appearing at distal parts of the tree or shrub contain the active pathogenic bacterial cells.

General Characteristics of *Pseudomonas savastanoi* pv. *savastanoi*

A member of the class *Gammaproteobacteria*, *Pseudomonas savastanoi* pv. *savastanoi* is a Gram-negative, rod-shaped, aerobic bacterium with polar flagella. Physiologically, the organism is negative for arginine dihydrolase and cytochrome oxidase activities. It does not oxidize glucose to 2-ketogluconate but utilizes glucose aerobically. Cells do not produce levan. In the presence of arginine, the bacterium can grow anaerobically because ATP is generated via fermentation of this substrate. The bacterial cells secrete a diffusible yellow-green pigment that fluoresces in ultraviolet light when grown on an appropriate medium, e.g., King's medium B (King et al., 1954). A semiselective medium, PVF-1, is used to enrich *Pseudomonas savastanoi* pv. *savastanoi* isolates from diseased plant samples (Surico and Lavermicocca, 1989). Optimal growth occurs at 26–28°C.

Although *Pseudomonas savastanoi* pv. *savastanoi* strains isolated from diseased olive trees are unable to grow on sucrose as the sole carbon source, strains isolated from oleander knots can utilize sucrose as the only carbon source (Wilson et al.,

1972). The phytohormones IAA and cytokinin are required for knot production. IAA is produced from tryptophan by its conversion to indoleacetamide, which in turn is converted into IAA (Yamada et al., 1985). Two genes, *iaaM*, encoding tryptophan monooxygenase, and *iaaH*, encoding indoleacetamide hydrolase, are arranged as an operon that is harbored in virulent strains of *Pseudomonas savastanoi* pv. *savastanoi* (Gaffney et al., 1990). Interestingly, *Pseudomonas savastanoi* pv. *savastanoi* can conjugate IAA with lysine to form 3-indoleacetyl-ε-L-lysine (IAA-lysine). The enzyme that converts IAA to IAA-lysine is (indole-3-acetyl)-L-lysine synthetase and is encoded by the *iaaL* gene that is located on plasmid pIAA1. Sequence analysis revealed that *iaaM* and *iaaH* have significant similarity with homologous genes of other plant-associated bacteria, whereas no nucleotide homologies have been found with the *iaaL* gene. Interestingly, most ash isolates of *Pseudomonas savastanoi* pv. *savastanoi* produce low levels of IAA and no cytokinin (Iacobellis et al., 1998).

Olive strains of this pathogen harbor the *iaaM* and adjacent *iaaH* genes as an operon in their chromosome, while the same operon is located on a plasmid, pIAA1, in oleander strains (Palm et al., 1989). Enzymes for biosynthesis of cytokinin in the form of *trans*-zeatin and isopentenyladenine are encoded by the *ptz* and *ipt* genes, respectively. The genes are located respectively on the plasmid pIAA1 or pCK1 (Comai and Kosuge, 1980; Powell and Morris, 1986; Akiyoshi et al., 1987). Sequence homologies occur between the *ptz* and *ipt* genes of *Pseudomonas savastanoi* pv. *savastanoi* and the *tzs* and *tmr* genes of *Agrobacterium tumefaciens*, an observation that suggests that lateral gene transfer has occurred between these bacteria. On the other hand, these homologous genes might have been acquired independently from a common ancestor through lateral gene transfer events. Comparative genetic analysis of 32 plasmids from 10 strains of *Pseudomonas savastanoi* pv. *savastanoi* revealed the presence of mosaic arrangements, direct repeats, and insertion elements, all of which suggest that recent lateral gene transfer events have taken place (Pérez-Martínez et al., 2008).

Host Range

The host range of *Pseudomonas savastanoi* pv. *savastanoi* is limited to members of the family Oleaceae. This pathogen infects olive (*Olea europaea* L.), oleander (*Nerium oleander* L.), Japanese privet (*Ligustrum japonicum* Thunb.), jasmine (*Jasminum* spp.), goldenbells (*Forsythia viridissima* Lindl.), and mock privet (*Phillyrea latifolia* L.). Infection of ash is typified by wartlike overgrowths rather than by tumorlike growths.

Cultivated olive varieties (cultivars) known to be susceptible to *Pseudomonas savastanoi* pv. *savastanoi* are Columbella, Nevadillo Blanco, Manzanillo, Sevillano, and Ascolano. The cultivar Mission is somewhat less susceptible than these varieties. The Portuguese cultivar Galega Vulgar is semiresistant.

History of Olive Knot

Olive knot was observed in biblical times (see Chapter 1). The bacterial cause of olive knot was demonstrated in 1881 by Luigi Savastano, who worked at the School of Agriculture in Portici and did his bacteriological studies in a lab set up in the Zoological Laboratory in Naples, Italy (Savastano, 1886). In 1890, N. B. Pierce (Pierce, 1891) reported observing bacteria in olive stem tissue surrounded by abnormally enlarged host cells. In 1898, F. T. Bioletti (Bioletti, 1898) observed olive knot in Merced County, California. By 1907, R. E. Smith (Smith, 1907) reported the disease in the Sacramento Valley of California. The pathogen isolated by Savastano was initially named *Bacillus oleae*. This name was first changed to *Bacillus oleae tuberculosis* and subsequently changed to *Bacterium savastanoi* by E. F. Smith in 1908 (Smith, 1908).

Symptomatology

Tumor induction by *Pseudomonas savastanoi* pv. *savastanoi* proceeds in successive steps and initial symptoms are difficult to assess. The site of infection usually is a wound, such as a leaf scar. Colonization of the wound site takes place and water-soaking occurs around the infected site. This is followed by hypertrophic growth of the host cells around the infection site. Cells nearest the bacterial biofilm (surface-attached communities of bacterial cells) undergo hypertrophy first and this cell response progressively extends to more distal locations of the host tissue. This is followed by hyperplastic growth and the formation of swollen outgrowths that develop into tumorous tissue (Fig. 8-5). The pathogen moves systemically up and down the infected stem or branch. Localized injuries induce formation of additional tumors. Frequently, leaf scars and bud scars are initial sites of infection with the appearance of localized tumors. Tumors continue to grow as the plant ages. Old

Fig. 8-5. Olive knot on young branches of an olive tree.

trees display long-standing tumors, but it is often difficult to isolate *Pseudomonas savastanoi* pv. *savastanoi* from them.

Disease Cycle

Fresh inoculum is not present on the surface of the knots during hot and dry periods of the year, but *Pseudomonas savastanoi* pv. *savastanoi* appears on the surface of the knots when wet. An aqueous film is essential for the exudation of the bacterial cells. Infection spread through splashing rain aided by wind is facilitated during the wet periods of December, January, and February. Wounds, including those made by hail, frost, windblown sand, and man-made pruning, are primary sites of infection. Strong winds aid in defoliating portions of the tree, leaving bountiful leaf scar wounds for entry of *Pseudomonas savastanoi* pv. *savastanoi*. Natural abscission of blossom racemes leaves raceme scars, which also can be sites of infection. Once infection is established, the bacterial cells move through the lacticifers to distal parts of the host. Even the roots of the olive tree form knots.

Pathogenicity Tests

Young, vigorously growing olive or oleander plants are used for pathogenicity tests. Overnight cultures of *Pseudomonas savastanoi* pv. *savastanoi* grown in medium 523 (Kado and Heskett, 1970) are harvested by centrifugation (6,000 × *g*, 10 min), resuspended in distilled water, and injected into the cortex with a fine hypodermic needle. Alternatively, inoculum may be delivered as a droplet placed on the surface of a leaf scar made by severing a leaf petiole close to the shoot. The pathogen can be readily and specifically detected by culture enrichment of cells derived from diseased tissue followed by polymerase chain amplification of the *iaaL* gene using primers IAALF (5'-GGCACCAGCGGCAACATCAA-3') and IAALR (5'-CGCCCTCGGAACTGCCATAC-3') (Penyalver et al., 2000).

Pathogenicity Factors

Pseudomonas savastanoi pv. *savastanoi* produces both IAA and cytokinin. These phytohormones serve as the primary components in the virulence of this pathogen. Tryptophan is the substrate required for the production of IAA. It is converted into indole-3-acetamide (IAM) by tryptophan-2-monooxygenase, which is encoded by the *iaaM* gene. IAM is then converted into IAA and ammonia by indoleacetamide hydrolase, which is encoded by the *iaaH* gene. In oleander strains of *Pseudomonas savastanoi* pv. *savastanoi*, the *iaaM* and *iaaH* genes, which are arranged as an operon, are located on a 52-kb plasmid, such as pIAA1 (Comai and Kosuge, 1980), and the 73-kb plasmid pIAA2 (Surico, 1986; Soby et al., 1994), whereas these genes are chromosomally located in olive strains of the pathogen (Comai et al., 1982). The expression of *iaaM* and *iaaH* is constitutive. The addition of L-tryptophan increases IAA production but not the IAA-producing enzymes themselves (Smidt and Kosuge, 1978; Surico et al., 1985). IAA is further metabolized by the pathogen, conjugating IAA with lysine to form IAA lysine. The conjugate formation is catalyzed by the (indole-3-acetyl)-L-lysine synthetase that is encoded by *iaaH* located on plasmid pIAA1 (Glass and Kosuge, 1986).

Besides IAA, *Pseudomonas savastanoi* pv. *savastanoi* also produces cytokinins. At least four cytokinins, zeatin, zeatin riboside, 1'-methylzeatin, and 1''-methylzeatin riboside, are synthesized, with zeatin being the most abundantly produced (Surico and Iacobellis, 1992). Isopentenyl phosphotransferase encoded by the *ipt* gene is the principal enzyme that catalyzes the transfer of an isoprenoid moiety derived from mevalonic acid via dimethylallyl pyrophosphate to 5'-adenosine monophosphate, generating isopentenyl monophosphate (iPMP). This latter compound is further hydroxylated to form zeatin and zeatin riboside. The *ipt* (also called *ptz*) gene is located on a 42-kb plasmid (MacDonald et al., 1986; Powell and Morris, 1986). The gene is strikingly homologous to the *tzs* and *ipt* genes of *Agrobacterium tumefaciens* (Powell and Morris, 1986).

Along with the phytohormone production required for virulence, *Pseudomonas savastanoi* pv. *savastanoi* is equipped with a type III secretion system (TTSS) that directly delivers effector proteins into the host cell (Hueck, 1998; Galan and Collmer, 1999; Aldridge and Hughes, 2001; Plano et al., 2001). The TTSS is also employed in bacterial flagellar assembly (Macnab, 1999). The product of the *virPphA* gene homolog found in *Pseudomonas savastanoi* pv. *savastanoi* is likely to be one of the effector proteins (Jackson et al., 2002). The effector VirPphA shares 52% amino acid identity with AvrPtoB produced by *Pseudomonas syringae* pv. *tomato.* This effector protein suppresses programmed cell death, thus facilitating host susceptibility to the pathogen (Abramovitch et al., 2003). The roles in pathogenesis of many other effector proteins, which have been identified on the basis of their ability to elicit the hypersensitive reaction of resistant hosts, remain poorly understood (Abramovitch et al., 2003).

In addition to IAA, which is required for virulence, the membrane localized regulator, MorA, controls the timing of flagellar development and governs chemotaxis leading to biofilm formation (Choy et al., 2004). Biofilms are formed during colonization of the leaf scars and other wounds made on the host.

Doubtless other genes involved in virulence, such as those required for growth on specific substrates of members of the family Oleaceae, will be revealed through current genome-based analyses. Conserved among all bacterial pathogens are the essential genes required for chromosomal integrity and perpetuation as well as the customary housekeeping genes (Galperin et al., 2006). However, the most important genes are those that confer on the pathogen its ability to occupy, survive, and proliferate in a given host. Many of these pathogenesis-related and virulence genes are on extrachromosomal elements, such as plasmids, or on intrachromosomal, specific pathogenicity islands. The ability of *Pseudomonas savastanoi* pv. *savastanoi* to proliferate and move in the laticifers suggests that latex-based precursors as well as latex itself might be utilized by this pathogen. Clearly, because of its relatively narrow host range, *Pseudomonas savastanoi* pv. *savastanoi* harbors specific substrate utilization genes to effectively colonize the host.

Gypsophila Gall

Pantoea agglomerans pv. *gypsophilae* causes gall formation on baby's breath plants (*Gypsophila paniculata* L.), a perennial European herb grown outdoors or under glass for cut flowers (Cooksey, 1986). The host plant is vegetatively propagated by rooting stem cuttings. The cut ends of the stem cuttings are the sites of entry and colonization of *Pantoea agglomerans* pv. *gypsophilae*, forming galls on the cut ends. A related pathovar is *Pantoea agglomerans* pv. *betae*, which causes galls on beet (*Beta vulgaris* L.), wisteria (*Wisteria* spp.), Douglas-fir (*Pseudotsuga taxifolia* (Poir.) Britton), and cranberry (*Vaccinium macrocarpon* Aiton) (Manulis and Barash, 2003a; Manulis and Barash, 2003b).

General Characteristics of *Pantoea agglomerans* pv. *gypsophilae*

Pantoea agglomerans pv. *gypsophilae* is a member of the class *Gammaproteobacteria* in the family *Enterobacteriaceae.* Its cells are Gram negative, measuring 1 × 3 μm when grown on rich media, and produce peritrichous flagella. It is cytochrome oxidase negative, reduces nitrate to nitrite, produces β-galactosidase, and is facultatively anaerobic. *Pantoea agglomerans* pv. *gypsophilae* grows in minimal medium composed of glucose, ammonium chloride, potassium phosphate, and magnesium sulfate. Casein hydrolysate added to the medium stimulates growth. Colonies on medium 523 (Kado and Heskett, 1970) or other rich media are pigmented light to deep yellow, smooth with entire margins, and semitranslucent. Optimum growth occurs between 28 and 30°C.

Host Range

The host range is restricted to *Gypsophila* spp. (Volcani, 1985). *Gypsophila* species are used in commercial cut flower production in California and Florida (United States), Israel, Taiwan, Japan, and to a limited extent, Scotland (Graham and Quinn, 1974; Cooksey, 1986; Hseu and Hsu, 1998).

History of Gypsophila Gall

The gall disease of *Gypsophila paniculata* L. was first reported in 1932 by Nellie A. Brown (Brown, 1932), who isolated and classified the yellow-pigmented bacterium as *Bacterium gypsophilae.* Because of its tumor-inducing activity, *Bacterium gypsophilae* was later changed to *Agrobacterium gypsophilae* by H. J. Conn (Conn, 1942) in the appendix to the genus *Agrobacterium* in the sixth edition of *Bergey's Manual of Determinative Bacteriology* (cited in Starr and Weiss, 1943). Conn had received a culture of the bacterium from N. A. Brown, Washington, DC. Being a facultative anaerobe, *Agrobacterium gypsophilae* did not possess the characteristics of the genus *Agrobacterium* but rather those of the members of the family *Enterobacteriaceae.* The species was hence moved into the genus *Erwinia* as *Erwinia herbicola* (Graham and Quinn, 1974; Miller et al., 1981) and later as the pathovar *Erwinia herbicola* pv. *gypsophilae* (Cooksey, 1986). The yellow-pigmented members of the family *Enterobacteriaceae* were all moved into the genus *Pantoea* within this family (Gavini et al., 1989). *Erwinia herbicola* pv. *gypsophilae* was subsequently changed to *Pantoea agglomerans* pv. *gypsophilae.*

Fig. 8-6. *Pantoea agglomerans* pv. *gypsophilae*-induced galls on cuttings of *Gypsophila paniculata* L. (Courtesy Isaac Barash)

Symptomatology

Pantoea agglomerans pv. *gypsophilae* induces gall formation at freshly made wounds of the stem cuttings of *Gypsophila paniculata* L. used for vegetative propagation (Fig. 8-6).

Disease Cycle

Galled *Gypsophila* cuttings and infected mother plants are the primary sources of inoculum. *Pantoea agglomerans* pv. *gypsophilae* is disseminated from infected plants via cutting tools and splashing water. Recycled irrigation water can serve as a reservoir of the pathogen; as the inoculum builds up, more and more cuttings become infected. The practice of placing freshly cut, bundled shoots in water contaminated with *Pantoea agglomerans* pv. *gypsophilae* effectively disseminates the pathogen to all the cuttings. Apparently, *Pantoea agglomerans* pv. *gypsophilae* moves from the galls to the shoots via the xylem vessels (Manulis and Barash, 2003b). Strict sanitation practices are mandatory for controlling the disease.

Pathogenicity Tests

Relatively young, vigorous plants of *Gypsophila paniculata* L., grown under glass, are used as the source of cuttings for pathogenicity tests. Freshly taken stem cuttings are dipped into inoculum prepared by suspending agar colonies of *Pantoea agglomerans* pv. *gypsophilae* in 0.01 M phosphate buffer, pH 7.0, and these are planted in potting mix for rooting. The formation of galls at the ends of the cuttings takes about 10 days.

Pathogenicity Factors

Pathogenicity and virulence (the degree of pathogenesis) of *Pantoea agglomerans* pv. *gypsophilae* are conferred by products encoded by genes on a large resident plasmid (ca. 150 kb) called pPATH (Manulis et al., 1991; Lichter et al., 1995). A 135-kb indigenous pPATH from strain Pag824-1 has been sequenced (Barash et al., 2005). It contains a pathogenicity island of about 75 kb that bears a functional *hrp* gene cluster (Mor et al., 2001). Both IAA and cytokinin are produced by this pathogen, utilizing the pPATH genes involved in the biosynthesis of these phytohormones (Clark et al., 1993; Guo et al., 2001; Barash and Manulis-Sasson, 2007). Like the IAA-producing pathway of *Pseudomonas savastanoi* pv. *savastanoi*, tryptophan-2-monooxygenase and indoleacetamide hydrolase are encoded by the *iaaM* and *iaaH* genes on pPATH. In addition, the cytokinin-producing gene *ipt* (or *etz*) encoding isopentenyl transferase is also located on pPATH (Guo et al., 2001). The pPATH pathogenicity island contains the *hrp* and *hrc* gene clusters that encode type III virulence effectors. Interspersed among these virulence effector genes are insertion sequences (Guo et al., 2002).

The pPATH pathogenicity island contains open reading frames (ORFs) with homologs to HrcJ, HrcV, HrcN, HrpO, HrpP, $HrcQ_A$, $HrcQ_B$, HrcR, HrcS, HrcT, and HrcU (Nizan et al., 1997). The conserved *hrc* genes are homologs of the genes of the TTSS found in many Gram-negative pathogens (see Chapter 10). For example, AvrPphd, HopPmaB, and HopPmaH are common to *Pantoea agglomerans* pv. *gypsophilae* and *Pseudomonas syringae* pathovars. Mutations in the *hrc* genes completely abolish the ability of *Pantoea agglomerans* pv. *gypsophilae* to induce galls (Nizan et al., 1997). Regulation of the *hrp* operon is through the activation of HrpL in the presence of sigma factor σ^{54} (Nissan et al., 2005). Some interesting effectors are translocated by the TTSS. PthG, HsvB, and HsvG are host range determinants (Ezra et al., 2000; Nissan et al., 2006). The PthG protein has a dual function in that it elicits the hypersensitive response in beet species, such as Avr of *Pseudomonas syringae* pathovars, and serves as a virulence factor in *Gypsophila* species. A mutation in PthG extends the host range of *Pantoea agglomerans* pv. *gypsophilae* to cause galls on beet. HsvG confers host specificity on both *Pantoea agglomerans* pv. *gypsophilae* and *Pantoea agglomerans* pv. *betae* (Valinsky et al., 2002). Both HsvB and HsvG are transcriptional activators that contain direct repeats and nuclear localization signals. These direct repeats were discovered to determine host specificity (Nissan et al., 2006).

Fasciation

Fasciation is characterized by the excessive and abnormal proliferation of leaves, flowers, and shoots, including gall formations on leaves and at wounds on stems on the plant host as a result of hyperplastic growth of the affected tissues. The disease is caused by *Rhodococcus fascians*, a member of the family *Nocardiaceae* in the class *Actinobacteria*. As with other pathogens that cause overgrowth diseases, *Rhodococcus fascians* produces indole-3-acetic acid and cytokinins. The disease is particularly important in the cut flower industry.

General Characteristics of *Rhodococcus fascians*

As a member of the family *Nocardiaceae*, *Rhodococcus fascians* is Gram positive and nonmotile with relatively non-pleomorphic but slightly curved rod-shaped cells that are either linked into chains forming friable branched hyphae or vary from cocci to long, unsegmented filaments. Colonies are slow growing on nutrient agar and are round, convex, and opaque. Best growth is obtained when arginine, biotin, histidine, and thiamine are present in the growth medium. *Rhodococcus fascians* is unable to utilize starch and is nitrate reductase negative. It utilizes urea, hydrolyzes esculin and allantoin, and produces catalase. Optimum growth temperature is 26°C. Owing to their capacity to produce carotenoid-like pigments, colonies range in color from yellow to orange, depending on their age. They grow on mannitol agar that is composed of (per liter) 10 g of mannitol, 1 g of yeast extract, 0.5 g of dipotassium phosphate, 0.5 g of ammonium sulfate, 0.2 g of sodium chloride, 0.2 g of magnesium sulfate, 3 g of calcium carbonate, 0.01 g of thiamine hydrochloride, and 15 g of agar.

Host Range

The host range of *Rhodococcus fascians* is broad and includes certain monocots as well as many dicots. *Rhodococcus fascians* has been isolated from sweet pea (*Lathyrus odoratus* L.), florist's chrysanthemum (*Chrysanthemum morifolium* (Ramat.) Hemsl.), chrysanthemum (*Chrysanthemum indicum* L.), jade (*Crassula* sp.), geranium (*Pelargonium* × *hortorum* L. H. Bailey), strawberry (*Fragaria chiloensis* (L.) Mill. var. *ananassa* L. H. Bailey), mullein (*Verbascum thapsus* L.), wallflower (*Cheiranthus cheiri* L.), dahlia (*Dahlia variabilis* (Willd.) Desf.), tobacco (*Nicotiana tabacum* L.), primrose (*Primula* sp.), gladiolus (*Gladiolus* sp.), coral bells (*Heuchera* sp.), beet (*Beta vulgaris* L.), phlox (*Phlox paniculata* L.), potato (*Solanum tuberosum* L.), petunia (*Petunia* × *atkinsiana* D. Don ex Louden), carnation (*Dianthus caryophyllus* L.), stock (*Matthiola* sp.), storkbill (*Pelargonium* sp.), snapdragon (*Antirrhinum majus* L.), asparagus (*Asparagus officinalis* L.), lily (*Lilium regale* E. H.

Wilson), phlox (*Phlox* sp.), azalea (*Rhododendron* sp.), and polyanthus (*Primula elatior* (L.) Hill) (Baker, 1950; Mohanty, 1951; Lacey, 1955).

History of Fasciation

Fasciation of sweet pea was first reported by P. E. Tilford (Tilford, 1936) and Margaret Lacey (Lacey, 1936). The disease was also called leafy gall, witches'-broom, and cauliflower depending on the host; cauliflower, for example, was a fasciation disease of strawberry (Lacey, 1936; Mohanty, 1951). Because of the variation in overgrowth symptoms, which include prolific floral, leaf, and stem production in a dense unorganized form, the disease is best called fasciation as originally defined by Tilford and Lacey. The causal organism was named *Phytomonas fascians*. Lacey (Lacey, 1939) then changed its name to *Bacterium fascians*, which was then changed to *Corynebacterium fascians* (Dowson, 1942). Owing to its general resemblance to actinomycetes, *Corynebacterium fascians* was reclassified as *Rhodococcus fascians* (Goodfellow, 1984).

Fasciation continues to sporadically make its appearance in areas where floriculture is concentrated. For example, fasciation of carnation, stock, and chrysanthemum periodically appears in Santa Cruz County, California, while the fasciation disease of strawberry is seen from time to time in Monterey County, California.

Symptomatology

Overgrowth symptoms vary depending on the host and the environment under which the plant host is grown. Multiple dwarfed flowers, leaves, and stems originating from a single meristematic region or a wound site are typical of fasciation (Fig. 8-7). Galls may form at leaf margins and along veins. Experimental inoculations on leaves or stems usually produce galls followed by the formation of leaves on the galls themselves. Dwarfing of the plant is often observed. Application of an antibiotic that kills *Rhodococcus fascians* results in remediation of the disease and normal outgrowth of shoots. Severely infected plants occasionally show a thick taproot with few lateral roots.

Fig. 8-7. Fasciation of carnation from Santa Cruz, California (left). Healthy plant (right).

Disease Cycle

Rhodococcus fascians is an epiphytic pathogen and thus is disseminated by splashing water in the form of rain or sprinkler irrigation (Baker, 1950). Moisture abundance is important for dissemination. Regions of low humidity with semiarid conditions are not conducive to the disease (Baker, 1950). The disease may be seedborne since *Rhodococcus fascians* can be isolated from seeds derived from diseased plants. In fields where fasciation occurs, pockets of endemic residence of *Rhodococcus fascians* can be found. The bacterial cells often are concentrated in various leaf and meristematic cavities. The pathogen is opportunistic and colonizes intercellular spaces when entry is gained through wounds (Cornelis et al., 2001). The disease is spread from infected plants mainly through rain and irrigation and by handling during the harvest of flowers. Once colonized and under appropriate conditions, which remain unclear (Goethals et al., 2001), fasciation of various types is induced. Both abiotic stress on the host and ensuing signals trigger the release of pathogenicity factors from *Rhodococcus fascians*.

Pathogenicity Tests

A number of plants can serve as the test host for *Rhodococcus fascians*. In general, young seedlings germinated from surface-sterilized seeds are used. Sweet pea seeds are surface sterilized by soaking for 3 min in 70% ethanol, germinated in dishes on moist paper towels, and planted into moist sand. Inoculations are performed with 50 μl of turbid inoculum (ca. 10^6 CFU/ml) placed directly on the uncut apical meristem. Fasciation appears after about 1 week of incubation.

Pathogenicity Factors

Both auxin and cytokinins contribute to fasciation. The cytokinin synthase gene of the *fas* operon is located on a conjugative linear plasmid of 180 kb (Crespi et al., 1992). Cytokinin is produced by the product of the *ipt* gene homolog (ORF4) of the six-gene *fas* operon. ORF1 encodes a homolog of cytochrome oxidase P450, ORF2 encodes a product with certain similarities to ferredoxin, ORF3 encodes a product similar to the β-subunit of pyruvate dehydrogenase, ORF5 encodes a cytokinin oxidase homolog, and ORF6 encodes a glutathione *S*-transferase-like homolog (Goethals et al., 2001). These products are thought to modulate cytokinin activity and therefore virulence. Interestingly, the *fas* operon also occurs on a large (about 325 kb), mobile pathogenicity island in *Streptomyces turgidiscabies* (Kers et al., 2005). The pathogenicity island has virulence-associated genes and genes that encode thaxtomin, which inhibits cellulose biosynthesis. *Rhodococcus fascians* and *Streptomyces turgidiscabies* are Gram-positive organisms, each equipped with a homologous *fas* operon arranged collinearly. Both are soil inhabitants and are in close association with each other, thus having ample opportunities for genetic exchange.

Expression of the *fas* locus requires the *att* locus. Sequence analysis of the *att* region of the plasmid mentioned above has revealed the presence of 10 ORFs, *attX*, *attA*, *attB*, *attC*, *attD*, *attE*, *attF*, *attG*, *attH*, and *attR* (Maes et al., 2001). The *att* locus plays an important regulatory role in virulence, mediated through the synthesis of uncharacterized inducing factors that are present only in gall extracts. Induction of the *fas* locus requires the presence of inducing factors (Maes et al., 2001) and is controlled at both the transcriptional and the translational levels (Temmerman et al., 2000). Besides plasmid genes involved in virulence, a chromosomal gene called *vicA* also affects the virulence of *Rhodococcus fascians* (Vereecke et al., 2002). This gene encodes a malate synthase and is required for efficacious growth of hyperplastic tissues.

Novel Gall Diseases

From the analyses of various gall-forming bacteria, it is clear that they have acquired and share genes that encode auxin and cytokinin. The acquisition of these "oncogenes" is likely mediated through horizontal gene transfer mechanisms, such as promiscuous conjugal transfer, integrative phages, and conjugative transposons. Novel gall diseases caused by bacteria will likely be discovered. For example, a gall disease of himeyuzuriha (*Daphniphyllum teijsmanni* Zoll.) is caused by *Pseudomonas syringae* pv. *daphniphylli* (Ogimi et al., 1990). Galls are formed on the trunks and branches of himeyuzuriha trees in various regions of Japan. This gall-producing pathogen is highly species specific. Inoculation tests on 70 other tree species were negative.

Aster Yellows

Aster yellows diseases are caused by members of the genus '*Candidatus* Phytoplasma', composed of at least 75 species (*Candidatus* nomen species). Symptoms begin with a general yellowing of the foliage followed by growth distortions, such as the loss of apical dominance that results in shoot and floral proliferation into a witches'-broom appearance. The conversion of petals and sepals of flowers into additional flowers or into leaflike structures is called phyllody (Fig. 8-8), similar to the structures in Figure 8-7. The yellowing leads to a general decline of the plant, culminating in death. Because aster yellows occurs in a range of domesticated plants, the disease remains economically important.

General Characteristics of '*Candidatus* Phytoplasma asteris'

'*Candidatus* Phytoplasma' species are members of the bacterial class *Mollicutes* that lack the characteristic rigid peptidoglycan and cell wall of most bacteria. They are Gram-negative pleomorphic cells with a small AT-rich genome (580–1,350 kb) and low guanine-plus-cytosine (GC) content. Currently, they are not culturable in vitro and are considered obligate parasites that reside in the plant phloem. In planta,

Fig. 8-8. Aster yellows blossom phyllody (right) compared with a normal healthy flower (left). (Photo by Jeff Hall)

they appear as 50- to 1,000-nm amorphous bodies (Doi et al., 1967). Phytoplasmas are able to replicate in the plant host and in the insects that vector them. The aster yellows group phytoplasma called '*Candidatus* Phytoplasma fraxini' (Griffiths et al., 1999) remains latent in various weeds (e.g., dandelion and thistle; Table 8-4 shows a complete list). The pathogen is picked up by and replicated in the aster leafhoppers *Macrosteles fascifrons* (Stål) and *Macrosteles quadrilineatus* Forbes and transmitted to nearby crops much like the phytoplasma of X-disease (Raine and Forbes, 1969; Purcell et al., 1981). After 8–10 days, the aster leafhopper becomes infectious and can transmit this organism to healthy plants for the remainder of its life span. Interestingly, the aster leafhopper lives longer when exposed to aster yellows phytoplasma (Beanland et al., 2000). Extension of longevity of the infectious leafhopper suggests that the pathogen and insect host have coexisted through evolution. Phytoplasmas have apparently evolved from acholeplasma ancestors (Lim and Sears, 1992; Gundersen et al., 1994) and their evolution has not stopped, apparently because they lack DNA repair mechanisms. Eventually, the ever-changing environment and lifestyle can convert the pathogen into an endosymbiont through mutator alleles and horizontal gene transfer (Taddel et al., 1997) (see Chapter 2). In response to changing environments, bacterial metabolic networks also evolve by the uptake of peripheral genes and operons through lateral gene transfer (Pál et al., 2005).

Host Range

The aster yellows phytoplasma has a very broad and extensive host range covering at least 60 plant families. Examples are listed in Table 8-4. '*Candidatus* Phytoplasma asteris' is a member of the aster yellows group of phytoplasmas that cause more than 100 hyperplastic overgrowth diseases. The wide host range reflects a highly diverse and widespread phytoplasma group. In particular, all plants of the family Asteraceae appear to be susceptible and therefore the disease name aster yellows appears most appropriate (Kunkel, 1926). The extremely broad host range indicates a lack of host specificity of the aster yellows pathogen. The organism appears to be nonfastidious since it has a broad utilization of nutrients in the phloem of plants as well as the ability to utilize nutrients in different leafhopper vectors. The absence of one or more two-component signal transduction systems in the phytoplasma, which are usually needed to activate housekeeping genes in planta for Gram-negative bacteria, likely evolved to allow the phytoplasma to exist on a diverse food base.

History of Aster Yellows

Aster yellows was first reported and described by Kunkel in 1926 (Kunkel, 1926) when he noted the unusual yellowing and multiple floral formations in aster plants (Fig. 8-8). Originally, aster yellows was thought to be caused by a virus since no organism could be cultured in vitro and since aster leafhoppers, other leafhoppers, and grafting efficiently transmitted the disease. The western aster yellows strain was readily transmitted from aster to celery by leafhoppers (Kunkel, 1932). The remission of disease symptoms by infiltrating tetracycline into diseased plants prompted cytological inspections of infected plants (Ishiie et al., 1967). Electron microscopic examination of diseased tissues of several yellows-type diseased plants by Doi and coworkers (Doi et al., 1967) revealed the presence of pleomorphic bodies in the

Table 8-4. Example of the Wide Range of Hosts of Phytoplasma Strains of the Aster Yellows Group

Host	Host Family	Reference
Shallot (*Allium ascalonicum* L.)	Alliaceae	U.S. Dep. Agric., 1960
Onion (*Allium cepa* L.)	Alliaceae	U.S. Dep. Agric., 1960; Khadhair et al., 2002
Garlic (*Allium sativum* L.)	Alliaceae	U.S. Dep. Agric., 1960; Khadhair et al., 2002
Rough pigweed (*Amaranthus retroflexus* L.)	Amaranthaceae	U.S. Dep. Agric., 1960
Madagascar periwinkle (*Catharanthus roseus* (L.) G. Don)	Apocynaceae	Torres et al., 2004
Periwinkle (*Vinca major* L.)	Apocynaceae	U.S. Dep. Agric., 1960
Date palm (*Phoenix dactylifera* L.)	Arecaceae	Alhudalb et al., 2007
Milkweed (*Asclepias* spp.)	Asclepiadaceae	U.S. Dep. Agric., 1960
Asparagus (*Asparagus officinalis* L.)	Asparagaceae	Fránová and Petrzik, 2009
Common ragweed (*Ambrosia artemisiifolia* L.)	Asteraceae	U.S. Dep. Agric., 1960
Chamomile (*Anthemis arvensis* L.)	Asteraceae	U.S. Dep. Agric., 1960
Mayweed (*Anthemis cotula* L.)	Asteraceae	U.S. Dep. Agric., 1960
Lazy daisy (*Aphanostephus humilis* (Benth.) A. Gray)	Asteraceae	O'Mara et al., 1993
Aster (*Aster* spp.)	Asteraceae	Kunkel, 1926
Beggars-ticks (*Bidens frondosa* L.)	Asteraceae	U.S. Dep. Agric., 1960
Bur marigold (*Bidens pilosa* L.)	Asteraceae	U.S. Dep. Agric., 1960
Swan River daisy (*Brachycome iberidifolia* Benth.)	Asteraceae	U.S. Dep. Agric., 1960
Bracted strawflower (*Bracteantha bracteata* (Vent.) Anderb. & Haegi)	Asteraceae	U.S. Dep. Agric., 1960
Pot marigold (*Calendula officinalis* L.)	Asteraceae	U.S. Dep. Agric., 1960
Basket flower (*Centaurea americana* Nutt.)	Asteraceae	U.S. Dep. Agric., 1960
Cornflower (*Centaurea cyanus* L.)	Asteraceae	U.S. Dep. Agric., 1960
Mums (*Chrysanthemum morifolium* (Ramat.) Hemsl.) and all other *Chrysanthemum* spp.	Asteraceae	U.S. Dep. Agric., 1960
Endive (*Cichorium endivia* L.)	Asteraceae	U.S. Dep. Agric., 1960
Chicory (*Cichorium intybus* L.)	Asteraceae	U.S. Dep. Agric., 1960
Horseweed (*Conyza canadensis* (L.) Cronquist)	Asteraceae	O'Mara et al., 1993
Tickseed (*Coreopsis grandiflora* Hogg ex Sweet)	Asteraceae	U.S. Dep. Agric., 1960
Tickseed (*Coreopsis lanceolata* L.)	Asteraceae	U.S. Dep. Agric., 1960
Cosmos (*Cosmos bipinnatus* Cav.)	Asteraceae	U.S. Dep. Agric., 1960
Artichoke (*Cynara scolymus* L.)	Asteraceae	U.S. Dep. Agric., 1960
Dahlia (*Dahlia variabilis* (Willd.) Desf.)	Asteraceae	U.S. Dep. Agric., 1960
Cape marigold (*Dimorphotheca aurantiaca* DC.)	Asteraceae	U.S. Dep. Agric., 1960
Daisy fleabane (*Erigeron annuus* (L.) Pers.)	Asteraceae	U.S. Dep. Agric., 1960
Horseweed (*Erigeron canadensis* L.)	Asteraceae	U.S. Dep. Agric., 1960
Streamside fleabane (*Erigeron glabellus* Nutt.)	Asteraceae	U.S. Dep. Agric., 1960
Gaillardia (*Gaillardia aristata* Pursh)	Asteraceae	U.S. Dep. Agric., 1960
Quickweed (*Galinsoga parviflora* Cav.)	Asteraceae	U.S. Dep. Agric., 1960
Cudweed (*Gnaphalium decurrens* Ives)	Asteraceae	U.S. Dep. Agric., 1960
Yellowdicks (*Helenium amarum* (Raf.) H. Rock var. *amarum* (Raf.) H. Rock)	Asteraceae	U.S. Dep. Agric., 1960
Sneezeweed (*Helenium autumnale* L.)	Asteraceae	U.S. Dep. Agric., 1960
Rosilla (*Helenium puberulum* DC.)	Asteraceae	O'Mara et al., 1993
Sunflower (*Helianthus annuus* L.)	Asteraceae	U.S. Dep. Agric., 1960
Coastal tarweed (*Hemizonia corymbosa* (DC.) Torr. & A. Gray)	Asteraceae	O'Mara et al., 1993
Garden lettuce (*Lactuca sativa* L.) and related lettuce species	Asteraceae	Severin, 1929
Prickly lettuce (*Lactuca serriola* L.)	Asteraceae	U.S. Dep. Agric., 1960; Koike and Davis, 2002
Fall dandelion (*Leontodon autumnalis* L.)	Asteraceae	U.S. Dep. Agric., 1960
Common tarweed (*Madia elegans* D. Don ex Lindl.)	Asteraceae	U.S. Dep. Agric., 1960

(continued)

Table 8-4. Example of the Wide Range of Hosts of Phytoplasma Strains of the Aster Yellows Group *(continued)*

Host	Host Family	Reference
Pineapple weed (*Matricaria matricarioides* (Less.) Porter)	Asteraceae	U.S. Dep. Agric., 1960
Guayule (*Parthenium argentatum* A. Gray)	Asteraceae	U.S. Dep. Agric., 1960
Congress grass (*Parthenium hysterophorus* L.)	Asteraceae	Raj et al., 2008b
Oxtongue (*Picris echioides* L.)	Asteraceae	O'Mara et al., 1993
Stevia (*Piqueria trinervia* Cav.)	Asteraceae	U.S. Dep. Agric., 1960
Texas dandelion (*Pyrrhopappus multicaulis* DC.)	Asteraceae	O'Mara et al., 1993
Black-eyed Susan (*Rudbeckia hirta* L.)	Asteraceae	U.S. Dep. Agric., 1960
Black salsify (*Scorzonera hispanica* L.)	Asteraceae	U.S. Dep. Agric., 1960
Dusty miller (*Senecio cineraria* DC.)	Asteraceae	U.S. Dep. Agric., 1960
Ragwort (*Senecio vulgaris* L.)	Asteraceae	U.S. Dep. Agric., 1960
Sowthistle (*Sonchus arvensis* L.)	Asteraceae	U.S. Dep. Agric., 1960; Koike and Davis, 2002
Aztec marigold (*Tagetes erecta* L.)	Asteraceae	U.S. Dep. Agric., 1960
French marigold (*Tagetes patula* L.)	Asteraceae	U.S. Dep. Agric., 1960
Common dandelion (*Taraxacum officinale* F. H. Wigg.)	Asteraceae	U.S. Dep. Agric., 1960; Koike and Davis, 2002
Salsify (*Tragopogon porrifolius* L.)	Asteraceae	U.S. Dep. Agric., 1960
Golden crownbeard (*Verbesina encelioides* (Cav.) Benth. & Hook. f. ex A. Gray)	Asteraceae	O'Mara et al., 1993
Zinnia (*Zinnia elegans* Jacq.)	Asteraceae	Severin, 1929
Begonia (*Begonia semperflorens* Link & Otto)	Begoniaceae	O'Mara et al., 1993
Forget-me-not (*Myosotis scorpioides* L.)	Boraginaceae	U.S. Dep. Agric., 1960
Edging lobelia (*Lobelia erinus* L.)	Campanulaceae	O'Mara et al., 1993
Hemp (*Cannabis sativa* L.)	Cannabaceae	Raj et al., 2008a
Papaya (*Carica papaya* L.)	Caricaceae	O'Mara et al., 1993
Sweet William (*Dianthus barbatus* L.)	Caryophyllaceae	U.S. Dep. Agric., 1960
Carnation (*Dianthus caryophyllus* L.)	Caryophyllaceae	U.S. Dep. Agric., 1960
Baby's breath (*Gypsophila paniculata* L.)	Caryophyllaceae	U.S. Dep. Agric., 1960
Corn spurry (*Spergula arvensis* L.)	Caryophyllaceae	U.S. Dep. Agric., 1960; O'Mara et al., 1993
Common chickweed (*Stellaria media* (L.) Vill.)	Caryophyllaceae	U.S. Dep. Agric., 1960
Prickly Russian thistle (*Salsola kali* L. subsp. *tragus* (L.) Čelak.)	Chenopodiaceae	Koike and Davis, 2002
Spinach (*Spinacia oleracea* L.)	Chenopodiaceae	U.S. Dep. Agric., 1960
Horseradish (*Armoracia rusticana* G. Gaertn., B. Mey. & Scherb.)	Cruciferae	O'Mara et al., 1993
Bird rape (*Brassica campestris* L.)	Cruciferae	U.S. Dep. Agric., 1960
India mustard (*Brassica juncea* (L.) Czern.)	Cruciferae	U.S. Dep. Agric., 1960
Cauliflower (*Brassica oleracea* L. var. *botrytis* L.)	Cruciferae	O'Mara et al., 1993
Cabbage (*Brassica oleracea* L. var. *capitata* L.)	Cruciferae	O'Mara et al, 1993
Sprouting broccoli (*Brassica oleracea* L. var. *italica* Plenk)	Cruciferae	O'Mara et al. 1993
Pe-tsai (*Brassica pekinensis* (Lour.) Rupr.)	Cruciferae	U.S. Dep. Agric., 1960
False flax (*Camelina sativa* (L.) Crantz)	Cruciferae	Khadhair et al., 2001
Shepherd's purse (*Capsella bursa-pastoris* (L.) Medik.)	Cruciferae	U.S. Dep. Agric., 1960
Wallflower (*Cheiranthus cheiri* L.)	Cruciferae	U.S. Dep. Agric., 1960
Sweet alyssum (*Lobularia maritima* (L.) Desv.)	Cruciferae	U.S. Dep. Agric., 1960
Garden radish (*Raphanus sativus* L.)	Cruciferae	U.S. Dep. Agric., 1960
Marsh cress (*Rorippa islandica* (Oeder) Borbás)	Cruciferae	U.S. Dep. Agric., 1960
London rocket (*Sisymbrium irio* L.)	Cruciferae	O'Mara et al., 1993
Winter squash (*Cucurbita maxima* Duchesne)	Cucurbitaceae	U.S. Dep. Agric., 1960
Winter crookneck squash (*Cucurbita moschata* Duchesne)	Cucurbitaceae	U.S. Dep. Agric., 1960
Pumpkin (*Cucurbita pepo* L.)	Cucurbitaceae	U.S. Dep. Agric., 1960

(continued)

Table 8-4. Example of the Wide Range of Hosts of Phytoplasma Strains of the Aster Yellows Group *(continued)*

Host	Host Family	Reference
Summer squash (*Cucurbita pepo* L. var. *melopepo* (L.) Alef.)	Cucurbitaceae	U.S. Dep. Agric., 1960
False hemp (*Datisca cannabina* L.)	Datiscaceae	O'Mara et al., 1993
Fuller's teasel (*Dipsacus fullonum* L.)	Dipsacaceae	U.S. Dep. Agric., 1960
Sweet scabious (*Scabiosa atropurpurea* L.)	Dipsacaceae	U.S. Dep. Agric., 1960
Rottlebox (*Crotalaria tetragona* Andrews)	Fabaceae	Baiswar et al., 2010
Alfileria (*Erodium cicutarium* (L.) L'Hér.)	Geraniaceae	U.S. Dep. Agric., 1960
Whitestem filaree (*Erodium moschatum* (L.) L'Hér. ex Aiton)	Geraniaceae	U.S. Dep. Agric., 1960
Didymocarpus (*Didymocarpus horsfieldii* Schniz.)	Gesneriaceae	O'Mara et al., 1993
Gloxinia (*Sinningia speciosa* (Lodd.) Hiern)	Gesneriaceae	U.S. Dep. Agric., 1960
Cultivated oat (*Avena sativa* L.)	Gramineae	O'Mara et al., 1993
Barley (*Hordeum vulgare* L.)	Gramineae	Chivkowski, 1964
Gladiolus (*Gladiolus hortulanus* Bailey)	Iridaceae	U.S. Dep. Agric., 1960
Henbit (*Lamium amplexicaule* L.)	Labiatae	U.S. Dep. Agric., 1960
Azure sage (*Salvia azurea* Lam.)	Labiatae	U.S. Dep. Agric., 1960
Medic (*Medicago arabica* (L.) Huds.)	Leguminosae	U.S. Dep. Agric., 1960
Clover (*Trifolium* spp.)	Leguminosae	U.S. Dep. Agric., 1960
Sting lily (*Blumenbachia hieronymi* Urb.)	Loasaceae	O'Mara et al., 1993
Chile nettle (*Cajophora lateritia* Klotzsch)	Loasaceae	O'Mara et al., 1993
Cheeseweed mallow (*Malva parviflora* L.)	Malvaceae	U.S. Dep. Agric., 1960
Common mallow (*Malva rotundifolia* L.)	Malvaceae	U.S. Dep. Agric., 1960
Ash (*Fraxinus excelsior* L.)	Oleaceae	Kamińska and Berniak, 2008
Red ribbons (*Clarkia concinna* (Fisch. & C. A. Mey.) Greene)	Onagraceae	O'Mara et al., 1993
Elegant clarkia (*Clarkia unguiculata* Lindl.)	Onagraceae	U.S. Dep. Agric., 1960; O'Mara et al., 1993
Willow herb (*Epilobium* spp.)	Onagraceae	U.S. Dep. Agric., 1960
Tall willow herb (*Epilobium brachycarpum* C. Presl)	Onagraceae	O'Mara et al., 1993
California willow herb (*Epilobium ciliatum* Raf. subsp. *ciliatum* Raf.)	Onagraceae	O'Mara et al., 1993
Biennial beeblossom (*Gaura biennis* L.)	Onagraceae	U.S. Dep. Agric., 1960
Lindheimer's beeblossom (*Gaura lindheimeri* Engelm. & A. Gray)	Onagraceae	U.S. Dep. Agric., 1960; O'Mara et al., 1993
Farewell-to-spring (*Godetia grandiflora* Lindl.)	Onagraceae	U.S. Dep. Agric., 1960
California poppy (*Eschscholzia californica* Cham.)	Papaveraceae	U.S. Dep. Agric., 1960
Sesame (*Sesamum indicum* L.)	Pedaliaceae	Khan et al., 2007
Black pepper (*Piper nigrum* L.)	Piperaceae	Adkar-Purushothama et al., 2009
Bracted plantain (*Plantago aristata* Michx.)	Plantaginaceae	U.S. Dep. Agric., 1960
Plantain (*Plantago lanceolata* L.)	Plantaginaceae	Fránová and Šimková, 2009
Common plantain (*Plantago major* L.)	Plantaginaceae	O'Mara et al., 1993
Annual statice (*Limonium sinuatum* (L.) Mill.)	Plumbaginaceae	O'Mara et al., 1993
Globe gilia (*Gilia capitata* Sims)	Polemoniaceae	O'Mara et al., 1993
Annual phlox (*Phlox drummondii* Hook.)	Polemoniaceae	U.S. Dep. Agric., 1960; O'Mara et al., 1993
Black bindweed (*Polygonum convolvulus* L.)	Polygonaceae	U.S. Dep. Agric., 1960
Garden sorrel (*Rumex acetosa* L.)	Polygonaceae	U.S. Dep. Agric., 1960
Sheep sorrel (*Rumex acetosella* L.)	Polygonaceae	O'Mara et al., 1993
Rock purslane (*Calandrinia grandiflora* Lindl.)	Portulacaceae	O'Mara et al., 1993
Purslane (*Portulaca oleracea* L.)	Portulacaceae	O'Mara et al., 1993
Scarlet pimpernel (*Anagallis arvensis* L.)	Primulaceae	U.S. Dep. Agric., 1960
Primrose (*Primula polyantha* Mill.)	Primulaceae	U.S. Dep. Agric., 1960
Poppy anemone (*Anemone coronaria* L.)	Ranunculaceae	U.S. Dep. Agric., 1960

(continued)

Table 8-4. Example of the Wide Range of Hosts of Phytoplasma Strains of the Aster Yellows Group *(continued)*

Host	Host Family	Reference
Rocket larkspur (*Delphinium ajacis* L.)	Ranunculaceae	U.S. Dep. Agric., 1960
Delphinium (*Delphinium* × *cultorum* Voss)	Ranunculaceae	O'Mara et al., 1993
Love-in-a-mist (*Nigella damascena* L.)	Ranunculaceae	O'Mara et al., 1993
Persian buttercup (*Ranunculus asiaticus* L.)	Ranunculaceae	U.S. Dep. Agric., 1960
Strawberry (*Fragaria chiloensis* (L.) Mill.)	Rosaceae	U.S. Dep. Agric., 1960
Avens (*Geum chiloense* Balb. ex Ser.)	Rosaceae	U.S. Dep. Agric., 1960
Rose (*Rosa* spp.)	Rosaceae	Kamińska et al., 2001
Blackberry (*Rubus fruticosus* agg. L.)	Rosaceae	Reeder et al., 2010
Lombardy poplar (*Populus nigra* L. 'Italica')	Salicaceae	Musič et al., 2008
Cloven-lip toad flax (*Linaria bipartita* (Vent.) Willd.)	Scrophulariaceae	O'Mara et al., 1993
Oldfield toad flax (*Linaria canadensis* (L.) Dum. Cours.)	Scrophulariaceae	O'Mara et al., 1993
Scarlet monkey-flower (*Mimulus cardinalis* Douglas ex Benth.)	Scrophulariaceae	O'Mara et al., 1993
Common monkey-flower (*Mimulus guttatus* DC.)	Scrophulariaceae	O'Mara et al., 1993
European speedwell (*Veronica beccabunga* L.)	Scrophulariaceae	O'Mara et al., 1993
Persian speedwell (*Veronica persica* Poir.)	Scrophulariaceae	O'Mara et al., 1993
Tomato (*Lycopersicon esculentum* Mill. var. *esculentum* Mill.)	Solanaceae	O'Mara et al., 1993
Indian tobacco (*Nicotiana quadrivalvis* Pursh var. *quadrivalvis* Pursh)	Solanaceae	U.S. Dep. Agric., 1960
Peasants' tobacco (*Nicotiana rustica* L.)	Solanaceae	U.S. Dep. Agric., 1960
Garden petunia (*Petunia* × *atkinsiana* D. Don ex Louden)	Solanaceae	U.S. Dep. Agric., 1960
Painted tongue (*Salpiglossis sinuata* Ruiz & Pav.)	Solanaceae	U.S. Dep. Agric., 1960
Scarlet eggplant (*Solanum integrifolium* Poir.)	Solanaceae	U.S. Dep. Agric., 1960
Black nightshade (*Solanum nigrum* L.)	Solanaceae	O'Mara et al., 1993
Potato (*Solanum tuberosum* L.)	Solanaceae	U.S. Dep. Agric., 1960
Garden nasturtium (*Tropaeolum majus* L.)	Tropaeolaceae	U.S. Dep. Agric., 1960
Dill (*Anethum graveolens* L.)	Umbelliferae	U.S. Dep. Agric., 1960
Celery (*Apium graveolens* L.)	Umbelliferae	Severin, 1929
Celeriac (*Apium graveolens* L. var. *rapaceum* (Mill.) Gaudin)	Umbelliferae	U.S. Dep. Agric., 1960
Caraway (*Carum carvi* L.)	Umbelliferae	U.S. Dep. Agric., 1960
Water hemlock (*Cicuta* spp.)	Umbelliferae	U.S. Dep. Agric., 1960
Coriander (*Coriandrum sativum* L.)	Umbelliferae	O'Mara et al., 1993
Carrot (*Daucus carota* L. subsp. *sativus* (Hoffm.) Arcang.)	Umbelliferae	U.S. Dep. Agric., 1960
Parsnip (*Pastinaca sativa* L.)	Umbelliferae	U.S. Dep. Agric., 1960
Nettle (*Urtica californica* Greene)	Urticaceae	O'Mara et al., 1993

phloem elements. Previously, careful cytological examination of the aster yellows leafhopper vector *Cicadula sexnotata* Fall. failed to reveal any histological or cytological difference between "virus"-free and "virus"-bearing individuals (Dobroscky, 1929). Later, electron microscopic studies of the leafhopper *Macrosteles fascifrons* (Stål) revealed mycoplasma-like bodies in its saliva (Hirumi and Maramorosch, 1969; Raine and Forbes, 1969). Additional antibiotic treatment-remission studies on aster yellows and chemotherapeutic applications to transmitting leafhoppers were conducted by several groups (Davis et al., 1968; Freitag and Smith, 1969). All of these efforts led to the conclusion that aster yellows was caused by a wall-less, pleomorphic bacterium rather than by a virus.

Symptomatology

Aster yellows symptoms begin with a semi-water-soaked appearance of the veins of the leaves. The interveinal regions of the infected leaves soon become chlorotic, while the veins (vascular elements) remain green. Infection of the apical region causes stunting and initiates phyllody of the flower heads (Fig. 8-8).

Pathogenicity Tests

Since in vitro culture of the aster yellows phytoplasma has so far not been possible, plant grafting and use of leafhopper vectors, such as the aster leafhopper *Macrosteles quadrilineatus* Forbes, to inoculate indicator plants appear to be reasonable but tedious alternatives. Both grafting and transmission by dodder are used to experimentally infect indicator plants, such as Madagascar periwinkle (*Catharanthus roseus* (L.) G. Don) (Kamińska et al., 2001). Earlier leafhopper-vectored transmission studies have been carried out between aster and celery (Severin, 1929).

Diagnosis for the presence of the phytoplasma can be performed using nested polymerase chain reaction (PCR) amplification of the total DNA extracted from growing tips of plants suspected of bearing the phytoplasma, using one of several procedures employing two universal primer pairs (Gundersen and Lee, 1996; Schneider et al., 1997; Khadhair et al., 2002). Alternatively, a simple DNA extraction procedure (Gundersen et al., 1996) is used to provide templates for PCR, using the universal 16S rRNA phytoplasma primer pair R16R2/F2 (Lee et al., 1993). For woody plants, a useful "pencil-sharpener" DNA extraction procedure has been described (Green et al., 1999). To reduce host or leafhopper genomic DNA background noise, host-free phytoplasmal genomic DNA can be isolated from the vector's saliva (Melamed et al., 2003). The resulting PCR products are reamplified using the group-specific rRNA primer pair R16(I)R1/F1 (Lee et al., 1994) or using the specific aster yellows primers P1/Ayint (Torres et al., 2004).

Pathogenicity Factors

There is a paucity of information on the mechanism of pathogenesis and on virulence genes of the aster yellows phytoplasmas. Partly this is due to the lack of a method for culturing the organisms in vitro. Transposon mutagenesis to help identify virulence determinants is therefore difficult. The lack of a bacterial cell wall indicates that phytoplasmas evolved along a secondary evolutionary branch of the domain *Bacteria*. They also lack a DNA repair system, which suggests that phytoplasmas need to evolve and adapt quickly as the environment changes. These lower bacteria also are devoid of two-component signal transduction systems to detect environmental cues for long-term survival (Hoch and Silhavy, 1995; Inouye and Dutta, 2003). The lack of type II, III, and IV secretion systems to deliver pathogenicity effectors reflects evolution toward an endosymbiotic state for this group of pathogens. Clearly, there are numerous infected plants that remain asymptomatic (see Chapter 9).

Comparative genomic sequence studies that can filter out usual housekeeping genes will reveal the presence of novel genes that can be examined closely for potential roles in pathogenicity. The growth distortions conferred on the plant host by the aster yellows phytoplasma suggests that some of the virulence genes encode plant growth hormones, such as cytokinins. Analyses of the genomes of the aster yellows phytoplasmas may reveal the presence of a homolog of the isopentenyl transferases encoded by the *ipt* gene found in most of the bacterial pathogens described in this chapter.

REFERENCES

Abramovitch, R. B., Kim, Y.-J., Chen, S., Dickman, M. B., and Martin, G. B. 2003. *Pseudomonas* type III effector AvrPtoB induces plant disease susceptibility by inhibition of host programmed cell death. EMBO J. 22:60-69.

Adkar-Purushothama, C. R., Casati, P., Quaglino, F., Durante, G., and Bianco, P. A. 2009. First report of a '*Candidatus* Phytoplasma asteris'-related strain associated with a yellows disease of black pepper (*Piper nigrum*) in India. Plant Pathol. 58:789.

Akiyoshi, D. E., Regier, D. A., and Gordon, M. P. 1987. Cytokinin production by *Agrobacterium* and *Pseudomonas* spp. J. Bacteriol. 169:4242-4248.

Aldridge, P., and Hughes, K. T. 2001. How and when are substrates selected for type III secretion? Trends Microbiol. 9:209-214.

Alhudalb, K., Arocha, Y., Wilson, M., and Jones, P. 2007. Identification and molecular characterization of a phytoplasma associated with Al-Wijam disease of date palm in Saudi Arabia. Arab J. Plant Prot. 25:116-122.

Anand, A., Krichevsky, A., Schornack, S., Lahaye, T., Tzfira, T., Tang, Y., Citovsky, V., and Mysore, K. S. 2007. *Arabidopsis* VirE2 interacting protein 2 is required for *Agrobacterium* T-DNA integration in plants. Plant Cell 19:1695-1708.

Anand, V. K., and Heberlein, G. T. 1977. Crown gall tumorigenesis in potato tuber tissue. Am. J. Bot. 64:153-158.

Anderson, L. B., Hertzel, A. V., and Das, A. 1996. *Agrobacterium tumefaciens* VirB7 and VirB9 form a disulfide-linked protein complex. Proc. Natl. Acad. Sci. U.S.A. 93:8889-8894.

Ark, P. A., and Schroth, M. N. 1958. Use of slices of carrot and other fleshy roots to detect crown gall bacteria in soil. Plant Dis. Rep. 42:1279-1281.

Ark, P. A., and Thompson, J. P. 1960. Susceptibility of *Artemisia vulgaris* and *Helianthus tuberosus* to crown gall, *Agrobacterium tumefaciens*. Plant Dis. Rep. 44:102-103.

Atmakuri, K., Ding, Z., and Christie, P. J. 2003. VirE2, a type IV secretion substrate, interacts with the VirD4 transfer protein at cell poles of *Agrobacterium tumefaciens*. Mol. Microbiol. 49:1699-1713.

Atmakuri, K., Cascales, E., and Christie, P. J. 2004. Energetic components VirD4, VirB11 and VirB4 mediate early DNA transfer reactions required for bacterial type IV secretion. Mol. Microbiol. 54:1199-1211.

Babic, A., Lindner, A. B., Vulic, M., Stewart, E. J., and Radman, M. 2008. Direct visualization of horizontal gene transfer. Science 319:1533-1536.

Baek, S.-H., Hartsock, A., and Shapleigh, J. P. 2008. *Agrobacterium tumefaciens* C58 uses ActR and FnrN to control *nirK* and *nor* expression. J. Bacteriol. 190:78-86.

Bailey, S., Ward, D., Middleton, R., Grossmann, J. G., and Zambryski, P. C. 2006. *Agrobacterium tumefaciens* VirB8 structure reveals potential protein-protein interaction sites. Proc. Natl. Acad. Sci. U.S.A. 103:2582-2587.

Baiswar, P., Arocha, Y., Chandra, S., and Ngachan, S. V. 2010. First report of '*Candidatus* Phytoplasma asteris' associated with witches' broom of *Crotalaria tetragona* in India. Plant Pathol. 59:397.

Baker, K. F. 1950. Bacterial fasciation disease of ornamental plants in California. Plant Dis. Rep. 34:121-126.

Banfield, W. M. 1935. Studies in cellular pathology. I. Effects of cane gall bacteria upon gall tissue cells of the black raspberry. Bot. Gaz. 97:193-239.

Barash, I., and Manulis-Sasson, S. 2007. Virulence mechanisms and host specificity of gall-forming *Pantoea agglomerans*. Trends Microbiol. 15:538-545.

Barash, I., Panijel, M., Gurel, F., Chalupowicz, L., and Manulis, S. 2005. Transformation of *Pantoea agglomerans* into a tumorigenic pathogen. Pages 10-19 in: Proc. Int. Conf. Plant-Microbe Interactions: Endophytes and Biocontrol Agents, 1st. S. Sovari and O. Toldi, eds. Saariselka, Lapland, Finland.

Baron, C., Domke, N., Beinhofer, M., and Hapfelmeier, S. 2001. Elevated temperature differentially affects virulence, VirB protein accumulation, and T-pilus formation in different *Agrobacterium tumefaciens* and *Agrobacterium vitis* strains. J. Bacteriol. 183:6852-6862.

Bayer, M., Eferl, R., Zellnig, G., Teferle, K., Dijkstra, A., Koraimann, G., and Hogenauer, G. 1995. Gene 19 of plasmid R1 is required for both efficient conjugative DNA transfer and bacteriophage R17 infection. J. Bacteriol. 177:4279-4288.

Beanland, L., Hoy, C. W., Miller, S. A., and Nault, L. R. 2000. Influence of aster yellows phytoplasma on the fitness of aster leafhopper (Homoptera: Cicadellidae). Ann. Entomol. Soc. Am. 93:271-276.

Bercetche, J., Chriqui, D., Adam, S., and David, C. 1987. Morphogenetic and cellular reorientations induced by *Agrobacterium rhizogenes* (strains 1855, 2659 and 8196) on carrot, pea and tobacco. Plant Sci. 52:195-210.

Berger, B., and Christie, P. J. 1994. Genetic complementation analysis of the *Agrobacterium tumefaciens virB* operon: *virB2* Through *virB11* are essential virulence genes. J. Bacteriol. 176:3646-3660.

Bioletti, F. T. 1898. The olive knot. Calif. Agric. Exp. Stn. Bull. 120:1-11.

Bouzar, H., and Moore, L. W. 1987. Isolation of different *Agrobacterium* biovars from a natural oak savanna and tallgrass prairie. Appl. Environ. Microbiol. 53:717-721.

Braun, A. C., and White, P. R. 1943. Bacteriological sterility of tissues derived from secondary crown-gall tumors. Phytopathology 33:85-100.

Bravo-Angel, A. M., Gloeckler, V., Hohn, B., and Tinland, B. 1999. Bacterial conjugation protein MobA mediates integration of complex DNA structures into plant cells. J. Bacteriol. 181:5758-5765.

Brencic, A., Xia, Q., and Winans, S. C. 2004a. *virA* Of *Agrobacterium tumefaciens* is an intradimer transphosphorylase and can actively block *vir* gene expression in the absence of phenolic signals. Mol. Microbiol. 52:1349-1362.

Brencic, A., Eberhard, A., and Winans, S. C. 2004b. Signal quenching, detoxification and mineralization of *vir* gene-inducing phenolics by the VirH2 protein of *Agrobacterium tumefaciens*. Mol. Microbiol. 51:1103-1115.

Brown, N. A. 1932. A gall similar to crown gall produced on *Gypsophila* by a new bacterium. J. Agric. Res. 48:1099-1112.

Cangelosi, G. A., Ankenbauer, R. G., and Nester, E. W. 1990. Sugars induce the *Agrobacterium* virulence genes through a periplasmic binding protein and a transmembrane signal protein. Proc. Natl. Acad. Sci. U.S.A. 87:6708-6712.

Cardarelli, M., Marlot, D., Pomponi, M., Spano, L., Capone, I., and Costantino, P. 1987. Induction and growth properties of carrot roots with different complements of *Agrobacterium rhizogenes* T-DNA. Plant Mol. Biol. 13:43-52.

Carle, S. A., Hao, G., Zheng, D., Sanni-Sanoussi, T., Talarico, M. J., Hilton, J. S., and Burr, T. J. 2008. A gene cluster in *Agrobacterium vitis* homologous to polyketide synthase operons is associated with grape necrosis and hypersensitive response induction on tobacco. FEMS Microbiol. Lett. 289:90-96.

Cascales, E., and Christie, P. J. 2004. Definition of a bacterial type IV secretion pathway for a DNA substrate. Science 304:1170-1173.

Cavara, F. 1897. Tubercolosi della vite. Intorno alla eziologia de alcune malattie di plante cultivate. Stn. Sper. Agrar. Ital. 30:483-487.

Charles, T. C., and Nester, E. W. 1993. A chromosomally encoded two-component sensory transduction system is required for virulence of *Agrobacterium tumefaciens*. J. Bacteriol. 175:6614-6625.

Charles, T. C., and Nester, E. W. 1994. Defining the contribution of the *Agrobacterium* chromosome in crown gall tumorigenesis. Pages 639-649 in: Molecular Mechanisms of Bacterial Virulence. C. I. Kado and J. H. Crosa, eds. Kluwer Academic, Dordrecht, the Netherlands.

Chen, L., Chen, Y., Wood, D. W., and Nester, E. W. 2002. A new type IV secretion system promotes conjugal transfer in *Agrobacterium tumefaciens*. J. Bacteriol. 184:4838-4845.

Chilton, M.-D. 1982. Integration and transcription of Ti plasmid fragments. Pages 299-319 in: Molecular Biology of Plant Tumors. G. Kahl and J. S. Schell, eds. Academic Press, New York.

Chilton, M.-D., Drummond, M. H., Merlo, D. J., Sciaky, D., Montoya, A. L., Gordon, M. P., and Nester, E. W. 1977. Stable incorporation of plasmid DNA into higher plant cells: The molecular basis of crown gall tumorigenesis. Cell 11:263-271.

Chilton, M.-D., Drummond, M. H., Merlo, D. J., and Sciaky, D. 1978. Highly conserved DNA of Ti plasmids overlaps T-DNA maintained in plant tumors. Nature 275:147-149.

Chivkowski, L. N. 1964. The reaction of barley varieties to aster yellows virus. Can. J. Bot. 43:373-378.

Choy, W.-K., Zhou, L., Syn, C. K.-C., Zhang, L.-H., and Warup, S. 2004. MorA defines a new class of regulators affecting flagellar development and biofilm formation in diverse *Pseudomonas* species. J. Bacteriol. 186:7221-7228.

Christie, P. J., and Vogel, J. P. 2000. Bacterial type IV secretion: Conjugation systems adapted to deliver effector molecules to host cells. Trends Microbiol. 8:354-360.

Christie, P. J., Ward, J. E., Winans, S. C., and Nester, E. W. 1988. The *Agrobacterium tumefaciens virE2* gene product is a single-stranded–DNA-binding protein that associates with T-DNA. J. Bacteriol. 170:2659-2667.

Chuchue, T., Tanboon, W., Prapagdee, B., Dubbs, J. M., Vattanaviboon, P., and Mongkolsuk, S. 2006. *ohrR* And *ohr* are the primary sensor/regulator and protective genes against organic hydroperoxide stress in *Agrobacterium tumefaciens*. J. Bacteriol. 188:842-851.

Citovsky, V., Wong, M. L., and Zambryski, P. 1989. Cooperative interaction of *Agrobacterium* VirE2 protein with single-stranded DNA: Implications for the T-DNA transfer process. Proc. Natl. Acad. Sci. U.S.A. 86:1193-1197.

Citovsky, V., Kapelnikov, A., Oliel, S., Zakai, N., Rojas, M. R., Gilbertson, R. L., Tzira, T., and Loyter, A. 2004. Protein interactions involved in nuclear import of the *Agrobacterium* VirE2 protein *in vivo* and *in vitro*. J. Biol. Chem. 279:29528-29533.

Citovsky, V., Kozlovsky, S. V., Lacroix, B., Zaltsman, A., Dafny-Yelin, M., Vyas, S., Tovkach, A., and Tzfira, T. 2007. Biological systems of the host cell involved in *Agrobacterium* infection. Cell Microbiol. 9:9-20.

Clark, E., Manulis, S., Ophir, Y., Barash, I., and Gafni, Y. 1993. Cloning and characterization of *iaaM* and *iaaH* from *Erwinia herbicola* pathovar *gypsophilae*. Phytopathology 83:234-240.

Clément, B., Perot, J., Geoffroy, P., Legrand, M., Zon, J., and Otten, L. 2007. Abnormal accumulation of sugars and phenolics in tobacco roots expressing the *Agrobacterium* T-*6b* oncogene and the role of these compounds in *6b*-induced growth. Mol. Plant-Microbe Interact. 20:53-62.

Close, T. J., Tait, R. C., and Kado, C. I. 1985. Regulation of the Ti plasmid virulence genes by a chromosomal locus of *Agrobacterium tumefaciens*. J. Bacteriol. 164:774-781.

Close, T. J., Tait, R. C., Rempel, H. C., Hirooka, T., Kim, L., and Kado, C. I. 1987a. Molecular characterization of the *virC* genes of the Ti plasmid. J. Bacteriol. 169:2336-2344.

Close, T. J., Rogowsky, P. M., Kado, C. I., Winans, S. C., Yanofsky, M. F., and Nester, E. W. 1987b. Dual control of *Agrobacterium tumefaciens* Ti plasmid virulence genes. J. Bacteriol. 169:5113-5118.

Comai, L., and Kosuge, T. 1980. Involvement of plasmid deoxyribonucleic acid in indoleacetic acid synthesis in *Pseudomonas savastanoi*. J. Bacteriol. 143:950-957.

Comai, L., Surico, G., and Kosuge, T. 1982. Relation of plasmid DNA to indoleacetic acid production in different strains of *Pseudomonas syringae* pv. *savastanoi*. J. Gen. Microbiol. 128:2157-2163.

Conn, H. J. 1942. Validity of the genus *Alcaligenes*. J. Bacteriol. 44:353-360.

Cooksey, D. A. 1986. Galls of *Gypsophila paniculata* caused by *Erwinia herbicola*. Plant Dis. 70:464-468.

Cooley, M. B., D'Souza, M. R., and Kado, C. I. 1991. The *virC* and *virD* operons of the *Agrobacterium* Ti plasmid are regulated by the *ros* chromosomal gene: Analysis of the cloned *ros* gene. J. Bacteriol. 1173:2608-2618.

Cornelis, K., Ritsema, T., Nijsse, J., Holsters, M., Goethals, K., and Jaziri, M. 2001. The plant pathogen *Rhodococcus fascians* colonizes the exterior and interior of the aerial parts of plants. Mol. Plant-Microbe Interact. 14:599-608.

Crespi, M., Messens, E., Caplan, A. B., Van Montagu, M., and Desomer, J. 1992. Fasciation induction by the phytopathogen *Rhodococcus fascians* depends upon a linear plasmid encoding a cytokinin synthase gene. EMBO J. 11:795-804.

Damiano, C., and Monticelli, S. 1998. *In vitro* fruit tree rooting by *Agrobacterium rhizogenes* wild type infection. EJB Electronic J. Biotechnol. 1:189-195.

Dang, T. A. T., and Christie, P. J. 1997. The VirB4 ATPase of *Agrobacterium tumefaciens* is a cytoplasmic membrane protein exposed at the periplasmic surface. J. Bacteriol. 179:453-462.

Dang, T. A., Zhou, X. R., Graf, B., and Christie, P. J. 1999. Dimerization of the *Agrobacterium tumefaciens* VirB4 ATPase and the effect of ATP-binding cassette mutations on the assembly and function of the T-DNA transporter. Mol. Microbiol. 32:1239-1253.

Das, A., and Xie, Y.-H. 2000. The *Agrobacterium* T-DNA transport pore proteins VirB8, VirB9, and VirB10 interact with one another. J. Bacteriol. 182:758-763.

Davis, R. E., Whitcomb, R. F., and Steere, R. L. 1968. Remission of aster yellows disease by antibiotics. Science 161:793-795.

De Cleene, M., and De Ley, J. 1976. The host range of crown gall. Bot. Rev. 42:389-466.

Deng, W., Chen, L., Wood, D. W., Metcalfe, T., Liang, X., Gordon, M. P., Comai, L., and Nester, E. W. 1998. *Agrobacterium* VirD2 protein interacts with plant host cyclophilins. Proc. Natl. Acad. Sci. U.S.A. 95:7040-7045.

Deng, W., Chen, L., Peng, W.-T., Liang, X., Sekiguchi, S., Gordon, M. P., Comai, L., and Nester, E. W. 1999. VirE1 is a specific molecular chaperone for the exported single-stranded-DNA-binding protein VirE2 in *Agrobacterium*. Mol. Microbiol. 31:1795-1807.

Depicker, A., Van Montagu, M., and Schell, J. 1978. Homologous DNA sequences in different Ti plasmids are essential for oncogenicity. Nature 275:150-152.

de Ropp, R. S. 1947. The growth promoting and tumefacient factors of bacteria-free crown gall tumor tissue. Am. J. Bot. 34:248-261.

Dobroscky, I. D. 1929. Is the aster-yellows virus detectable in its insect vector? Phytopathology 19:1009-1015.

Doi, Y., Teranaka, M., Yora, K., and Asuyama, H. 1967. *Mycoplasma* or PLT-group-like microorganisms found in the phloem elements of plants infected with mulberry dwarf, potato witches-broom, aster yellows, or paulownia witches-broom. Ann. Phytopathol. Soc. Jpn. 33:259-266.

Doty, S. L., Chang, M., and Nester, E. W. 1993. The chromosomal virulence gene, *chvE*, of *Agrobacterium tumefaciens* is regulated by a LysR family member. J. Bacteriol. 175:7880-7886.

Douglas, C. J., Staneloni, R. J., Rubin, R. A., and Nester, E. W. 1985. Identification and genetic analysis of an *Agrobacterium tumefaciens* chromosomal virulence region. J. Bacteriol. 161:850-860.

Dowson, W. J. 1942. On the generic name of the Gram-positive bacterial plant pathogens. Trans. Br. Mycol. Soc. 25:311-314.

Duckely, M., Oomen, C., Axtheim, F., van Gelder, P., Waksman, G., and Engel, A. 2005. The VirE1VirE2 complex of *Agrobacterium tumefaciens* interacts with single-stranded DNA and forms channels. Mol. Microbiol. 58:1130-1142.

Dunoyer, P., Himber, C., and Voinnet, O. 2006. Induction, suppression and requirement of RNA silencing pathways in virulent *Agrobacterium tumefaciens* infections. Nature Genet. 38:258-263.

Dűrrenberger, F., Crameri, A., Hohn, B., and Koukolíková-Nicola, Z. 1989. Covalently bound VirD2 protein of *Agrobacterium tumefaciens* protects the T-DNA from exonucleolytic degradation. Proc. Natl. Acad. Sci. U.S.A. 86:9154-9158.

Eiamphungporn, W., Charoenlap, N., Vattanaviboon, P., and Mongkolsuk, S. 2006. *Agrobacterium tumefaciens soxR* is involved in superoxide stress protection and also directly regulates superoxide-inducible expression of itself and a target gene. J. Bacteriol. 188:8669-8673.

Eisenbrandt, R., Kalkum, M., Lai, E.-M., Lurz, R., Kado, C. I., and Lanka, E. 1999. Conjugative pili of IncP plasmids, and the Ti plasmid T pilus are composed of cyclic subunits. J. Biol. Chem. 274:22548-22555.

Escudero, J., den Dulk-Ras, A., Regensburg-Tuïnk, T. J. G., and Hooykaas, P. J. J. 2003. VirD4-independent transformation by CloDF13 evidences an unknown factor required for the genetic colonization of plants via *Agrobacterium*. Mol. Microbiol. 47:891-901.

Ezra, D., Barash, I., Valinsky, L., and Manulis, S. 2000. The dual function in virulence and host range restriction of a gene isolated from $pPATH_{Ehg}$ plasmid of *Erwinia herbicola* pv. *gypsophilae*. Mol. Plant-Microbe Interact. 13:683-692.

Fabre, E., and Dunal, F. 1853. Observations sur les maladies régnantes de la vigne. Bull. Soc. Cent. Agric. Dép. Hérault 40:46.

Fernandez, D., Spudich, G. M., Zhou, X. R., and Christie, P. J. 1996. The *Agrobacterium tumefaciens* VirB7 lipoprotein is required for stabilization of VirB proteins during assembly of the T-complex transport apparatus. J. Bacteriol. 178:3168-3176.

Filichkin, S. A., and Gelvin, S. B. 1993. Formation of a putative relaxation intermediate during T-DNA processing directed by the *Agrobacterium tumefaciens* VirD1,D2 endonuclease. Mol. Microbiol. 8:915-926.

Fránová, J., and Petrzik, K. 2009. *Asparagus officinalis*: A new host of '*Candidatus* Phytoplasma asteris'. J. Phytopathol. 158:317-320.

Fránová, J., and Šimková, M. 2009. Association of '*Candidatus* Phytoplasma asteris' with yellowing and phyllody of *Plantago lanceolata*. Folia Microbiol. 54:469-472.

Freitag, J. H., and Smith, S. H. 1969. Effects of tetracyclines on symptom expression and leafhopper transmission of aster yellows. Phytopathology 59:1820-1823.

Fründt, C., Meyer, A. D., Ichikawa, T., and Meins, F., Jr. 1998. Evidence for the ancient transfer of Ri plasmid T-DNA genes between bacteria and plants. Pages 94-117 in: Horizontal Gene Transfer. M. Syvanen and C. I. Kado, eds. Chapman and Hall, London.

Gaffney, T. D., da Costa e Silva, O., Yamada, T., and Kosuge, T. 1990. Indoleacetic acid operon of *Pseudomonas syringae* subsp. *savastanoi*: Transcription analysis and promoter identification. J. Bacteriol. 172:5593-5601.

Galan, J. E., and Collmer, A. 1999. Type III secretion machines: Bacterial devices for protein delivery into host cells. Science 284:1322-1328.

Galperin, M. Y., Moroz, O. V., Wilson, K. S., and Murzin, A. G. 2006. House cleaning, a part of good housekeeping. Mol. Microbiol. 59:5-19.

Gao, R., and Lynn, D. G. 2005. Environmental pH sensing: Resolving the VirA/VirG two-component system inputs for *Agrobacterium* pathogenesis. J. Bacteriol. 187:2182-2189.

Gavini, F., Mergaert, J., Beji, A., Mielcarek, C., Izard, D., Kersters, K., and DeLey, J. 1989. Transfer of *Enterobacter agglomerans* (Beijerinck 1888) Ewing and Fife 1972 to *Pantoea* gen. nov. as *Pantoea agglomerans* comb. nov. and description of *Pantoea dispersa* sp. nov. Int. J. Syst. Bacteriol. 39:337-345.

Gelvin, S. B. 2003. *Agrobacterium*-mediated plant transformation: The biology behind the "gene-jockeying" tool. Microbiol. Mol. Biol. Rev. 67:16-37.

Ghai, J., and Das, A. 1989. The *virD* operon of *Agrobacterium tumefaciens* Ti plasmid encodes a DNA-relaxing enzyme. Proc. Natl. Acad. Sci. U.S.A. 86:3109-3113.

Glass, N. L., and Kosuge, T. 1986. Cloning of the gene for indoleacetic acid-lysine synthetase from *Pseudomonas syringae* subsp. *savastanoi*. J. Bacteriol. 166:598-603.

Goethals, K., Vereecke, D., Jaziri, M., Van Montagu, M., and Holsters, M. 2001. Leafy gall formation by *Rhodococcus fascians*. Annu. Rev. Phytopathol. 39:27-52.

Goldmann, A., Tempé, J., and Morel, G. 1968. Quelques particularités de diverses souches d'*Agrobacterium tumefaciens*. C. R. Seances Soc. Biol. Ses Fil. 162:623-631.

Goldmann, A., Thomas, D. W., and Morel, G. 1969. Sur la structure de la nopaline metabolite anormal de certaines tumeurs de crown-gall. C. R. Acad. Sci. Paris 268:852-854.

Goodfellow, M. 1984. Reclassification of *Corynebacterium fascians* (Tilford) Dowson in the genus *Rhodococcus*, as *Rhodococcus fascians* comb. nov. Syst. Appl. Microbiol. 5:225-229.

Goodner, B., Hinkle, G., Gattung, S., Miller, N., Blanchard, M., Qurollo, B., Goldman, B. S., Cao, Y., Askenazi, M., Halling, C., Mullin, L., Houmiel, K., Gordon, J., Vaudin, M., Lartchouk, O., Epp, A., Liu, F., Wollam, C., Allinger, M., Doughty, D., Scott, C., Lappas, C., Markelz, B., Flanagan, C., Crowell, C., Gurson, J., Lomo, C., Sear, C., Strub, G., Cielo, C., and Slater, S. 2001. Genome sequence of the plant pathogen and biotechnology agent *Agrobacterium tumefaciens* C58. Science 294:2323-2329.

Graham, D. C., and Quinn, C. E. 1974. Identification of *Agrobacterium gypsophilae* strains NCPPB 179 and NCPPB 1948 as *Erwinia herbicola*. Int. J. Syst. Bacteriol. 24:238-241.

Gray, J., Wang, J., and Gelvin, S. B. 1992. Mutation of the *miaA* gene of *Agrobacterium tumefaciens* results in reduced *vir* gene expression. J. Bacteriol. 174:1086-1098.

Green, M. J., Thompson, D. A., and MacKenzie, D. J. 1999. Easy and efficient DNA extraction from woody plants for the detection of phytoplasmas by polymerase chain reaction. Plant Dis. 83:482-485.

Griffiths, H. M., Sinclair, W. A., Smart, C. D., and Davis, R. E. 1999. The phytoplasma associated with ash yellows and lilac witches'-broom: '*Candidatus* Phytoplasma fraxini'. Int. J. Syst. Bacteriol. 49:1605-1614.

Grimsley, N., Hohn, B., Ramos, C., Kado, C., and Rogowsky, P. 1989. DNA transfer from *Agrobacterium* to *Zea mays* or *Brassica* by agroinfection is dependent on bacterial virulence functions. Mol. Gen. Genet. 217:309-316.

Gundersen, D. E., and Lee, I.-M. 1996. Ultrasensitive detection of phytoplasmas by nested-PCR assays using two universal primer pairs. Phytopathol. Mediterr. 35:114-151.

Gundersen, D. E., Lee, I.-M., Rehner, S. A., Davis, R. E., and Kingsbury, D. T. 1994. Phylogeny of mycoplasmalike organisms (phytoplasmas): A basis for their classification. J. Bacteriol. 176:5244-5254.

Gundersen, D. E., Lee, I.-M., Schaff, D. A., Harrison, N. A., Chang, C. J., Davis, R. E., and Kingbury, D. T. 1996. Genomic diversity and differentiation among phytoplasma strains in 16S rRNA group I (aster yellows and related phytoplasmas) and III (X-disease and related phytoplasmas). Int. J. Syst. Bacteriol. 46:64-75.

Guo, M., Manuli, S., Barash, I., and Lichter, A. 2001. The operon for cytokinin biosynthesis of *Erwinia herbicola* pv. *gypsophilae* contains two promoters and is plant induced. Can. J. Microbiol. 47:1126-1131.

Guo, M., Manulis, S., Mor, H., and Barash, I. 2002. The presence of diverse IS elements and an *avrPphD* homologue that acts as a virulence factor on the pathogenicity plasmid of *Erwinia herbicola* pv. *gypsophilae*. Mol. Plant-Microbe Interact. 15:709-716.

Guo, M., Hou, Q., Hew, C. L., and Pan, S. Q. 2007. *Agrobacterium* VirD2-binding protein is involved in tumorigenesis and redundantly encoded by conjugative transfer gene clusters. Mol. Plant-Microbe Interact. 20:1201-1212.

Hamilton, R. H., and Fall, M. Z. 1971. The loss of tumor-initiating ability in *Agrobacterium tumefaciens* by incubation at high temperature. Experientia 27:229-230.

Hansen, G., and Chilton, M. D. 1999. Lessons in gene transfer to plants by a gifted microbe. Curr. Topics Microbiol. Immunol. 240:21-57.

Hao, G., Zhang, H., Zheng, D., and Burr, T. J. 2005. *luxR* Homolog *avhR* in *Agrobacterium vitis* affects the development of a grape-specific necrosis and a tobacco hypersensitive response. J. Bacteriol. 187:185-192.

Hapfelmeier, S., Domke, N., Zambryski, P. C., and Baron, C. 2000. VirB6 is required for stabilization of VirB5 and VirB3 and formation of VirB7 homodimers in *Agrobacterium tumefaciens*. J. Bacteriol. 182:4505-4511.

Hedgcock, G. G. 1905. Some of the results of three years' experiments with crown gall. Science 22:120-122.

Hendrickson, A. A., Baldwin, I. L., and Riker, A. J. 1934. Studies on certain physiological characters of *Phytomonas tumefaciens*, *Phytomonas rhizogenes* and *Bacillus radiobacter*. J. Bacteriol. 28:597-618.

Henze, M. 1908. Chemische Untersuchungen an Octopoden. Hoppe-Seylers Zs. Physiol. Chem. 55:433-444.

Henze, M. 1911. Ueber das Vorkommen des Betains be Chephalopoden. Hoppe-Seylers Zs. Physiol. Chem. 70:253-255.

Hewitt, W. B. 1938. Leaf-scar infection in relation to the olive-knot disease. Hilgardia 12:41-66.

Hirooka, T., and Kado, C. I. 1986. Location of the right boundary of the virulence region on *Agrobacterium tumefaciens* plasmid pTiC58 and a host-specifying gene next to the boundary. J. Bacteriol. 168:237-243.

Hirooka, T., Rogowsky, P. M., and Kado, C. I. 1987. Characterization of the *virE* locus of *Agrobacterium tumefaciens* plasmid pTiC58. J. Bacteriol. 169:1529-1536.

Hirota, Y. 1960. The effect of acridine dyes on mating type factors in *E. coli.* Proc. Natl. Acad. Sci. U.S.A. 46:57-64.

Hirota, Y., and Lijima, T. 1957. Acriflavine as an effective agent for eliminating F-factor in *Escherichia coli* K-12. Nature 180:655-656.

Hirumi, H., and Maramorosch, K. 1969. Mycoplasma-like agents of aster yellows in the salivary gland of leafhopper vectors. (Abstr.) Phytopathology 59:399-400.

Hoch, J. A., and Silhavy, T. J., eds. 1995. Two-Component Signal Transduction. ASM Press, Washington, DC.

Hodges, L. D., Cuperus, J., and Ream, W. 2004. *Agrobacterium rhizogenes* GALLS protein substitutes for *Agrobacterium tumefaciens* single-stranded DNA-binding protein VirE2. J. Bacteriol. 186:3065-3077.

Hodges, L. D., Vergunst, A. C., Neal-McKinney, J., den Dulk-Ras, A., Moyer, D. M., Hooykaas, P. J. J., and Ream, W. 2006. *Agrobacterium rhizogenes* GALLS protein contains domains for ATP binding, nuclear localization and type IV secretion. J. Bacteriol. 188:8222-8230.

Hooykaas, P. J. J., Hofker, M., den Dulk-Ras, H., and Schilperoort, R. A. 1984. A comparison of virulence determinants in an octopine Ti plasmid, a nopaline Ti plasmid, and an Ri plasmid by complementation analysis of *Agrobacterium tumefaciens* mutants. Plasmid 11:195-205.

Höppner, C., Liu, Z., Domke, N., Binns, A. N., and Baron, C. 2004. VirB1 orthologs from *Brucella suis* and pKM101 complement defects of the lytic transglycosylase required for efficient type IV secretion from *Agrobacterium tumefaciens.* J. Bacteriol. 186:1415-1422.

Howard, E. A., Zupan, J. R., Citovsky, V., and Zambryski, P. C. 1992. The VirD2 protein of *A. tumefaciens* contains a C-terminal bipartite nuclear localization signal: Implications for nuclear uptake of DNA in plant cells. Cell 68:109-118.

Hseu, S. H., and Hsu, S. T. 1998. Occurrence of bacterial gall of *Gypsophila paniculata* caused by *Erwinia herbicola* pv. *gypsophilae* in Taiwan. Plant Prot. Bull. 40:329-338.

Huang, M. W., Cangelosi, G. A., Halperin, W., and Nester, E. W. 1990. A chromosomal *Agrobacterium tumefaciens* gene required for effective plant signal transduction. J. Bacteriol. 172:1814-1822.

Huang, Y., Morel, P., Powell, B., and Kado, C. I. 1990. VirA, a coregulator of Ti-specified virulence genes, is phosphorylated in vitro. J. Bacteriol. 172:1142-1144.

Hueck, C. J. 1998. Type III protein secretion systems in bacterial pathogens of animals and plants. Microbiol. Mol. Biol. Rev. 62:379-433.

Humann, J., Andrews, S., and Ream, W. 2006. VirE1-mediated resistance to crown gall in transgenic *Arabidopsis thaliana.* Phytopathology 96:105-110.

Iacobellis, N. S., Caponero, A., and Evidente, A. 1998. Characterization of *Pseudomonas syringae* ssp. *savastanoi* strains isolated from ash. Plant Pathol. 47:73-83.

Iñon de Iannino, N., and Ugalde, R. A. 1989. Biochemical characterization of avirulent *Agrobacterium tumefaciens chvA* mutants: Synthesis and excretion of β-(1,2) glucan. J. Bacteriol. 171:2842-2849.

Inouye, M., and Dutta, R., eds. 2003. Histidine Kinases in Signal Transduction. Academic Press, San Diego, CA.

Ishiie, T., Doi, Y., Yora, K., and Asuyama, H. 1967. Suppressive effects of antibiotics of tetracycline group on symptom development of mulberry dwarf disease. Ann. Phytopathol. Soc. Jpn. 33:267-275.

Jackson, R. W., Mansfield, J. W., Ammouneh, H., Dutton, L. C., Wharton, B., Ortiz-Barredo, A., Arnold, D. L., Tsiamis, G., Sesma, A., Butcher, D., Boch, J., Kim, Y. J., Martin, G. B., Tegli, S., Murillo, J., and Vivian, A. 2002. Location and activity of members of a family of virPphA homologues in pathovars of *Pseudomonas syringae* and *P. savastanoi.* Mol. Plant Pathol. 3:205-216.

Jakubowski, S. J., Krishnamoorthy, V., and Christie, P. J. 2003. *Agrobacterium tumefaciens* VirB6 protein participates in formation of VirB7 and VirB9 complexes required for type IV secretion. J. Bacteriol. 185:2867-2878.

Jakubowski, S. J., Cascales, E., Krishnamoorthy, V., and Christie, P. J. 2005. *Agrobacterium tumefaciens* VirB9, an outer-membrane-associated component of a type IV secretion system, regulates substrate selection and T-pilus biogenesis. J. Bacteriol. 187:3486-3495.

Jamet, A., Kiss, E., Batut, J., Puppo, A., and Hérouart, D. 2005. The *katA* catalase gene is regulated by OxyR in both free-living and symbiotic *Sinorhizobium meliloti*. J. Bacteriol. 187:376-381.

Jarchow, E., Grimsley, N. H., and Hohn, B. 1991. *virF*, The host-range-determining virulence gene of *Agrobacterium tumefaciens*, affects T-DNA transfer to *Zea mays*. Proc. Natl. Acad. Sci. U.S.A. 88:10426-10430.

Jasper, F., Koncz, C., Schell, J., and Steinbiss, H.-H. 1994. *Agrobacterium* T-strand production *in vitro*: Sequence-specific cleavage and 5' protection of single-stranded DNA templates by purified VirD2 protein. Proc. Natl. Acad. Sci. U.S.A. 91:694-698.

Jin, S., Prusti, R. K., Roitsch, T., Ankenbauer, R. G., and Nester, E. W. 1990a. Phosphorylation of the VirG protein of *Agrobacterium tumefaciens* by the autophosphorylated VirA protein: Essential role in biological activity of VirG. J. Bacteriol. 172:4945-4950.

Jin, S., Roitsch, T., Christie, P. J., and Nester, E. W. 1990b. The regulatory VirG protein specifically binds to a *cis*-acting regulatory sequence involved in transcriptional activation of *Agrobacterium tumefaciens* virulence genes. J. Bacteriol. 172:531-537.

Jones, A. L., Shirasu, K., and Kado, C. I. 1994. The product of the *virB4* gene of *Agrobacterium tumefaciens* promotes accumulation of VirB3 protein. J. Bacteriol. 176:5255-5261.

Jones, A. L., Lai, E.-M., Shirasu, K., and Kado, C. I. 1996. VirB2 is a processed pilin-like protein encoded by the *Agrobacterium tumefaciens* Ti plasmid. J. Bacteriol. 178:5706-5711.

Judd, P. K., Kumar, R. B., and Das, A. 2005. The type IV secretion apparatus protein VirB6 of *Agrobacterium tumefaciens* localizes to a cell pole. Mol. Microbiol. 55:115-124.

Kado, C. I. 1998. *Agrobacterium*-mediated horizontal gene transfer. Genet. Eng. 20:1-24.

Kado, C. I. 2002a. Horizontal transmission of genes by *Agrobacterium* species. Pages 45-50 in: Horizontal Gene Transfer, 2nd ed. M. Syvanen and C. I. Kado, eds. Academic Press, San Diego, CA.

Kado, C. I. 2002b. Negative transcriptional regulation of virulence and oncogenes of the Ti plasmid by Ros bearing a conserved C2H2-zinc finger motif. Plasmid 48:179-185.

Kado, C. I., and Heskett, M. G. 1970. Selective media for isolation of *Agrobacterium*, *Corynebacterium*, *Erwinia*, *Pseudomonas*, and *Xanthomonas*. Phytopathology 60:969-976.

Kalogeraki, V. S., and Winans, S. C. 1995. The octopine-type Ti plasmid pTiA6 of *Agrobacterium tumefaciens* contains a gene homologous to the chromosomal virulence gene *acvB*. J. Bacteriol. 177:892-897.

Kalogeraki, V. S., and Winans, S. C. 1998. Wound-released chemical signals may elicit multiple responses from an *Agrobacterium tumefaciens* strain containing an octopine-type Ti plasmid. J. Bacteriol. 180:5660-5667.

Kalogeraki, V. S., Zhu, J., Eberhard, A., Madsen, E. L., and Winans, S. C. 1999. The phenolic *vir* gene inducer ferulic acid is O-demethylated by the VirH2 protein of an *Agrobacterium tumefaciens* Ti plasmid. Mol. Microbiol. 34:512-522.

Kamińska, M., and Berniak, H. 2008. The occurrence of '*Candidatus* Phytoplasma asteris' in *Fraxinus excelsior* trees in Poland. Phytopathol. Pol. 49:35-41.

Kamińska, M., Dziekanowaka, D., and Rudzińska-Langwald, A. 2001. Detection of phytoplasma infection in rose, with degeneration symptoms. J. Phytopathol. 149:3-10.

Kamoun, S., Cooley, M. B., Rogowsky, P. M., and Kado, C. I. 1989. Two chromosomal loci involved in production of exopolysaccharide in *Agrobacterium tumefaciens*. J. Bacteriol. 171:1755-1759.

Kerr, A. 1969. Transfer of virulence between isolates of *Agrobacterium*. Nature 223:1175-1176.

Kerr, A. 1971. Acquisition of virulence by non-pathogenic isolates of *Agrobacterium radiobacter*. Physiol. Plant Pathol. 1:241-246.

Kers, J. A., Cameron, K. D., Joshl, M. V., Bukhalid, R. A., Morello, J. E., Wach, M. J., Gibson, D. M., and Loria, R. 2005. A large, mobile pathogenicity island confers plant pathogenicity on *Streptomyces* species. Mol. Microbiol. 55:1025-1033.

Khadhair, A. H., Tewari, J. P., Howard, R. J., and Paul, V. H. 2001. Detection of aster yellows phytoplasma in false flax based on PCR and RFLP. Microbiol. Res. 156:179-184.

Khadhair, A. H., Evans, I. R., and Choban, B. 2002. Identification of aster yellows phytoplasma in garlic and green onion. Microbiol. Res. 157:161-167.

Khan, M. S., Raj, S. K., and Snehi, S. K. 2007. First report of '*Candidatus* Phytoplasma asteris' affecting sesame cultivation in India. J. Plant Pathol. 89:301-305.

King, E. O., Ward, M. K., and Raney, D. E. 1954. Two simple media for the demonstration of pyocyanin and fluorescin. J. Lab. Clin. Med. 44:301-307.

Koenig, R. L., Morris, R. O., and Polacco, J. C. 2002. tRNA is the source of low-level *trans*-zeatin production in *Methylobacterium* spp. J. Bacteriol. 184:1832-1842.

Koike, S. T., and Davis, R. M. 2002. UC IPM Pest Management Guidelines: Lettuce. Univ. Calif. Agric. Nat. Res. Publ. 3450.

Koraimann, G. 2003. Lytic transglycosylases in macromolecular transport systems of Gram-negative bacteria. Cell Mol. Life Sci. 60:2371-2388.

Koukolíková-Nicola, Z., Raineri, D., Stephens, K., Ramos, C., Tinland, B., Nester, E. W., and Hohn, B. 1993. Genetic analysis of the *virD* operon of *Agrobacterium tumefaciens*: A search for functions involved in transport of T-DNA into the plant cell nucleus and in T-DNA integration. J. Bacteriol. 175:723-731.

Kumar, R. B., and Das, A. 2001. Functional analysis of the *Agrobacterium tumefaciens* T-DNA transport pore protein VirB8. J. Bacteriol. 183:3636-3641.

Kumar, R. B., Xie, Y. H., and Das, A. 2000. Subcellular localization of the *Agrobacterium tumefaciens* T-DNA transport pore proteins: VirB8 is essential for the assembly of the transport pore. Mol. Microbiol. 36:608-617.

Kumar, S. V., and Rajam, M. U. 2005. Polyamines enhance *Agrobacterium tumefaciens vir* gene induction and T-DNA transfer. Plant Sci. 118:475-480.

Kunkel, L. O. 1926. Studies on aster yellows. Am. J. Bot. 13:646-705.

Kunkel, L. O. 1932. Celery yellows of California not identical with the aster yellows of New York. Contrib. Boyce Thompson Inst. 4:405-414.

Lacey, M. S. 1936. Studies in bacteriosis. XXII. The isolation of a bacterium associated with 'fasciation' of sweet peas, 'cauliflower' of strawberry plants and 'leaf gall' of various plants. Ann. Appl. Biol. 23:302-310.

Lacey, M. S. 1939. Studies in bacteriosis. XXIV. Studies on a bacterium associated with leafy galls, fasciations and cauliflower disease of various plants. Ann. Appl. Biol. 26:262-278.

Lacey, M. S. 1955. The cytology and relationships of *Corynebacterium fascians*. Trans. Br. Mycol. Soc. 38:49-58.

Lacroix, B., Vaidya, M., Tzfira, T., and Citovsky, V. 2005. The VirE3 protein of *Agrobacterium* mimics a host cell function required for plant genetic transformation. EMBO J. 24:428-437.

Lai, E.-M., and Kado, C. I. 1998. Processed VirB2 is the major subunit of the promiscuous pilus of *Agrobacterium tumefaciens*. J. Bacteriol. 180:2711-2717.

Lai, E.-M., and Kado, C. I. 2000. The T-pilus of *Agrobacterium tumefaciens*. Trends Microbiol. 8:8361-8369.

Lai, E.-M., Chesnokova, O., Banta, L. M., and Kado, C. I. 2000. Genetic and environmental factors affecting T-pilin export and T-pilus biogenesis in relation to flagellation of *Agrobacterium tumefaciens*. J. Bacteriol. 182:3705-3716.

Lai, E.-M., Eisenbrandt, R., Kalkum, M., Lanka, E., and Kado, C. I. 2002. Biogenesis of T pili in *Agrobacterium tumefaciens* requires precise VirB2 propilin cleavage and cyclization. J. Bacteriol. 184:327-330.

Lee, I.-M., Hammond, R. W., Davis, R. E., and Gundersen, D. E. 1993. Universal amplification and analysis of pathogen 16S rDNA for classification and identification of mycoplasmalike organisms. Phytopathology 83:834-842.

Lee, I.-M., Gundersen, D. E., Hammond, R. W., and Davis, R. E. 1994. Use of mycoplasma-like organism (MLO) group-specific oligonucleotide primers for nested-PCR assays to detect mixed-MLO infections in a single host plant. Phytopathology 84:559-566.

Lee, Y. W., Ha, U. H., Sim, W. S., and Nester, E. W. 1998. Characterization of an unusual sensor gene (*virA*) of *Agrobacterium*. Gene 210:307-314.

Lemcke, K., and Schmulling, T. 1998. Gain of function assays identify non-*rol* genes from *Agrobacterium rhizogenes* TL-DNA that alter plant morphogenesis or hormone sensitivity. Plant J. 15:423-433.

Leroux, B., Yanofsky, M. F., Winans, S. C., Ward, J. E., Ziegler, S. F., and Nester, E. W. 1987. Characterization of the *virA* locus of *Agrobacterium tumefaciens*: A transcriptional regulator and host range determinant. EMBO J. 6:849-856.

Levine, M. 1919. Studies on plant cancers. I. The mechanism of the formation of the leafy crown gall. Bull. Torrey Bot. Club 46:447-452.

Lichter, A., Manulis, S., Sagee, O., Gafni, Y., Gray, J., Meilan, R., Morris, R. O., and Barash, I. 1995. Production of cytokinins by *Erwinia herbicola* pv. *gypsophilae* and isolation of a locus conferring cytokinin biosynthesis. Mol. Plant-Microbe Interact. 8:114-121.

Lim, P. O., and Sears, B. B. 1992. Evolutionary relationships of a plant-pathogenic mycoplasmalike organism and *Acholeplasma laidlawii* deduced from two ribosomal protein gene sequences. J. Bacteriol. 174:2606-2611.

Lin, B.-C., and Kado, C. I. 1977. Studies on *Agrobacterium tumefaciens*. VII. Avirulence induced by temperature and ethidium bromide. Can. J. Microbiol. 23:1554-1561.

Lin, T.-S., and Kado, C. I. 1993. The *virD4* gene is required for virulence while *virD3* and *orf5* are not required for virulence of *Agrobacterium tumefaciens*. Mol. Microbiol. 9:803-812.

Lioret, C. 1956. Sur la mise en evidence d'un acide amine non identifié particulier aux tissues de crown-gall. Bull. Soc. Fr. Physiol. Vég. 2:76.

Lippincott, J. A., and Lippincott, B. B. 1967. Time required for tumour initiation by *Agrobacterium tumefaciens* on pinto bean leaves. Nature 213:596-598.

Liu, Z., Jacobs, M., Schaff, D. A., McCullen, C. A., and Binns, A. N. 2001. ChvD, a chromosomally encoded ATP-binding cassette transporter-homologous protein involved in regulation of virulence gene expression in *Agrobacterium tumefaciens*. J. Bacteriol. 183:3310-3317.

MacDonald, E. M. S., Powell, G. K., Regier, D. A., Glass, N. L., Roberto, F., Kosuge, T., and Morris, R. O. 1986. Secretion of zeatin, ribosylzeatin, and ribosyl-1''-methylzeatin by *Pseudomonas savastanoi*. Plant Physiol. 82:742-750.

Machón, C., Rivas, S., Albert, A., Goñi, F. M., and de la Cruz, F. 2002. TrwD, the hexameric traffic ATPase encoded by plasmid R388, induces membrane destabilization and hemifusion of lipid vesicles. J. Bacteriol. 184:1661-1668.

Macnab, R. M. 1999. The bacterial flagellum: Reversible rotary propeller and type III export apparatus. J. Bacteriol. 181:7149-7153.

Maes, T., Vereecke, D., Ritsema, T., Cornelis, K., Thu, H. N. T., van Montagu, M., Holsters, M., and Goethals, K. 2001. The *att* locus of *Rhodococcus fascians* strain D188 is essential for full virulence on tobacco through the production of an autoregulatory compound. Mol. Microbiol. 42:13-28.

Magrelli, A., Langenkemper, K., Dehio, C., Schell, J., and Spena, A. 1994. Splicing of the *rolA* transcript of *Agrobacterium rhizogenes* in *Arabidopsis*. Science 266:1986-1988.

Mantis, N. J., and Winans, S. C. 1993. The chromosomal response regulatory gene *chvI* of *Agrobacterium tumefaciens* complements an *Escherichia coli phoB* mutation and is required for virulence. J. Bacteriol. 175:6626-6636.

Manulis, S., and Barash, I. 2003a. *Pantoea agglomerans* pvs. *gypsophilae* and *betae*, recently evolved pathogens? Mol. Plant Pathol. 4:307-314.

Manulis, S., and Barash, I. 2003b. Molecular basis for transformation of an epiphyte into a gall-forming pathogen as exemplified by *Erwinia herbicola* pv. *gypsophilae*. Pages 19-52 in: Plant-Microbe Interactions, Vol. 6. G. Stacey and N. T. Keen, eds. American Phytopathological Society, St. Paul, MN.

Manulis, S., Gafni, Y., Clark, E., Zutra, D., Ophir, Y., and Barash, I. 1991. Identification of a plasmid DNA probe for detection of strains of *Erwinia herbicola* pathogenic on *Gypsophila paniculata*. Phytopathology 81:54-57.

Marks, J. R., Lynch, T. J., Karlinsey, J. E., and Thomashow, M. F. 1987. *Agrobacterium tumefaciens* virulence locus *pscA* is related to the *Rhizobium meliloti exoC* locus. J. Bacteriol. 169:5835-5837.

Matthysse, A. G., and McMahan, S. 2001. The effect of the *Agrobacterium tumefaciens attR* mutation on attachment and root colonization differs between legumes and other dicots. Appl. Environ. Microbiol. 67:1070-1075.

Matthysse, A. G., Yarnall, H., Boles, S. B., and McMahan, S. 2000. A region of the *Agrobacterium tumefaciens* chromosome containing genes required for virulence and attachment to host cells. Biochim. Biophys. Acta 1490:208-212.

Matthysse, A. G., Jaeckel, P., and Jeter, C. 2008. *attG* And *attC* mutations of *Agrobacterium tumefaciens* are dominant negative mutations that block attachment and virulence. Can. J. Microbiol. 54:241-247.

Melamed, S., Tanne, E., Ben-Haim, R., Edelbaum, O., Yogev, D., and Sela, H. 2003. Identification and characterization of phytoplasmal genes, employing a novel method of isolating phytoplasmal genomic DNA. J. Bacteriol. 185:6513-6521.

Melchers, L. S., Regensburg-Tuïnk, T. J., Bourret, R. B., Sedee, N. J., Schilperoort, R. A., and Hooykaas, P. J. J. 1989. Membrane topology and functional analysis of the sensory protein VirA of *Agrobacterium tumefaciens*. EMBO J. 8:1919-1925.

Melchers, L. S., Maroney, M. J., den Dulk-Ras, A., Thompson, D. V., van Vuuren, H. A. J., Schilperoort, R. A., and Hooykaas, P. J. J. 1990. Octopine and nopaline strains of *Agrobacterium tumefaciens* differ in virulence; Molecular characterization of the *virF* locus. Plant Mol. Biol. 14:249-259.

Ménagé, A., and Morel, G. 1964. Sur la presence d'octopine dans les tissues de crown-gall. C. R. Acad. Sci. Paris 259:4795-4796.

Miller, H. J., Quinn, C. E., and Graham, D. C. 1981. A strain of *Erwinia herbicola* pathogenic on *Gypsophila paniculata*. Neth. J. Plant Pathol. 87:167-172.

Mohanty, U. 1951. *Corynebacterium fascians* (Tilford) Dowson; Its morphology, physiology, nutrition and taxonomic position. Trans. Br. Mycol. Soc. 34:23-34.

Mor, H., Manulis, S., Zuck, M., Nizan, R., Coplin, D. L., and Barash, I. 2001. Genetic organization of the *hrp* gene cluster and *dspAE/BF* operon in *Erwinia herbicola* pv. *gypsophilae*. Mol. Plant-Microbe Interact. 14:431-436.

Morel, G. 1956. Métabolisme de l'arginine par les tissues de crown-gall de topinambour. Bull. Soc. Fr. Physiol. Veg. 2:75.

Morel, P., Powell, B. S., Rogowsky, P. M., and Kado, C. I. 1989. Characterization of the *virA* virulence gene of the nopaline plasmid, pTiC58, of *Agrobacterium tumefaciens*. Mol. Microbiol. 3:1237-1246.

Moriuchi, H., Okamoto, C., Nishihama, R., Yamashita, I., Machida, Y., and Tanaka, N. 2004. Nuclear localization and interaction of RolB with plant 14-3-3 proteins correlates with induction of adventitious roots by the oncogene *rolB*. Plant J. 38:260-275.

Morizawa, K. 1927. Über die Extraktivstoff von *Oktopus oktopodia*. Acta Sch. Med. Univ. Imp. Kyoto 9:285-298.

Mushegian, A. R., Fullner, K. J., Koonin, E. V., and Nester, E. W. 1996. A family of lysozyme-like virulence factors in bacterial pathogens of plants and animals. Proc. Natl. Acad. Sci. U.S.A. 93:7321-7326.

Musič, M. S., Novokmet, M., and Novak, R. 2008. Distribution of phytoplasma diseases in the Lombardy poplar tree population of Zagreb urban area. Acta Bot. Croat. 67:119-130.

Nissan, G., Manulis, S., Weinthal, D. M., Sessa, G., and Barash, I. 2005. Analysis of promoters recognized by HrpL, an alternative σ-factor protein from *Pantoea agglomerans* pv. *gypsophilae*. Mol. Plant-Microbe Interact. 18:634-643.

Nissan, G., Manulis-Sasson, S., Weinthal, D., Mor, H., Sessa, G., and Barash, I. 2006. The type III effectors HsvG and HsvB of gall-forming *Pantoea agglomerans* determine host specificity and function as transcriptional activators. Mol. Microbiol. 61:1118-1131.

Nizan, R., Barash, I., Valinsky, L., Lichter, A., and Manulis, S. 1997. The presence of *hrp* genes on the pathogenicity-associated plasmid of the tumorigenic bacterium *Erwinia herbicola* pv. *gypsophilae*. Mol. Plant-Microbe Interact. 10:677-682.

Offringa, R., van den Elzen, P. J. M., and Hooykaas, P. J. J. 1992. Gene targeting in plants using *Agrobacterium* vector system. Transgenic Res. 1:114-123.

Ogimi, C., Kubo, Y., Higuchi, H., and Takikawa, Y. 1990. Bacterial gall disease of himeyuzuriha (*Daphniphyllum teijsmanni* Z.) caused by *Pseudomonas syringae* pv. *daphniphylli* pv. nov. J. Jpn. For. Soc. 72:17-22.

O'Mara, J., Bauernfeind, R., Stevens, A., Gast, K. L. B., and Stevens, S. 1993. Aster yellows. Coop. Ext. Serv. MF-1086. Kansas State University, Manhattan, KS.

Otten, L., Piotrowiak, G., Hooykaas, P., Dubois, M., Szegedi, E., and Schell, J. 1985. Identification of an *Agrobacterium tumefaciens* pTiB6S3 *vir* region fragment that enhances the virulence of pTiC58. Mol. Gen. Genet. 199:189-193.

Pál, C., Papp, B., and Lercher, M. J. 2005. Adaptive evolution of bacterial metabolic networks by horizontal gene transfer. Nat. Genet. 37:1372-1375.

Palm, C. J., Gaffney, T., and Kosuge, T. 1989. Cotranscription of genes encoding indoleacetic acid production in *Pseudomonas syringae* subsp. *savastanoi*. J. Bacteriol. 171:1002-1009.

Palumbo, J. D., Phillips, D. A., and Kado, C. I. 1998. Characterization of a new *Agrobacterium tumefaciens* strain from alfalfa (*Medicago sativa* L.). Arch. Microbiol. 169:381-386.

Pan, S. Q., Jin, S., Boulton, M. I., Hawes, M., Gordon, M. P., and Nester, E. W. 1995. An *Agrobacterium* virulence factor encoded by a Ti plasmid gene or a chromosomal gene is required for T-DNA transfer into plants. Mol. Microbiol. 17:259-269.

Panek, H. R., and O'Brien, M. R. 2004. KatG is the primary detoxifier of hydrogen peroxide produced by aerobic metabolism in *Bradyrhizobium japonicum*. J. Bacteriol. 186:7874-7880.

Pansegrau, W., Schoumacher, F., Hohn, B., and Lanka, E. 1993. Site-specific cleavage and joining of single-stranded DNA by VirD2 protein of *Agrobacterium tumefaciens* Ti plasmids: Analogy to bacterial conjugation. Proc. Natl. Acad. Sci. U.S.A. 90:11538-11542.

Pantoja, M., Chen, L., Chen, Y., and Nester, E. W. 2002. *Agrobacterium* type IV secretion is a two-step process in which export substrates associate with the virulence protein VirJ in the periplasm. Mol. Microbiol. 45:1325-1335.

Pazour, G. J., and Das, A. 1990. *virG*, an *Agrobacterium tumefaciens* transcriptional activator, initiates translation at a UUG codon and is a sequence-specific DNA-binding protein. J. Bacteriol. 174:4169-4174.

Pelczar, P., Kalck, V., Gomez, D., and Hohn, B. 2004. *Agrobacterium* proteins VirD2 and VirE2 mediate precise integration of synthetic T-DNA complexes in mammalian cells. EMBO Rep. 5:632-637.

Peng, W.-T., Banta, L. M., Charles, T. C., and Nester, E. W. 2001. The *chvH* locus of *Agrobacterium* encodes a homologue of an elongation factor involved in protein synthesis. J. Bacteriol. 183:36-45.

Penyalver, R., Garcia, A., Ferrer, A., Bertolini, E., and Lopéz, M. M. 2000. Detection of *Pseudomonas savastanoi* pv. savastanoi in olive plants by enrichment and PCR. Appl. Environ. Microbiol. 66:2673-2677.

Pérez-Martínez, I., Zhao, Y., Murillo, J., Sundin, G. W., and Ramos, C. 2008. Global genomic analysis of *Pseudomonas savastanoi* pv. savastanoi plasmids. J. Bacteriol. 190:625-635.

Perry, K. L., and Kado, C. I. 1982. Characteristics of Ti plasmids from broad-host-range and ecologically specific biotype 2 and 3 strains of *Agrobacterium tumefaciens*. J. Bacteriol. 151:343-350.

Petit, A., Delhaye, S., Tempé, J., and Morel, G. 1970. Recherches sur les guanidines des tissues de crown gall. Mise en evidence d'une relation biochimique spécifique entre les souches d'*Agrobacterium* et les tumeurs qu'elles induisent. Physiol. Veg. 8:205-213.

Pierce, N. B. 1891. Tuberculosis of the olive. J. Mycol. 6:148-153.

Plano, G. V., Day, J. B., and Ferracci, F. 2001. Type III export: New uses for an old pathway. Mol. Microbiol. 40:284-293.

Powell, B. S., and Kado, C. I. 1990. Specific binding of VirG to the *vir* box requires a *C*-terminal domain and exhibits a minimum concentration threshold. Mol. Microbiol. 4:2159-2166.

Powell, G. K., and Morris, R. O. 1986. Nucleotide sequence and expression of a *Pseudomonas savastanoi* cytokinin biosynthetic gene: Homology with *Agrobacterium tumefaciens tmr* and *tzs* loci. Nucleic Acids Res. 14:2555-2565.

Purcell, A. H., Richardson, J., and Finlay, A. 1981. Multiplication of the agent of X-disease in a non-vector leafhopper *Macrosteles fascifrons*. Ann. Appl. Biol. 99:283-289.

Raina, S., Riana, R., Venkatesh, T. V., and Das, H. K. 1995. Isolation and characterization of a locus from *Azospirillum brasilense* Sp7 that complements the tumorigenic defect of *Agrobacterium tumefaciens chvB* mutant. Mol. Plant-Microbe Interact. 8:322-326.

Raine, J., and Forbes, A. R. 1969. Mycoplasma-like bodies in the saliva of the leafhopper *Macrosteles fascifrons* (Stal) (Homoptera: Cicadellidae). Can. J. Microbiol. 15:1105-1107.

Raj, S. K., Snehi, S. K., Khan, M. S., and Kumar, S. 2008a. '*Candidatus* Phytoplasma asteris' (group 16SrI) associated with a witches' broom disease of *Cannabis sativa* in India. Plant Pathol 57:1173.

Raj, S. K., Khan, M. S., Snehi, S. K., Kumar, S., Mall, S., and Rao, G. P. 2008b. First report of phytoplasma '*Candidatus* phytoplasma asteris' (16Srl) from *Parthenium hysterophorus* L. showing symptoms of virescence and witches'-broom in India. Australas. Plant Dis. Notes 3:44-45.

Rashkova, S., Zhou, X.-R., Chen, J., and Christie, P. J. 2000. Self-assembly of the *Agrobacterium tumefaciens* VirB11 traffic ATPase. J. Bacteriol. 182:4137-4145.

Reeder, R., Kelly, P. L., and Arocha, Y. 2010. '*Candidatus* Phytoplasma asteris' identified in blackberry (*Rubus fruticosus* agg.) in the United Kingdom. Plant Pathol. 59:394.

Regensburg-Tuïnk, A. J. G., and Hooykaas, P. J. J. 1993. Transgenic *N. glauca* plants expressing bacterial virulence gene *virF* are converted into hosts for nopaline strains of *A. tumefaciens*. Nature 363:69-70.

Relić, B., Andjelković, M., Rossi, L., Nagamine, Y., and Hohn, B. 1998. Interaction of the DNA modifying proteins VirD1 and VirD2 of *Agrobacterium tumefaciens*: Analysis by subcellular localization in mammalian cells. Proc. Natl. Acad. Sci. U.S.A. 95:9105-9110.

Riker, A. J., Banfield, W. M., Wright, W. H., Keitt, G. W., and Sagen, H. E. 1930. Studies on infectious hairy-root of nursery apple trees. J. Agric. Res. 41:507-540.

Rogowsky, P. M., Powell, B. S., Shirasu, K., Lin, T.-S., Morel, P., Zyprian, E. M., Steck, T. R., and Kado, C. I. 1990. Molecular characterization of the *vir* regulon of *Agrobacterium tumefaciens*: Complete nucleotide sequence and gene organization of the 28.63-kbp regulon cloned as a single unit. Plasmid 23:85-106.

Rong, L. R., Karcher, S. J., and Gelvin, S. B. 1991. Genetic and molecular analyses of *picA*, a plant-inducible locus on the *Agrobacterium tumefaciens* chromosome. J. Bacteriol. 173:5110-5120.

Rong, L., Carpita, N. C., Mort, A., and Gelvin, S. B. 1994. Soluble cell wall compounds from carrot roots induce the *picA* and *pgl* loci of *Agrobacterium tumefaciens*. Mol. Plant-Microbe Interact. 7:6-14.

Roset, M. S., Ciocchini, A. E., Ugalde, R. A., and Iñon de Iannino, N. 2004. Molecular cloning and characterization of *cgt*, the *Brucella abortus* cyclic beta-1,2-glucan transporter gene, and its role in virulence. Infect. Immun. 72:2263-2271.

Rossi, L., Hohn, B., and Tinland, B. 1996. Integration of complete transferred DNA units is dependent on the activity of virulence E2 protein of *Agrobacterium tumefaciens*. Proc. Natl. Acad. Sci. U.S.A. 93:126-130.

Saenkham, P., Eiamphungporn, W., Farrand, S. K., Vattanaviboon, P., and Mongkolsuk, S. 2007. Multiple superoxide dismutases in *Agrobacterium tumefaciens*: Functional analysis, gene regulation, and influence on tumorigenesis. J. Bacteriol. 189:8807-8817.

Sagulenko, E., Sagulenko, V., Chen, J., and Christie, P. J. 2001. Role of *Agrobacterium* VirB11 ATPase in T-pilus assembly and substrate selection. J. Bacteriol. 183:5813-5825.

Sagulenko, V., Sagulenko, E., Jakubowski, S., Spudich, E., and Christie, P. J. 2001. VirB7 lipoprotein is exocellular and associates with the *Agrobacterium tumefaciens* T pilus. J. Bacteriol. 183:3642-3651.

Savastano, L. 1886. Les maladies de l'olvier; hyperplasies et tumeurs. C. R. Acad. Agric. (France) 103:1278.

Schmidt-Eisenlohr, H., Domke, N., Angerer, C., Wanner, G., Zambryski, P. C., and Baron, C. 1999. Vir proteins stabilize VirB5 and mediate its association with T pilus of *Agrobacterium tumefaciens*. J. Bacteriol. 181:7485-7492.

Schneider, B., Gibb, K. S., and Seemuller, F. 1997. Sequence and RFLP analysis of the elongation factor Tu gene used in differentiation and classification of phytoplasmas. Microbiology 143:3381-3389.

Sen, P., Pazour, G. J., Anderson, D., and Das, A. 1989. Cooperative binding of the VirE2 protein to single-stranded DNA. J. Bacteriol. 171:2573-2580.

Severin, H. H. P. 1929. Yellows disease of celery, lettuce and other plants transmitted by *Cicadula sexnotata* (Fall.). Hilgardia 3:543-582.

Shamaei-Tousi, A., Cahill, R., and Frankel, G. 2004. Interaction between protein subunits of the type IV secretion system of *Bartonella henselae*. J. Bacteriol. 186:4796-4801.

Sheikholeslam, S., Okubara, P. A., Lin, B.-C., Dutra, J. C., and Kado, C. I. 1978. Large and small plasmids in tumorigenic and cured non-tumorigenic *Agrobacterium tumefaciens* and in *Agrobacterium radiobacter*. Pages 132-135 in: Microbiology—1978. D. Schlessinger, ed. American Society of Microbiology, Washington, DC.

Sheikholeslam, S., Lin, B.-C., and Kado, C. I. 1979. Multiple-size plasmids in *Agrobacterium radiobacter* and *A. tumefaciens*. Phytopathology 69:54-58.

Shimoda, N., Toyoda-Yamamoto, A., Nagamine, J., Usami, S., Katayama, M., Sakagami, Y., and Machida, Y. 1990. Control of expression of *Agrobacterium vir* genes by synergistic actions of phenolic signal molecules and monosaccharides. Proc. Natl. Acad. Sci. U.S.A. 87:6684-6688.

Shirasu, K., and Kado, C. I. 1993a. The *virB* operon of the *Agrobacterium tumefaciens* virulence regulon has sequence similarities to B, C and D open reading frames downstream of the pertussis toxin-operon and to the DNA transfer-operons of broad-host-range conjugative plasmids. Nucleic Acids Res. 21:353-354.

Shirasu, K., and Kado, C. I. 1993b. Membrane location of the Ti plasmid VirB proteins involved in the biosynthesis of a pilin-like conjugative structure on *Agrobacterium tumefaciens*. FEMS Microbiol. Lett. 111:287-294.

Shirasu, K., Koukolikova-Nicola, Z., Hohn, B., and Kado, C. I. 1994. An inner-membrane-associated virulence protein essential for T-DNA transfer from *Agrobacterium tumefaciens* to plants exhibits ATPase activity and similarities to conjugative transfer genes. Mol. Microbiol. 11:581-588.

Siegler, E. A. 1929. The woolly-knot type of crown gall. J. Agric. Res. 39:427-450.

Simone, M., McCullen, C. A., Stahl, L. E., and Binns, A. N. 2001. The carboxy-terminus of VirE2 from *Agrobacterium tumefaciens* is required for its transport to host cells by the *virB*-encoded type IV transport system. Mol. Microbiol. 41:1283-1293.

Slater, S. C., Goldman, B. S., Goodner, B., Setubal, J. C., Farrand, S. K., Nester, E. W., Burr, T. J., Banta, L., Dikerman, A. W., Paulsen, I., Otten, L., Suen, G., Welch, R., Almeida, N. F., Arnold, F., Burton, O. T., Du, Z., Ewing, A., Godsy, E., Heisel, S., Houmiel, K. L., Jhaveri, J., Lu, J., Miller, N. M., Norton, S., Chen, Q., Phoolcharoen, W., Ohlin, V., Ondrusek, D., Pride, N., Stricklin, S. L., Sun, J., Wheeler, C., Wilson, L., Zhu, H., and Wood, D. W. 2009. Genome sequences of three *Agrobacterium* biovars help elucidate the evolution of multichromosome genomes in bacteria. J. Bacteriol. 191:2501-2511.

Smidt, M., and Kosuge, T. 1978. The role of indole-3-acetic acid accumulation by alpha methyl tryptophan resistant mutants of *Pseudomonas savastanoi* in gall formation on oleanders. Physiol. Plant Pathol. 13:203-214.

Smith, C. O. 1912. Further proof of the cause and infectiousness of crown gall. Univ. Calif. Agric. Exp. Stn. Bull. 235:531-557.

Smith, E. F. 1908. Recent studies of the olive-tubercule organism. U.S. Dep. Agric. Bur. Plant Ind. Bull. 131:25-43.
Smith, E. F. 1921. Effect of crowngall inoculations on Bryophyllum. J. Agric. Res. 21:593-597.
Smith, E. F., and Townsend, C. O. 1907. A plant-tumor of bacterial origin. Science 25:671-673.
Smith, R. E. 1907. Report of the plant pathologist. Calif. Agric. Exp. Stn. Bull. 184:252.
Soby, S., Kirkpatrick, B., and Kosuge, T. 1994. Characterization of high-frequency deletions in the *iaa*-containing plasmid, pIAA2, of *Pseudomonas syringae* pv. *savastanoi*. Plasmid 31:21-30.
Spena, A., Schműlling, T., Koncz, C., and Schell, J. S. 1987. Independent and synergistic activity of *rolA*, *B* and *C* loci in stimulating abnormal growth in plants. EMBO J. 6:3891-3899.
Spudich, G., Fernandez, D., Zhou, X.-R., and Christie, P. J. 1996. Intermolecular disulfide bonds stabilize VirB7 homodimers and VirB7/VirB9 heterodimers during biogenesis of the *Agrobacterium tumefaciens* T-complex transport pore. Proc. Natl. Acad. Sci. U.S.A. 93:7512-7517.
Starr, M. P., and Weiss, J. E. 1943. Growth of phytopathogenic bacteria in a synthetic asparagin medium. Phytopathology 33:314-318.
Steck, T. R., Lin, T.-S., and Kado, C. I. 1990. VirD2 gene product from the nopaline plasmid pTiC58 has at least two activities required for virulence. Nucleic Acids Res. 18:6953-6958.
Stockwell, V. O., Moore, L. W., and Loper, J. E. 1993. Fate of *Agrobacterium radiobacter* K84 in the environment. Appl. Environ. Microbiol. 59:2112-2120.
Suksomtip, M., Liu, P., Anderson, T., Tungpradabkul, S., Wood, D. W., and Nester, E. W. 2005. Citrate synthase mutants of *Agrobacterium* are attenuated in virulence and display reduced *vir* gene induction. J. Bacteriol. 187:4844-4852.
Sundberg, C. D., and Ream, W. 1999. The *Agrobacterium tumefaciens* chaperone-like protein, VirD1, interacts with VirE2 at domains required for single-stranded DNA binding and cooperative interaction. J. Bacteriol. 181:6850-6855.
Surico, G. 1986. Indoleacetic acid and cytokinins in the olive knot disease. An overview of their role and their genetic determinants. Pages 315-322 in: Biology and Molecular Biology of Plant-Pathogen Interactions. NATO ASI Ser. H, Vol. 1. J. Bailey, ed. Springer-Verlag, Berlin.
Surico, G., and Iacobellis, N. S. 1992. Phytohormones and olive knot disease. Pages 209-228 in: Molecular Signals in Plant-Microbe Communications. D. P. S. Verma, ed. CRC Press, Boca Raton, FL.
Surico, G., and Lavermicocca, P. 1989. A semiselective medium for the isolation of *Pseudomonas syringae* pv. *savastanoi*. Phytopathology 79:185-190.
Surico, G., Iacobellis, N. S., and Sisto, A. 1985. Studies on the role of indole-3-acetic acid and cytokinins in the formation of knots on olive and oleander plants by *Pseudomonas syringae* pv. *savastanoi*. Physiol. Plant Pathol. 26:309-320.
Suzuki, K., Iwata, K., and Yoshida, K. 2001. Genome analysis of *Agrobacterium tumefaciens*: Construction of physical maps for linear and circular chromosomal DNAs, determination of copy number ratio and mapping of chromosomal virulence genes. DNA Res. 8:141-152.
Swart, S., Smit, G., Lugtenberg, B. J. J., and Kijne, J. W. 1993. Restoration of attachment, virulence and nodulation of *Agrobacterium tumefaciens chvB* mutants by rhicadhesin. Mol. Microbiol. 19:597-605.
Taddel, F., Matic, I., Godelle, B., and Radman, M. 1997. To be a mutator, or how pathogenic and commensal bacteria can evolve rapidly. Trends Microbiol. 5:427-428.
Tanaka, N., Fujikawa, Y., Aly, M. A. M., Saneoka, H., Fujita, K., and Yamashita, I. 2001. Proliferation and *rol* gene expression in hairy root lines of Egyptian clover (*Trifolium alexandrinum* L.). Plant Cell Tissue Organ. Cult. 66:175-182.
Temmerman, W., Vereecke, D., Dreesen, R., Van Montagu, M., Holsters, M., and Goethals, K. 2000. Leafy gall formation is controlled by *fasR*, an AraC-type regulatory gene in *Rhodococcus fascians*. J. Bacteriol. 182:5832-5840.
Tempé, J., and Schell, J. 1977. Is crown gall a natural instance of gene transfer? Pages 416-420 in: Translation of Natural and Synthetic Polynucleotides. A. B. Legocki, ed. University of Agriculture, Poznan, Poland.

Tilford, P. E. 1936. Fasciation of sweet peas caused by *Phytomonas fascians* n. sp. J. Agric. Res. 53:383-394.

Tinland, B., and Hohn, B. 1995. Recombination between prokaryotic and eukaryotic DNA: Integration of *Agrobacterium tumefaciens* T-DNA into the plant genome. Genet. Eng. 17:209-229.

Tolin, S. A., and Lacy, G. A. 2004. Viral, bacterial, and phytoplasmal diseases of soybean. Pages 765-819 in: Soybeans: Improvement, Production, and Uses, 3rd ed. H. R. Boerma and J. E. Specht, eds. American Society of Agronomy, Crop Science Society of America, Soil Science Society of America, Madison, WI.

Toro, N., Datta, A., Carmi, O. A., Young, C., Prusti, R. K., and Nester, E. W. 1989. The *Agrobacterium tumefaciens virC1* gene product binds to overdrive, a T-DNA transfer enhancer. J. Bacteriol. 171:6845-6849.

Torres, L., Galdeano, E., Docampo, D., and Conci, L. 2004. Characterization of an aster yellows phytoplasma associated with Catharanthus little leaf in Argentina. J. Plant Pathol. 86:209-214.

Unger, L., Ziegler, S. F., Huffman, G. A., Knauf, V. C., Peet, R., Moore, L. W., Gordon, M. P., and Nester, E. W. 1985. New class of limited-host-range *Agrobacterium* mega-tumor-inducing plasmids lacking homology to the transferred DNA of a wide-host-range, tumor-inducing plasmid. J. Bacteriol. 164:723-730.

United States Department of Agriculture. 1960. Index of plant diseases in the United States. U.S. Dep. Agric. Agric. Handb. 165.

Urbanczyk, H., Suzuki, K., Yoshida, K., and Kondo, K. 2003. Physical and gene maps of *Agrobacterium* biovar 2 strains and their relationship to biovar 1 chromosomes. Microbiology 149:3035-3042.

Valinsky, L., Barash, I., Chalupowicz, L., Ezra, D., and Manulis, S. 2002. Regulation of *hsvG*, a host-specific virulence gene from *Erwinia herbicola* pv. *gypsophilae*. Physiol. Mol. Plant Pathol. 60:19-29.

Van Larebeke, N., Genetello, C., Schell, J., Schilperoort, R. A., Hermans, A. K., Hernalsteens, J. P., and Van Montagu, M. 1975. Acquisition of tumour-inducing ability by non-oncogenic bacteria as a result of plasmid transfer. Nature 255:742-743.

Van Veen, R. J. M., den Dulk-Ras, H., Schilperoort, R. A., and Hooykaas, P. J. J. 1987. Chromosomal nodulation genes: Sym-plasmid containing *Agrobacterium* strains need chromosomal virulence genes (*chvA* and *chvB*) for nodulation. Plant Mol. Biol. 8:105-108.

Vereecke, D., Cornelis, K., Temmerman, W., Jaziri, M., Van Montagu, M., Holsters, M., and Goethals, K. 2002. Chromosomal locus that affects pathogenicity of *Rhodococcus fascians*. J. Bacteriol. 184:1112-1120.

Vogel, A. M., and Das, A. 1992a. Mutational analysis of *Agrobacterium tumefaciens virD2*: Tyrosine 29 is essential for endonuclease activity. J. Bacteriol. 174:303-308.

Vogel, A. M., and Das, A. 1992b. The *Agrobacterium tumefaciens virD3* gene is not essential for tumorigenicity on plants. J. Bacteriol. 174:5161-5164.

Volcani, Z. 1985. Bacterial Diseases of Plants in Israel. Agricultural Research Organization, The Volcani Center, Bet Dagan, Israel.

Ward, D. V., Draper, O., Zupan, J. R., and Zambryski, P. C. 2002. Peptide-linkage mapping of the *Agrobacterium tumefaciens vir*-encoded type IV secretion system reveals protein subassemblies. Proc. Natl. Acad. Sci. U.S.A. 99:11493-11500.

Ward, E. R., and Barnes, W. M. 1988. VirD2 protein of *Agrobacterium tumefaciens* is very tightly linked to the 5' end of T-strand DNA. Science 242:927-930.

Watson, B., Currier, T. C., Gordon, M. P., Chilton, M.-D., and Nester, E. W. 1975. Plasmid required for virulence of *Agrobacterium tumefaciens*. J. Bacteriol. 123:255-264.

Wehling, M. D., Guo, M., Fu, Z. Q., and Alfano, J. R. 2004. The *Pseudomonas syringae* HopPtoV protein is secreted in culture and translocated into plant cells via the type III protein secretion system in a manner dependent on the ShcV type III chaperone. J. Bacteriol. 186:3621-3630.

Weller, S. A., Stead, D. E., O'Neill, T. M., Hargreaves, D., and McPherson, G. M. 2000. Rhizogenic *Agrobacterium* biovar 1 and cucumber root mat in the UK. Plant Pathol. 49:43-50.

White, F. F., Taylor, B. H., Huffman, G. A., Gordon, M. P., and Nester, E. W. 1985. Molecular and genetic analysis of the transferred DNA regions of the root-inducing plasmid of *Agrobacterium rhizogenes.* J. Bacteriol. 164:3-44.

White, P. R., and Braun, A. C. 1942. A cancerous neoplasm of plants. Autonomous bacteria-free crown-gall tissue. Cancer Res. 2:597-617.

Wilson, E. E., and Magie, A. R. 1964. Systemic invasion of the host plant by the tumor-inducing bacterium, *Pseudomonas savastanoi.* Phytopathology 54:576-579.

Wilson, E. E., Heskett, M. G., Johnson, M. L., and Kosuge, T. 1972. Metabolic behavior of *Pseudomonas savastanoi* isolates from olive and oleander on certain carbohydrate and amino substrates. Phytopathology 62:349-355.

Wirawan, I. G. P., Kang, H. W., and Kojima, M. 1993. Isolation and characterization of a new chromosomal virulence gene of *Agrobacterium tumefaciens.* J. Bacteriol. 175:3208-3212.

Wise, A. A., Voinov, L., and Binns, A. N. 2005. Intersubunit complementation of sugar signal transduction in VirA heterodimers and posttranslational regulation of VirA activity in *Agrobacterium tumefaciens.* J. Bacteriol. 187:213-223.

Wise, A. A., Fang, F., Lin, Y.-H., He, F., Lynn, D. G., and Binns, A. N. 2010. The receiver domain of hybrid histidine kinase VirA: An enhancing factor for *vir* gene expression in *Agrobacterium tumefaciens.* J. Bacteriol. 192:1534-1542.

Wood, D. W., Setubal, J. C., Kaul, R., Monks, D. E., Kitajima, J. P., Okura, V. K., Zhou, Y., Chen, L., Wood, G. E., Almeida, N. F., Jr., Woo, L., Chen, Y., Paulsen, I. T., Eisen, J. A., Karp, P. D., Bovee, D., Sr., Chapman, P., Clendenning, J., Deatherage, G., Gillet, W., Grant, C., Kutyavin, T., Levy, R., Li, M. J., McClelland, E., Palmieri, A., Raymond, C., Rouse, G., Saenphimmachak, C., Wu, Z., Romero, P., Gordon, D., Zhang, S., Yoo, H., Tao, Y., Biddle, P., Jung, M., Krespan, W., Perry, M., Gordon-Kamm, B., Liao, L., Kim, S., Hendrick, C., Zhao, Z. Y., Dolan, M., Chumley, F., Tingey, S. V., Tomb, J. F., Gordon, M. P., Olson, M. V., and Nester, E. W. 2001. The genome of the natural genetic engineer *Agrobacterium tumefaciens* C58. Science 294:2317-2323.

Wright, W. H., Hendrickson, A. A., and Riker, A. J. 1930. Studies on the progeny of single-cell isolations from the hairy-root and crown-gall organisms. J. Agric. Res. 41:541-547.

Wu, H.-Y., Chung, P.-C., Shih, H.-W., Wen, S.-R., and Lai, E.-M. 2008. Secretome analysis uncovers an Hcp-family protein secreted via a type VI secretion system in *Agrobacterium tumefaciens.* J. Bacteriol. 190:2841-2850.

Xu, X. Q., and Pan, S. Q. 2000. An *Agrobacterium* catalase is a virulence factor involved in tumorigenesis. Mol. Microbiol. 35:407-414.

Xu, X. Q., Li, L. P., and Pan, S. Q. 2001. Feedback regulation of an *Agrobacterium* catalase gene *katA* involved in *Agrobacterium*-plant interaction. Mol. Microbiol. 42:645-657.

Yamada, T., Palm, C. J., Brooks, B., and Kosuge, T. 1985. Nucleotide sequence of the *Pseudomonas savastanoi* indoleacetic acid genes and homology with *Agrobacterium tumefaciens* T-DNA. Proc. Natl. Acad. Sci. U.S.A. 82:6522-6526.

Yanofsky, M. F., Porter, S. G., Young, C., Albright, L. M., Gordon, M. P., and Nester, E. W. 1986. The *virD* operon of *Agrobacterium tumefaciens* encodes a site-specific endonuclease. Cell 47:471-477.

Yeo, H.-J., Yuan, Q., Beck, M. R., Baron, C., and Waksman, G. 2003. Structural and functional characterization of the VirB5 protein from the type IV secretion system encoded by the conjugative plasmid pKM101. Proc. Natl. Acad. Sci. U.S.A. 100:15947-15952.

Young, C., and Nester, E. W. 1988. Association of the VirD2 protein with the 5' end of T strands in *Agrobacterium tumefaciens.* J. Bacteriol. 170:3367-3374.

Yuan, Z.-C., Zaheer, R., and Finan, T. M. 2005. Phosphate limitation induces catalase expression in *Sinorhizobium meliloti, Pseudomonas aeruginosa* and *Agrobacterium tumefaciens.* Mol. Microbiol. 58:877-894.

Yuan, Z.-C., Liu, P., Saenkham, P., Kerr, K., and Nester, E. W. 2008. Transcriptome profiling and functional analysis of *Agrobacterium tumefaciens* reveals a general conserved response to acidic conditions (pH 5.5) and a complex acid-mediated signaling involved in *Agrobacterium*-plant interactions. J. Bacteriol. 190:494-507.

Yusibov, V. M., Steck, T. R., Gupta, V., and Gelvin, S. B. 1994. Association of single-stranded transferred DNA from *Agrobacterium tumefaciens* with tobacco cells. Proc. Natl. Acad. Sci. U.S.A. 91:2994-2998.

Zaenen, I., Van Larebeke, H., Teuchy, H., Van Montagu, M., and Schell, J. 1974. Supercoiled circular DNA in crown-gall inducing *Agrobacterium* strains. J. Mol. Biol. 86:109-127.

Zhao, Z.-M., Sagulenko, E., Ding, Z. Y., and Christie, P. J. 2001. Activities of VirE1 and the VirE1 secretion chaperone in export of the multifunctional VirE2 effector via an *Agrobacterium* type IV secretion pathway. J. Bacteriol. 183:3855-3865.

Zhou, X.-R., and Christie, P. J. 1999. Mutagenesis of the *Agrobacterium* VirD2 single-stranded DNA-binding protein identifies regions required for self-association and interaction with VirE1 and a permissive site for hybrid protein construction. J. Bacteriol. 181:4342-4352.

Zhou, Z., Sagulenko, E., Ding, Z., and Christie, P. J. 2001. Activities of *virE1* and the VirE1 secretion chaperone in export of the multifunctional VirE2 effector via an *Agrobacterium* type IV secretion pathway. J. Bacteriol. 183:3855-3865.

Zhu, J., Oger, P. M., Schrmmeijer, B., Hooykaas, P. J. J., Farrand, S. K., and Winans, S. C. 2000. The bases of crown gall tumorigenesis. J. Bacteriol. 182:3885-3895.

Zorreguieta, A., Geremia, R. A., Cavaignac, S., Cangelosi, G. A., Nester, E. W., and Ugalde, R. A. 1988. Identification of the product of an *Agrobacterium tumefaciens* chromosomal virulence gene. Mol. Plant-Microbe Interact. 1:121-127.

Zupan, J. R., Citovsky, V., and Zambryski, P. 1996. *Agrobacterium* VirE2 protein mediates nuclear uptake of single-stranded DNA in plant cells. Proc. Natl. Acad. Sci. U.S.A. 93:2392-2397.

Zupan, J., Muth, T. R., Draper, O., and Zambryski, P. 2000. The transfer of DNA from *Agrobacterium tumefaciens* into plants: A feast of fundamental insights. Plant J. 23:11-28.

Zupan, J., Hackworth, C. A., Aguilar, J., Ward, D., and Zambryski, P. 2007. VirB1* promotes T-pilus formation in the *vir*-type IV secretion system of *Agrobacterium tumefaciens*. J. Bacteriol. 189:6551-6563.

Zyprian, E., and Kado, C. I. 1990. *Agrobacterium*-mediated plant transformation by novel mini-T vectors in conjunction with a high copy *vir* region helper plasmid. Plant Mol. Biol. 15:345-356.

Asymptomatic and Latent Infections

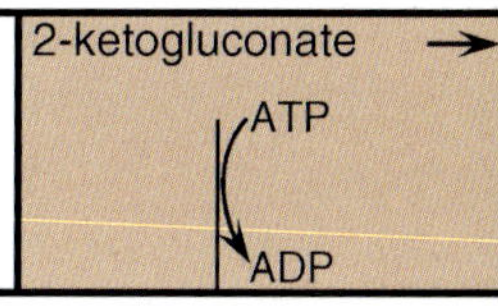

9

As discussed in the first half of Chapter 2, a steady state between the pathogen and its host plant is achieved when the pathogen resides in equilibrium with surrounding host tissues and causes very little, if any, physiological disturbance to the cells of the host. This host–pathogen interaction moves toward an endophytic, commensal interrelationship and the plant harboring the pathogen remains asymptomatic. However, such plants can begin to show symptoms when there is a shift in this equilibrium, as for example, when there is a significant change in temperature, moisture, or humidity or when the plant is altered genetically for horticultural purposes. Environmental or genetic changes induce visible disease symptoms because the pathogen reestablishes its original parasitic form. The virulence expression of the pathogen is dependent upon the particular host environment. When the pathogen is isolated from an asymptomatic host and introduced into a new host, strong pathogenic reactions may be observed, as discussed below.

There are many cases of latent infections by pathogens such as viruses, viroids, fungi, oomycetes, and bacteria. Most bacterial pathogens can be isolated and cultured but certain '*Candidatus* Phytoplasma' species, which often are associated with asymptomatic native hosts, are nonculturable. For detection purposes, they can be introduced via vector transmission into cultivated species that develop severe disease symptoms. For example, the asymptomatic wild grapevine *Vitis riparia* Michx. harbors phytoplasma strains that cause severe disease symptoms in cultivated grapevine (*Vitis vinifera* L.) (Davis and Sinclair, 1998). Also, trees of black locust (*Robinia pseudoacacia* L.) in the eastern United States exhibit no apparent disease symptoms unless the tree is cut to leave a stump; then, shoots emerge from the stump and roots, producing symptoms characteristic of the black locust witches'-broom, a phytoplasma-caused disease. Likewise, when pathogenic bacteria are isolated, cultured, and inoculated into a domestically cultivated host, rapid necrosis and yellowing of the inoculated area can occur. Rapid killing of surrounding tissues provides outward appearance of necrosis characteristic of the hypersensitive response in nonhost plants. Table 9-1 lists examples of pathogenic bacteria that reside latently in wild plants that can serve as reservoirs of inoculum that can move into cultivated plants.

Asymptomatic plants are usually overlooked and pass inspection as they are transported across national borders or awarded plant and seed-piece certifications. They are the primary way in which new diseases are introduced into disease-free regions and new territories. Asymptomatic plants present a serious problem in disease containment.

Pink Disease of Pineapple

Disease control is often difficult when crops remain asymptomatic in the field. For example, the pink disease of pineapple, caused by *Pantoea citrea* (now called

Table 9-1. Bacterial Pathogens Found in Latently Infected Asymptomatic Plants

Pathogen	Host	Symptoms Induced or Lack of Symptoms	Reference
Clavibacter michiganensis subsp. *sepedonicus*	Solanaceous weeds; sugar beet (*Beta vulgaris* L.)	Asymptomatic in wild solanaceous plants	Bugbee et al., 1987; Van der Wolf et al., 2005
Clavibacter michiganensis subsp. *michiganensis*	Solanaceous weeds; tomato (*Lycopersicon esculentum* Mill. var. *esculentum* Mill.)	Asymptomatic in young tomato transplants and solanaceous weeds; vascular wilt and fruit spotting induced	Gitaitis et al., 1991; Kaneshiro et al., 2006
Clavibacter xyli subsp. *cynodontis*	Bermuda grass (*Cynodon dactylon* (L.) Pers.)	Xylem-inhabiting endophyte of Bermuda grass (*Cynodon dactylon* (L.) Pers.); colonizes corn (*Zea mays* L.)	Kostka et al., 1988
Xylella fastidiosa	More than 114 genera of wild herbaceous and woody plants, including monocots	Latent endophytic associations in wild hosts; leaf tissue, berry, and stem necrosis on cultivated grapevines	Hopkins and Purcell, 2002
'*Candidatus* Phytoplasma phoenicium'	Almond (*Prunus amygdalus* Batsch)	Latent in certain cultivated varieties	Verdin et al., 2003
Phytoplasma of grapevine	Wild grapevine (*Vitis vinifera* L.)	Asymptomatic in wild species; wilt and collapse in cultivated hosts	Davis and Sinclair, 1998
Phormium yellow leaf phytoplasma	Strawberry (*Fragaria chiloensis* (L.) Mill.)	Asymptomatic in cv. Pajaro	Andersen et al., 1998
Eastern black locust phytoplasma	Black locust (*Robinia pseudoacacia* L.)	Asymptomatic unless truncated	Davis and Sinclair, 1998
Xanthomonas campestris pv. *pelargonii*	Geranium (*Pelargonium* spp.)	Symptoms masked by cool conditions; leaf spots, stem rots, and wilting induced by high temperatures and humidity	Taylor et al., 1996
Xanthomonas fragariae	Strawberry (*Fragaria chiloensis* (L.) Mill.)	Angular leaf spots, sepal blackening, vascular oozing	Mahuku and Goodwin, 1997; Moltmann and Zimmerman, 2005
Xanthomonas axonopodis pv. *dieffenbachiae*	Anthurium (*Anthurium andraeanum* Linden ex André)	Asymptomatic microplants	Norman and Alvarez, 1994; Norman and Alvarez, 1996
Xanthomonas campestris pv. *campestris*	Cabbage (*Brassica oleracea* L. var. *capitata* L.)	Asymptomatic under cool conditions; black rot symptoms induced by warm conditions and high humidity	Williams, 1980
Ralstonia solanacearum race 3, biovars 1 and 2	Geranium (*Pelargonium* × *hortorum* L. H. Bailey); potato (*Solanum tuberosum* L.); many vegetable crops and weeds	Southern wilt, leaf yellowing, vascular browning	Priou et al., 2000; Maciel et al., 2001; Swanson et al., 2005
Agrobacterium tumefaciens	Rose (*Rosa* spp.)	Endophytic, systemic, latent infection; galls produced by injury	Marti et al., 1999
Erwinia amylovora	Apple (*Malus* spp.); pear (*Pyrus* spp.)	Systemic, blackening necrosis of stems when induced	Bonn, 1979; Calzolari et al., 1982; Steiner and Suleman, 1993; Crepel et al., 1996; Momol et al., 1998

(continued)

Table 9-1. Bacterial Pathogens Found in Latently Infected Asymptomatic Plants *(continued)*

Pathogen	Host	Symptoms Induced or Lack of Symptoms	Reference
Curtobacterium flaccumfaciens pv. *poinsettiae*	Poinsettia (*Euphorbia pulcherrima* Willd. ex Klotzsch)	Leaf spots, water-soaked lesions on stems and petioles induced by warm conditions and high moisture levels	Raabe et al., 2002
Pantoea citrea (now called *Tatumella citrea*)	Pineapple (*Ananas comosus* (L.) Merr.)	Asymptomatic in the field; internal pink coloration of fruit induced by heating (canning)	Kado, 2003
Erwinia tracheiphila	Watermelon (*Citrullus vulgaris* Schrad.)	Systemically latent in 10-leaf-stage watermelon	Watterson et al., 1971
Agrobacterium tumefaciens	Orchid protocorms (*Dendrobium* spp.); tobacco seedlings (*Nicotiana* spp.); chrysanthemum (*Chrysanthemum morifolium* (Ramat.) Hemsl.); grape (*Vitis* spp.); marguerite daisy (*Argyranthemum frutescens* (L.) Sch. Bip.); raspberry (*Rubus* spp.)	Asymptomatic T-DNA transformation or systemic resident	Nan et al., 1997; Nan et al., 1998; Kado, 2001; Brencic et al., 2005
Agrobacterium rhizogenes	Carrot (*Daucus carota* L.); wild tobacco (*Nicotiana glauca* Graham); petunia (*Petunia* × *atkinsiana* D. Don ex Louden)	T-DNA fingerprints in genome	Kado, 2002; Suzuki et al., 2002

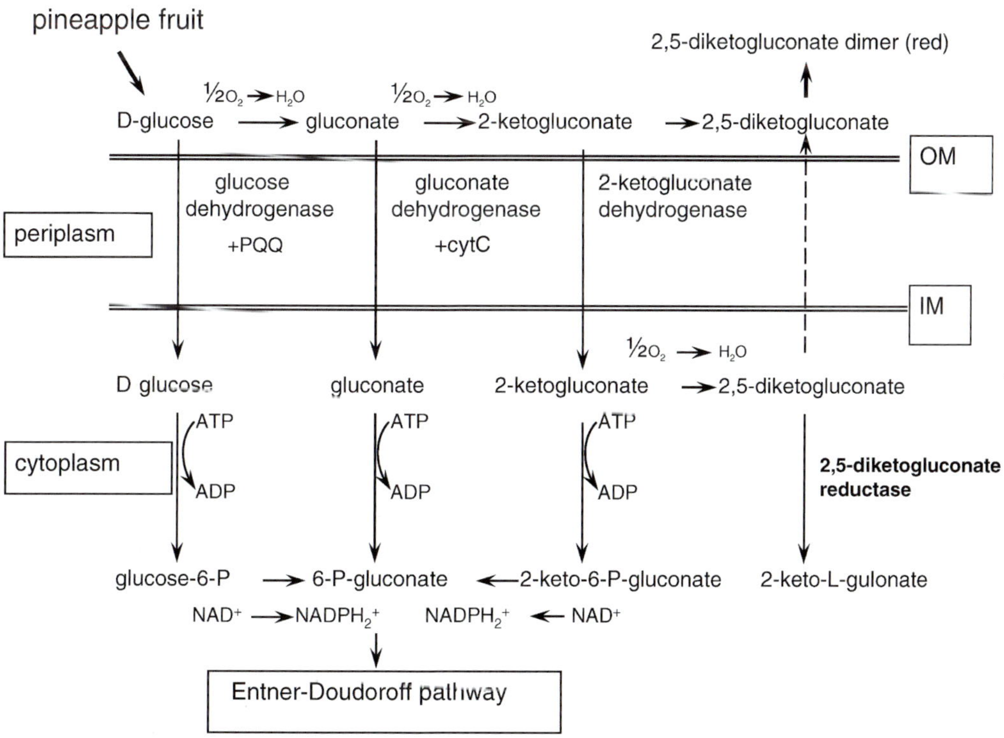

Fig. 9-1. Biochemical pathway leading to pink color formation from the oxidation of D-glucose by *Pantoea citrea* (now called *Tatumella citrea*).

Tatumella citrea) (Cha et al., 1997; Brady et al., 2009), is likely spread by insects and possibly by vegetative propagation. The disease is not conspicuously obvious in the field but prominently appears when the fruit is harvested and canned (see Fig. 4-14). Pink disease symptoms are produced by heating as usually carried out during the sterilization step in the fruit canning process. Cans from each lot are screened randomly for pink disease. Owing to zero tolerance for the disease, the detection of one can with discolored fruit results in the destruction of the entire lot of canned fruit (one lot equals 32,000 cans of processed fruit) (Kado, 2003). Obviously, pink disease is an economically important problem in the pineapple industry. In the fruit, *Pantoea citrea* converts glucose into 2,5-diketogluconate, which upon heating forms a deep red compound. The biochemical pathway of 2,5-diketogluconate production has been worked out (Pujol and Kado, 2000) (Fig. 9-1). Given that processed pineapple is the third most consumed fruit worldwide (pineapple is the fifth most consumed fruit in the United States), pink disease may be one of the most costly plant bacterial diseases. However, because of its sporadic incidence, the disease is not widely reported.

Pierce's Disease and Related Diseases

Pierce's disease, almond leaf scorch, phony peach, plum leaf scald, citrus variegated chlorosis, and a number of other diseases caused by *Xylella fastidiosa* are cases in which strong disease symptoms occur when the pathogen is introduced into hosts in which the pathogen's environment in planta has been affected by genetic selection for domestication from the wild plant species. Typical examples are the French hybrid grapevines that react strongly to *Xylella fastidiosa.* There are at least 114 genera of wild plants that naturally harbor *Xylella fastidiosa* and yet remain symptomless (Table 9-1). The bacterial cells in the xylem are acquired through the feeding of leafhoppers that vector the cells of *Xylella fastidiosa.* When the bacteria are introduced to other hosts and to cultivated species, the disease becomes obvious. Colonization of the xylem by *Xylella fastidiosa* requires a number of genes whose identity remains unknown. The *pglA* gene that encodes polygalacturonase appears to play a role in xylem colonization of *Vitis vinifera* L. grapevines (Roper et al., 2007). The *pilY1* and *pilB* genes contribute to the twitching motility of a type IV pilus that appears to be needed for cell motility against the flow direction of the xylem sap (De La Fuente et al., 2007). Multidrug resistance efflux systems in *Xylella fastidiosa* are required for survival and pathogenicity in *Vitis vinifera* L. (Reddy et al., 2007). Pierce's disease and similar diseases are typical examples of steady-state, commensal habitation of the pathogen in most of its hosts in an asymptomatic condition. Interestingly, the glassy-winged sharpshooter (*Homalodisca coagulata* (Say)) vector harbors two endosymbionts (*Baumannia cicadellinicola* and *Sulcia muelleri*) that provide the insect vector with vitamins and cofactors and with essential amino acids, respectively, that are absent in the xylem fluid (Wu et al., 2006).

Asymptomatic Crown Gall

Agrobacterium tumefaciens systemically resides in the xylem vessels and intercellular spaces of chrysanthemum, grapevine, marguerite daisy, and raspberry in the absence of any visible crown gall symptoms (Miller, 1975; Tarbah and Goodman,

1987; Bishop et al., 1988; Burr et al., 1988; Jones and Raju, 1988; Kado, 2001). Crown gall tumors are produced when the systemically infected plants are injured. Tumors appear at wound sites caused by pruning, removal of apical buds (pinching), and frost injury. *Agrobacterium tumefaciens* isolated from systemically infected symptomless plant material causes tumors on indicator plants, such as *Kalanchoë* spp. and Jimson weed (*Datura stramonium* L.). *Agrobacterium tumefaciens* can transform certain plants in the absence of any visible tumor symptoms. For example, *Dendrobium* orchid protocorm-like bodies can be transformed by *Agrobacterium tumefaciens* without producing tumors (Nan et al., 1998). The protocorm-like bodies naturally produce the *vir* gene inducer coniferyl alcohol (Nan et al., 1997). Tobacco seedlings can also be transformed by *Agrobacterium tumefaciens* without visible tumor induction (Brencic et al., 2005). The seedlings produce syringate and vanillate as *vir* gene inducers.

Systemic invasion of the above-named plants usually begins at the roots. Bacteria are known to localize at the origin of secondary roots and below the outer parenchyma layer of naturally exfoliated cells (Huang, 1986). They colonize the root surfaces and form benign biofilms (Matthysse and McMahan, 1998). *Agrobacterium rhizogenes* causes hairy root in a number of herbaceous and woody crop plants. Its relationship with plants likely evolved millions of years ago (Suzuki et al., 2002), since remnants of its association with plants remain as T-DNA fingerprints in symptomless carrot, wild tobacco, and petunia (Fründt et al., 1998; Kado, 2002).

Protective Plant-Pathogenic Endophytes

Wild plant species can harbor endophytic bacteria, some of which appear to protect the plant from pathogens. The endophytes apparently protect the plant by naturally secreting antifungal agents, such as lipopeptides (iturins), and antibacterial compounds, such as cyclic peptides (e.g., bacitracin). *Bacillus subtilis* apparently protects silver and sugar maple seedlings (*Acer saccharinum* L. and *Acer saccharophorum* K. Koch, respectively) against *Verticillium dahliae* Kleb. (Hall and Davis, 1990). The *Bacillus subtilis* appears to inhabit the xylem elements. *Clavibacter xyli* subsp. *cynodontis* resides as an endophyte in the xylem of Bermuda grass (*Cynodon dactylon* (L.) Pers.) and can be transmitted to numerous plant species, including corn. Colonization in corn (*Zea mays* L.) by this organism has little effect on growth and yield under normal growing conditions (Prunier et al., 1990). *Clavibacter xyli* subsp. *cynodontis* can be used as a biocontrol organism that can be genetically engineered to express bactericides, fungicides, and pesticidal toxins, such as the delta-toxin of *Bacillus thuringiensis* (Kostka et al., 1988). Concerns that this approach might lead to indiscriminate dissemination of *Clavibacter xyli* subsp. *cynodontis* by xylem-feeding and chewing insects were minimized by studies showing that the various xylem-feeding and chewing insects tested did not serve as vectors of this bacterium (Barbehenn and Purcell, 1993).

More than one species of bacteria have been found living as endophytes in cotton (Misaghi and Donndelinger, 1990). They include species of the genera *Pectobacterium* and *Bacillus.*

Symptomless "diseases" are likely to be more widespread than realized currently. Given that the direction of pathogenesis is for the good of the pathogen itself, rapid killing of the host is therefore not in the best interest of the pathogen for its long-term survival. Surveys of asymptomatic plants will doubtless reveal the presence of endophytic pathogens, some of whose characteristics will be much like those already known and some of which will have novel ways of interacting with their plant hosts.

REFERENCES

Andersen, M. T., Longmore, J., Liefting, L. W., Wood, G. A., Sutherland, P. W., Beck, D. L., and Forster, R. L. S. 1998. Phormium yellow leaf phytoplasma is associated with strawberry lethal yellows disease in New Zealand. Plant Dis. 82:606-609.

Barbehenn, R. V., and Purcell, A. H. 1993. Factors limiting the transmission of a xylem-inhabiting bacterium *Clavibacter xyli* subsp. *cynodontis* to grasses by insects. Phytopathology 83:859-863.

Bishop, A. L., Katz, B. H., and Burr, T. J. 1988. Infection of grapevines by soilborne *Agrobacterium tumefaciens* biovar 3 and population dynamics in host and nonhost rhizospheres. Phytopathology 78:945-948.

Bonn, W. G. 1979. Fire blight bacteria in symptomless dormant apple and pear buds. Can. J. Plant Pathol. 1:61-62.

Brady, C. L., Venter, S. N., Cleenwerck, I., Vandemeulebroecke, K., de Vos, P., and Coutinho, T. A. 2009. Transfer of *Pantoea citrea*, *Pantoea punctata* and *Pantoea terrea* to the genus *Tatumella* emend. as *Tatumella citrea* comb. nov., *Tatumella punctata* comb. nov. and *Tatumella terrea* (Kageyama *et al.* 1992) comb. nov. and description of *Tatumella morbirosei* sp. nov. Int. J. Syst. Evol. Microbiol. First published on August 4, 2009, as doi: 10.1099/ijs.0.012070-0.

Brencic, A., Angert, E. R., and Winans, S. C. 2005. Unwounded plants elicit *Agrobacterium vir* gene induction and T-DNA transfer: Transformed plant cells produce opines yet are tumour free. Mol. Microbiol. 57:1522-1531.

Bugbee, W. M., Gudmestad, N. C., Secor, G. A., and Nolte, P. 1987. Sugar beet as a symptomless host for *Corynebacterium sepedonicum*. Phytopathology 77:765-770.

Burr, T. J., Katz, B. H., Bishop, A. L., Myeres, C. A., and Mittak, V. L. 1988. Effect of shoot age and tip culture propagation on grapes on systemic infestation by *Agrobacterium tumefaciens* biovar 3. Am. J. Enol. Vitic. 39:67-70.

Calzolari, A., Peddes, P., Mazzuchi, U., Mori, P., and Garzena, C. 1982. Occurrence of *Erwinia amylovora* in buds of asymptomatic apple plants in commerce. J. Phytopathol. 103:156-162.

Cha, J.-S., Pujol, C., Ducusin, A. R., Macion, E. A., Hubbard, C. H., and Kado, C. I. 1997. Studies on *Pantoea citrea*, the causal agent of pink disease of pineapple. J. Phytopathol. 145:313-320.

Crepel, C., Greenen, J., and Maes, M. 1996. The latent survival of *Erwinia amylovora* in hibernating shoots. Acta Hortic. 411:21-25.

Davis, R. E., and Sinclair, W. A. 1998. Phytoplasma identity and disease etiology. Phytopathology 88:1372-1376.

De La Fuente, L., Burr, T. J., and Hoch, H. C. 2007. Mutations in type I and type IV pilus biosynthetic genes affect twitching motility rates in *Xylella fastidiosa*. J. Bacteriol. 189:7507-7510.

Fründt, C., Meyer, A. D., Ichikawa, T., and Meins, F., Jr. 1998. Evidence for the ancient transfer of Ri plasmid T-DNA genes between bacteria and plants. Pages 94-117 in: Horizontal Gene Transfer. M. Syvanen and C. I. Kado, eds. Chapman and Hall, London.

Gitaitis, R. D., Beaver, R. W., and Voloudakis, A. E. 1991. Detection of *Clavibacter michiganensis* subsp. *michiganensis* in symptomless tomato transplants. Plant Dis. 75:834-838.

Hall, T. J., and Davis, W. E. E. 1990. Survival of *Bacillus subtilis* in silver and sugar maple seedlings over a two-year period. Plant Dis. 74:608-609.

Hopkins, D. L., and Purcell, A. H. 2002. *Xylella fastidiosa*: Cause of Pierce's disease of grapevine and other emergent diseases. Plant Dis. 86:1056-1066.

Huang, J.-S. 1986. Ultrastructure of bacterial penetration in plants. Annu. Rev. Phytopathol. 24:141-157.

Jones, J. B., and Raju, B. C. 1988. Systemic movement of *Agrobacterium tumefaciens* in symptomless stem tissue of *Chrysanthemum morifolium*. Plant Dis. 72:51-54.

Kado, C. I. 2001. Agrobacteria. Pages 18-20 in: Encyclopedia of Plant Pathology. O. C. Maloy and T. D. Murray, eds. John Wiley & Sons, New York.

Kado, C. I. 2002. Horizontal transmission of genes by *Agrobacterium* species. Pages 45-50 in: Horizontal Gene Transfer, 2nd ed. M. Syvanen and C. I. Kado, eds. Academic Press, San Diego, CA.

Kado, C. I. 2003. Pink disease of pineapple. APS*net* Feature. American Phytopathological Society, St. Paul, MN. www.APSnet.org/online/feature/pineapple.

Kaneshiro, W. S., Mizumoto, C. Y., and Alvarez, A. M. 2006. Differentiation of *Clavibacter michiganensis* subsp. *michiganensis* from seed-borne saprophytes using ELISA, Biolog and 16SrDNA sequencing. Eur. J. Plant Pathol. 116:45-56.

Kostka, S. J., Resser, P. W., and Miller, D. P. 1988. Experimental host range of *Clavibacter xyli* subsp. *cynodontis* (Cxc) and a Cxc/*Bacillus thuringiensis* recombinant (Cxc/Bt). (Abstr.) Phytopathology 78:1540.

Maciel, J. L. N., Duarte, V., Silveira, J. R. P., and Van der Sand, S. T. 2001. Freqüencia de biovares de *Ralstonia solanacearum* em differentes cultivares e épocas de cultivo de batat no rio grande do sui. Fitopatol. Bras. 26:144-153.

Mahuku, G. S., and Goodwin, P. H. 1997. Presence of *Xanthomonas fragariae* in symptomless strawberry crowns in Ontario detected using a nested polymerase chain reaction (PCR). Can. J. Plant Pathol. 19:366-370.

Marti, R., Cubero, J., Daza, A., Piquer, J., Salcedo, C. I., Morente, C., and Lopez, M. M. 1999. Evidence of migration and endophytic presence of *Agrobacterium tumefaciens* in rose plants. Eur. J. Plant Pathol. 105:39-50.

Matthysse, A. G., and McMahan, S. 1998. Root colonization by *Agrobacterium tumefaciens* is reduced in *cel*, *attB*, *attD*, and *attR* mutants. Appl. Environ. Microbiol. 64:2341-2345.

Miller, H. N. 1975. Leaf, stem, crown, and root galls induced in chrysanthemum by *Agrobacterium tumefaciens*. Phytopathology 65:805-811.

Misaghi, I. J., and Donndelinger, C. R. 1990. Endophytic bacteria in symptom-free cotton plants. Phytopathology 80:808-811.

Moltmann, E., and Zimmerman, C. 2005. Detection of *Xanthomonas fragariae* in symptomless strawberry plants by nested PCR. EPPO Bull. 35:53-54.

Momol, M. T., Norelli, J. L., Piccioni, D. E., Momol, E. A., Gustafson, H. L., Cummins, J. N., and Aldwinckle, H. S. 1998. Internal movement of *Erwinia amylovora* through symptomless apple scion tissues into the rootstock. Plant Dis. 82:646-650.

Nan, G.-L., Tang, C.-S., Kuehnle, A. R., and Kado, C. I. 1997. *Dendrobium* orchids contain an inducer of *Agrobacterium* virulence genes. Physiol. Mol. Plant Pathol. 51:391-399.

Nan, G.-L., Kuehnle, A. R., and Kado, C. I. 1998. Transgenic *Dendrobium* orchid through *Agrobacterium*-mediated transformation. Malay. Orchid Rev. 32:93-96.

Norman, D. J., and Alvarez, A. M. 1994. Latent infections of in vitro anthurium caused by *Xanthomonas campestris* pv. *dieffenbachiae*. Plant Cell Tissue Organ. Cult. 39:55-61.

Norman, D. J., and Alvarez, A. M. 1996. Monitoring the spread of *Xanthomonas campestris* pv. *dieffenbachiae* introduced from symptomless *Anthurium* cuttings into production fields. J. Am. Soc. Hortic. Sci. 121:582-585.

Priou, S., Torres, R., Villar, A., Martinez, J., Barrea, O., Gutarrra, L., and de Mendiburu, F. 2001. Optimization of sample size for the detection of latent infection by *Ralstonia solanacearum* in potato seed tubers in the highlands of Peru and Bolivia. Pages 145-153 in: Scientist and Farmer: Partners in Research for the 21st Century. CIP Program Report 1999-2000. Centro Internacional de la Papa (International Potato Centre), Lima, Peru.

Prunier, J. P., Kostka, S. J., and Labonne, G. 1990. *Clavibacter xyli* subsp. *cynodontis*: An endophytic bacterium in corn. Pages 305-310 in: Proc. Int. Conf. Plant Pathogenic Bacteria, 7th. Z. Klement, ed. Akadémia Kiado, Budapest, Hungary.

Pujol, C. J., and Kado, C. I. 2000. Genetic and biochemical characterization of the pathway in *Pantoea citrea* leading to pink disease of pineapple. J. Bacteriol. 182:2230-2237.

Raabe, R. D., Grebus, M. E., Wilen, C. A., and McCain, A. H. 2002. UC IPM Pest Management Guidelines: Floriculture and Ornamental Nurseries. Univ. Calif. Agric. Nat. Resour. Publ. 392.

Reddy, J. D., Reddy, S. L., Hopkins, D. L., and Gabriel, D. W. 2007. TolC is required for pathogenicity of *Xylella fastidiosa* in *Vitis vinifera* grapevines. Mol. Plant-Microbe Interact. 20:403-410.

Roper, M. C., Greve, L. C., Warren, J. G., Labavitch, J. M., and Kirkpatrick, B. C. 2007. *Xylella fastidiosa* requires polygalacturonase for colonization and pathogenicity in *Vitis vinifera* grapevines. Mol. Plant-Microbe Interact. 20:411-419.

Steiner, P. W., and Suleman, P. 1993. The systemic nature of fire blight: Implications for control. Penn. Fruit News 73:16-22.

Suzuki, K., Yamashita, I., and Tanaka, N. 2002. Tobacco plants were transformed by *Agrobacterium rhizogenes* infection during their evolution. Plant J. 32:775-787.

Swanson, J. K., Yao, J., Tans-Kersten, J., and Allen, C. 2005. Behavior of *Ralstonia solanacearum* race 3 biovar 2 during latent and active infection of geranium. Phytopathology 95:136-143.

Tarbah, F., and Goodman, R. N. 1987. Systemic spread of *Agrobacterium tumefaciens* biovar 3 in the vascular system of grapes. Phytopathology 77:915-920.

Taylor, R. J., Vanette, J. R., and Lamey, H. A. 1996. Leaf spot and stem rot (bacterial blight) of geranium (*Pelargonium* spp.). N. D. State Univ. Ext. Bull. 1996.

Van der Wolf, J. M., van Beckhoven, J. R. C. M., Hukkanen, A., Karjalainen, R., and Müller, P. 2005. Fate of *Clavibacter michiganensis* ssp. *sepedonicus*, the causal organism of bacterial ring rot of potato, in weeds and field crops. J. Phytopathol. 153:358-365.

Verdin, E., Pascal, S., Danet, J.-L., Choueiri, E., Jreijiri, F., El Zammar, S., Gélie, B., Bové, J. M., and Garnier, M. 2003. '*Candidatus* Phytoplasma phoenicium' sp. nov., a novel phytoplasma associated with an emerging lethal disease of almond trees in Lebanon and Iran. Int. J. Syst. Evol. Microbiol. 53:833-838.

Watterson, J. C., Williams, P. H., and Durbin, R. D. 1971. Response of cucurbits to *Erwinia tracheiphila*. Plant Dis. Rep. 55:816-819.

Williams, P. H. 1980. Black rot: A continuing threat to world crucifers. Plant Dis. 64:736-742.

Wu, D., Daugherty, S. C., Van Aken, S. E., Pai, G. H., Watkins, K. L., Khouri, H., Tallon, L. J., Zaborsky, J. M., Dunbar, H. E., Tran, P. L., Moran, N. A., and Eisen, J. A. 2006. Metabolic complementarity and genomics of the dual bacterial symbiosis of sharpshooters. PLoS Biol. 4:e188.

Molecular Mechanisms of Virulence and Pathogenesis

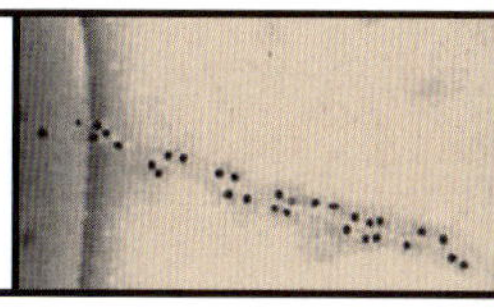

10

Gene activation is central for plant pathogens to sense, swim toward, colonize, and infect their hosts. In this chapter, this basic premise is categorically described. The chapter is not meant to be a comprehensive review but rather provide principles and examples of each pattern of behavior as conferred through gene regulation and expression.

Recognition and Chemotaxis to the Host

Many plant-pathogenic bacteria can recognize the presence of their host plants through the perception of plant-generated chemical signals. Signals perceived by the bacterial cells are transduced through the bacterial membranes to set off specific motility events whereby the pathogen follows a chemical gradient of the signal until it reaches its host plant. The bacterial cells move through the gradient by means of either flagella or pili. Various chemicals, such as products of the phenylpropanoid-flavonoid pathway, serve as specific signals that are recognized by the pathogen. In the case of the crown gall pathogen, *Agrobacterium tumefaciens* recognizes methoxyphenols, such as sinapinic and cinnamic acids, acetosyringone, and coniferyl alcohol, derived from the lignin biosynthetic pathway produced by the plant. In this bacterium, a transmembrane-docked histidine kinase called VirA aids in perceiving the methoxyphenols potentiated by simple sugars and low pH and in activating the response regulator VirG (see Chapter 8). Acidic environments of around pH 5.5 have been shown to induce as many as 78 genes and to repress 74 genes in *Agrobacterium tumefaciens* (Yuan et al., 2008). Likewise, *Sinorhizobium meliloti* possesses nodulation (*nod*) genes that respond to flavonoids, such as luteolin, and other inducers as well as to low-pH environments around alfalfa roots.

In addition, members of the phytochrome sensor histidine kinase family (HWE histidine kinases) (Inouye and Dutta, 2003) are present in *Agrobacterium tumefaciens, Pseudomonas syringae*, and *Xanthomonas campestris* (Karniol and Vierstra, 2004). The presence of HWE histidine kinases provides extensive metabolic flexibility and allows the organism to expeditiously associate with their plant host. It is quite likely that plant pathogens are equipped with a variety of mechanisms to intercommunicate with each other as well as with their plant host. Such "chemical conversations" span species and kingdoms (Bassler and Losick, 2006).

Not only do pathogens recognize plant-generated chemical signals but they also recognize self-generated signals. One mechanism that is widespread among both Gram-negative and Gram-positive bacteria is their ability to communicate and coordinate behavior via signaling molecules, a trait that is dependent on cell population size (figuratively, a process called quorum sensing) (Magnuson et al., 1994; Håvarstein and Morrison, 1999; Nes and Eijsink, 1999; Novick, 1999; Risøen et al., 2000; Fuqua et al., 2001; Miller and Bassler, 2001; Whitehead et al., 2001; von Bodman et al., 2003; Newton and Fray, 2004; Visick and Fuqua, 2005). More than one quorum-sensing system can be present in a given species (Hoang et al., 2004; Lupp and Ruby, 2005). In this process, as they grow, bacteria regulate gene expression by sensing the concentration of their own diffusible signal molecule. Each bacterial cell produces a small amount of a quorum-sensing signal molecule, which becomes sufficiently concentrated as the numbers of bacteria increase.

In Gram-positive bacteria, among various functions, quorum sensing regulates virulence. The quorum-sensing diffusible signal is a small peptide called an inducing peptide, or lantibiotic, that binds to a histidine protein kinase in a phosphorelay regulator of transcription. Inducing peptides are members of bacterial pheromones. Pheromones exert control over many cellular processes, such as conjugation, sporulation, virulence, and cell division. *Staphylococcus aureus* uses thiolactone peptides to control the expression of many virulence factors, including toxins, proteases, and adhesion (Novick, 1999). An 11-residue peptide lactone induces virulence factors, such as gelatinase and serine protease, in *Enterococcus faecalis* (Nakayama et al., 2001) (Fig. 10-1).

In Gram-negative bacteria the best characterized signal molecules are the *N*-acyl-homoserine lactones (AHLs) (Fig. 10-2). Quorum-sensing systems are tuned to preferentially synthesize, and respond to, specific AHLs having an acyl side group that may range in length from four to sixteen carbons, may contain a degree of unsaturation, and may be modified with a carbonyl or hydroxyl functional group at carbon position 3. Two genes, *I* (autoinducer synthesis) and *R* (detection), encode an AHL synthase for the production of AHL and an AHL-responsive transcription regulator, respectively. Examples are TraI and TraR of *Agrobacterium tumefaciens* and SinI and ExpR of *Sinorhizobium meliloti*. In the absence of AHL, the R protein, such as TraR, is rapidly degraded by proteolysis in *Agrobacterium tumefaciens* (Chai and Winans, 2005). Ligation to its amino terminus by AHL stabilizes TraR as a homodimer (White and Winans, 2005). Each bacterial cell only produces small amounts of AHL, and as the bacterial population grows, AHL accumulates in the extracellular environment to a threshold concentration that leads to the AHL-mediated gene expression. Any AHL entering the bacterial cytosol is degraded by a lactonase, such as the lactonase called BlcC in *Agrobacterium tumefaciens*. Interestingly, mutation in the *blcC* gene had no significant effect on the induction of or continued transfer of pTiC58 from donors in any stage of growth, including the stationary phase (Khan and Farrand, 2009). The AHL autoregulates its own production (a process called autoinduction).

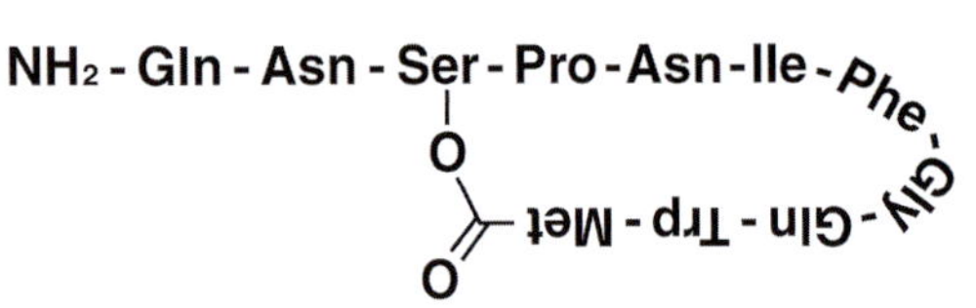

Fig. 10-1. Chemical structure of a peptide lactone pheromone.

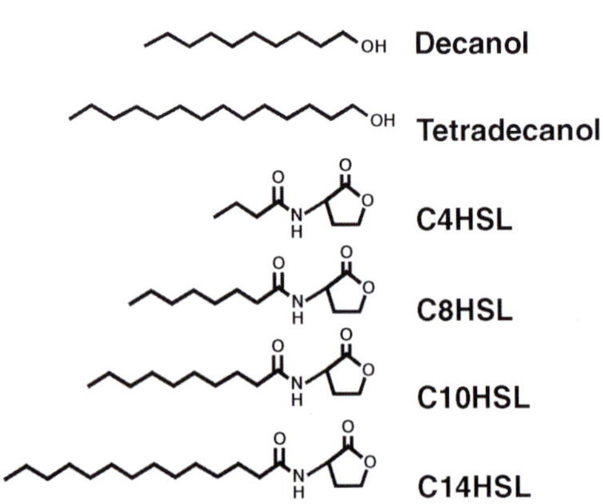

Fig. 10-2. Chemical structures of acylated C4-, C8-, C10-, and C14-homoserine lactones and their alcohol precursors decanol (*n*-decyl alcohol) and tetradecanol (myristic alcohol).

AHL is produced and used by many Gram-negative plant pathogens and endosymbionts, including *Agrobacterium tumefaciens*, *Pseudomonas syringae* pathovars, *Xanthomonas campestris* pathovars, *Burkholderia cepacia*, *Pantoea stewartii* subsp. *stewartii*, *Tatumella citrea*, *Erwinia amylovora*, *Pectobacterium carotovorum*, and *Sinorhizobium meli-*

loti. More than one LuxR-type response regulator can be encoded by the bacterial genome, e.g., *Sinorhizobium meliloti* (Patankar and González, 2009). AHL is not produced by *Xanthomonas oryzae* pv. *oryzae*. Instead it produces a modular protein called OryR that has an AHL domain and a helix-turn-helix domain, both of which are typical of the LuxR-family subgroup of quorum-sensing regulators (Fuqua et al., 2001; Ferluga et al., 2007). OryR protein is solubilized by a rice signal molecule (RSM) that is present in high concentration in the xylem sap of rice plants (Ferluga and Venturi, 2009). The chemical structure of RSM is unknown but the molecular size is less than 1 kDa.

Quorum sensing appears to be essential for efficacious survival and colonization in planta, including biofilm maturation that involves regulating twitching motility, conjugative transfer of plasmids such as the Ti plasmid, type III secretion pathway regulon expression, and exopolysaccharide production and nodulation (Jones et al., 1993; Shaw et al., 1997; Huber et al., 2001; He et al., 2003; Marketon et al., 2003; Venturi et al., 2004; Bleves et al., 2005; McGowan et al., 2005; Quiñones et al., 2005; Patankar and González, 2009; M. Schuster, personal communication). There is some question on whether or not twitching motility is directly regulated by quorum sensing in *Pseudomonas aeruginosa* (Beatson et al., 2002).

The quorum-sensing signaling molecule *N*-(3-oxo-hexanoyl)-L-homoserine lactone, produced by *Pectobacterium carotovorum* subsp. *carotovorum*, is required for the production of a β-lactam antibiotic called carbapenem and for the production of pectin lyases, cellulases, and proteases, which are required for efficacious survival and for optimal virulence (McGowan et al., 2005). *Pectobacterium carotovorum* subsp. *carotovorum* encodes ExpR, a homolog of LuxR that activates the expression of a global negative regulator that binds RNA (Cui et al., 2005). In *Pectobacterium atrosepticum*, the expression of a gene, designated *svx*, is regulated by quorum sensing. This gene encodes a 67-kDa Svx protein that is secreted and posttranslationally modified. Although its precise function is unknown in conferring virulence on *Pectobacterium atrosepticum*, Svx is a homolog of AvrXca, which is produced in *Xanthomonas campestris* pv. *raphani*. *Xanthomonas campestris* requires the quorum-sensing gene cluster *rpf*, which encodes sensors of the diffusible signal factor and is required for virulence (Dow et al., 2003). The HD-GYP domain regulator RpfG in this organism regulates the synthesis of extracellular virulence factors and negatively regulates biofilm formation (Dow et al., 2003). The recognition of the diffusible signal factor is somehow coordinated with the degradation of cyclic di-GMP via a two-component RpfC (histidine kinase) and the RpfG response regulator (Ryan et al., 2006). In *Pantoea stewartii* subsp. *stewartii*, the LuxR-type quorum-sensing transcription factor EsaR functions as a repressor of exopolysaccharide synthesis, which is required for Stewart's wilt in corn (Minogue et al., 2005). In *Burkholderia cepacia*, quorum sensing regulates the expression of extracellular proteins, siderophores, and biofilm maturation (Aguilar et al., 2003). *Erwinia amylovora* harbors *luxI* and *lusR* homologs of *expI* and *expR*, *hslI* and *hslR* genes of *Pectobacterium carotovorum*, and *expI* and *expR* genes of *Dickeya chrysanthemi*. These genes are used for AHL synthesis and activation in these plant pathogens (Molina et al., 2005). AHL in *Erwinia amylovora* regulates the expression of virulence effectors and subsequent symptom development.

Interestingly, cross-domain (bacteria to eukaryote) signaling occurs between *Xanthomonas campestris*, which produces *cis*-11-methyl-2-dodecenoic acid, and the fungus *Candida albicans* (C. P. Robin) Berkhout, affecting its filamentation (Wang et al., 2004). This diffusible chemical signals the expression of virulence (*vir*) genes and biofilm dispersal (Y.-W. He et al., 2006). Moreover, plants and animals have innate defense mechanisms for disrupting bacterial quorum sensing by producing compounds that mimic AHL signals (Givskov et al., 1996; Gao et al., 2003). In the case of *Pseudomonas aeruginosa*, the AHL signals enter the eukaryotic cells and activate transcription factors (Williams et al., 2004). Hence, the chemical cross talk between plants and their associated bacteria appears to reflect a highly evolved and complicated biological system that seems to benefit both parties of the interaction. Clearly, quorum sensing is used to synchronize the production of virulence factors in pathogens utilizing this signaling system. More than 350 genes in *Pseudomonas aeruginosa* have been observed to be regulated by quorum sensing (Dong et al., 2005).

Regulation of Virulence and Pathogenicity

As described above, plant-pathogenic bacteria can sense the presence of their host or hosts through signature molecules specifically produced by host tissues. Several methoxyphenolic compounds or phenolic β-glycosides in combination with mono- or disaccharides induce the expression of virulence and phytotoxin-producing genes in *Agrobacterium* species and *Pseudomonas syringae* pathovars, respectively. In *Agrobacterium tumefaciens,* a two-component regulatory system composed of a VirA membrane sensory protein (a histidine kinase) and VirG transcriptional activator initiate the expression of virulence genes on the Ti plasmid. In *Pseudomonas syringae* pv. *syringae*, the production of the phytotoxin syringomycin is induced by flavonoid glycosides, such as arbutin (4-hydroxyphenyl-β-D-glucopyranoside), quercetin (2-(3,4-dihydroxyphenyl)-3,5,7-trihydroxy-4*H*-1-benzopyran-4-one), and D-fructose, compounds naturally present in cherry tissues (Mo and Gross, 1991) and perceived by the *syrB1* and *sypA* gene products (Quigley and Gross, 1994).

The expression of virulence and pathogenic effectors is dependent on the faithful transcription of the genes encoding these proteins. Transcription in bacteria is catalyzed by a multisubunit RNA polymerase that is composed of a core enzyme $\beta\beta'\alpha_2$ and a sigma (σ) factor. Sigma factors are required for the faithful initiation of transcription at the promoter site. They are members of a family of relatively small, dissociable subunits that bind to the core subunits to form an initiation-specific RNA polymerase holoenzyme. Thus, each holoenzyme specifically recognizes promoters bearing a consensus sequence determined by the bound sigma factor.

In *Erwinia, Dickeya, Pseudomonas, Pectobacterium*, and *Pantoea* species, an alternate sigma factor (σ^{54}) designated HrpL (or sigma L) is complexed with its RNA polymerase core enzyme to recognize specific promoters and initiate the expression of hypersensitive response and pathogenicity (*hrp*) response genes, pathogenicity genes, and genes encoding the type III secretion machinery of the *hrp* regulon (Xiao et al., 1994; Wei and Beer, 1995; Collmer et al., 2000; Wei et al., 2000; Frederick et al., 2001; Mor et al., 2001; Chatterjee et al., 2002; Jock et al., 2003; Losada et al., 2004; Nissan et al., 2005; Ferreira et al., 2006). With *Pseudomonas syringae* pv. *tomato*, HrpL initiates the expression of *hrp, hrc, hop*, and *avr* genes and of genes encoding the type III secretion machinery (discussed below). Pathogenicity is conferred by a gene cluster located in the *hrp* pathogenicity island (Lindgren et al., 1986; Hutcheson, 1999; Alfano et al., 2000). HrpL recognizes and binds to a consensus *cis* element known as the *hrp* box, which is present in the promoters of these genes or operons. The consensus sequence of the *hrp* box is 5'-5XT/GGGAACC/T16XCCAC2XA2X-3', in which X is any deoxynucleotide (Innes et al., 1993; Shen and Keen, 1993; Xiao and Hutcheson, 1994; Fouts et al., 2002; Zwiesler-Vollick et al., 2002).

The expression of genes encoding the type III secretion system requires HrpR, HrpS, and HrpL (Xiao et al., 1994; Grimm et al., 1995; Hutcheson et al., 2001) and involves the two-component response regulator RhpR and a sensory kinase, RhpS (Xiao et al., 2007). HrpR and HrpS are distant members of the enhancer-binding family of proteins (Grimm and Panopoulos, 1989; Xiao et al., 1994; Deng et al., 1998). HrpR is required for the transcriptional activation of *hrpS*. HrpS is a positive regulator of *hrpL*. Both *hrpS* and *hrpL* bear a consensus σ^{54} sequence (Grimm and Panopoulos, 1989; Xiao et al., 1994). At least 24 genes/operons are activated through this regulatory cascade in *Pseudomonas syringae* strains (Zwiesler-Vollick et al., 2002).

In *Pantoea agglomerans* pv. *gypsophilae*, the *hrp* and *hrc* genes are also regulated by HrpS, HrpR, and HrpL. HrpL recognizes a *hrp* box that is 52% similar to that recognized by HrpL of *Pseudomonas syringae* pv. *tomato* (Nissan et al., 2005). An additional two-component regulatory system involves HrpX and HrpY. HrpY regulates HrpS and its own production. In *Pantoea agglomerans* pv. *gypsophilae* and *Pantoea stewartii* subsp. *stewartii*, transphosphorylation of HrpY by HrpX activates transcription of *hrpS* (Merighi et al., 2003; Nizan-Koren et al., 2003). HrpS activates *hrpL* and triggers a positive autoregulatory loop that increases the transcription levels of *hrpX* and *hrpY* as well as of *hrpS* itself (Merighi et al., 2005).

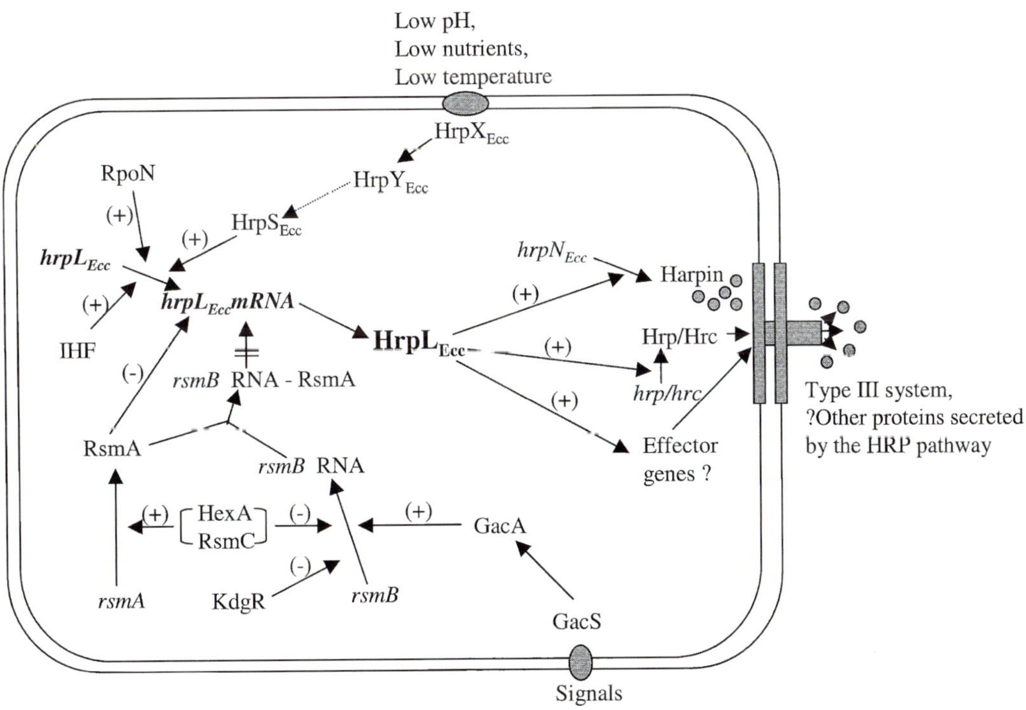

Fig. 10-3. Signal circuits and cascades following detection of environmental or host signals. (Reprinted from Chatterjee et al., 2002)

In *Pectobacterium carotovorum* subsp. *carotovorum*, full expression of the *hrpL* gene requires several regulatory players. They include σ^{54} (RpoN) and HrpS. HrpS together with σ^{54} activate *hrpL* transcription as well as RsmA, rsmB RNA, GacA (a response regulator of a two-component system), KdgR (an IclI-type repressor), HexA (an LysR-type repressor), and RsmC (a putative transcriptional adapter) (Chatterjee et al., 2002). GacA is activated by transphosphorylation by GacS (a histidine kinase) and initiates the expression of *hrp* genes and extracellular enzymes. The entire length of GacA is required to interact with itself and with GacS (Workentine et al., 2009). A summary of complex regulatory circuits following the perception of host or environmental signals are shown in Figure 10-3.

Delivery of Pathogenicity Factors into Plants

Efficacious establishment of infection by plant-pathogenic bacteria utilizes various influx and efflux systems as well as enzyme- and effector-protein translocation pathways. The influx systems are used for the uptake of essential components needed for survival, which are usually generated by the enzymatic breakdown of complex substrates into smaller compounds accessible by the influx system. The efflux systems are used to void the bacterial cell of toxic compounds and antimetabolites, including pH differences and induced innate host defense compounds, such as hydrogen peroxide, nitric oxide, and secondary metabolites (phytoalexins and phenolics) (discussed below).

For the secretion of polypeptide virulence effectors, there is a wide range of secretion pathways currently classified as type I, II, III, IV, V, and VI secretion systems. The secretion machineries of these secretion systems utilize the inner and outer membranes of the bacterial envelope. Hence, only Gram-negative bacteria can be equipped with these secretion pathways. It is not known how many secretion complexes are present in one bacterium during active secretion, nor is the stoichiometry of proteins in the complex known.

Type I Secretion

Type I secretion involves protein translocation from the cytoplasm across the inner and outer membranes without interaction in the periplasm. The type I export mechanism requires only three proteins and does not generate periplasmic intermediates (Koronakis et al., 1991) (Fig. 10-4). The secreted proteins utilize an uncleaved carboxyl-terminal secretion signal, and the last 60 C-terminal amino acid residues of the protein are sufficient for secretion (Binet et al., 1997). Type I secretion system (T1SS) components are a group of structurally and functionally related protein secretion complexes that recognize polypeptides without a recognizable N-terminal signal peptide. Instead, the polypeptides are recognized through their C-terminal domain. The type I secretion apparatus is composed of a membrane ATPase (the ATP-binding cassette [ABC] protein), a membrane fusion protein that belongs to the membrane fusion protein family, and an outer membrane peptide. The ABC transporter containing these membrane proteins can translocate a wide variety of substrates across extra- and intracellular membranes. A metalloprotease C of *Dickeya chrysanthemi* is translocated by this secretion pathway (Binet and Wandersman, 1995). Other substrates include metabolic products, lipids, sterols, toxins, and drugs.

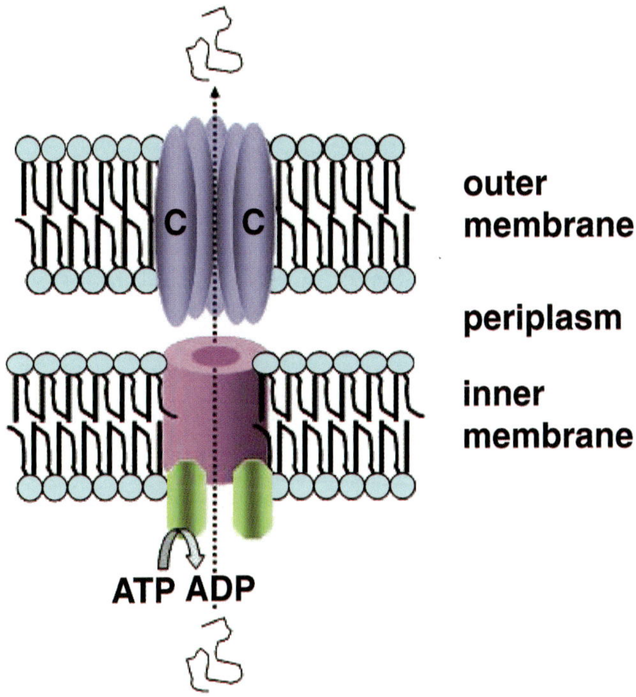

Fig. 10-4. Diagram of the type I transmembrane secretion apparatus. ATP hydrolysis is used to drive the protein substrates across the inner and outer membranes in one step without pausing in the periplasm and without removing any part of the substrate. Three export components are a traffic ATPase and an accessory protein, which together form an energized substrate-specific complex in the inner membrane, and a third protein, a member of the TolC family, in the outer membrane. TolC (C in the diagram) forms a homotrimeric hollow tapered cylinder (140 Å in length) that is composed of a 100-Å-long α-helical barrel anchored in the outer membrane by a 40-Å-long β-barrel. This composite structure forms the equatorial domain. The cylinder forms a conduit of 35 Å diameter facing outward and with decreasing diameter toward the periplasmic end (Andersen et al., 2000; 2002).

Polypeptides that are secreted via the type I secretion apparatus are translocated from the cytosol, without a periplasmic intermediate, through the inner and outer membranes out into the extracellular environment. The secreted polypeptides can be toxins or enzymes, such as hemolysin of *Escherichia coli*, repeat toxin (RTX) family (based on a series of glycine-aspartic acid-rich nonapeptide repeats found in the carboxyl-terminal third of the toxin protein) of *Bordetella pertussis*, and proteases of *Dickeya chrysanthemi* (Létoffé et al., 1990; Delepelaire and Wandersman, 1991; Ghigo and Wandersman, 1994; Martis et al., 1999) and *Pseudomonas aeruginosa* (Duong et al., 2001). Based on genome sequence analyses, *Agrobacterium tumefaciens*, *Burkholderia cepacia*, *Dickeya chrysanthemi*, *Leifsonia xyli*, *Pectobacterium carotovorum*, *Pseudomonas syringae* pathovars, *Ralstonia solanacearum*, *Xanthomonas axonopodis* pathovars, *Xanthomonas campestris*, *Xanthomonas oryzae* pv. *oryzae*, and *Xylella fastidiosa* are equipped with T1SSs.

Interestingly, the common feature that secreted proteins form one or more intramolecular disulfide bonds in their native structures is not applicable to 22 known proteins secreted by the T1SS out of *Dickeya chrysanthemi* (Palacios et al., 2001). In contrast, in the type II secretion pathway (described in the next section), disulfide bond formation is essential for the secretion of pectate lyases and a cellulase in *Dickeya chrysanthemi* (Bortoli-German et al., 1994; Shevchik et al., 1995). Disulfide bonds are important for the structure and stability of numerous extracellular proteins. It is well established that the formation of disulfide bonds is an assisted process that occurs in the periplasm and is catalyzed by the thiol-disulfide oxidoreductase DsbA (Bardwell et al., 1991; Kamitani et al., 1992). Although there are examples in which disulfide bond formation appears not to be needed for secretion, as for secretion of pullulanase in *Klebsiella* species (Sauvonnet and Pugsley, 1998), in general, secretion depends on the presence of the *dsbA* gene (Urban et al., 2001).

Type II Secretion

The type II secretion system (T2SS), conserved among Gram-negative bacteria, is characterized by a two-step protein translocation process: protein processing through the inner (cytoplasmic) membrane, followed by protein modifications to exit the outer membrane (Fig. 10-5). The major route of protein translocation across the inner membrane is the secretion (Sec) pathway, to which substrates are targeted by amino-terminal signal peptides. Protein translocation by the Sec pathway is by an amino- to carboxyl-terminal threading mechanism driven by SecA-dependent ATP hydrolysis (Pugsley, 1993). The cytoplasmic proteins to be secreted by this pathway contain a known recognizable motif in the N-terminus called signal peptides. The PelB amino-terminal signal sequence of pectate lyase from *Pectobacterium carotovorum* has proven generally useful in gene fusions used in protein secretion studies (Liebeton et al., 2001). These amino-terminal signal peptides are cleaved off by a protease after translocation of the peptide across the inner membrane. Translocation across the inner membrane is dependent on integral inner-membrane secretory proteins called Sec proteins. The processed protein is then folded and released into the periplasmic space. Proteins to be secreted require binding to SecB in a posttranslational reaction (Fekkes and Driessen, 1999). The translocase machinery is a protein-conducting pore that receives and translocates secretory proteins. The translocase consists of an integral inner-membrane, oligomeric protein complex composed of SecY, SecE, and SecG and peripheral membrane ATPase SecA that catalyzes the polypeptide transfer through the translocase. ATP causes the insertion and deinsertion of SecA at the translocase, an

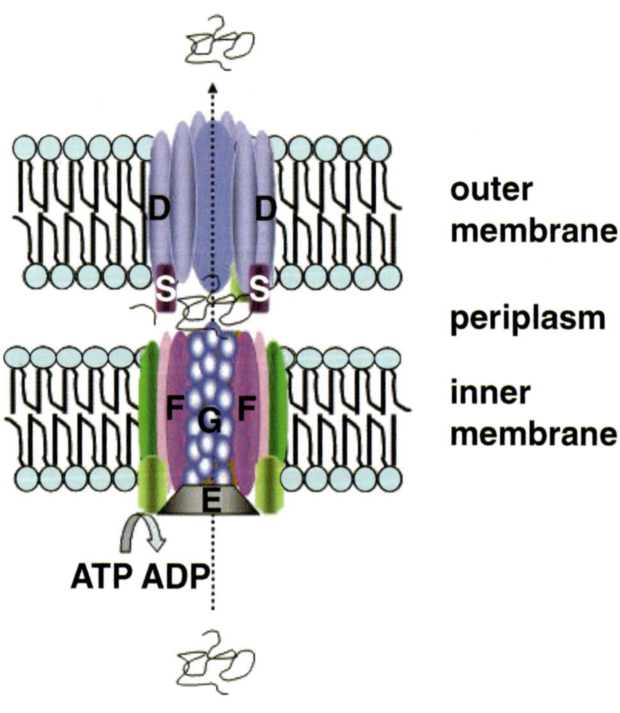

Fig. 10-5. Diagram of the type II secretion apparatus. After passage through the inner membrane, protein substrates make an intermediate stop in the periplasm for processing (cleavage off of the signal peptide) before delivery through the outer membrane. The apparatus is composed of at least 15 proteins. The "pore" in the outer membrane is composed of protein D forming a large oligomer of 12–14 subunits. The piluslike structure is composed of protein G subunits. The pilus may act as a piston to push substrates out to the exterior (Sandkvist, 2001). Protein E makes up the basal body with ATPase activity facing into the cytoplasm. Protein F is an inner-membrane protein. Protein S is a lipoprotein. Proteins D, E, F, and G are present in *Pectobacterium carotovorum*, *Dickeya chrysanthemi*, and *Xanthomonas campestris* (Sandkvist, 2001).

action that ratchets the translocating peptide across the inner membrane and into the periplasm.

The type II secretion machinery also includes a pore in the outer membrane that is composed of 12–14 subunits of protein D, which belongs to a large family of homologous proteins called secretins. In *Pseudomonas aeruginosa*, the type II secretion machinery is made with 12–15 such proteins (called Xcp or Gsp, depending on the organism) (Filloux et al., 1998; Gérard-Vincent et al., 2002). XcpA is involved in the processing and, together with XcpT, the synthesis of a piluslike structure called pseudopilin. These appendages are composed of individual fibrils assembled together in a bundle structure (Durand et al., 2003). XcpR is a cytoplasmic membrane-associated protein that is a member of the trafficking ATPase family (Ball et al., 1999). Members of this family of ATPases are involved in DNA uptake, type IV pili assembly, and type II and type IV secretion pathways. More than one T2SS can occur in *Pseudomonas* species (Ball et al., 2002). In the case of *Dickeya chrysanthemi* pathogenesis, the PelB enzyme binds to protein D, specifically to the N-terminal region during secretion (Shevchik et al., 1997). Protein D is stabilized by a lipoprotein called protein S, which binds to the C-terminal region of protein D. This pathogen is well known for secreting pectate lyases in culture (Dahler et al., 1990). Similarly, the polygalacturonase PehA is targeted by the T2SS at the carboxyl-terminal residues 84–135, and at 342–369 in *Pectobacterium carotovorum* (Palomäki et al., 2002). In *Erwinia amylovora*, levansucrase is secreted by the T2SS (Geier and Geider, 1993). With *Xanthomonas campestris* pv. *campestris*, a cytoplasmic membrane protein called XpsN binds to protein XpsD (Lee et al., 2000). XpsN is required for XpsL–XpsM complex formation (Lee et al., 2001). XpsL, XpsM, and XpsN span the inner membrane with a large carboxyl-terminal periplasmic domain (Lee et al., 2004). At least 15 proteins make up the type II secretion machinery of this organism. In *Xanthomonas oryzae* pv. *oryzae*, the *hrp* regulatory genes *hrpXo* and *trh* mediate expression of *hrp* genes and type II- and type III-secreted proteins, such as a cysteine protease (CysP2) (Furutani et al., 2004) and cellulase (ClsA) and cellobiosidase (CbsA) (Tsuge et al., 2006; Jha et al., 2007).

Pectobacterium carotovorum secretes a number of virulence factors via the type II secretion pathway. Cellulases, proteases, and pectate lyases, including the Svx protein, are secreted (Dahler et al., 1990; Hinton et al., 1990; Walker et al., 1994; Corbett et al., 2005). The production of these virulence factors is also additionally regulated by quorum sensing (see Recognition and Chemotaxis to the Host above). A variety of virulence factors and degradative enzymes are secreted by the type II secretion pathway.

Type III Secretion

A number of plant-pathogenic bacteria use one or more type III secretion systems (T3SSs) to alter host cell machinery and cause disease. T3SSs (Desvaux et al., 2006) are found in a variety of bacterial pathogens, including major human pathogens (*Salmonella*, *Shigella*, *Yersinia*, and *Pseudomonas* species and enteropathogenic *Escherichia coli*), and are composed of at least 40 genes, making their temporal and spatial regulation very complex (Hueck, 1998; Aizawa, 2001; Francis et al., 2004). They occur in endosymbionts, such as *Rhizobium* species (Deakin et al., 2005) and *Sinorhizobium* species (Krishnan, 2002), and even in obligate intracellular pathogens, such as *Chlamydia* species (Fields et al., 2003; Ouellette et al., 2005). In *Rhizobium* sp. NGR234, the T3SS-dependent pilus is composed of NopA and associated proteins NopB and NopX (Saad et al., 2008). Synthesis and assembly of the secretion apparatus is induced by change in pH (Daefler, 1999; van Dijk et al., 1999), presence of flavonoids (Deakin et al., 2005), and low calcium and high magnesium concentrations (Linares et al., 2005), even at low bacterial cell numbers (Genin et al., 2005). The right temperature and pH are critical for the operation of the T3SS. Among pathogens bearing needle secretion structures, which are usually human pathogens, the T3SSs can also be activated by direct physical contact with host cells. In *Shigella* species, a protein complex at the tip of the needle, consisting of IpaB and IpaD, is directly involved in sensing host cells and regulating secretion (Veenendaal et al., 2007). The length of the needle is determined by a protein called Spa32 (Botteaux et al., 2008).

With *Pseudomonas syringae* pathovars, pH 6.0 and 20°C promote the secretion of HrmA and AvrPto (van Dijk et al., 1999). Various environmental and host factors play a role in regulating the expression of the T3SS, including a two-component sensory system involving an environmental sensor, RhpS, and a response regulator, RhpR (Xiao et al., 2007). Given the diverse array of inducers of T3SSs, other environmental stimuli that induce T3SSs will likely be discovered as studies continue with this interesting system. For example, there are at least 15 positive and negative regulators of the T3SS in strains of enterohemorrhagic *Escherichia coli* (Iyoda and Watanabe, 2005). It should be stressed that more than one T3SS, some of which are contact mediated, can operate in a given bacterial species (Young and Young, 2002).

In plant pathogens, the type III secretion apparatus is encoded by *hrp* genes. In *Xanthomonas vesicatoria* and likely other xanthomonads, the *hrpE* locus encodes Hrp pili, which are required for type III-dependent protein secretion (Weber et al., 2005). Also, the *hpaC* gene in this organism is required for the translocation of the effector proteins XopJ and XopF1 into plant cells but not for HrpE pilus formation (Büttner et al., 2006). A homolog of HpaC, designated HpaP in *Xanthomonas oryzae* pv. *oryzae*, plays a role in recruiting substrates for type III secretion in this organism (Furutani et al., 2009). In *Pseudomonas syringae* pv. *tomato*, the Hrp pilus is synthesized by the deposition of HrpA pilin subunits at the distal end of the growing pilus (Li et al., 2002). In *Ralstonia solanacearum*, the HrpY pilin subunits require HrpV and HrpX proteins in the assembly of the Hrp pilus (Van Gijsegem et al., 2002). The HrpY pilus apparently serves as the conduit (utilizing the lumen of the pilus) for PopA, PopB, and PopC virulence effectors for this organism. PopF1 is homologous to HrpF and functions as a translocator.

The type III secretion apparatus is usually encoded by a set of genes often on a plasmid or within a chromosomal pathogenicity island. The secretion apparatus is composed of a macromolecular complex spanning both the inner and outer membranes. In *Pseudomonas syringae* pv. *tomato*, a 6- to 8-nm-diameter pilus extends well out of the cell (Li et al., 2002) (Fig. 10-6). A lumen running the length of the Hrp pilus serves as the conduit for translocating presumably unfolded HrpA as well as HrpZ (Li et al., 2002) and HrpW and AvrPto (Wei et al., 2000). The Hrp pilus extends from the bacterial cell surface several micrometers (Brown et al., 2001). In contrast, the T3SS needle observed in mammalian/human pathogens (e.g., *Yersinia* and *Shigella* species and *Escherichia coli*) measures 7–8 nm in diameter with an average length ranging from 40 to 192 nm (Sekiya et al., 2001; Magdalena et al., 2002; Tamano et al., 2002; reviewed in Minamino and Pugsley, 2005). The needle is mostly composed of many copies of a single, small (<10 kDa) protein. In *Yersinia* species, the needle is composed of YscF and three translocators, LcrV, YopB, and YopD. YopB and YopD contain hydrophobic domains. LcrV localizes at the tip of the YscF needle and serves as the locus for the assembly of the translocation pore.

Also, a new class of type III secretion apparatus first discovered in enteropathogenic *Escherichia coli* is present in plant pathogens. The T3SS filamentous apparatus is composed of translocator protein EspA subunits that form a hollow filamentous conduit for translocating virulence determinants (Daniell et al., 2003; Crepin et al., 2005). The filament is 12 nm in diameter with a central channel of 2.5 nm that translocates EspA as well as virulence determinants (Daniell et al., 2001; Daniell et al., 2003; Crepin et al., 2005). Of the transmembrane secretion apparatus, the *Pseudomonas syringae* HrcQ$_B$ protein that belongs to the conserved core of the type III secretion apparatus has been examined at the 50- to 128-residue carboxyl-terminal region by x-ray crystallography to reveal a potential C-ring structure (a multiprotein, ring-shaped assembly) like that of flagellar C-ring structure (Fadouloglou et al., 2004). Indeed, the type III protein secretion system and bacterial flagella-producing system are likely derived from a common ancestor since these systems share at least nine homologous constituents (Fig. 10-7). Both systems apparently arose from a simpler primordial secretion system.

All T3SSs mediate the injection of virulence proteins (called type III effectors) as well as nodulation outer proteins (Nops) (Marie et al., 2003; Deakin et al., 2005) into the host cell, where they manipulate host cell function through sets of signal cascades to promote infection (Cornelis and Van Gijsegem, 2000; Collmer et al., 2002; Greenberg and Vinatzer, 2003; Jin et al., 2003; Alfano and Collmer, 2004; Chang et al., 2004; Mudgett, 2005; Tang et al., 2006; Nimchuk et al., 2007). The proteins secreted by this pathway apparently lack an identifiable signal sequence. However, based on recent studies, there appears to be amino-terminal amino acid

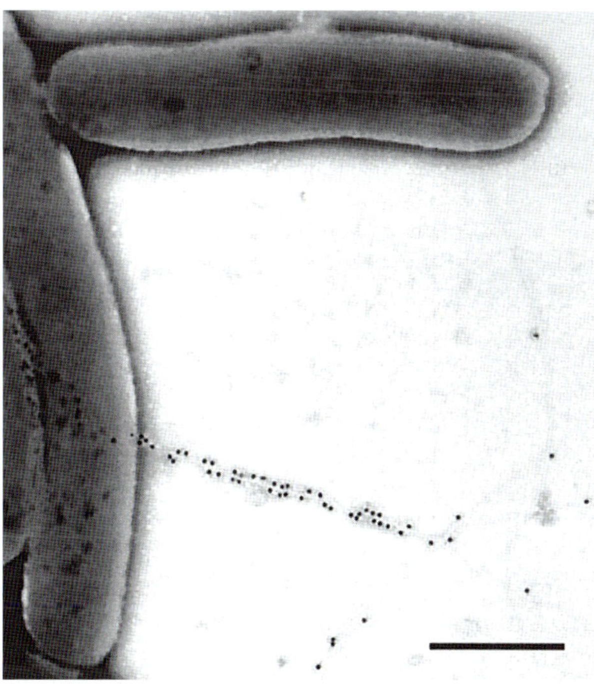

Fig. 10-6. An Hrp pilus decorated by immunogold HrpA antibody. Bar = 0.5 μm. (Courtesy Martin Romantschuk; reprinted, by permission from Macmillan Publishers Ltd., from Li et al., 2002)

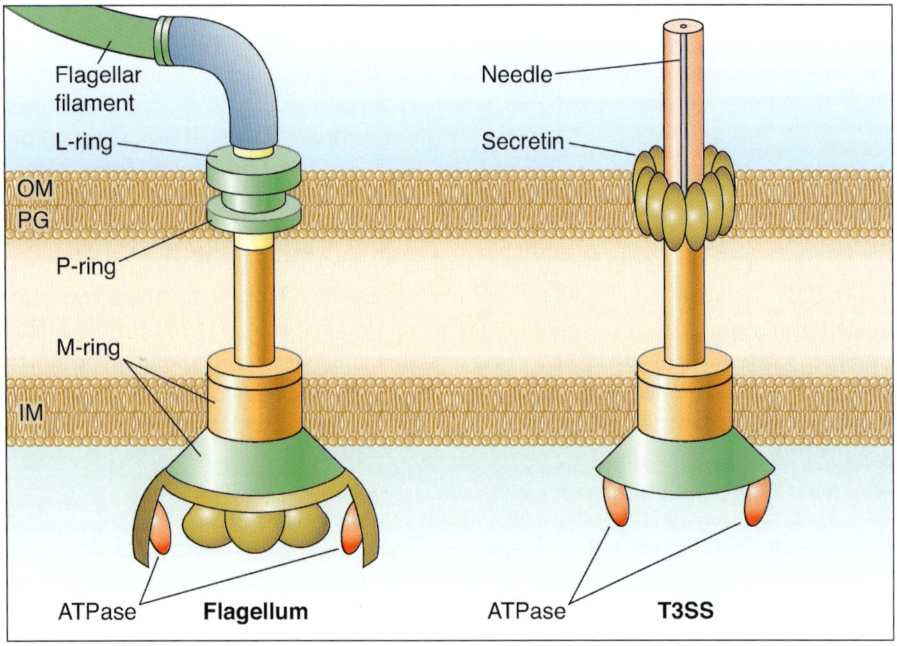

Fig. 10-7. Comparative schematic of the flagellum system and the type III secretion system (T3SS). Homologous structures in the basal region of both structures are similarly colored. Abbreviations: IM, inner (cytoplasmic) membrane; PG, peptidoglycan cell wall; and OM, outer (lipopolysaccharide-containing) membrane. ATP hydrolysis is catalyzed by FliI ATPases that are regulated by FliH to provide the energy required for protein export and assembly of these systems. (Courtesy Milton Saier; reprinted, by permission from the American Society for Microbiology and Penumbra Design, from Wong et al., 2007)

sequence dependency for processing and secretion (Karavolos et al., 2005). In fact, the amino-terminal amino acid sequence must have amphipathic sequences containing four or five serine residues (Lloyd et al., 2002).

Although the precise functions of a majority of type III effectors in plant pathogens remain unknown, their numbers are impressive. In *Ralstonia solanacearum*, more than 40 effector proteins (possibly as many as 95) are translocated via the type III secretion pathway (Cunnac et al., 2004a; Cunnac et al., 2004b; Mukaihara et al., 2004). Likewise, *Pseudomonas syringae* pv. *tomato* harbors 48 or more type III effectors (Collmer et al., 2002; Buell et al., 2003; Alfano and Collmer, 2004; Schechter et al., 2004; Grant et al., 2006; Cunnac et al., 2009). With the growing number of effectors and translocators being identified in *Pseudomonas syringae* pathovars, a new set of guidelines have been proposed to unify the nomenclature of Hrp outer protein (Hop) effectors and translocators (Lindeberg et al., 2005). Thus far, these effectors are thought to each have a minor, but additive, effect on pathogenicity. Hence, a repertoire of diverse effectors are produced by these pathovars, which includes *Pseudomonas syringae* pv. *tomato* (Cunnac et al., 2009). Although numerous effectors are produced, many are subject to Lon protease-mediated degradation if the corresponding protective chaperones are not present. For example, AvrPto produced by *Pseudomonas syringae* has a half-life of 8 min, but in a Lon-deficient mutant strain, its half-life was 30 min (Losada and Hutcheson, 2005). The chaperones HopPsyB1, HopPsyV1, and HopPtoM provided at least tenfold more stability to effectors (Losada and Hutcheson, 2005). The half-life of HrpPsyR was 3 min in wild-type *Pseudomonas syringae*, whereas its half-life was extended to 35 min in a Lon-deficient mutant strain (Bretz et al., 2002). Interestingly, strains deficient in the production of Lon protease do not express a functional T3SS (Jackson et al., 2004). Additional *Pseudomonas syringae* Hop protein effectors appear to have corresponding chaperones to protect the effector from protease degradation and to facilitate type III-mediated translocation (Guo et al., 2005). With *Pseudomonas syringae*, a number of effectors appear to suppress so-called plant "immunity" or basal resistance. How these effectors cause the suppression remains unknown. These effector proteins contribute to pathogenicity by interfering with plant defense signal transduction (Mudgett, 2005; Tang et al., 2006). Effectors such as NolX, which confers host-range specificity on *Sinorhizobium fredii* and is a homolog of HrpF of *Xanthomonas vesicatoria*, remain uncharacterized for biochemical/structural function (Krishnan, 2002). Other effectors, such as PthA produced by *Xanthomonas citri* pv. *citri*, have a major effect and expression in citrus leaf cells and produce almost all the symptoms of citrus canker disease (Duan et al., 1999). Interestingly, gene *hssB3.0* from this pathovar confers host-specific suppression of virulence (Shiotani et al., 2007).

Most studies therefore have focused on avirulence proteins because their phenotypic activity is easily scored by the hypersensitive response (HR) in resistant hosts or in nonhost plants (White et al., 2000). For example, the biochemical/enzymatic function of AvrXa21 of *Xanthomonas oryzae* pv. *oryzae* is unknown. This avirulence protein requires modification and activation by an ATP sulfurylase and an adenosine-5'-phosphosulfate kinase (Shen et al., 2002). Avirulence effector proteins, designated as AvrBs2, AvrBs3, and AvrXv4 in *Xanthomonas vesicatoria*, are secreted and translocated into host plants via a type III secretion pathway (Casper-Lindley et al., 2002; Szurek et al., 2002; Roden et al., 2004). AvrBs3 is localized in the nucleus of *Nicotiana benthamiana*, whereas AvrBsl is observed in the cytoplasm (Gürlebeck et al., 2009). AvrXv4 is a member of the YopJ family and, like YopJ of *Yersinia pestis*, it mimics Ulpl protease activity in eukaryotic cells (Orth et al., 2000). AvrXv4 encodes an active small ubiquitin-like modifier (SUMO) protein that localizes in the plant cytoplasm (Roden et al., 2004). SUMO is a protein of 97 amino acids that is structurally similar to ubiquitin and therefore covalently attaches to certain lysine residues of specific target proteins. Besides AvrXv4, other type III effectors have been assigned functions, namely, XopD (Hotson et al., 2003), AvrPphB (Shao et al., 2003a), HopPtoM (Badel et al., 2003), AvrRpt2 (Axtell et al., 2003), and CysP2 (Furutani et al., 2004). Other type III effectors are XopE1 and XopE2, which belong to the HopX family and possess a conserved N-myristoylation motif (Thieme et al., 2007). HrpX proteins of *Xanthomonas oryzae* pathovars and of *Xanthomonas campestris* pathovars were shown previously to contain a protein myristoylation consensus domain whose consequence may be a loss of recognition by host defense and surveillance systems (Kamdar et al., 1993). HrpX is highly conserved among all *Xanthomonas* species (Oku et al., 1995). Also, a predicted myristoylation sequence has been reported in members of the HopZ family (except HopZ3) in *Pseudomonas syringae* (Lewis et al., 2008). HopZ1, which is also a member of this family, has allelic variants that are differentially recognized by plant resistance systems (Zhou et al., 2009). Amino acid sequence alignment and residue substitution analyses of these variants revealed that amino acid residues between positions 63 and 203 and the cysteine residue at position 141 are critical for triggering and recognition in the HR reaction. A putative peptidyl-prolyl *cis-trans* isomerase that appears to be involved in pathogenesis is encoded by *Xanthomonas campestris* pv. *campestris* (Zang et al., 2007). The amino acid sequence of this enzyme has 49% similarity to a macrophage infectivity potentiator (Mip) protein of *Legionella pneumophila*. A type III effector, named $AvrAC_{Xcc8004}$, produced by *Xanthomonas campestris* pv. *campestris*, contains a leucine-rich repeat domain and appears to be recognized by vascular tissues of *Arabidopsis thaliana* ecotype Col-O (Xu et al., 2008). This effector is encoded by a chromosomal region whose guanine-plus-cytosine (GC) content is 57 mol %, whereas the genomic average GC content is 64.94 mol %. A lateral gene transfer event might have led to the acquisition of this avirulence gene (Xu et al., 2008).

In *Ralstonia solanacearum*, the RipT effector belongs to the YopT family of cysteine proteases (Cunnac et al., 2004b), which are cysteine proteases whose precise function besides targeted protein modification remains to be elucidated (Hotson and Mudgett, 2004).

For translocation through the type III secretion system (Fig. 10-8), certain protein effectors appear to be identified at a signal within the 5'-untranslated region of their mRNA. This untranslated region contains a stem-loop structure traversing the Shine-Dalgarno box and the translational start AUG codon of the transcribed product as demonstrated for AvrPto of *Pseudomonas syringae* pv. *tomato* (Anderson et al., 1999). However, recent studies on the RNA targeting concept have shown that it is not the RNA that is involved but that the first 15 amino-terminal amino acid residues of the effector are recognized and required for translocation (Karavolos

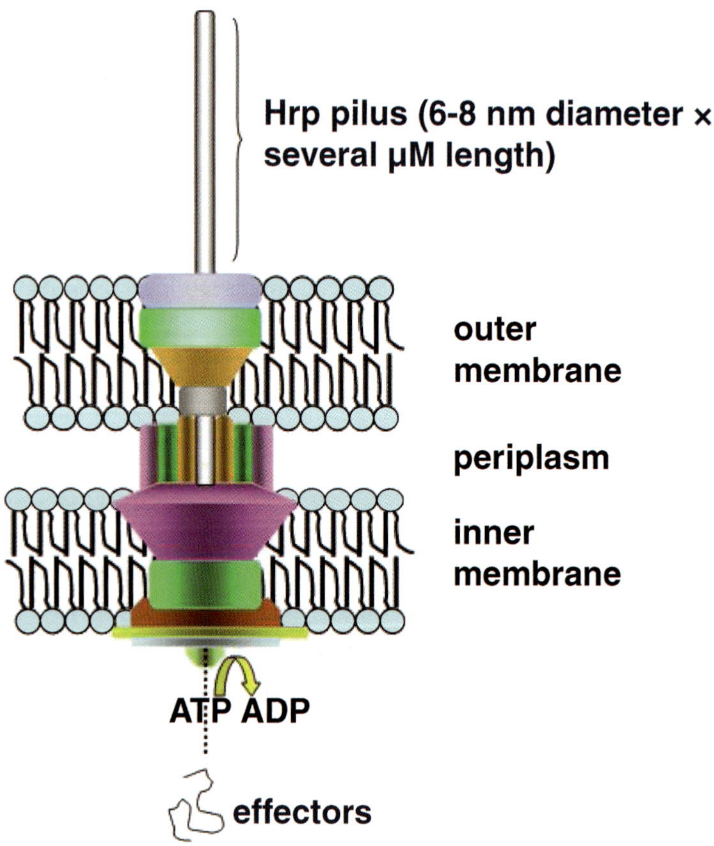

Fig. 10-8. Diagram of the type III secretion apparatus. The energy derived from ATP hydrolysis mediates substrate secretion across the inner and outer membranes. The protein effectors are unfolded to be exported through the 2- to 3-nm channel of the Hrp pilus. The assembly of the type III secretion system in *Salmonella*, *Shigella*, and *Escherichia* species requires at least 25 proteins and that the core of the secretion apparatus or "injectisome" does not exceed nine proteins (Henderson et al., 2004; Cornelis, 2006).

et al., 2005). Thus, further in-depth studies on fine mapping the signal domains in effector proteins of the T3SS will be needed to pinpoint the critical amino acid residues needed for substrate recognition of this secretion system. Work along these lines showed that there indeed are critical amino acid residues in AvrPto at positions 21 and 30 that eliminate function (Goss et al., 2004).

In *Pseudomonas syringae*, the AvrPphB cysteine protease specifically cleaves a PBS1 kinase of *Arabidopsis thaliana* (L.) Heyhn. (Shao et al., 2003a). The x-ray crystal structure of AvrPphB revealed a central antiparallel β-sheet, with α-helices packing on both sides of the sheet to form a two-lobed structure. The core of this structure resembles the papainlike cysteine proteases (Zhu et al., 2004). HopPtoN is a cysteine protease produced by *Pseudomonas syringae* pv. *tomato* that is thought to suppress the HR in both compatible and incompatible interactions (López-Solanilla et al., 2004). AvrRpt2 of *Pseudomonas syringae* pv. *tomato* is 28 kDa and is processed into a 21-kDa product whose carboxyl-terminal portion is a cysteine protease that targets *Arabidopsis* RIN4 (Axtell et al., 2003; Coaker et al., 2005). The cysteine protease is activated upon folding, facilitated by a cyclophilin (Coaker et al., 2005). Proper folding is required for autocatalytic processing of the cysteine protease. Besides cysteine proteases, HopPtoD2 produced by *Pseudomonas syringae* pv. *tomato* appears to possess tyrosine phosphatase activity that is needed for virulence (Bretz et al., 2003; Espinosa et al., 2003). Earlier studies have shown that the gene product of *avrPto* interacts directly with a serine/threonine protein kinase encoded by the tomato *R* gene (Loh and Martin, 1995; Frederick et al., 1998).

Interestingly, in the mammalian system, the *Yersinia* T3SS exports a set of proteins in the Yop and Lcr families, such as YopH and LcrV, which insert into the host cell membrane and cytoplasm. YopH and YspP are tyrosine phosphatases that uncouple multiple signal transduction pathways in the host cell and dephosphorylate several focal adhesion proteins (Black and Bliska, 1997; Persson et al., 1997; Matsumoto and Young, 2006). Tyrosine phosphatase disrupts host signal transduction processes involved in bacterial killing (Bliska et al., 1991; Anderson et al., 1996). Substrate recognition by YopH involves two substrate-targeting sites that function in a cooperative manner (Ivanov et al., 2005). Analysis of the crystal structure of the substrate-binding domain of YopH suggests that the enzyme functions in a monomeric state after it is assisted for translocation through the type III secretion pathway by a chaperone, SycH (Evdokimov et al., 2001; Smith et al., 2001). LcrV is a multifunctional protein (V antigen) with antihost and positive regulatory roles on Yop secretion. Clearly, the Yop proteins and LcrV act in concert to thwart innate cellular immune responses to *Yersinia* species (Cornelis, 2002; Ramamurthi and Schneewind, 2002; Cornelis, 2006). Two tyrosine phosphatases in *Mycobacterium tuberculosis* are essential for virulence and may operate similar to YopH (Koul et al., 2000).

It appears likewise that type III effectors, such as cysteine proteases and tyrosine phosphatases of plant pathogens, are used to circumvent, avoid, and inactivate innate cellular defense responses (Espinosa and Alfano, 2004; P. He et al., 2006). HopPtoD2 of *Pseudomonas syringae* tyrosine phosphatase inhibits MAPK kinase in the host (Espinosa and Alfano, 2004). Hence, it is not too surprising that plant pathogens are also equipped to secrete tyrosine phosphatases via the type III secretion pathway. In general, however, bacteria do not contain tyrosine phosphatase (Guan and Dixon, 1990). The function of each identified effector for each bacterial plant pathogen will be forthcoming in the near future as these enzymes are isolated and characterized. Besides circumventing innate immunity by protein digestion in the plant, cysteine proteases and, likely, tyrosine phosphatases produced by the plant pathogen may mimic plant cysteine proteases and tyrosine phosphatases that are generally used in housekeeping functions, e.g., protein turnover, suppressing signal transduction pathways. Indeed, the structure and catalytic domains of these enzymes are very similar to those of eukaryotic enzymes (Guan and Dixon, 1990; Zhang and Dixon, 1994; Fauman and Saper, 1996; Forsyth et al., 2004).

Cationic antimicrobial peptides (defensins) are major innate immunity compounds. Bacterial surface changes are affected by cysteine proteases to prevent the action of cationic antimicrobial peptides. The general opinion is that tyrosine phosphatase effectors injected by the pathogen may remove phosphates from phosphorylated tyrosines of signaling proteins that play a crucial role in plant defense responses. In addition, effectors that cause irreversible damage to cellular membranes can lead to rapid cell death. Effectors such as phospholipase A, which indeed causes membrane damage (Sato and Frank, 2004), can likely be effectors injected by bacterial plant-pathogenic species. Phospholipases are important virulence factors in a number of diverse bacterial pathogens (Songer, 1997; Sato et al., 2005). Also, effectors that are cysteine-rich proteins may eventually be discovered to be produced by xylem-invading bacterial pathogens. For example, a 12-kDa cysteine-rich *avr* gene product is produced in a xylem-invading fungal plant pathogen to prevent premature necrosis (Rep et al., 2004). Likewise, the effector AvrPtoB produced by *Pseudomonas syringae* pv. *tomato* inhibits programmed cell death (apoptosis) (Abramovitch et al., 2003). Suppression of a set of plant host genes that encode cell wall-based defense materials, such as callose, phenolics, and hydroxyproline-rich glycoproteins, has been reported for *Xanthomonas vesicatoria* (Brown et al., 1995) and *Pseudomonas syringae* pv. *tomato* (Hauck et al., 2003).

Type IV Secretion

Type IV secretion systems (T4SSs) transport protein, nucleoprotein, and nucleic acid substrates through specific channeling complexes incorporated into the outer membrane and cell envelope. The highly conserved T4SSs have been identified in both plant and mammalian pathogens. The T4SS may have been adapted ancestrally from a plasmid conjugation system, whereby both DNA and protein are conjugatively transferred to the recipient from the donor strain. This secretion system can also include pathogen-based effectors that are injected into eukaryotic cells. Moreover, with the T4SS, DNA can be taken from the extracellular environment or from the recipient cell and used in intracellular survival (Pansegrau and Lanka, 1996; Covacci et al., 1999; Sexton and Vogel, 2002; Cascales and Christie, 2003; Liu and Binns, 2003; Backert et al., 2005; Segal et al., 2005; Slater et al., 2009). Like the T3SSs, T4SSs are used by a variety of bacterial pathogens to transfer bacterial proteins and DNA into eukaryotic hosts (Fig. 10-9). All *Brucella* species require T4SSs for intracellular survival and persistent infection. At least two proteins, VceA (105 amino acids) and VceC (418 amino acids), are translocated into J774 macrophages (de Jong et al., 2008). Of the plant pathogens, *Agrobacterium tumefaciens* harbors the well-characterized prototypical T4SS used to transfer the T-DNA into plant cells to cause crown gall (see Chapter 8 for details). This organism has served as the model for *virB* operon-encoded type IV secretion in other pathogens (Zupan et al., 1998; Lai and Kado, 2000). The best-characterized members of the T4SS family are those involved in bacterial conjugation in which broad-host-range plasmids are transferred between Gram-negative bacteria (Llosa et al., 2002). Genome

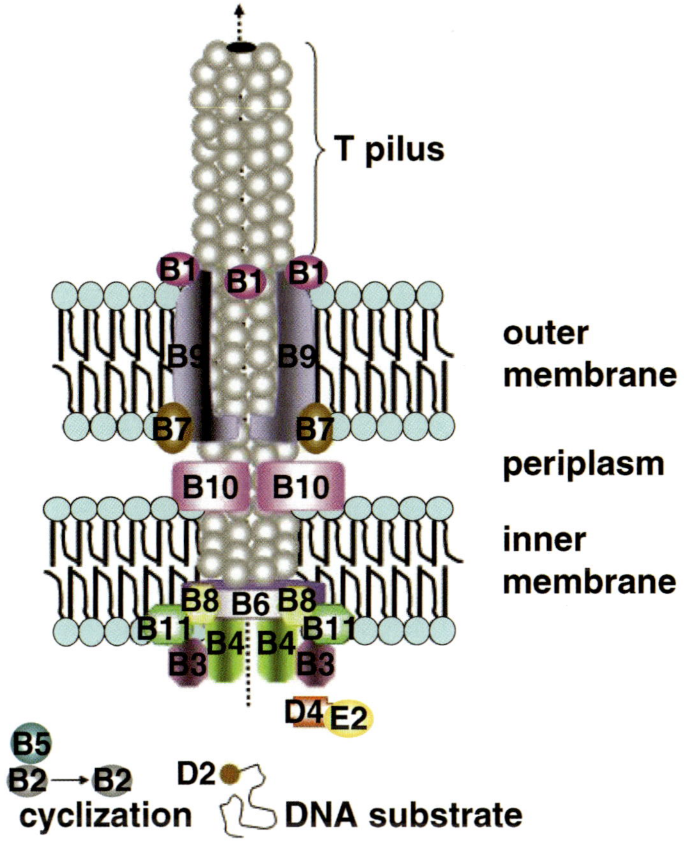

Fig. 10-9. Diagram of the type IV secretion apparatus. The apparatus is composed of VirB proteins that are described in Table 8-2. The hydrolysis of ATP generates the energy for substrate transfer out of the cell. VirB4, VirD4, and VirB11 are ATPases. VirB5 is a chaperone and VirB1* (not shown) is required for processing the VirB2 pilin subunit of the T pilus (Zupan et al., 2007). The VirE2 substrate docks via an arginine-rich, carboxyl-terminal domain at the VirD4 receptor to mediate translocation (Christie et al., 2007).

and plasmid sequences of *Agrobacterium tumefaciens*, *Agrobacterium rhizogenes*, *Agrobacterium vitis*, *Xanthomonas* species, and *Xylella fastidiosa* have revealed the presence of type IV secretion components. Their functional presence and the role of the Ti plasmid-based T4SS in *Agrobacterium tumefaciens* are relatively well characterized. *Xanthomonas citri* pv. *citri* contains two T4SS loci; one on the chromosome and the other on a 64-kb pXAC64 plasmid (da Silva et al., 2002). These loci are VirB orthologs, including a VirB7 protein discovered recently in *Xanthomonas citri* pv. *citri* (Alegria et al., 2005). T4SS homologs have been observed on plasmid pXAC64 in *Xanthomonas axonopodis* pv. *aurantifolii*, which may prove to be a self-mobilizing plasmid (Brunings and Gabriel, 2003). A marked plasmid derivative of pXcB transfers by conjugation, presumably via the type IV secretion system, between two phylogenetically distinct pathovars, *Xanthomonas axonopodis* pv. *aurantifolii* to *Xanthomonas citri* pv. *citri*, in planta (El Yacoubi et al., 2007). Likewise, *Xanthomonas vesicatoria* can conjugatively transfer in planta chromosomal sequences among different *Xanthomonas axonopodis* strains (Basim et al., 1999). In *Sinorhizobium meliloti* strain 1021, the T4SS is not required for intracellular symbiosis; it is only required for conjugation (Jones et al., 2007).

More than one type IV secretion pathway can occur in a given pathogen, and a pathogen can be also equipped with type II and type III as well as type IV secretion systems. *Burkholderia cepacia* is one example. Genes encoding the T4SS components can be found on plasmids and on chromosomes such as that of *Burkholderia cepacia* (see Chapter 6). In the human pathogen *Actinobacillus actinomycetemcomi-*

tans, two T4SSs are present. They are located on the *Actinobacillus actinomycetemcomitans* chromosome and on a resident plasmid (Novak et al., 2001). In the human ulcer-producing pathogen *Helicobacter pylori*, there are four T4SSs, namely, ComB, CagPAI, Tfs3, and Rlx2 (Backert et al., 2005). The growing body of evidence indicates that T4SS loci are highly conserved and that more than one T4SS has been widely dispersed among various Gram-negative bacteria via lateral gene transfer, clustered on pathogenicity islands and on relaxase-bearing clusters. For example, *Spiroplasma kunkelii* harbors a 14.6-kb plasmid, pSKU146, that apparently encodes a T4SS. Plasmid pSKU146 contains open reading frames (ORFs) that encode proteins having similarity to conjugative elements and components of the bacterial type IV secretion apparatus (Davis et al., 2005). The deduced amino acid sequence encoded by ORF10 possesses a signal peptide (amino acid residues 1–34) and a single transmembrane region (residues 62–80). The plasmid also encodes a 96-kDa adhesin (SkARP1) that may play a critical role in bacterial cell attachment to the gut of its insect vectors. ORF12 encodes a putative membrane-bound ATPase with a VirB4 domain. This putative gene encoding a conjugal transfer protein is among the first to be found in the class *Mollicutes*. The ORF12 protein contains a signal peptide (residues 1–49), two transmembrane regions (residues 50–72 and 79–93), and a low complexity region (residues 244–253). The presence of a signal peptide and transmembrane regions is uncharacteristic of VirB4 proteins and may indicate that the ORF12 protein carries out additional functions (Davis et al., 2005). The T4SS can also be found in *Mesorhizobium loti*, which uses the secretion pathway to translocate at least two effector proteins, Msi059 and Msi061, to plant cells as well as to yeast cells (Hubber et al., 2004). Interestingly, Msi059 has sequence similarity to a cysteine protease of the C48 family of peptidases that includes the ubiquitin-like protein protease Ulp1 and the type III effector XopD of *Xanthomonas vesicatoria* (Hubber et al., 2004). The second effector protein, Msi061, has sequence similarity to VirF of *Agrobacterium* species.

Agrobacterium tumefaciens, *Agrobacterium rhizogenes*, and *Agrobacterium vitis* are not the only pathogens that secrete DNA (as T-DNA). *Neisseria gonorrhoeae* uses the T4SS to secrete chromosomal DNA that is active in the transformation of recipient gonococci (Hamilton et al., 2005).

Basically, T4SSs are used to either transfer protein effectors into host cells or to take up protein and DNA (retrotransfer) for environmental adaptation (Kado, 2002). Interestingly, T4SSs operate to move substrates in a bidirectional manner (Knipers et al., 1998; Backert and Meyer, 2006). That is, certain T4SSs function to deliver or take up protein or nucleic acid substrates (Heinemann and Ankenbauer, 1993; Heinemann et al., 1996; Kado, 2002; Liu and Binns, 2003). The *Agrobacterium* Ti and Ri plasmid T4SSs appear to have evolved to transfer protein and nucleic acid substrates. The T-DNA is the prime example of a substrate that is transferred into host cells, where it becomes integrated into the nuclear genome and T-DNA oncogenic loci are expressed, culminating in the formation of crown gall tumors. In *Neisseria gonorrhoeae*, the T4SS appears to be used to benefit the pathogen by providing transforming DNA to recipients for antigenic variation to avoid the immunodefense system of the host (Hamilton et al., 2005).

An integral part of the T4SS is the biosynthesis of the T pilus whose processed subunit is encoded by the *virB2* gene of the virulence regulon located on the Ti plasmid of *Agrobacterium tumefaciens* (Fig. 10-10). The *virB2* gene sequence shows similarities to TraA of *Escherichia coli* plasmid F and to TrbC of the PilW operon of plasmid R388 (Shirasu and Kado, 1993a). The striking similarities between VirB2 and TraA in their amino acid sequence, their protein processing into a 7.2-kDa subunit, and their location in the bacterial cell brought forth the hypothesis that *virB2* encodes a VirB2 pilin subunit used in the transfer of the T-DNA (Shirasu and Kado, 1993b). The hypothesis was substantiated by the isolation of a 10-nm-diameter T pilus formed by induced *Agrobacterium tumefaciens* cells and encoded by the *virB2* gene (Lai and Kado, 1998). The diameter of the T pilus is clearly larger and distinct from the common pilus of 3 nm diameter (Fullner et al., 1996). The 3-nm common pilus is produced by both virulent and avirulent strains of *Agrobacterium tumefaciens* (Lai et al., 2000). Like the F pilus of *Escherichia coli* (Babic et al., 2008), the T pilus with its 2-nm lumen is likely the conduit through which the T-DNA is transferred into the host plant cell (Lai and Kado, 2000; Lai and Kado, 2002). Interestingly, deconvolution fluorescence microscopic studies on *vir*-induced *Agrobacterium tumefaciens* cells revealed T pili at multiple sites on the cell envelope (Aguilar et al., 2010).

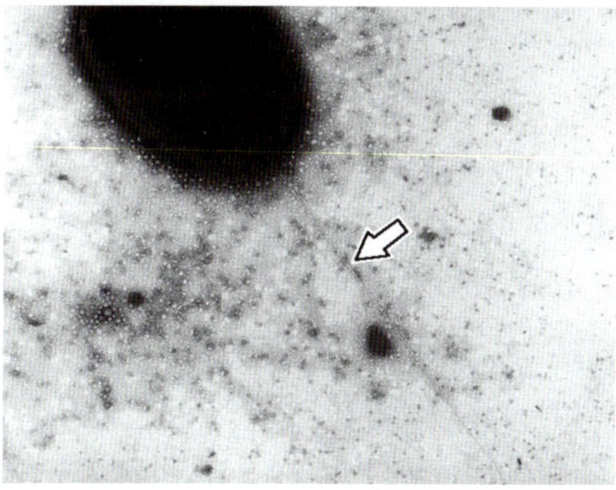

Fig. 10-10. *Agrobacterium tumefaciens* cell with polar T pilus of 10 nm diameter (arrow). (Photo by Erh-Min Lai)

Type V Secretion

The type V secretion system (T5SS) (Fig. 10-11), known as the autotransported protein system, is composed of proteins that are subclassified as either an autotransporter or a two-partner secretion pathway family member. Characteristic of this family is the capacity of each member to mediate its own translocation across the bacterial outer membrane. The Sec system is used to mediate translocation across the inner membrane (see Type II Secretion above). Autotransporter proteins use a C-terminal translocator domain to mediate secretion of an N-terminal effector domain across the outer membrane to the bacterial surface in a process requir-

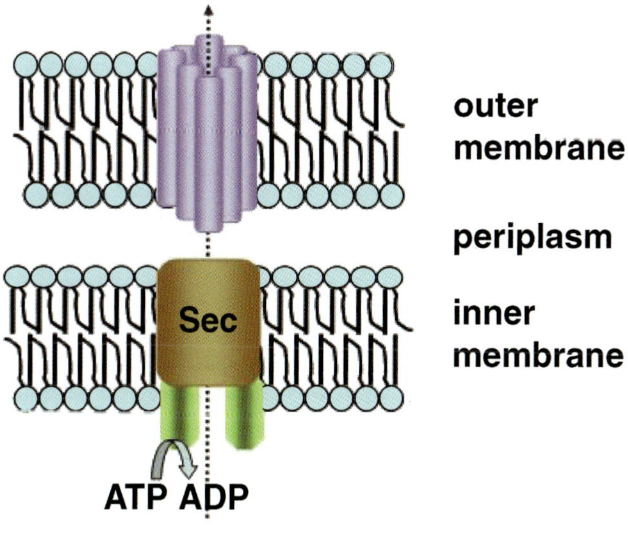

Fig. 10-11. Schematic of the type V secretion apparatus. At least three types of protein substrates are transported: autotransporter proteins (type Va); two-partner system proteins (type Vb); and type Vc, members of the AT-2 family of proteins (Henderson et al., 2004). Effector proteins are composed of an unusual extended signal sequence that is cleaved in the periplasm and a β-domain that inserts into the outer membrane to form the pore. After formation of the β-barrel structure, the passenger domain inserts into the pore and is translocated to the bacterial cell surface (Henderson et al., 2004).

ing no accessory proteins. Proteins belonging to the two-partner secretion pathway differ from autotransporters in that the effector and translocator domains are expressed as separate proteins (Henderson et al., 2004). Unlike the type I–IV secretion machineries, only a single gene encodes the secreted protein and the complete machinery required for secretion.

All autotransporter proteins consist of a modular structure composed of three domains. The signal sequence (also called signal peptide or leader sequence) is present at the N-terminal end of the protein and allows targeting of the protein to the inner membrane for type II export into the periplasm. The subsequent domain is the passenger domain, which confers the diverse effector functions on the various autotransporters. The last main domain, located at the C-terminal end of the protein, is the translocation unit, consisting of a short linker region having an α-helical secondary structure and a β-domain that adopts a β-barrel structure when embedded in the outer membrane and facilitates translocation of the passenger domain through the outer membrane (Henderson et al., 1998). Since no specific energy coupling or accessory factors seem to be required for the translocation process, the proteins secreted in this fashion received the name of autotransporters.

Hence, members of the family of type V-secreted virulence factors comprise three functional domains in a single autoexported protein, namely, a signal sequence, a passenger domain, and a β-strand domain (Henderson et al., 2004). Export of this family of proteins through the inner membrane occurs via the Sec pathway and is coupled to the cleavage of a 25- to 30-amino acid signal peptide at its amino terminus. After entering the periplasm, the carboxyl-terminal domain of the processed protein or effector mediates the passage of the protein through the β-protein pore in the outer membrane (Fig. 10-11). No energy coupling or accessory factors are known to be required for the protein translocation process.

Virulence factors such as immunoglobulin A (IgA) β-protease of *Neisseria gonorrhoeae* (Pohlner et al., 1987), AIDA-I adhesin of *Escherichia coli* (Benz and Schmidt, 1989), IcsA (VirG), an actin polarizer of *Shigella* species (Suzuki et al., 1995), and VacA of *Helicobacter pylori* (Cover, 1996) are members of the autotransporter family of the T5SS. They represent members of the largest group of secreted proteins in Gram-negative bacteria. Among plant-pathogenic bacteria, *Xylella fastidiosa* contains a homolog of the *Escherichia coli mttC* gene that is involved in type V secretion (Keen et al., 2000).

Type VI Secretion

The virulence-associated secretion (VAS) system of *Vibrio cholerae* transports proteins to the exterior of the bacterial cell by a mechanism that does not require hydrophobic N-terminal signal sequences. Several *vas* gene products constitute the structural components of the VAS secretion apparatus that delivers effector proteins via the type VI secretion system (T6SS) (Pukatzki et al., 2006). The VAS pathway mediates *Vibrio cholerae* cytotoxicity toward mammalian macrophages and amoebae in the genus *Dictyostelium*. Several Gram-negative bacterial plant pathogens and endosymbionts, such as *Agrobacterium tumefaciens* and *Rhizobium leguminosarum*, respectively, harbor genes homologous to *vas* genes (Roest et al., 1997; Bladergroen et al., 2003; Das and Chaudhuri, 2003). The T6SS is used by *Agrobacterium tumefaciens* to secrete a hemolysin-regulated protein (Hcp) that is required for efficient tumorigenesis (Wu et al., 2008). T6SSs will undoubtedlessly be discovered in other Gram-negative plant pathogens (Fig. 10-12).

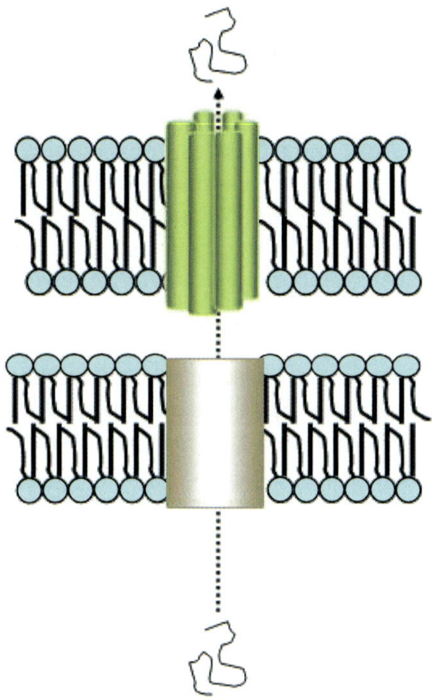

Fig. 10-12. Putative diagram of the type VI secretion apparatus encoded by the virulence-associated secretion (*vas*) genes (Pukatzki et al., 2006).

vasA is a homolog of *impG* of *Rhizobium leguminosarum* (Bladergoen et al., 2003; Das and Chaudhuri, 2003). This gene is involved in mediating legume root infection. The inner-membrane spanning protein is ClpV1 in *Pseudomonas aeruginosa*. The outer-membrane spanning protein is the VasA/Hcp1 hexameric oligomer that has been observed by electron microscopy and analyzed by crystallography (Mougous et al., 2006).

Twin-Arginine Translocation Protein Secretion System

In the sections above, six protein/nucleic acid translocation pathways (types I–VI) are described. There is an additional pathway used by bacteria called the twin-arginine translocation (Tat) secretion system, which is found in the cytoplasmic membranes of most bacteria and in the energy-transducing membranes of plant organelles (Settles and Martienssen, 1998; Berks et al., 2000; Berks et al., 2005; Müller and Klösgen, 2005). The Tat secretion pathway transports proteins in a Sec-independent manner. Such proteins are synthesized with amino-terminal signal peptides that contain a distinctive invariant peptide sequence of SRRxFLK, which contains the twin-arginine amino acid motif. Proteins bearing twin-arginine signal peptides are directed posttranslationally to the membrane-embedded Tat secretion system that transports fully folded proteins across the inner membrane using energy provided by the transmembrane pH gradient (Berks et al., 2000). The core of the Tat transporter is a large oligomeric complex of TatA, TatB, and TatC proteins. The TatA protein provides the protein-conducting channel within the complex through which folded proteins are transported. The Tat secretion pathway is structurally and mechanistically related to the ΔpH-dependent pathway in chloroplast thylakoid membranes (Cline et al., 1993; Müller and Klösgen, 2005). In

Escherichia coli, there are at least 26 Tat substrates, each of which possesses single transmembrane α-helices at the extreme carboxyl terminus. This portion is permanently integrated into the inner membrane (Hatzixanthis et al., 2003). The Tat substrate TorA of *Escherichia coli* is compatible with Tat secretion systems from *Agrobacterium tumefaciens* (Oates et al., 2003). Studies on a Tat secretion system in *Agrobacterium tumefaciens* showed that inactivation of this system abolishes anaerobic respiration and chemotaxis and attenuates virulence (Ding and Christie, 2003). Substrates of the Tat secretion system were found recently in *Pseudomonas syringae* pv. *tomato* (Ferreira et al., 2006). Phospholipases were found to be substrates that are translocated across the outer membrane by a *Pseudomonas syringae* T2SS encoded by the *gsp* operon (Bronstein et al., 2005). This indicated that the Tat secrection system and T2SS coordinate the secretion of these substrates. In general, soluble periplasmic proteins that are exported by the Tat secretion pathway are recognized by their distinct cleavable amino-terminal signal peptides in their cognate substrate proteins (Robinson and Bolhuis, 2001). Signal peptides of Tat-dependent precursor proteins are optimally adapted only to their cognate export apparatus, which is in marked contrast to the protein processing system of the Sec-dependent protein translocation pathway. As more work becomes focused on the Tat secretion pathway, additional plant-pathogenic bacteria, such as *Pseudomonas syringae* (Caldelari et al., 2006) and *Ralstonia solanacearum* (González et al., 2007), will be found to be using this pathway to deliver effectors used in virulence and pathogenesis.

Colonization of the Plant Host

Microscopic examinations of infection sites show that pathogenic bacterial cells first adhere to a plant substratum as a monolayer, which is followed by the formation of microcolonies, usually in the intercellular space. Microcolony formation is followed by the appearance of macrocolonies, namely, biofilm maturation, that measure 100 μm or more in thickness. *Dickeya chrysanthemi* forms aggregates that are assisted by cellulose and depend on the T3SS (Yap et al., 2005). Macrocolonies are separated by fluid-filled channels surrounded by protein-polysaccharide matrices. Host surface attachment by *Agrobacterium tumefaciens* apparently involves the synthesis of cellulose fibrils and cyclic β-glucans encoded by *cel* and by *chvA* and *chvB*, respectively (Matthysse, 1983; Breedveld and Miller, 1994). The architecture and maturation rate of the biofilm formation of *Agrobacterium tumefaciens* were found to be regulated by SinR, a transcriptional regulator superimposed by the fumarate and nitrate reduction (FNR)-type global transcriptional regulator of anaerobically expressed genes (Ramey et al., 2004). Cellulose production aids in the biofilm formation needed for the survival of *Salmonella enteritidis*, but cellulose production is not involved in the virulence of this enteric pathogen (Solano et al., 2002). Two operons, bcsABZC and bcsEFG, are responsible for cellulose biosynthesis.

To return from the biofilm to the planktonic state as free-living cells, rhamnolipids are required for detachment. For example, *Pseudomonas aeruginosa* produces rhamnolipids to facilitate detachment of cells from the biofilm (Boles et al., 2005), affect surface motility (Deziel et al., 2003), maintain open biofilm channels (Davey et al., 2003), change cell surface polarity (Zhang and Miller, 1994), assimilate hydrocarbons (Beal and Betts, 2000), and disrupt host defenses (Read et al., 1992). With *Pseudomonas fluorescens*, rhamnolipid biosurfactant facilitates soft rotting of broccoli and cauliflower (Hernandez-Anguiano et al., 2004).

Some plant pathogens, such as *Pseudomonas syringae* pv. *syringae* and *Xanthomonas campestris* pv. *campestris*, enter their hosts (beans and cauliflower/cabbage, respectively) through natural openings, such as stomata and hydathodes. Cells of these pathogens multiply and ingress further through intercellular spaces or xylem elements. Real-time ingression through the xylem by cells of *Xanthomonas campestris* pv. *campestris* that have been tagged with a constitutive luciferase gene set reveals passive translocation rather than assistance by flagella (Kamoun and Kado, 1990; McElhaney et al., 1998). Cells of this pathogen move passively, presumably by diffusion in the plant xylem. Thus, forward movement of these cells in the xylem is not due to motility or chemotaxis since the bacterial cells are nonmotile and nonchemotactic in the host plant (Kamoun and Kado, 1990). The growth rate in the xylem is approximately 4×10^7 cells per 72 h (Kamoun and Kado, 1990). Likewise, *Ralstonia solanacearum*, which also colonizes xylem vessel elements, becomes nonmotile after invading the xylem (Tans-Kersten et al., 2001). *Dickeya chrysanthemi* pv. *chrysanthemi* uses genes of the *phoP-phoQ* operon to aid in growth and survival at low pH (pH of less than 5.5) in planta and to induce further production of polygalacturonase (Liama-Palacios et al., 2003). In *Pseudomonas syringae* pv. *tomato*, the induction of a number of normally silent bacterial genes occurs during colonization of the host. These genes are upregulated by HrpL (an alternative RNA polymerase sigma factor) (Boch et al., 2002).

Circumventing Resistance Barriers

Avirulence Effectors Countering Innate Immunity

Invasive plant-pathogenic bacteria are equipped with an arsenal of products that help evade innate and induced host defense responses. Innate immunity depends on germline-encoded receptors that have evolved to recognize highly conserved pathogen-associated molecular patterns (PAMPs). These receptors are termed pattern recognition receptors (PRRs). Host defenses are activated by two distinct classes of elicitors: general elicitors and specific elicitors. General elicitors are bacterial lipopolysaccharides, phospholipids, peptidoglycans, flagellins, proteins such as elongation factors, double-stranded nucleic acids, cell wall pieces, and methylated DNA that can bind to host PRRs and trigger host defenses (Kunze et al., 2004; Ingle et al., 2006). Specific elicitors are perceived by very specific receptors that are often host-cultivar specific. One of the effectors injected by *Pseudomonas syringae* via the type III secretion apparatus is an ADP-ribosylase that ribosylates RNA-binding proteins, the results of which suppress plant defenses (Fu et al., 2007).

During the colonization of host tissues, e.g., leaf tissues, the pathogen injects (via the secretion systems described above) several proteins into the apoplastic space of the colonized leaves. These effector proteins, which include proteins that suppress defense reactions and promote bacterial survival, include race-specific elicitor proteins encoded by avirulence (*avr*) genes. Avr proteins usually induce defense responses, such as the well-known HR, if the host plant carries the corresponding *R* gene. This is known as an incompatible host–pathogen interaction, whereby the pathogen produces an Avr effector protein that is recognized by a cognate *R* gene product in the resistant cultivated variety (Keen, 1990; Martin et al., 2003). *R* genes encode extracellular, plasma membrane-anchored, receptor proteins bearing leucine-rich repeats (Belkhadir et al., 2004). R proteins are thought to recognize the modification of the virulence target by the Avr protein. The simple model that

R proteins elicit HR in response to cell death suppression activity, which is only observed in the absence of the cognate R protein, has been replaced by a more complex model, called the trump model (Abramovitch and Martin, 2005). The trump model includes a trump (T) factor that would act on the cognate R protein to generate an HR-based immunity in response to a cell death suppressor. Certain effector proteins, such as HopPtoN, suppress cell death in both compatible and incompatible interactions (López-Solanilla et al., 2004).

Innate immunity defenses are triggered by PAMPs (Abramovitch et al., 2006; Chisholm et al., 2006; Jones and Dangl, 2006). These defenses are regulated by interconnected chemical signaling networks that utilize ethylene, jasmonic acid, and salicylic acid (Kunkel and Brooks, 2002; Thatcher et al., 2005). PAMPs activate mitogen-activated protein kinases (MAPKs) and the production of reactive oxygen species (ROS) by the host. Superoxide dismutases produced by the pathogen help reduce the effects of ROS. Such virulence factors can inhibit PAMP-elicited basal defenses (Espinosa and Alfano, 2004). Recruitment of host factors by the pathogen is a highly evolved means of circumventing innate defenses. For example, the *Agrobacterium* VirE2-binding protein VIP1 produced by the plant host appears to be a substrate of MAPK3 that phosphorylates VIP1 at serine position 79 of the VIP1

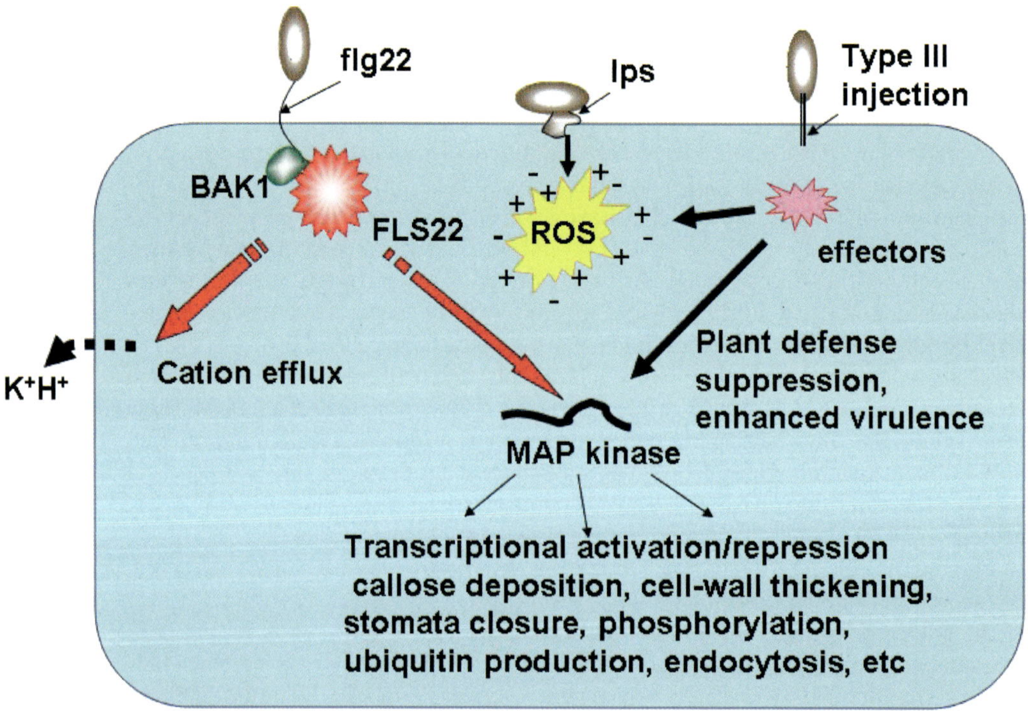

Fig. 10-13. Plant innate immunity responses to a pathogen and counteractions by the pathogen. The plant perceives various bacterial components, such as the receptor complex FLS22 and BAK1 to flagellar proteins (flg22), that trigger a cascade of defense responses (Chinchilla et al., 2007a,b; Heese et al., 2007). These include cation efflux that changes the external environment, production of reactive oxygen species (ROS), localized cell death (hypersensitive response), callose formation, cell wall thickening, stomata closure, endocytosis, phosphorylation, and ubiquitin production, all activated by mitogen-activated protein (MAP) kinases (Anderson et al., 2006; Robatzek et al., 2006). To counter these defenses, the pathogen secretes effectors that suppress or inactivate plant cell defense activities.

peptide in order to translocate the T-DNA strand-VirE2 complex into the plant cell nucleus (Djamei et al., 2007). *Agrobacterium tumefaciens* therefore has recruited MAPK3 to facilitate nuclear localization and eventual integration of the T-DNA strand in the host chromosome. In *Arabidopsis thaliana*, an elongation factor, Tu, perceives and binds to receptor kinases, the action of which reduces *Agrobacterium*-mediated transformation efficiency (Zipfel et al., 2006).

Figure 10-13 summarizes the known mechanisms that induce and circumvent (suppress) innate immunity in plants.

Importance of Cysteine Proteases in Establishing Infection

Cysteine proteases are represented by papain, calpain, caspases, clostripain, gingipain, cathepsins, streptopain, and staphopains. Although relatively new to discovery in plant pathogens, many cysteine protease effectors have been recognized for more than two decades as critical components in the parasitic egress and invasion of the malaria parasite, *Plasmodium falciparum* Welch (Rosenthal and Nelson, 1992; Greenbaum et al., 2002; Wickham et al., 2003). Also, *Trypanosoma cruzi* Chagas, the causal organism of Chagas disease, produces a cysteine protease effector called cruzipain (Cazzulo, 2002), and cysteine proteases produced by the amoebic dysentery protozoan parasite *Entamoeba histolytica* Schaudinn are essential for establishing a niche in the human host (Li et al., 1995; Ankri et al., 1998; Zhang et al., 2000; Bruchhaus et al., 2003). In fact, *Entamoeba histolytica* uses cysteine proteases to evade host defenses by inhibiting phagocytosis (Ankri et al., 1998; Bruchhaus et al., 2003). In bacterial systems, the cysteine protease streptopain in *Streptococcus pyogenes* has played a major role in human pathogenesis by cleaving the cytokine interleukin-1 beta (Kapur et al., 1993). *Porphyromonas gingivalis*, the etiological organism of periodontitis, produces cysteine proteases, collectively referred to as gingipains, to degrade several host proteins, including hemoglobin, to obtain iron and establish infection (Sroka et al., 2001). In *Yersinia* species, the YopT type III effector is a cysteine protease that cleaves a GTPase (Shao et al., 2003b). So, it is not too surprising to find cysteine proteases playing an important role in plant pathogenesis. For example, the effectors AvrXv4, AvrPphb, AvrRpt2, and XopD have cysteine protease functions (Hotson and Mudgett, 2004). Because cysteine proteases have been identified as effectors injected into plant cells by the T3SS, their role is likely to be targeting defense-related proteins. Caspases are cysteine proteases that trigger apoptosis and play a critical role in disease production in humans. Bacterial proteases also degrade antibacterial peptides produced by the human host (Schmidtchen et al., 2002). Cysteine protease inhibitors produced by Gram-positive and Gram-negative bacteria are likely used to protect cytoplasmic and cellular membrane proteins against the secreted proteases (Kagawa et al., 2005; Potempa et al., 2005). Hence, additional focused studies on cysteine proteases produced by plant pathogens will help identify the elaborate biochemical pathways used by these pathogens in plant pathogenesis.

Neutralizing Reactive Oxygen Species

When perturbed, signaled, or damaged, lysosomes in the host produce ROS as a host defense response. Pathogens equipped with an ROS detoxification pathway can effectively evade this defense. The human pathogen *Mycobacterium tuberculosis* produces catalase encoded by a dual-regulated *katG* gene, which is an important virulence determinant (Master et al., 2001; Ng et al., 2004). The 54-kDa KatA protein of *Enterococcus faecalis* contains a protoheme IX group (Frankenberg et al., 2002). The acquisition of heme is required for the production of active catalase. Among plant pathogens, *Agrobacterium tumefaciens* harbors the *katG* gene that encodes catalase to hydrolyze toxic peroxides (Xu and Pan, 2000; Xu et al., 2001) and is induced by low phosphate concentration (Yuan et al., 2005). Both *Sinorhizobium meliloti* and *Bradyrhizobium japonicum* also contain *katG*, whose product detoxifies hydrogen peroxide (Panek and O'Brian, 2004; Jamet et al., 2005). A catalase-peroxidase encoded by *katG* in *Xanthomonas campestris* pv. *campestris* appears to be required for virulence by conferring resistance to the detrimental effects of the hydrogen peroxide defense system (Jittawuttipoka et al., 2009). ROS is also detoxified by an alkyl-hydroperoxide reductase encoded by the *ahpC* gene in *Xanthomonas campestris* pv. *phaseoli* (Sukchawalit et al., 2001). The expression of both *katG* and *ahpC* are regulated by the hydrogen peroxide-induced transcription factor OxyR (Zheng et al., 2001). Hydrogen peroxide activates OxyR by the formation of an intramolecular disulfide bond. Oxidized OxyR in turn activates the transcription of *katG* and *ahpC* among other reductases. Interestingly, a lipopolysaccharide with a glucose, galacturonic acid, and 3-deoxy-D-manno-oct-2-ulosonic acids core produced by *Xanthomonas campestris* pv. *campestris* induces an oxidative burst in tobacco cell cultures (Braun et al., 2005). It is curious why this signal is maintained for recognition by the host plant. Other aerobic plant-pathogenic bacteria are likely to degrade $H_2O_2^-$ to water and molecular oxygen by the sequential action of superoxide dismutase and catalase, as well as of alkyl-hydroperoxide reductases. The latter enzyme expression is regulated by OhrR in *Xanthomonas campestris* pv. *phaseoli* (Panmanee et al., 2002). In addition, an alkyl-hydroperoxide reductase encoded by *ahpC* inactivates linoleic hydroperoxides in *Xanthomonas campestris* pv. *phaseoli* (Klomsiri et al., 2005). Interestingly, a *Xanthomonas* species encodes RoxA, a rubber oxygenase that cleaves poly-(*cis*-1,4-isoprene), the natural latex from *Hevea brasiliensis* (Willd. ex A. Juss.) Müll. Arg. (Braaz et al., 2005). *Agrobacterium tumefaciens* possesses three iron-containing superoxide dismutases called SodBI, SodBII, and SodBIII, all of which play a role in virulence (Saenkham et al., 2007).

Detoxification of Phytoalexins

Phytoalexins are antimicrobial compounds generally considered as secondary metabolites that are produced by the plant in response to pathogen invasion (Dixon, 1986; Dixon, 2001). Phytoalexins accumulate rapidly in reaction to avirulent pathogen races but not to virulent types (Darvill and Albersheim, 1984; Dixon, 1986). Some of these compounds restrict or inhibit the growth of bacteria by hydrogen bonding interactions with vital proteins, such as enzymes (Field and Lettinga, 1992). Virulent races of pathogens induce plant defense responses at later stages of infection. They are able to detoxify phytoalexins by a variety of mechanisms that include modification of these compounds followed by their metabolism as well as efficient efflux systems that prevent intracellular accumulation of toxic compounds. Efflux pumps have broad substrate specificity and expel a wide range of noxious compounds from the cell directly into the extracellular environment, bypassing the periplasm (Zgurskaya and Nikaido, 1999). An efflux pump that is induced by isoflavonoids was found in *Agrobacterium tumefaciens* (Palumbo et al., 1998). The pump greatly reduces the cellular accumulation of isoflavonoids that is produced by alfalfa roots as exudates. The efflux system is induced by various isoflavonoids and a chalcone and is encoded by the *ifeABR* locus. Mutations in *ifeABR* reduced the ability of *Agrobacterium tumefaciens* to colonize the root system. An efflux pump encoded by a putative *acrB* operon is present in *Erwinia amylovora* (Burse et al., 2004a). Disruption of the efflux pump by mutation in *acrB* dramatically reduces the virulence of *Erwinia amylovora*. The apple phytoalexins phloretin, quercetin, catechin, and naringenin reduced the growth of the *acrB* mutant compared with its wild-type strain (Burse et al., 2004b). *Rhizobium etli* possesses a flavanoid-induced *rmrAB* operon that encodes a multidrug resistance pump (Gonzalez-Pasayo and Martinez-Romero, 2000). Mutations in *rmrAB* diminished the root-colonizing ability of *Rhizobium etli* and increased sensitivity to naringenin and coumarate. *Dickeya chrysanthemi* uses an efflux system that is activated by a combination of salicylic acid, *t*-cinnamic acid, and benzoic acid (Ravirala et al., 2007).

Plants can counteract the bacterial efflux system by producing inhibitors of the multidrug resistance pumps (Tegos et al., 2002). The phytoalexin camalexin (3-thiazol-2'-yl-indole) (Zhao and Last, 1996; Glawischnig et al., 2004) was found to accumulate in *Arabidopsis thaliana* when inoculated with *Pseudomonas syringae* pv. *maculicola* (Glazebrook and Ausubel, 1994). *Arabidopsis* mutants defective in camalexin biosynthesis enhanced the sensitivity to *Pseudomonas syringae* pv. *maculicola*. Interestingly, camalexin is not required for resistance to avirulent *Pseudomonas syringae* strains in *Arabidopsis thaliana* (Glazebrook and Ausubel, 1994).

Small Interfering RNAs Suppression

Bacteria that incorporate parts of their genome into plant cells are equipped with mechanisms to suppress the small interfering RNAs (siRNAs) produced by the plant's defensive action. The current established example is *Agrobacterium tumefaciens*, which was shown to suppress siRNAs whose production is induced by the incorporation of multiple copies of the T-DNA into the plant genome (Dunoyer et al., 2006). The tumor-inducing pathogen triggers the formation of siRNAs 3 days after inoculation, but these siRNAs are no longer detected in mature crown gall tumors. The formation of microRNAs (miRNAs) (see Chapter 12), which are required for tumor development presumably by a degradation pathway of target mRNAs, induces formation in the tumor of siRNAs that are subsequently degraded by uncharacterized *Agrobacterium*-mediated factor or factors. The factor or factors might be a novel ribonuclease.

Interestingly, RNA viruses suppress crown gall tumor formation, the mechanism of which could be a circulating siRNA induced by the virus (Saedi et al., 1979). Cowpea (*Vigna unguiculata* (L.) Walp.) seedlings failed to develop tumors after being inoculated with *Agrobacterium tumefaciens*, if 1 day before bacterial inoculation the primary leaves of the cowpea were inoculated with *Cowpea mosaic virus*, *Tobacco mosaic virus*, or *Southern bean mosaic virus*.

Nonproteinaceous Virulence Factors

Since research efforts have given much emphasis to virulence and avirulence genes and the contribution of their protein/peptide products in pathogenicity, there is a paucity of information on small molecules manufactured and used by pathogens for pathogenesis. Clearly there are bacterial chemicals produced that are injurious or lethal to the host. For example, the versatile pathogen *Pseudomonas aeruginosa*, which exhibits virulence toward *Arabidopsis thaliana*, was shown to use small molecule poisons, such as cyanide, phenazines, and rhamnolipid biosurfactant, to facilitate pathogenesis (Britigan et al., 1999; Olvera et al., 1999; Blumer and Haas, 2000; Gallagher and Manoil, 2001). The *hcnABC* operon of *Pseudomonas aeruginosa* encodes hydrogen cyanide (HCN) synthase whose promoter is controlled by an anaerobic regulator and by two quorum-sensing regulators (Pessi and Haas, 2000). It is therefore apparent that studies are needed to identify bacterial chemicals that are used in plant pathogenesis. A number of bacterial plant pathogens produce peptide- or protein-containing toxins, such as tabtoxin, phaseolotoxin, coronatine, syringomycin, tagetitoxin, and rhizobitoxine, that were discussed in other chapters of this book and reviewed elsewhere (Mitchell, 1981; Aguilera et al., 2007). The mechanisms of action of these toxins have been studied. For example, tagetitoxin was found specifically to inhibit the action of RNA polymerase III in eukaryotic 5S and tRNA gene transcription (Steinberg et al., 1990) and of chloroplast RNA polymerase (Mathews and Durbin, 1990). Likewise, coronatine induces effects on photosynthetic machinery and on ROS in modulating necrotic cell death during bacterial speck disease of tomato (Ishiga et al., 2009). Currently, there is a paucity of data on bacterial chemicals that play a role in plant pathogenesis. Such information should become available as additional functional genomic studies help identify operons that encode bacterial chemicals that contribute to disease development in host plants.

REFERENCES

Abramovitch, R. B., and Martin, G. B. 2005. AvrPtoB: A bacterial type III effector that both elicits and suppresses programmed cell death associated with plant immunity. FEMS Microbiol. Lett. 245:1-8.

Abramovitch, R. B., Kim, Y.-J., Chen, S., Dickman, M. B., and Martin, G. B. 2003. *Pseudomonas* type III effector AvrPtoB induces plant disease susceptibility by inhibition of host programmed cell death. EMBO J. 22:60-69.

Abramovitch, R. B., Anderson, J. C., and Martin, G. B. 2006. Bacterial elicitation and evasion of plant innate immunity. Nat. Rev. Mol. Cell. Biol. 7:601-611.

Aguilar, C., Friscina, A., Devescovi, G., Kojic, M., and Venturi, V. 2003. Identification of quorum-sensing-regulated genes of *Burkholderia cepacia*. J. Bacteriol. 185:6456-6462.

Aguilar, J., Zupan, J., Cameron, T. A., and Zambryski, P. C. 2010. *Agrobacterium* type IV secretion system and its substrates form helical arrays around the circumference of *virulence*-induced cells. Proc. Natl. Acad. Sci. U.S.A. 107:3758-3763.

Aguilera, S., López-López, K., Nieto, Y., Garcidueñas-Piña, R., Hernández-Guzman, G., Hernández-Flores, J. L., Murillo, J., and Alvarez-Morales, A. 2007. Functional characterization of the gene cluster from *Pseudomonas syringae* pv. phaseolicola NPS3121 involved in synthesis of phaseolotoxin. J. Bacteriol. 189:2834-2843.

Aizawa, S. I. 2001. Bacterial flagella and type III secretion systems. FEMS Microbiol. Lett. 202:157-164.

Alegria, M. C., Souza, D. P., Andrade, M. O., Docena, C., Khater, L., Ramos, C. H. I., da Silva, A. C. R., and Farah, C. S. 2005. Identification of new protein-protein interactions involving the products of the chromosome- and plasmid-encoded type IV secretion loci of the phytopathogen *Xanthomonas axonopodis* pv. citri. J. Bacteriol. 187:2315-2325.

Alfano, J. R., and Collmer, A. 2004. Type III secretion system effector proteins; Double agents in bacterial disease and plant defense. Annu. Rev. Phytopathol. 42:385-414.

Alfano, J. R., Charkowski, A. O., Deng, W.-L., Badel, J. L., Petnicki-Ocwieja, T., van Dijk, K., and Collmer, A. 2000. The *Pseudomonas syringae* Hrp pathogenicity island has a tripartite mosaic structure composed of a cluster of type III secretion genes bound by exchangeable effector and conserved effector loci that contribute to parasitic fitness and pathogenicity in plants. Proc. Natl. Acad. Sci. U.S.A. 97:4856-4861.

Andersen, C., Hughes, C., and Koronakis, V. 2000. Chunnel vision: Export and efflux through bacterial channel-tunnels. EMBO Rep. 1:313-318.

Andersen, C., Koronakis, E., Bokma, E., Eswaran, J., Humphreys, D., Hughes, C., and Koronakis, V. 2002. Transition to the open state of the TolC periplasmic tunnel entrance. Proc. Natl. Acad. Sci. U.S.A. 99:11103-11108.

Anderson, D. M., Fouts, D. E., Collmer, A., and Schneewind, O. 1999. Reciprocal secretion of proteins by the bacterial type III machines of plant and animal pathogens suggests universal recognition of mRNA targeting signals. Proc. Natl. Acad. Sci. U.S.A. 96:12839-12843.

Anderson, J. C., Pascuzzi, P. E., Xiao, F., Sessa, G., and Martin, G. B. 2006. Host-mediated phosphorylation of type III effector AvrPto promotes *Pseudomonas* virulence and avirulence in tomato. Plant Cell 18:502-514.

Anderson, K., Carballeira, N., Magnusson, K. E., Persson, C., Stendahl, O., Wolf-Watz, H., and Fallman, M. 1996. YopH of *Yersinia pseudotuberculosis* interrups early phosphotyrosine signaling associated with phagocytosis. Mol. Microbiol. 20:1057-1069.

Ankri, S., Stolarsky, T., and Mirelman, D. 1998. Antisense inhibition of expression of cysteine proteinases does not affect *Entamoeba histolytica* cytopathic or haemolytic activity but inhibits phagocytosis. Mol. Microbiol. 28:777-785.

Axtell, M. J., Chisholm, S. T., Dahlbeck, D., and Staskawicz, B. J. 2003. Genetic and molecular evidence that the *Pseudomonas syringae* type III effector protein AvrPrt2 is a cysteine protease. Mol. Microbiol. 49:1537-1546.

Babic, A., Lindner, A. B., Vulic, M., Stewart, E. J., and Radman, M. 2008. Direct visualization of horizontal gene transfer. Science 319:1533-1536.

Backert, S., and Meyer, T. F. 2006. Type IV secretion systems and their effectors in bacterial pathogenesis. Curr. Opin. Microbiol. 9:207-217.

Backert, S., Kwok, T., and König, W. 2005. Conjugative plasmid DNA transfer in *Helicobacter pylori* mediated by chromosomally encoded relaxase and TraG-like proteins. Microbiology 151:3493-3503.

Badel, J. L., Nomura, K., Bandyopadhyay, S., Shimizu, R., Collmer, A., and He, S. Y. 2003. *Pseudomonas syringae* pv. *tomato* DC3000 HopPtoM (CEL ORF3) is important for lesion formation but not growth in tomato and is secreted and translocated by the Hrp type III secretion system in a chaperone-dependent manner. Mol. Microbiol. 49:1239-1251.

Ball, G., Chapon-Herve, V., Bleves, S., Michel, G., and Bally, M. 1999. Assembly of XcpR in the cytoplasmic membrane is required for extracellular protein secretion in *Pseudomonas aeruginosa*. J. Bacteriol. 181:383-388.

Ball, G., Durand, E., Lazdunski, A., and Filloux, A. 2002. A novel type II secretion system in *Pseudomonas aeruginosa*. Mol. Microbiol. 43:475-485.

Bardwell, J. C., McGovern, K., and Beckwith, J. 1991. Identification of a protein required for disulfide bond formation in vivo. Cell 67:581-589.

Basim, H., Stall, R. E., Minsavage, G. V., and Jones, J. B. 1999. Chromosomal gene transfer by conjugation in the plant pathogen *Xanthomonas axonopodis* pv. *vesicatoria*. Phytopathology 89:1044-1049.

Bassler, B. L., and Losick, R. 2006. Bacterially speaking. Cell 125:237-246.

Beal, R., and Betts, W. B. 2000. Role of rhamnolipid biosurfactants in the uptake and mineralization of hexadecane in *Pseudomonas aeruginosa*. J. Appl. Microbiol. 89:158-168.

Beatson, S. A., Whitchurch, C. B., Semmler, A. B. T., and Mattick, J. S. 2002. Quorum sensing is not required for twitching motility in *Pseudomonas aeruginosa*. J. Bacteriol. 184:3598-3604.

Belkhadir, Y., Subramaniam, R., and Dangl, J. L. 2004. Plant disease resistance protein signaling: NBS LRR proteins and their partners. Curr. Opin. Plant Biol. 7:391-399.

Benz, I., and Schmidt, M. A. 1989. Cloning and expression of an adhesin (AIDA-I) involved in diffuse adherence of enteropathogenic *Escherichia coli*. Infect. Immun. 57:1506-1511.

Berks, B. C., Sargent, F., and Palmer, T. 2000. The Tat protein export pathway. Mol. Microbiol. 35:260-274.

Berks, B. C., Palmer, T., and Sargent, F. 2005. Protein targeting by the bacterial twin-arginine translocation (Tat) pathway. Curr. Opin. Microbiol. 8:174-181.

Binet, R., and Wandersman, C. 1995. Protein secretion by hybrid bacterial ABC-transporters: Specific functions of the membrane ATPase and the membrane fusion protein. EMBO J. 14:2298-2306.

Binet, R., Letoffe, S., Ghigo, J. M., Delepelaire, P., and Wandersman, C. 1997. Protein secretion by gram-negative bacterial ABC transporters—A review. Gene 192:7-11.

Black, D. S., and Bliska, J. B. 1997. Identification of p130Cas as a substrate of *Yersinia* YopH (Yop51), a bacterial protein tyrosine phosphatase that translocates into mammalian cells and targets focal adhesions. EMBO J. 16:2730-2744.

Bladergroen, M. R., Badelt, K., and Spaink, H. P. 2003. Infection-blocking genes of a symbiotic *Rhizobium leguminosarum* strain that are involved in temperature-dependent protein secretion. Mol. Plant-Microbe Interact. 16:53-64.

Bleves, S., Soscia, C., Nogueira-Orlandi, P., Lazdunski, A., and Filloux, A. 2005. Quorum sensing negatively controls type III secretion regulon expression in *Pseudomonas aeruginosa* PAO1. J. Bacteriol. 187:3898-3902.

Bliska, J. B., Guan, K., Dixon, J. E., and Falkow, S. 1991. Tyrosine phosphatase hydrolysis of host proteins by an essential *Yersinia* virulence determinant. Proc. Natl. Acad. Sci. U.S.A. 88:1187-1191.

Blumer, C., and Haas, D. 2000. Mechanism, regulation, and ecological role of bacterial cyanide biosynthesis. Arch. Microbiol. 173:170-177.

Boch, J., Joardar, V., Gao, L., Robertson, T. L., Lim, M., and Kunkel, B. N. 2002. Identification of *Pseudomonas syringae* pv. *tomato* genes induced during infection of *Arabidopsis thaliana*. Mol. Microbiol. 44:73-88.

Boles, B. R., Thoendel, M., and Singh, P. K. 2005. Rhamnolipids mediate detachment of *Pseudomonas aeruginosa* from biofilms. Mol. Microbiol. 57:1210-1223.

Bortoli-German, I., Brun, E., Py, B., Chippaux, M., and Barras, F. 1994. Periplasmic disulphide bond formation is essential for cellulase secretion by the plant pathogen *Erwinia chrysanthemi*. Mol. Microbiol. 11:545-553.

Botteaux, A., Sani, M., Kayath, C. A., Boekema, E. J., and Allaoui, A. 2008. Spa32 interaction with the inner-membrane Spa40 component of the type III secretion of *Shigella flexneri* is required for the control of the needle length by a molecular tape measure mechanism. Mol. Microbiol. 70:1515-1528.

Braaz, R., Armbruster, W., and Jendrossek, D. 2005. Heme-dependent rubber oxygenase RoxA of *Xanthomonas* sp. cleaves the carbon backbone of poly(*cis*-1,4-isoprene) by a dioxygenase mechanism. Appl. Environ. Microbiol. 71:2473-2478.

Braun, S. G., Meyer, A., Holst, O., Pühler, A., and Niehaus, K. 2005. Characterization of the *Xanthomonas campestris* pv. *campestris* lipopolysaccharide substructures essential for elicitation of an oxidative burst in tobacco cells. Mol. Plant-Microbe Interact. 18:674-681.

Breedveld, M. W., and Miller, K. J. 1994. Cyclic β-glucans of members of the family *Rhizobiaceae*. Microbiol. Rev. 58:145-161.

Bretz, J., Losada, L., Lisboa, K., and Hutcheson, S. W. 2002. Lon protease functions as a negative regulator of type III protein secretion in *Pseudomonas syringae*. Mol. Microbiol. 45:397-409.

Bretz, J. R., Mock, N. M., Charity, J. C., Zeyad, S., Baker, C. J., and Hutcheson, S. W. 2003. A translocated protein tyrosine phosphatase of *Pseudomonas syringae* pv. *tomato* DC3000 modulates plant defence response to infection. Mol. Microbiol. 49:389-400.

Britigan, B. E., Railsback, M. A., and Cox, C. D. 1999. The *Pseudomonas aeruginosa* secretory product pyocyanin inactivates α1 protease inhibitor: Implications for the pathogenesis of cystic fibrosis lung disease. Infect. Immun. 67:1207-1212.

Bronstein, P. A., Marrichi, M., Cartinhour, S., Schneider, D. J., and DeLisa, M. P. 2005. Identification of a twin-arginine translocation system in *Pseudomonas syringae* pv. *tomato* DC3000 and its contribution to pathogenicity and fitness. J. Bacteriol. 187:8450-8461.

Brown, I., Mansfield, J., and Bonas, U. 1995. *hrp* Genes in *Xanthomonas campestris* pv. *vesicatoria* determine ability to suppress papilla deposition in pepper mesophyll cells. Mol. Plant-Microbe Interact. 8:825-836.

Brown, I. R., Mansfield, J. W., Taira, S., Roine, E., and Romantschuk, M. 2001. Immunocytochemical localization of HrpA and HrpZ supports a role for the Hrp pilus in the transfer of effector proteins from *Pseudomonas syringae* pv. *tomato* across the host plant cell wall. Mol. Plant-Microbe Interact. 14:394-404.

Bruchhaus, I., Loftus, B. J., Hall, N., and Tannich, E. 2003. The intestinal protozoan parasite *Entamoeba histolytica* contains 20 cysteine protease genes, of which only a small subset is expressed during in vitro cultivation. Eukaryotic Cell 2:501-509.

Brunings, A. M., and Gabriel, D. W. 2003. *Xanthomonas citri*: Breaking the surface. Mol. Plant Pathol. 4:141-157.

Buell, C. R., Joardar, V., Lindeberg, M., Selengut, J., Paulsen, L. T., Gwinn, M. L., Dodson, R. J., Deboy, R. T., Durkin, A. S., Kolonay, J. F., Madupu, R., Daugherty, S., Brinkac, L., Beanan, M. J., Haft, D. H., Nelson, W. C., Davidsen, T., Zafar, N., Zhou, L., Liu, J., Yuan, Q., Khouri, H., Fedorova, N., Tran, B., Russell, D., Berry, K., Utterback, T., Van Aken, S. E., Feldblyum, T. V., D'Ascenzo, M., Deng, W. L., Ramos, A. R., Alfano, J. R., Cartinhour, S., Chatterjee, A. K., Delaney, T. P., Lazarowitz, S. G., Martin, G. B., Schneider, D. J., Tang, X., Bender, C. L., White, O., Fraser, C. M., and Collmer, A. 2003. The complete genome sequence of the *Arabidopsis* and tomato pathogen *Pseudomonas syringae* pv. *tomato* DC3000. Proc. Natl. Acad. Sci. U.S.A. 100:10181-10186.

Burse, A., Weingart, H., and Ullrich, M. S. 2004a. The phytoalexin-inducible multidrug efflux pump AcrAB contributes to virulence in the fire blight pathogen, *Erwinia amylovora*. Mol. Plant-Microbe Interact. 17:43-54.

Burse, A., Weingart, H., and Ullrich, M. S. 2004b. NorM, an *Erwinia amylovora* multidrug efflux pump involved in in vitro competition with other epiphytic bacteria. Appl. Environ. Microbiol. 70:693-703.

Büttner, D., Lorenz, C., Weber, E., and Bonas, U. 2006. Targeting two effector protein classes to the type III secretion system by a HpaC- and HpaB-dependent protein complex from *Xanthomonas campestris* pv. *vesicatoria*. Mol. Microbiol. 59:513-527.

Caldelari, I., Mann, S., Crooks, C., and Palmer, T. 2006. The Tat pathway of the plant pathogen *Pseudomonas syringae* is required for optimal virulence. Mol. Plant-Microbe Interact. 19:200-212.

Cascales, E., and Christie, P. J. 2003. The versatile bacterial type IV secretion systems. Nat. Rev. Microbiol. 1:137-149.

Casper-Lindley, C., Dahlbeck, D., Clark, E. T., and Staskawicz, B. J. 2002. Direct biochemical evidence for type III secretion-dependent translocation of the AvrBs2 effector protein into plant cells. Proc. Natl. Acad. Sci. U.S.A. 99:8336-8341.

Cazzulo, J. J. 2002. Proteinases of *Trypanosoma cruji*: Potential targets for chemotherapy of Chagas disease. Curr. Topics Med. Chem. 2:1261-1271.

Chai, Y., and Winans, S. C. 2005. Amino-terminal protein fusions to the TraR quorum-sensing transcription factor enhance protein stability and autoinducer-independent activity. J. Bacteriol. 187:1219-1226.

Chang, J. H., Goel, A. K., Grant, S. R., and Dangl, J. L. 2004. Wake of the flood: Ascribing functions to the wave of type III effector proteins of phytopathogenic bacteria. Curr. Opin. Microbiol. 7:11-18.

Chatterjee, A., Cui, Y., and Chatterjee, A. K. 2002. Regulation of *Erwinia carotovora* $hrpL_{Ecc}$ ($sigma\text{-}L_{Ecc}$), which encodes an extracytoplasmic function subfamily of sigma factor required for expression of the Hrp regulon. Mol. Plant-Microbe Interact. 15:971-980.

Chinchilla, D., Zipfel, C., Robatzek, S., Kemmerling, B., Nürnberger, T., Jones, J. D. G., Felix, G., and Boller, T. 2007a. A flagellin-induced complex of the receptor FLS2 and BAK1 initiates plant defence. Nature 448:497-500.

Chinchilla, D., Boller, T., and Robatzek, S. 2007b. Flagellin signaling in plant immunity. Adv. Exp. Med. Biol. 598:358-371.

Chisholm, S. T., Coaker, G., Day, B., and Staskawicz, B. J. 2006. Host-microbe interactions: Shaping the evolution of the plant immune response. Cell 124:803-814.

Christie, P. J., Atmakuri, K., Jakubowski, S., Krihnamoorthy, V., Cascales, E., Bayliss, R., and Waksman, G. 2007. Requirements for receptor binding and DNA translocation through the *Agrobacterium tumefaciens* VirB/D4 type IV secretion system. (Abstr.) Plasmid 57:211.

Cline, K., Henry, R., Li, C., and Yuan, J. 1993. Multiple pathways for protein transport into or across the thylakoid membrane. EMBO J. 12:4105-4114.

Coaker, G., Falick, A., and Staskawicz, B. 2005. Activation of a phytopathogenic bacterial effector protein by a eukaryotic cyclophilin. Science 308:548-550.

Collmer, A., Badel, J. L., Charkowski, A. O., Deng, W.-L., Fouts, D. E., Ramos, A. R., Rehm, A. H., Anderson, D. M., Schneewind, O., van Dijk, K., and Alfano, J. R. 2000. *Pseudomonas syringae* Hrp type III secretion system and effector proteins. Proc. Natl. Acad. Sci. U.S.A. 97:8770-8777.

Collmer, A., Lindeberg, M., Petnicki-Ocwieja, T., Schneider, D. J., and Alfano, J. R. 2002. Genomic mining type III secretion system effectors in *Pseudomonas syringae* yields new picks for all TTSS prospectors. Trends Microbiol. 10:462-469.

Corbett, M., Virtue, S., Bell, K., Birch, P., Burr, T., Hyman, L., Lilley, K., Poock, S., Toth, I., and Salmond, G. 2005. Identification of a new quorum-sensing-controlled virulence factor in *Erwinia carotovora* subsp. *atroseptica* secreted via the type II targeting pathway. Mol. Plant-Microbe Interact. 18:334-342.

Cornelis, G. R. 2002. The *Yersinia* Ysc-Yop 'type III' weaponry. Nat. Rev. Mol. Cell. Biol. 3:742-752.

Cornelis, G. R. 2006. The type III secretion injectisome. Nat. Rev. Microbiol. 4:811-825.

Cornelis, G. R., and Van Gijsegem, F. 2000. Assembly and function of type III secretory systems. Annu. Rev. Microbiol. 54:735-774.

Covacci, A., Telford, J. L., del Giudice, G., Parsonnet, J., and Rappouli, R. 1999. *Helicobacter pylori* virulence and genetic geography. Science 284:1328-1333.

Cover, T. L. 1996. The vacuolating cytotoxin of *Helicobacter pylori*. Mol. Microbiol. 20:241-246.

Crepin, V. F., Shaw, R., Abe, C. M., Knutton, S., and Frankel, G. 2005. Polarity of enteropathogenic *Escherichia coli* EspA filament assembly and protein secretion. J. Bacteriol. 187:2881-2889.

Cui, Y., Chatterjee, A., Hasegawa, H., Dixit, V., Leigh, N., and Chatterjee, A. K. 2005. ExpR, a LuxR homolog of *Erwinia carotovora* subsp. *carotovora*, activates transcription of *rsmA*, which specifies a global regulatory RNA-binding protein. J. Bacteriol. 187:4792-4803.

Cunnac, S., Boucher, C., and Genin, S. 2004a. Characterization of the *cis*-acting regulatory element controlling HrpB-mediated activation of the type III secretion system and effector genes in *Ralstonia solanacearum*. J. Bacteriol. 186:2309-2318.

Cunnac, S., Occhialini, A., Barberis, P., Boucher, C., and Genin, S. 2004b. Inventory and functional analysis of the large Hrp regulon in *Ralstonia solanacearum*: Identification of novel effector proteins translocated to plant host cells through the type III secretion system. Mol. Microbiol. 53:115-128.

Cunnac, S., Lindeberg, M., and Collmer, A. 2009. *Pseudomonas syringae* type III secretion system effectors: Repertoires in search of functions. Curr. Opin. Microbiol. 12:53-60.

Daefler, S. 1999. Type III secretion by *Salmonella typhimurium* does not require contact with a eukaryotic host. Mol. Microbiol. 31:45-51.

Dahler, G. S., Barras, F., and Keen, N. T. 1990. Cloning of genes encoding extracellular metalloproteases from *Erwinia chrysanthemi* EC16. J. Bacteriol. 172:5803-5815.

Daniell, S. J., Takashashi, N., Wilson, R., Friedberg, D., Rosenshine, I., Booy, F. P., Shaw, R. K., Knutton, S., Frankel, G., and Aizawa, S.-I. 2001. The filamentous type III secretion translocon of enteropathogenic *Escherichia coli*. Cell. Microbiol. 3:864-871.

Daniell, S. J., Kocsis, E., Morris, E., Knutton, S., Booy, F. P., and Frankel, G. 2003. 3D structure of EspA filaments from enteropathogenic *Escherichia coli*. Mol. Microbiol. 49:301-308.

Darvill, A. G., and Albersheim, P. 1984. Phytoalexins and their elicitors—A defense against microbial infection in plants. Annu. Rev. Plant Physiol. 35:243-275.

Das, S., and Chaudhuri, K. 2003. Identification of a unique IAHP (IcmF associated homologous proteins) cluster in *Vibrio cholerae* and other proteobacteria through in silico analysis. In Silico Biol. 3:287-300.

da Silva, A. C., Ferro, J. A., Reinach, F. C., Farah, C. S., Furlan, L. R., Quaggio, R. B., Monteiro-Vitorello, C. B., Van Sluys, M. A., Aleida, N. F., Alves, L. M., Do Ammaral, A. M., Bertolini, M. C., Camargo, L. E., Camarotte, G., Cannavan, F., Cardozo, J., Chambergo, F., Ciapina, L. P., Cicarelli, R. M., Coutinho, L. L., Cursino-Santos, J. R., El-Dorry, H., Faraia, J. B., Ferreira, A. J., Ferreira, R. C., Ferro, M. I., Formighleri, E. F., Franco, M. C., Greggio, C. C., Gruber, A., Katsuyama, A. M., Kishi, L. T., Lette, R. P., Lemos, E. G., Lemos, M. V., Locali, E. C., Machado, M. A., Madeira, A. M., Martinez-Rossi, N. M., Matins, E. C., Meidanis, J., Menck, C. F., Miyaki, C. V., Moon, D. H., Moreira, L. M., Novo, M. T., Okura, V. K., Oliveira, M. C., Oleveira, V. R., Pereira, H. A., Rossi, A., Sena, J. A., Silva, C., de Souza, R. F., Spinola, L. A., Takita, M. A., Tamura, R. E., Teixeira, E. C., Tezza, R. I., Trindade dos Santos, M., Truffi, D., Tsai, S. M., White, F. F., Setubal, J. C., and Kitajima, J. P. 2002. Comparison of the genomes of two *Xanthomonas* pathogens with differing host specificities. Nature 417:459-463.

Davey, M. E., Calazza, N. C., and O'Toole, G. A. 2003. Rhamnolipid surfactant production affects biofilm architecture in *Pseudomnas aeruginosa* PA01. J. Bacteriol. 185:1027-1036.

Davis, R. E., Dally, E. L., Jomantiene, R., Zhao, Y., Roe, B., Lin, S., and Shao, J. 2005. Cryptic plasmid pSKU146 from the wall-less plant pathogen *Spiroplasma kunkelii* encodes an adhesin and components of a type IV translocation-related conjugation system. Plasmid 3:179-190.

Deakin, W. J., Marie, C., Saad, M. M., Krishnan, H. B., and Broughton, W. J. 2005. NopA is associated with cell surface appendages produced by the type III secretion system of *Rhizobium* sp. strain NGR234. Mol. Plant-Microbe Interact. 18:499-507.

de Jong, M. F., Sun, Y.-H., Den Hartigh, A. B., van Dijl, J. M., and Tsolis, R. M. 2008. Identification of VceA and VceC, two members of the VjbR regulon that are translocated into macrophages by the *Brucella* type IV secretion system. Mol. Microbiol. 70:1378-1396.

Delepelaire, P., and Wandersman, C. 1991. Characterization, localization and transmembrane organization of the three proteins PrtD, PrtE and Prtf necessary for protease secretion by the gram-negative bacterium *Erwinia chrysanthemi*. Mol. Microbiol. 5:2427-2434.

Deng, W. L., Preston, G., Collmer, A., Change, C.-J., and Huang, H.-C. 1998. Characterization of the *hrpC* and *hrpRS* operons of *Pseudomonas syringae* pathovars syringae, tomato and glycinea and analysis of the ability of *hrpF*, *hrpG*, *hrcC*, *hrpT*, and *hrpV* mutants to elicit the hypersensitive response and disease in plants. J. Bacteriol. 180:4523-4531.

Desvaux, M., Hébraud, M., Henderson, I. R., and Pallen, M. J. 2006. Type III secretion: What's in a name? Trends Microbiol. 14:157-160.

Deziel, E., Lepine, F., Milot, S., and Villemur, R. 2003. *rhlA* is required for the production of a novel biosurfactant promoting swarming motility in *Pseudomonas aeruginosa*: 3-(3-Hydroxyalkanoyloxy)alkanoic acids (HAAs), the precursors of rhamnolipids. Microbiology 149:2005-2013.

Ding, Z., and Christie, P. J. 2003. *Agrobacterium tumefaciens* twin-arginine-dependent translocation is important for virulence, flagellation, and chemotaxis but not type IV secretion. J. Bacteriol. 185:760-771.

Dixon, R. A. 1986. The phytoalexin response: Elicitation, signaling and the control of host gene expression. Biol. Rev. 61:239-291.

Dixon, R. A. 2001. Natural products and plant disease resistance. Nature 411:843-847.

Djamei, A., Pitzschke, A., Nakagami, H., Rajh, I., and Heribert, H. 2007. Trojan horse strategy in *Agrobacterium* transformation: Abusing MAPK defense signaling. Science 318:453-456.

Dong, Y. H., Zhang, X.-F., Xu, J.-L., Tan, A.-T., and Zhang, L.-H. 2005. VqsM, a novel AraC-type global regulator of quorum-sensing signaling and virulence in *Pseudomonas aeruginosa*. Mol. Microbiol. 58:552-564.

Dow, J. M., Crossman, L. C., Findlay, K., He, Y. Q., Feng, J. X., and Tang, J. L. 2003. Biofilm dispersal in *Xanthomonas campestris* is controlled by cell-cell signaling and is required for full virulence to plants. Proc. Natl. Acad. Sci. U.S.A. 100:10995-11000.

Duan, Y. P., Castañeda, A., Zhao, G., Erdos, G., and Gabriel, D. W. 1999. Expression of a single, host-specific, bacterial pathogenicity gene in plant cells elicits division, enlargement, and cell death. Mol. Plant-Microbe Interact. 12:556-560.

Dunoyer, P., Himber, C., and Voinnet, O. 2006. Induction, suppression and requirement of RNA silencing pathways in virulent *Agrobacterium* infections. Nat. Genet. 38:258-263.

Duong, F., Bonnet, E., Geli, V., Lazdunski, A., Murgier, M., and Filloux, A. 2001. The AprX protein of *Pseudomonas aeruginosa*: A new substrate for the Apr type I secretion system. Gene 262:147-153.

Durand, E., Bernadac, A., Ball, G., Lazdunski, A., Sturgis, J. N., and Filloux, A. 2003. Type II protein secretion in *Pseudomonas aeruginosa*: The pseudopilus is a multifibrillar and adhesive structure. J. Bacteriol. 185:2749-2758.

El Yacoubi, B., Brunings, A. M., Yuan, Q., Shankar, S., and Gabriel, D. W. 2007. In planta horizontal transfer of a major pathogenicity effector gene. Appl. Environ. Microbiol. 73:1612-1621.

Espinosa, A., and Alfano, J. R. 2004. Disabling surveillance: Bacterial type III secretion system effectors that suppress innate immunity. Cell. Microbiol. 6:1027-1040.

Espinosa, A., Guo, M., Tam, V. C., Fu, Z. Q., and Alfano, J. R. 2003. The *Pseudomonas syringae* type III-secreted protein HopPtoD2 possesses protein tyrosine phosphatase activity and suppresses programmed cell death in plants. Mol. Microbiol. 49:377-387.

Evdokimov, A. G., Tropea, J. E., Routzahn, K. M., Copeland, T. D., and Waugh, D. S. 2001. Structure of the N-terminal domain of *Yersinia pestis* YopH at 2.0 Å resolution. Acta Crystallogr. D. Biol. Crystallogr. 57:793-799.

Fadouloglou, V. E., Tampakaki, A. P., Glykos, N. M., Bastaki, M. N., Hadden, J. M., Phillips, S. E., Panopoulos, N. J., and Kokkinidis, M. 2004. Structure of $HrcQ_B$-C, a conserved component of the bacterial type III secretion systems. Proc. Natl. Acad. Sci. U.S.A. 101:70-75.

Fauman, E. B., and Saper, M. A. 1996. Structure and function of the protein tyrosine phosphatases. Trends Biochem. Sci. 21:413-417.

Fekkes, P., and Driessen, A. J. M. 1999. Protein targeting to the bacterial cytoplasmic membrane. Microbiol. Mol. Biol. Rev. 63:161-173.

Ferluga, S., and Venturi, V. 2009. OryR is a LuxR-family protein involved in interkingdom signaling between pathogenic *Xanthomonas oryzae* and rice. J. Bacteriol. 191:890-897.

Ferluga, S., Bigirimana, J., Hőfte, M., and Venturi, V. 2007. A LuxR homologue of *Xanthomonas oryzae* pv. *oryzae* is required for optimal rice virulence. Mol. Plant Pathol. 8:529-538.

Ferreira, A. O., Myers, C. R., Gordon, J. S., Martin, G. B., Vencato, M., Collmer, A., Wehling, M. D., Alfano, J. R., Moreno-Hagelsieb, G., Lamboy, W. F., DeClerck, G., Schneider, D. J., and Cartinhour, S. W. 2006. Whole-genome expression profiling defines the HrpL regulon of *Pseudomonas syringae* pv. *tomato* DC3000, allows de novo reconstruction of the Hrp *cis* element and identifies novel coregulated genes. Mol. Plant-Microbe Interact. 19:1167-1179.

Field, J. A., and Lettinga, G. 1992. Toxicity of tannic compounds to microorganisms. Pages 673-692 in: Plant Polyphenols. R. W. Hemmingway and P. E. Laks, eds. Plenum Press, New York.

Fields, K. A., Mead, D. J., Dooley, C. A., and Hackstadt, T. 2003. *Chlamydia trachomatis* type III secretion: Evidence for a functional apparatus during early-cycle development. Mol. Microbiol. 48:671-683.

Filloux, A., Michel, G., and Bally, M. 1998. GSP-dependent protein secretion in gram-negative bacteria: The Xcp system of *Pseudomonas aeruginosa*. FEMS Microbiol. Rev. 22:177-198.

Forsyth, C. M., Lemongello, D., LaCount, D. J., Friesen, P. D., and Fisher, A. J. 2004. Crystal structure of an invertebrate caspase. J. Biol. Chem. 279:7001-7008.

Fouts, D. E., Abramovitch, R. B., Alfano, J. R., Baldo, A. M., Buell, C. R., Cartinhour, S., Chatterjee, A. K., D'Ascenzo, M., Gwinn, M. L., Lazarowitz, S. G., Lin, N.-C., Martin, G. B., Rehm, A. H., Schneider, D. J., van Dijk, K., Tang, X., and Collmer, A. 2002. Genomewide identification of *Pseudomonas syringae* pv. *tomato* DC3000 promoters controlled by the HrpL alternative sigma factor. Proc. Natl. Acad. Sci. U.S.A. 99:2275-2280.

Francis, M. S., Schesser, K., Forsberg, Å., and Wolf-Watz, H. 2004. Type III secretion systems in animal- and plant-interacting bacteria. Pages 361-392 in: Cellular Microbiology, 2nd ed. P. Cossart, P. Boquet, S. Normark, and R. Rappuoli, eds. American Society for Microbiology Press, Washington, DC.

Frankenberg, L., Brugna, M., and Hederstedt, L. 2002. *Enterococcus faecalis* heme-dependent catalase. J. Bacteriol. 184:6351-6356.

Frederick, R. D., Thilmony, R. L., Sessa, G., and Martin, G. B. 1998. Recognition specificity for the bacterial avirulence protein AvrPto is determined by Thr-204 in the activation loop of the tomato Pto kinase. Mol. Cell. 2:241-245.

Frederick, R. D., Ahmad, M., Majerczak, D. R., Arroyo-Rodríguez, A. S., Manulis, S., and Coplin, D. L. 2001. Genetic organization of the *Pantoea stewartii* subsp. *stewartii hrp* gene cluster and sequence analysis of the *hrpA*, *hrpC*, *hrpN*, and *wtsE* operons. Mol. Plant-Microbe Interact. 14:1213-1222.

Fu, Z. Q., Guo, M., Jeong, B.-R., Tian, F., Elthon, T. E., Cerny, R. I., Stalger, D., and Alfano, J. R. 2007. A type III effector ADP-ribosylates RNA-binding proteins and quells plant immunity. Nature 447:284-288.

Fullner, K. J., Lara, J. C., and Nester, E. W. 1996. Pilus assembly by *Agrobacterium* T-DNA transfer genes. Science 273:1107-1109.

Fuqua, C., Parsek, M. R., and Greenberg, E. P. 2001. Regulation of gene expression by cell-to-cell communication: Acyl-homoserine lactone quorum sensing. Annu. Rev. Genet. 35:439-468.

Furutani, A., Tsuge, S., Ohnishi, K., Hikichi, Y., Oku, T., Tsuno, K., Inoue, Y., Ochiai, H., Kaku, H., and Kubo, Y. 2004. Evidence for HrpXo-dependent expression of type II secretory proteins in *Xanthomonas oryzae* pv. oryzae. J. Bacteriol. 186:1374-1380.

Furutani, A., Takaoka, M., Sanada, H., Noguchi, Y., Oku, T., Tsuno, K., Ochiai, H., and Tsuge, S. 2009. Identification of novel type III secretion effectors in *Xanthomonas oryzae* pv. *oryzae*. Mol. Plant-Microbe Interact. 22:96-106.

Gallagher, L. A., and Manoil, C. 2001. *Pseudomonas aeruginosa* PAO1 kills *Caenorhabditis elegans* by cyanide poisoning. J. Bacteriol. 183:6207-6214.

Gao, M., Teplitski, M., Robinson, J. B., and Bauer, W. D. 2003. Production of substances by *Medicago truncatula* that affect bacterial quorum sensing. Mol. Plant-Microbe Interact. 16:827-834.

Geier, G., and Geider, K. 1993. Characterization and influence on virulence of the levansucrase gene from the fireblight pathogen *Erwinia amylovora*. Physiol. Mol. Plant Pathol. 42:387-404.

Genin, S., Brito, B., Denny, T. P., and Boucher, C. 2005. Control of the *Ralstonia solanacearum* type III secretion system (Hrp) genes by the global virulence regulator PhcA. FEBS Lett. 579:2077-2081.

Gérard-Vincent, M., Robert, V., Ball, G., Bleves, S., Michel, F. P. F., Lazdunski, A., and Filloux, A. 2002. Identification of XcpP domains that confer functionality and specificity to the *Pseudomonas aeruginosa* type II secretion apparatus. Mol. Microbiol. 44:1651-1665.

Ghigo, J., and Wandersman, C. 1994. A carboxyl-terminal four-amino acid motif is required for secretion of the metalloprotease PtrG through the *Erwinia chrysanthemi* protease secretion pathway. J. Biol. Chem. 269:8979-8985.

Givskov, M., De Nys, R., Manefield, M., Gram, L., Maximilien, R., Eberl, L., Molin, S., Steinberg, P. D., and Kjelleberg, S. 1996. Eukaryotic interference with homoserine lactone-mediated prokaryotic signaling. J. Bacteriol. 178:6618-6622.

Glawischnig, E., Hansen, B. G., Olsen, C. E., and Halkier, B. A. 2004. Camalexin is synthesized from indole-3-actaldoxime, a key branching point between primary and secondary metabolism in *Arabidopsis*. Proc. Natl. Acad. Sci. U.S.A. 101:8245-8250.

Glazebrook, J., and Ausubel, F. M. 1994. Isolation of phytoalexin-deficient mutants of *Arabidopsis thaliana* and characterization of their interactions with bacterial pathogens. Proc. Natl. Acad. Sci. U.S.A. 91:8955-8959.

González, E. T., Brown, D. G., Swanson, J. K., and Allen, C. 2007. Using the *Ralsonia solanacearum* Tat secretome to identify bacterial wilt virulence factors. Appl. Environ. Microbiol. 73:3779-3786.

González-Pasayo, R., and Martínez-Romero, E. 2000. Multiresistance genes of *Rhizobium etli* CFN42. Mol. Plant-Microbe Interact. 13:572-577.

Goss, J. W., Sorg, J. A., Ramamurthi, K. S., Ton-That, H., and Schneewind, O. 2004. The secretion signal of YopN, a regulatory protein of the *Yersinia enterocolitica* type III secretion pathway. J. Bacteriol. 186:6320-6324.

Grant, S. R., Fisher, E. J., Chang, J. H., Mole, B. M., and Dangl, J. L. 2006. Subterfuge and manipulation: Type III effector proteins of phytopathogenic bacteria. Annu. Rev. Microbiol. 60:425-449.

Greenbaum, D. C., Baruch, A., Grainger, M., Bozdech, Z., Medzihradszky, K. F., and Engel, J. 2002. A role for the protease falcipain 1 in host cell invasion by the human malaria parasite. Science 298:2002-2006.

Greenberg, J. T., and Vinatzer, B. A. 2003. Identifying type III effectors of plant pathogens and analyzing their interaction with plant cells. Curr. Opin. Microbiol. 6:20-28.

Grimm, C., and Panopoulos, N. J. 1989. The predicted protein product of a pathogenicity locus from *Pseudomonas syringae* pv. *phaseolicola* is homologous to a highly conserved domain of several prokaryotic regulatory proteins. J. Bacteriol. 171:5031-5038.

Grimm, C., Aufsatz, W., and Panopoulos, N. J. 1995. The *hrpRS* locus of *Pseudomonas syringae* pv. *phaseolicola* constitutes a complex regulatory unit. Mol. Microbiol. 15:155-165.

Guan, K., and Dixon, J. E. 1990. Protein tyrosine phosphatase activity of an essential virulence determinant in *Yersinia*. Science 249:553-556.

Guo, M., Chancey, S. T., Tian, F., Ge, Z., Jamir, Y., and Alfano, J. R. 2005. *Pseudomonas syringae* type III chaperones ShcO1, ShcS1, and ShcS2 facilitate translocation of their cognate effectors and can substitute for each other in the secretion of HopO1-1. J. Bacteriol. 187:4257-4269.

Gürlebeck, D., Jahn, S., Gürlebeck, N., Szczesny, R., Szurek, B., Hahn, S., Hause, G., and Bonas, U. 2009. Visualization of novel virulence activities of *Xanthomonas* type III effectors AvrBs1, AvrBs3 and AvrBs4. Mol. Plant Pathol. 10:175-188.

Hamilton, H. L., Dominguez, N. M., Schwartz, K. J., Hackett, K. T., and Dillard, J. P. 2005. *Neisseria gonorrhoeae* secretes chromosomal DNA via a novel type IV secretion system. Mol. Microbiol. 55:1704-1721.

Hatzixanthis, K., Palmer, T., and Sargent, F. 2003. A subset of bacterial inner membrane proteins integrated by the twin-arginine translocase. Mol. Microbiol. 49:1377-1390.

Hauck, P., Thilmony, R., and He, S. Y. 2003. A *Pseudomonas syringae* type III effector suppresses cell wall-based extracellular defense in susceptible *Arabidopsis* plants. Proc. Natl. Acad. Sci. U.S.A. 100:8577-8582.

Håvarstein, L. S., and Morrison, D. A. 1999. Quorum-sensing and peptide pheromones in streptococcal competence for genetic transformation. Pages 9-26 in: Cell-Cell Signaling in Bacteria. G. M. Dunny and S. C. Winans, eds. American Society for Microbiology Press, Washington, DC.

He, P., Shan, L., Lin, N.-C., Martin, G. B., Kemmerling, B., Nürnberger, T., and Sheen, J. 2006. Specific bacterial suppressors of MAMP signaling upstream of MAPKKK in *Arabidopsis* innate immunity. Cell 125:563-575.

He, X., Chang, W., Pierce, D. L., Seib, L. O., Wagner, J., and Fuqua, C. 2003. Quorum sensing in *Rhizobium* sp. strain NGR234 regulates conjugal transfer (*tra*) gene expression and influences growth rate. J. Bacteriol. 185:809-822.

He, Y.-W., Xu, M., Lin, K., Ng, Y.-J. A., Wen, C.-M., Wang, L.-H., Liu, Z.-D., Zhang, H.-B., Dong, Y.-H., Dow, J. M., and Zhang, L.-H. 2006. Genome scale analysis of diffusible signal factor regulon in *Xanthomonas campestris* pv. *campestris*: Identification of novel cell-cell communication-dependent genes and functions. Mol. Microbiol. 59:610-622.

Heese, A., Hann, D. R., Gimenez-Ibanez, S., Jones, A. M. E., He, K., Li, J., Schroeder, J. I., Peck, S. C., and Rathjen, J. P. 2007. The receptor-like kinase SERK3/BAK1 is a central regulator of innate immunity in plants. Proc. Natl. Acad. Sci. U.S.A. 104:12217-12222.

Patankar, A. V., and González, J. E. 2009. An orphan LuxR homolog of *Sinorhizobium meliloti* affects stress adaptation and competition for nodulation. Appl. Environ. Microbiol. 75:946-955.

Persson, C., Carballeira, N., Wolf-Watz, H., and Falman, M. 1997. The PTPase YopH inhibits uptake of *Yersinia*, tyrosine phosphorylation of p130Cas and FAK, and the associated accumulation of these proteins in peripheral focal adhesions. EMBO J. 16:2307-2318.

Pessi, G., and Haas, D. 2000. Transcriptional control of the hydrogen cyanide biosynthetic genes *hcnABC* by the anaerobic regulator ANR and the quorum-sensing regulators LasR and RhlR in *Pseudomonas aeruginosa*. J. Bacteriol. 182:6940-6949.

Pohlner, J., Halter, R., Beyreuther, K., and Meyer, T. F. 1987. Gene structure and extracellular secretion of *Neisseria gonorrhoeae* IgA protease. Nature 325:458-462.

Potempa, J., Golonka, E., Filipek, R., and Shaw, L. N. 2005. Fighting an enemy within: Cytoplasmic inhibitors of bacterial cysteine proteases. Mol. Microbiol. 57:605-610.

Pugsley, A. P. 1993. The complete general secretory pathway in gram-negative bacteria. Microbiol. Rev. 57:50-108.

Pukatzki, S., Ma, A. T., Sturtevant, D., Krastins, B., Sarracino, D., Nelson, W. C., Heidelberg, J. F., and Mekalanos, J. J. 2006. Identification of a conserved bacterial protein secretion system in *Vibrio cholerae* using the *Dictyostelium* host model system. Proc. Natl. Acad. Sci. U.S.A. 103:1528-1533.

Quigley, N. B., and Gross, D. C. 1994. Syringomycin production among strains of *Pseudomonas syringae* pv. *syringae*: Conservation of the *syrB* and *syrD* genes and activation of phytotoxin production by plant signal molecules. Mol. Plant-Microbe Interact. 7:78-90.

Quiñones, B., Dulla, G., and Lindow, S. E. 2005. Quorum sensing regulates exopolysaccharide production, motility, and virulence in *Pseudomonas syringae*. Mol. Plant-Microbe Interact. 18:682-693.

Ramamurthi, K. S., and Schneewind, O. 2002. Type III protein secretion in *Yersinia* species. Annu. Rev. Cell Dev. Biol. 18:107-133.

Ramey, B. E., Matthysse, A. G., and Fuqua, C. 2004. The FNR-type transcriptional regulator SinR control maturation of *Agrobacterium tumefaciens* biofilms. Mol. Microbiol. 52:1495-1511.

Ravirala, R. S., Barabote, R. D., Wheeler, D. M., Reverchon, S., Tatum, O., Malouf, J., Liu, H., Pritchard, L., Hedley, P. E., Birch, P. R. J., Toth, I. K., Payton, P., and San Francisco, M. J. D. 2007. Efflux pump gene expression in *Erwinia chrysanthemi* is induced by exposure to phenolic acids. Mol. Plant-Microbe Interact. 20:313-320.

Read, R. C., Roberts, P., and Munro, N. 1992. Effect of *Pseudomonas aeruginosa* rhamnolipids on mucociliary transport and ciliary beating. J. Appl. Physiol. 72:2271-2277.

Rep, M., van der Does, H. C., Mijer, M., van Wijk, R., Houterman, P. M., Dekker, H. L., de Koster, C. G., and Cornelissen, B. J. C. 2004. A small, cysteine-rich protein secreted by *Fusarium oxysporum* during colonization of xylem vessels is required for 1-3-mediated resistance in tomato. Mol. Microbiol. 53:1373-1383.

Risøen, P. A., Brurberg, M. B., Eijsink, V. G. H., and Nes, I. F. 2000. Functional analysis of promoters involved in quorum sensing-based regulation of bacteriocin production in *Lactobacillus*. Mol. Microbiol. 37:619-628.

Robatzek, S., Chinchilla, D., and Boller, T. 2006. Ligand-induced endocytosis of the pattern recognition receptor FLS2 in *Arabidopsis*. Genes Dev. 20:537-542.

Robinson, C., and Bolhuis, A. 2001. Protein targeting by the twin-arginine translocation pathway. Nat. Rev. Mol. Cell Biol. 2:350-355.

Roden, J., Eardley, L., Hotson, A., Cao, Y., and Mudgett, M. B. 2004. Characterization of the *Xanthomonas* AvrXv4 effector, a SUMO protease translocated into plant cells. Mol. Plant-Microbe Interact. 17:633-643.

Roest, H. P., Mulders, I. H. M., Spaink, H. P., Wijffelman, C. A., and Lugtenberg, B. J. J. 1997. A *Rhizobium leguminosarum* biovar *trifolii* locus not localized on the Sym plasmid hinders effective nodulation on plants of the pea cross-inoculation group. Mol. Plant-Microbe. Interact. 10:938-941.

Rosenthal, P. J., and Nelson, R. G. 1992. Isolation and characterization of a cysteine proteinase gene of *Plasmodium falciparum*. Mol. Biochem. Parasitol. 51:143-152.

Ryan, R. P., Fouhy, Y., Lucey, J. F., and Dow, J. M. 2006. Cyclic di-GMP signaling in bacteria: Recent advances and new puzzles. J. Bacteriol. 188:8327-8334.

Saad, M. M., Staehelin, C., Broughton, W. J., and Deakin, W. J. 2008. Protein-protein interactions within type III secretion system-dependent pili of *Rhizobium* sp. strain NGR234. J. Bacteriol. 190:750-754.

Saedi, D., Bruening, G., Kado, C. I., and Dutra, J. C. 1979. Tumor induction by *Agrobacterium tumefaciens* prevented in *Vigna sinensis* seedlings systemically infected by ribonucleic acid viruses. Infect. Immun. 23:298-304.

Saenkham, P., Eiamphungporn, W., Farrand, S. K., Vattanaviboon, P., and Mongkolsuk, S. 2007. Multiple superoxide dismutases in *Agrobacterium tumefaciens*: Functional analysis, gene regulation, and influence on tumorigenesis. J. Bacteriol. 189:8807-8817.

Sandkvist, M. 2001. Biology of type II secretion. Mol. Microbiol. 40:271-283.

Sato, H., and Frank, D. W. 2004. ExoU is a potent intracellular phospholipase. Mol. Microbiol. 53:1279-1290.

Sato, H., Feix, J. B., Hillard, C. J., and Frank, D. W. 2005. Characterization of phospholipase activity of the *Pseudomonas aeruginosa* type III cytotoxin, ExoU. J. Bacteriol. 187:1192-1195.

Sauvonnet, N., and Pugsley, A. P. 1998. The requirement for DsbA in pullulanase secretion is independent of disulphide bond formation in the enzyme. Mol. Microbiol. 27:661-667.

Schechter, L. M., Roberts, K. A., Jamir, Y., Alfano, J. R., and Collmer, A. 2004. *Pseudomonas syringae* type III secretion system targeting signals and novel effectors studied with a Cya translocation reporter. J. Bacteriol. 186:543-555.

Schmidtchen, A., Frick, I.-M., Andersson, E., Tapper, H., and Björck, L. 2002. Proteinases of common pathogenic bacteria degrade and inactivate the antibacterial peptide LL-37. Mol. Microbiol. 46:157-168.

Segal, G., Feldman, M., and Zusman, T. 2005. The Icm/Dot type-IV secretion systems of *Legionella pneumophila* and *Coxiella burnetii*. FEMS Microbiol. Rev. 29:65-81.

Sekiya, K., Ohishi, M., Ogino, T., Tamano, K., Sasakawa, C., and Abe, A. 2001. Supermolecular structure of the enteropathogenic *Escherichia coli* type III secretion system and its direct interaction with the EspA-sheath-like structure. Proc. Natl. Acad. Sci. U.S.A. 98:11638-11643.

Settles, A. M., and Martienssen, R. 1998. Old and new pathways of protein export in chloroplasts and bacteria. Trends Cell Biol. 8:2494-2498.

Sexton, J. A., and Vogel, J. P. 2002. Type IVB secretion by intracellular pathogens. Traffic 3:178-185.

Shao, F., Golstein, C., Ade, J., Stoutemyer, M., Dixon, J. E., and Innes, R. W. 2003a. Cleavage of *Arabidopsis* PBS1 by a bacterial type III effector. Science 301:1230-1233.

Shao, F., Vacratsis, P. O., Bao, Z., Bowers, K. E., Fierke, C. A., and Dixon, J. E. 2003b. Biochemical characterization of the *Yersinia* YopT protease: Cleavage site and recognition elements in Rho GTPases. Proc. Natl. Acad. Sci. U.S.A. 100:904-909.

Shaw, P. D., Ping, G., Daly, S. L., Cha, C., Cronan, J. E., Jr., Rinehart, K. L., and Farrand, S. K. 1997. Detecting and characterizing *N*-acyl-homoserine lactone signal molecules by thin-layer chromatography. Proc. Natl. Acad. Sci. U.S.A. 94:6036-6041.

Shen, H., and Keen, N. T. 1993. Characterization of the promoter of avirulence gene D from *Pseudomonas syringae* pv. tomato. J. Bacteriol. 175:5916-5924.

Shen, Y., Sharma, P., da Silva, F. G., and Ronald, P. 2002. The *Xanthomonas oryzae* pv. *oryzae raxP* and *raxQ* genes encode an ATP sulphurylase and adenosine-5'-phophosulfate kinase that are required for AvrXa21 avirulence activity. Mol. Microbiol. 44:37-48.

Shevchik, V. E., Bortoli-German, I., Robert-Baudouy, J., Robinet, S., Barras, F., and Condemine, G. 1995. Differential effect of *dsbA* and *dsbC* mutations on extracellular enzyme secretion in *Erwinia chrysanthemi*. Mol. Microbiol. 16:745-753.

Shevchik, V. E., Robert-Baudouy, J., and Condemine, G. 1997. Specific interaction between OutD, an *Erwinia chrysanthemi* outer membrane protein of the general secretory pathway, and secreted proteins. EMBO J. 16:3007-3016.

Shiotani, H., Fujikawa, T., Ishihara, H., Tsuyumu, S., and Ozaki, K. 2007. A *pthA* homolog from *Xanthomonas axonopodis* pv. citri responsible for host-specific suppression of virulence. J. Bacteriol. 1890:3271-3279.

Shirasu, K., and Kado, C. I. 1993a. The *virB* operon of the *Agrobacterium tumefaciens* virulence regulon has sequence similarities to B, C and D open reading frames downstream of the pertussis toxin-operon and to the DNA transfer-operons of broad-host-range conjugative plasmids. Nucleic Acids Res. 21:353-354.

Shirasu, K., and Kado, C. I. 1993b. Membrane location of the Ti plasmid VirB proteins involved in the biosynthesis of a pilin-like conjugative structure on *Agrobacterium tumefaciens*. FEMS Microbiol. Lett. 111:287-294.

Slater, S. C., Goldman, B. S., Goodner, B., Setubal, J. C., Farrand, S. K., Nester, E. W., Burr, T. J., Banta, L., Dikerman, A. W., Paulsen, I., Otten, L., Suen, G., Welch, R., Almeida, N. F., Arnold, F., Burton, O. T., Du, Z., Ewing, A., Godsy, E., Heisel, S., Houmiel, K. L., Jhaveri, J., Lu, J., Miller, N. M., Norton, S., Chen, Q., Phoolcharoen, W., Ohlin, V., Ondrusek, D., Pride, N., Stricklin, S. L., Sun, J., Wheeler, C., Wilson, L., Zhu, H., and Wood, D. W. 2009. Genome sequences of three *Agrobacterium* biovars help elucidate the evolution of multichromosome genomes in bacteria. J. Bacteriol. 191:2501-2511.

Smith, C. L., Khandelwal, P., Keliikuli, K., Zulderweg, E. R. P., and Saper, M. A. 2001. Structure of the type III secretion and substrate-binding domain of *Yersinia* YopH phosphatase. Mol. Microbiol. 42:967-979.

Solano, C., Garcia, B., Valle, J., Berasain, C., Chigo, J.-M., Gamazo, C., and Lasa, I. 2002. Genetic analysis of *Salmonella enteritidis* biofilm formation: Critical role of cellulose. Mol. Microbiol. 43:793-808.

Songer, J. G. 1997. Bacterial phospholipases and their role in virulence. Trends Microbiol. 5:156-161.

Sroka, A., Sztukowska, M., Potempa, J., Travis, J., and Genco, C. A. 2001. Degradation of host heme proteins by lysine- and arginine-specific cysteine proteinases (gingipains) of *Porphyromonas gingivalis*. J. Bacteriol. 183:5609-5616.

Steinberg, T. H., Matthews, D. E., Durbin, R. D., and Burgess, R. R. 1990. Tagetitoxin: A new inhibitor of eukaryotic transcription by RNA polymerase III. J. Biol. Chem. 265:499-505.

Sukchawalit, R., Loprasert, S., Atichartpongkul, S., and Mongkolsuk, S. 2001. Complex regulation of the organic hydroperoxide resistance gene (*ohr*) from *Xanthomonas* involves Ohr, a novel organic peroxide-inducible negative regulator, and posttranscriptional modifications. J. Bacteriol. 183:4405-4412.

Suzuki, T., Lett, M.-C., and Sasakaw, C. 1995. Extracellular transport of VirG protein in *Shigella*. J. Biol. Chem. 270:30874-30880.

Szurek, B., Rossier, O., Hause, G., and Bonas, U. 2002. Type III-dependent translocation of the *Xanthomonas* AvrBs3 protein into the plant cell. Mol. Microbiol. 46:13-23.

Tamano, K., Katayama, E., Toyotome, T., and Sasakawa, C. 2002. *Shigella* Spa32 is an essential secretory protein for functional type III secretion machinery and uniformity of its needle length. J. Bacteriol. 184:1244-1252.

Tang, X., Xiao, Y., and Zhou, J.-M. 2006. Regulation of the type III secretion system in phytopathogenic bacteria. Mol. Plant-Microbe Interact. 19:1159-1166.

Tans-Kersten, J., Huang, H., and Allen, C. 2001. *Ralstonia solanacearum* needs motility for invasive virulence on tomato. J. Bacteriol. 183:3597-3605.

Tegos, G., Stermitz, F. R., Lomovskaya, O., and Lewis, K. 2002. Multidrug pump inhibitors uncover remarkable activity of plant antimicrobials. Antimicrob. Agents Chemother. 46:3133-3141.

Thatcher, L. F., Anderson, J. P., and Singh, K. B. 2005. Plant defence responses: What have we learnt from *Arabidopsis*? Funct. Plant Biol. 32:1-19.

Thieme, F., Szczesny, R., Urban, A., Kirchner, O., Hause, G., and Bonas, U. 2007. New type III effectors from *Xanthomonas campestris* pv. *vesicatoria* trigger plant reactions dependent on a conserved N-myristoylation motif. Mol. Plant-Microbe Interact. 20:1250-1261.

Tsuge, S., Nakayama, T., Terashima, S., Ochiai, H., Furutani, A., Oku, T., Tsuno, K., Kubo, Y., and Kaku, H. 2006. Gene involved in transcriptional activation of the *hrp* regulatory gene *hrpG* in *Xanthomonas oryzae* pv. oryzae. J. Bacteriol. 188:4158-4162.

Urban, A., Leipelt, M., Eggert, T., and Jaeger, K.-E. 2001. DsbA and DsbC affect extracellular enzyme formation in *Pseudomonas aeruginosa*. J. Bacteriol. 183:587-596.

van Dijk, K., Fouts, D. E., Rehm, A. H., Hill, A. R., Collmer, A., and Alfano, J. R. 1999. The Avr (effector) proteins HrmA (HopPsyA) and AvrPto are secreted in culture from *Pseudomonas syringae* pathovars via the Hrp (type III) protein secretion system in a temperature- and pH-sensitive manner. J. Bacteriol. 181:4790-4797.

Van Gijsegem, F., Vasse, J., De Rycke, R., Castello, P., and Boucher, C. 2002. Genetic dissection of the *Ralstonia solanacearum hrp* gene cluster reveals that the HrpV and HrpX proteins are required for Hrp pilus assembly. Mol. Microbiol. 44:935-946.

Veenendaal, A. K., Hodgkinson, J. L., Schwarzer, L., Stabat, D., Zenk, S. F., and Blocker, A. J. 2007. The type III secretion system needle tip complex mediates host cell sensing and translocon insertion. Mol. Microbiol. 63:1719-1730.

Venturi, V., Venuti, C., Devescovi, G., Lucchese, C., Fricina, A., Degrassi, G., Aguilar, C., and Mazzucchi, U. 2004. The plant pathogen *Erwinia amylovora* produces acyl-homoserine lactone signal molecules in vitro and in planta. FEMS Microbiol. Lett. 241:179-183.

Visick, K. L., and Fuqua, C. 2005. Decoding microbial chatter: Cell-cell communication in bacteria. J. Bacteriol. 187:5507-5519.

von Bodman, S. B., Bauer, W. D., and Coplin, D. L. 2003. Quorum sensing in plant-pathogenic bacteria. Annu. Rev. Phytopathol. 41:455-482.

Walker, D. S., Reeves, P. J., and Salmond, G. P. 1994. The major secreted cellulose, CelV, of *Erwinia carotovora* subsp. *carotovora* is an important soft rot virulence factor. Mol. Plant-Microbe Interact. 7:425-431.

Wang, L., He, Y., Gao, Y., Dong, J. E., He, C., Wang, S. X., Weng, L.-X., Xu, J.-L., Tay, L., Fang, R. X., and Zhang, L.-H. 2004. A bacterial cell-cell communication signal with cross kingdom structural analogs. Mol. Microbiol. 51:903-912.

Weber, E., Ojanen-Reuhs, T., Huguet, E., Hause, G., Romantschuk, M., Korhonen, T. K., Bonas, U., and Koebnik, R. 2005. The type III-dependent Hrp pilus is required for productive interaction of *Xanthomonas campestris* pv. *vesicatoria* with pepper host plants. J. Bacteriol. 187:2458-2468.

Wei, W., Plovanich-Jones, A., Deng, W.-L., Jin, Q.-L., Collmer, A., Huang, H.-C., and He, S. Y. 2000. The gene coding for the Hrp pilus structural protein is required for type III secretion of Hrp and Avr proteins in *Pseudomonas syringae* pv. *tomato*. Proc. Natl. Acad. Sci. U.S.A. 97:2247-2252.

Wei, Z. M., and Beer, S. V. 1995. *hrpL* activates *Erwinia amylovora hrp* gene transcription and is a member of the ECF subfamily of σ factors. J. Bacteriol. 177:6201-6210.

White, E. W., and Winans, S. C. 2005. Indentification of amino acid residues of the *Agrobacterium tumefaciens* quorum-sensing regulator traR that are critical for positive control of transcription. Mol. Microbiol. 55:1473-1486.

White, F. F., Yang, B., and Johnson, L. B. 2000. Prospects for understanding avirulence gene function. Curr. Opin. Plant Biol. 3:291-298.

Whitehead, N. A., Barnard, A. M. L., Slater, H., Simpson, N. J. L., and Salmond, G. P. C. 2001. Quorum-sensing in Gram-negative bacteria. FEMS Microbiol. Rev. 25:365-404.

Wickham, M. E., Culvenor, J. G., and Cowman, A. F. 2003. Selective inhibition of a two-step egress of malaria parasites from the host erythrocyte. J. Biol. Chem. 278:37658-37663.

Williams, S. C., Patterson, E. K., Carty, N. L., Grisold, J. A., Hamood, A. N., and Rumbaugh, K. P. 2004. *Pseudomonas aeruginosa* autoinducer enters and functions in mammalian cells. J. Bacteriol. 186:2281-2287.

Wong, T., Amidi, A., Dodds, A., Siddiqi, S., Wang, J., Yep, T., Tamang, D. G., and Saier, M. H., Jr. 2007. Evolution of the bacterial flagellum. Microbe 2:335-340.

Workentine, M. L., Chang, L., Ceri, H., and Turner, R. J. 2009. The GacS-GacA two-component regulatory system of *Pseudomonas flurorescens*: A bacterial two-hybrid analysis. FEMS Microbiol. Lett. 292:50-56.

Wu, H.-Y., Chung, P.-C., Shih, H.-W., Wen, S.-R., and Lai, E.-M. 2008. Secretome analysis uncovers an Hcp-family protein secreted via a type VI secretion system in *Agrobacterium tumefaciens*. J. Bacteriol. 190:2841-2850.

Xiao, Y., and Hutcheson, S. W. 1994. A single promoter sequence recognized by newly identified alternate sigma factor direct expression of pathogenicity and host range determinants in *Pseudomonas syringae*. J. Bacteriol. 176:3089-3091.

Xiao, Y., Heu, S., Yi, J., Lu, Y., and Hutcheson, S. W. 1994. Identification of a putative alternate sigma factor and characterization of a multicomponent regulatory cascade controlling the expression of *Pseudomonas syringae* pv. syringae Pss61 *hrp* and *hrmA* genes. J. Bacteriol. 176:1025-1036.

Xiao, Y., Lan, L., Yin, C., Deng, X., Baker, D., Zhou, J.-M., and Tang, X. 2007. Two-component sensor RhpS promotes induction of *Pseudomonas syringae* type III secretion system by repressing negative regulator RhpR. Mol. Plant-Microbe Interact. 20:223-234.

Xu, Q., and Pan, S. Q. 2000. An *Agrobacterium* catalase is a virulence factor involved in tumorigenesis. Mol. Microbiol. 35:407-414.

Xu, R.-Q., Blanvillain, S., Feng, J.-X., Jiang, B.-L., Li, X.-Z., Wei, H.-Y., Kroj, T., Lauber, E., Roby, D., Chen, B., He, Y.-Q., Lu, G.-T., Tang, D.-J., Vasse, J., Arlat, M., and Tang, J.-L. 2008. $AvrAC_{Xcc8004}$, a type III effector with a leucine-rich repeat domain from *Xanthomonas campestris* pathovar campestris confers avirulence in vascular tissues of *Arabidopsis thaliana* ecotype Col-0. J. Bacteriol. 190:343-355.

Xu, X. Q., Li, L. P., and Pan, S. Q. 2001. Feedback regulation of an *Agrobacterium* catalase gene katA involved in *Agrobacterium*-plant interaction. Mol. Microbiol. 42:645-657.

Yap, M.-N., Yang, C.-H., Barak, J. D., Jahn, C. E., and Charkowski, A. O. 2005. The *Erwinia chrysanthemi* type III secretion system is required for multicellular behavior. J. Bacteriol. 187:639-648.

Young, B. M., and Young, G. M. 2002. YplA is exported by the Ysc, Ysa, and flagellar type III secretion systems of *Yersinia enterocolitica*. J. Bacteriol. 184:1324-1334.

Yuan, Z.-C., Zaheer, R., and Finan, T. M. 2005. Phosphate limitation induces catalase expression in *Sinorhizobium meliloti*, *Pseudomonas aeruginosa* and *Agrobacterium tumefaciens*. Mol. Microbiol. 58:877-894.

Yuan, Z.-C., Liu, P., Saenkham, P., Kerr, K., and Nester, E. W. 2008. Transcriptome profiling and functional analysis of *Agrobacterium tumefaciens* reveals a general conserved response to acidic conditions (pH 5.5) and a complex acid-mediated signaling involved in *Agrobacterium*-plant interactions. J. Bacteriol. 190:494-507.

Zang, N., Tang, D.-J., Wei, M.-L., He, Y.-Q., Chen, B., Feng, J.-X., Xu, J., Gan, Y.-Q., Jiang, B.-L., and Tang, J.-L. 2007. Requirement of a *mip*-like gene for virulence in phytopathogenic bacterium *Xanthomonas campestris* pv. *campestris*. Mol. Plant-Microbe Interact. 20:21-30.

Zgurskaya, H. I., and Nikaido, H. 1999. Bypassing the periplasm: Reconstitution of the AcrAB multidrug efflux pump of *Escherichia coli*. Proc. Natl. Acad. Sci. U.S.A. 96:7190-7195.

Zhang, Y., and Miller, R. M. 1994. Effect of *Pseudomonas* rhamnolipid biosurfactant on cell hydrophobicity and biodegradation of octadecane. Appl. Environ. Microbiol. 60:2101-2106.

Zhang, Z.-Y., and Dixon, J. E. 1994. Protein tyrosine phosphatases: Mechanism of catalysis and substrate specificity. Adv. Enzymol. Relat. Areas Mol. Biol. 68:1-36.

Zhang, Z., Yan, L., Wang, L., Seydel, K. B., Li, E., Ankri, S., Mirelman, D., and Stanley, S. L., Jr. 2000. *Entamoeba histolytica* cysteine proteinases with interleukin-1 beta converting enzyme (ICE) activity cause intestinal inflammation and tissue damage in amoebiasis. Mol. Microbiol. 37:542-548.

Zhao, J., and Last, R. L. 1996. Coordinate regulation of the tryptophan biosynthetic pathway and indolic phytoalexin accumulation in *Arabidopsis*. Plant Cell 8:2235-2244.

Zheng, M., Wang, X., Doan, B., Lewis, K. A., Schneider, T. D., and Storz, G. 2001. Computation-directed identification of OxyR DNA binding sites in *Escherichia coli*. J. Bacteriol. 183:4571-4579.

Zhou, H., Morgan, R. L., Guttman, D. S., and Ma, W. 2009. Allelic variants of the *Pseudomonas syringae* type III effector HopZ1 are differentially recognized by plant resistance systems. Mol. Plant-Microbe Interact. 22:176-189.

Zhu, M., Shao, F., Innes, R. W., Dixon, J. E., and Xu, Z. 2004. The crystal structure of *Pseudomonas* avirulence protein AvrPphB: A papain-like fold with a distinct substrate-binding site. Proc. Natl. Acad. Sci. U.S.A. 101:302-307.

Zipfel, C., Kunze, G., Chinchilla, D., Caniard, A., Jones, J. D., Boller, T., and Felix, G. 2006. Perception of the bacterial PAMP EF-Tu by the receptor EFR restricts *Agrobacterium*-mediated transformation. Cell 125:749-760.

Zupan, J. R., Ward, D., and Zambryski, P. 1998. Assembly of the VirB transport complex for DNA transfer from *Agrobacterium tumefaciens* to plant cells. Curr. Opin. Microbiol. 1:649-655.

Zupan, J., Hackworth, C. A., Aguilar, J., Ward, D., and Zambryski, P. 2007. VirB1* promotes T-pilus formation in the *vir*-type IV secretion system of *Agrobacterium tumefaciens*. J. Bacteriol. 189:6551-6563.

Zwiesler-Vollick, J., Plovanich-Jones, A. E., Nomura, K., Bandyopadhyay, S., Joardar, V., Kunkel, B. N., and He, S. Y. 2002. Identification of novel *hrp*-regulated genes through functional genomic analysis of *Pseudomonas syringae* pv. *tomato* DC3000 genome. Mol. Microbiol. 45:1207-1218.

Epidemiology and Control of Bacterial Plant Diseases

11

Before focusing on the epidemiology of bacterial plant diseases, special attention should be given to the ecology of the bacterial pathogen in relationship to its physical environment. What is ecology? Ecology is the study of the interactions that organisms have with other organisms and with their environment.

Ecology and the Spread and Survival of Bacterial Pathogens

Most bacterial plant pathogens do not form spores and therefore do not depend on spores for their survival and dissemination. Their haploid status ensures clonal lineage and rapid propagation, and their microscopic size facilitates efficiency for dispersal via water, wind, and dust. Their survival in nature is highly dependent on cellular reserves, resiliency of their cell wall, and capsular material to protect them against cellular desiccation. Under many environmental conditions, the phylloplane can be considered a hostile habitat, since epiphytic bacteria are exposed to desiccation and solar ultraviolet radiation (Burrage, 1971; Sleesman and Leben, 1976; Lindow, 1991). Various pigments produced by, and high guanine-plus-cytosine (GC) content contained in, a number of these bacteria help shield the cells from harmful ultraviolet radiation of the sun (Singer and Ames, 1970). Many plant pathogens as well as many uncharacterized environmental bacteria have the ability to form biofilms on plants and on various sites, such as on clay minerals (e.g., kaolinites and montmorillonites), sand (e.g., silica sheets), and organic matter (e.g., humus). Pathogens possess genetic factors that allow them to adapt to microbial competition (e.g., microbial antibiosis) and to be carried, harbored, or both in insect vectors. They can be transported on infested/infected plant parts (as debris) and by insects. Their modes of survival vary widely; some seek to colonize the rhizoplane, while other plant-pathogenic bacteria reside in the phyllosphere (i.e., on leaves, stems, and shoots). Still other bacterial plant pathogens, such as *Xanthomonas axonopodis* pv. *malvacearum*, establish epiphytic residence in/on seeds, florets, immature buds, pollen, fruits, and nuts, as well as on vegetative cuttings and vegetative seed pieces.

The size of epiphytic populations of bacteria can vary substantially over time by the aging of the plant, the type or cultivar of the plant, plant nutrition and genetic background, and microbial competition and microbial interactions (Neal et al., 1973; Azad et al., 1985; Handelsman and Stabb, 1996; Hirano and Upper, 2000; Lindow and Brandl, 2003; Walker et al., 2003) that lead to habitat preferences (von Mering et al., 2007). Aboveground and belowground plant surfaces can be substantially colonized by plant-pathogenic bacteria. Their population size is dependent on moisture and available nutrients in the microenvironment as well as on their ability to survive against microbial competitors. On bean leaves, the cell numbers can be as high as 10^6–10^7 bacterial cells per square centimeter of leaf (Beattie and Lindow, 1995; Andrews and Harris, 2000; Hirano and Upper, 2000;

Lindow and Brandl, 2003). On pear leaves, *Pseudomonas syringae* populations can reach as high as 10^5 cells per gram fresh weight about 3 weeks following full bloom of 'Bartlett' pear trees. They also can reside in the rhizoplane in great numbers (10^6–10^7 cells per gram of root tissue) (Azad et al., 1985; van Loon et al., 1998; Gardener and Weller, 2001). Members of the genera *Agrobacterium*, *Burkholderia*, *Erwinia*, *Pantoea*, *Pectobacterium*, *Pseudomonas*, *Rhizobium*, and *Streptomyces* can be found as root inhabitants. Their existence and abundance is highly dependent on a number of variables that include adaptation to continued changes in the surrounding environment.

Successful establishment of disease in different agricultural production regions of the world has been witnessed in recent times because of the relatively high-speed transportation of inadvertently contaminated agriculturally based plant food and fiber products to worldwide markets. Of course, differing climates aid in precluding widespread establishment of certain plant bacterial diseases where pathogen survival is favored by either warm, humid conditions or cool temperate zone climates. In addition, disease establishment is precluded by the unavailability of the host plant to the pathogen. For example, pineapple and banana are produced in tropical regions, while wheat and soybean are grown in temperate zones. Pink disease of pineapple and Moko disease of banana are limited to tropical regions, while wheat streak and leaf blight of soybean occur in northern climate regions, such as in Saskatchewan, Canada. Tomatoes can be grown in subtropical climate zones, but their commercial propagation is limited in these parts of the world because they can be highly susceptible to *Ralstonia solanacearum*, fungal blights, and mildews, as well as to root-infecting oomycetes. All these factors that aid in the spread and establishment of bacterial plant pathogens are challenges faced in instituting efficient and effective control measures of bacterially caused diseases of fiber and consumable crops.

Epidemiology

Epidemiology is the study of the causes, frequency, and distribution of plant diseases in plant populations. The factors that cause and influence the spread and outcome of the disease are the basis for epidemiological studies. Within a plant population, assessments are made on the number of plants or plant parts that are affected and on the degree of injury (severity on tissues affected). The rate of disease spread in a plant population depends on a number of critical factors, particularly environmental influences; for example, wet versus dry conditions, whereby wet conditions promote growth and spread of the bacterial population, or hot versus cold conditions, whereby the growth of the bacterial population may be optimized by warmer conditions. These environmental factors can aid in disease forecasting; albeit, they are not absolute.

Plant diseases caused by bacterial pathogens are sustained by a number of factors that include ill-advised cultural practices, lax sanitary control measures, and inadequate diagnosis. Pathogenic bacterial cells have various dispensing instruments: infected vegetative planting material; infested seeds, soil, cultivating equipment, and harvesting machinery parts; recycled irrigation water; and vectoring insects.

Disease Cycles

Following the establishment and spread of infection on a susceptible cultivar or host, there can be repeated cycles of pathogen development. That is, the bacterial pathogen propagates in susceptible host tissues (and on host residues) to produce new inoculum that is dispersed to new susceptible sites to initiate new infections. Pathogens that produce only one cycle of development (one infection cycle) per crop cycle are called monocyclic. Pathogens that produce more than one infection per crop cycle are called polycyclic. The terms monocyclic and polycyclic are herein based on the number of crop cycles per year. One crop cycle per year can occur in temperate climatic zones, whereas multiple crop cycles can occur in subtropical and tropical zones.

Inoculum can build up when it is produced in one growing season and carried over to the next growing season. Inoculum buildup can occur in perennial plants, such as in orchards, pastures, lawns, and forests, as well as in annual crops grown in monoculture year after year, since inoculum produced in one growing season can carry over to the next. Tropical crops, such as coffee, banana, and pineapple, are grown year-round in tropical and subtropical regions without clear breaks in the growing season. Diseases of these crops are sustained nearly continuously over many years. Such conditions are termed polyetic epidemics (epiphytotics), regardless of whether or not the pathogen is monocyclic or polycyclic within a season.

Postharvest diseases and soilborne diseases tend to be monocyclic. In postharvest diseases, infections causing soft rots in storage may have already occurred either before harvest or during postharvest handling before the crop is placed into storage. As the rot progresses during storage, new inoculum is produced that can be spread when the infected crop is handled and if new sites of infection are produced through inadvertent wounding. Otherwise, there are no new infection cycles. Soft rot of potatoes, carrots, and bulb crops is usually initiated through infections that occurred during harvest and handling before the harvested crop was placed in storage and packed for shipment. Soft rot takes place during storage and transport to markets. Further handling of the stored crop provides new opportunities for inoculum to spread to fresh infection sites (wounds). Soft rot pathogens, such as *Pectobacterium* species, may not be strictly monocyclic but are constrained by the environment to produce a single cycle of infection. The same pathogens could produce polycyclic epidemics under conducive environments.

Soilborne diseases, such as vascular root infections, also produce one cycle of infection per crop cycle. Inoculum is dispersed in the soil by plowing and disking under infected crop residues. Infection occurs during the new crop cycle as the roots from new plantings come into contact with the inoculum present in the soil matrix and infected crop residues. The new inoculum that is produced remains confined on host debris until the pathogen-bearing debris is dispersed by plowing. Thus, there is only one cycle of infection per cropping cycle. For example, the bacterial wilt-producing pathogen *Ralstonia solanacearum* (race 1 and race 3) that resides in solanaceous crop residues and debris in soil can swim chemotactically toward the roots of newly planted hosts (see Chapter 10). Directed motility and not random motion is required for the virulence of *Ralstonia solanacearum* (Yao and Allen, 2006). As the roots ramify and extend out, they may come into contact with the pathogen. Successful infection ensures the production of fresh inoculum that awaits dispersal by cultivation practices. Cells of the pathogen are dispersed in the soil by irrigation and by plowing under infected crop residues. Infection of the next crop occurs through microscopic wounds developed during growth and differentiation of the roots. Because the new inoculum produced is not dispersed until the soil is again cultivated, there is only one complete infection cycle per cropping cycle.

Dissemination of Bacterial Pathogens

Besides dispersal of bacterial cells via an aqueous mode (e.g., wind-blown rain, irrigation water, and wash water), transmission of inoculum is effected through cultivation tools and harvesting equipment. For example, the claws of the walnut tree mechanical nut harvester become infested with inoculum after clamping and shaking a tree limb that is oozing *Brenneria rubrifaciens*, the cause of deep bark canker. As the harvester moves from tree to tree, it transmits this pathogen into fresh wounds made by the shaker on healthy trees in the orchard (Kado and Gardner, 1977) (see Chapter 5 and Fig. 5-7). Another example is the spread of alfalfa wilt in the field. The blades of the mower used in alfalfa fields can become contaminated with *Clavibacter michiganensis* subsp. *insidiosus* and generate new sites of infection via fresh wounds made by mowing (Malvick, 1988; Davis and Frate, 2006) (see Chapter 7).

Besides inoculum being spread by cultivational and harvesting activities, inoculum can be spread by flying insects. Clearly documented is the insect transmission of fire blight. Honey bees readily transmit *Erwinia amylovora* from blossom to blossom (see Chapters 1 and 5). Invasion and infection of pear blossoms are very rapid when this pathogen gains entry through the blossom nectaries, where β-glucosidase activity is very low and where hydrolysis of arbutin to the antibacterial compound hydroquinone is very low (Hildebrand and Schroth, 1964). Although crown gall is usually spread via infected planting stocks such as rooted grapevine cuttings (in which the *Agrobacterium* pathogen is a systemic resident) and through infested soil, insects occasionally have been documented as transmitting this disease. For example, the sweet potato whitefly, *Bemisia tabaci* (Gennadius), which is best known for transmitting members of the genus *Crinivirus*, has been shown to acquire and transmit *Agrobacterium tumefaciens* from tomato crown galls to healthy tomato seedlings (Zeidan and Czosnek, 1994). Although wind-blown rain is significant in dispersing bacteria, insects and pollen from infected catkins can spread *Xanthomonas arboricola* pv. *juglandis* to healthy walnut trees bearing catkins and can initiate new holdover infections of walnut blight (Rudolph, 1933; Schuster and Coyne, 1974).

Disease Control Management Principles

Strategies toward controlling bacterial plant diseases to a manageable level rely on several basic principles. They are eradication, exclusion, therapy, protection (prophylaxis, avoidance), and legal applications (regulation). Most of these principles are not meant to have a goal of zero tolerance for a given disease. Instead, they are meant to reduce disease incidence and severity down to a manageable and acceptable level.

Eradication

Eradication is undertaken to eliminate the pathogen after it has become established in the sphere where the host plant is growing. This principle is most successful when accompanied by a rigid quarantine. Such legal measures are not appropriate when the disease is fully endemic to the area. To eradicate a disease, infected crops are rogued and usually burned to reduce inoculum levels and to prevent further spread of the disease. Soilborne pathogen populations are reduced by treating infested soil with chemical and physical treatments, such as heat, steam, solarization, or volatile free-radical-producing compounds, such as methyl bromide and organic amendments. Steaming is regularly used in nurseries for eliminating bacteria as well as nematodes from the planting beds. Hot-water treatment of seeds of eggplant, pepper, tomato, carrot, spinach, lettuce, celery, cabbage, turnip, and radish has been used to eradicate bacterial pathogens (Miller and Lewis Ivey, 2005). Hot-water treatment of 44°C for 240 min minimized damage of nursery strawberry planting stocks but was lethal to *Xanthomonas fragariae*, the cause of strawberry angular leaf spot (Herder and Turechek, 2006).

Roguing is used for fire blight control, whereby infected twigs and branches of pear and apple trees are pruned, collected, and burned. More drastic roguing may be necessary in quarantined-crop-growing regions, for example, complete removal of citrus trees to prevent the epidemic spread of citrus canker (Gottwald et al., 2001). The recognition of fire blight in 1958 in three regions of Kent, England, resulted in the Fire Blight Disease Order of November 6, 1958. Approximately 2,600 trees were pulled and burned, including all beehives in the area (Lelliott, 1967). This action appeared to have been successful because no further infection has been found in the West Midlands since then. The roguing of cherry, apricot, and peach trees severely and systemically infected by *Pseudomonas syringae* pv. *syringae*, the cause of bacterial canker, is practiced in California and Canada (Fig. 11-1). The systemic infection of cherry trees has explained the lack of effective control from protective bactericides (Cameron, 1970). In some cases, entire orchards are cleared and replanted to another tree crop, such as walnuts. The roguing of bananas infected by *Ralstonia solanacearum* race 3 is commonly practiced in the Philippines (Buddenhagen, 1994). The rogued area is left fallow for 6–12 months.

Crop rotation is another alternative under this control principle. Sometimes a soilborne pathogen can be successfully eradicated by growing a nonhost for a

Fig. 11-1. Row of peach trees killed by *Pseudomonas syringae* that will be rogued.

period of time to eliminate the pathogen through sustaining a microbial population, such as *Streptomyces* species, that is antagonistic to the pathogen and produces lethal bactericides (antibiotics). Modification of the microbial environment provides a safe and nonchemical means of disease control. These efforts plus sanitation practices are essential for limiting the spread of disease.

Exclusion

Exclusion refers to the control of disease by excluding a pathogen from a region where it is not present. Distance and climatic differences, including mountains and other geographical barriers such as large rivers and lakes, between regions of the country provide adequate restrictions against some diseases. For example, in the United States, the control of ring rot of potato, caused by *Clavibacter michiganensis* subsp. *sepedonicus*, requires that potato seed pieces come from certified regions and be certified by testing. There are many varieties or cultivars of potato. Their certified seed is grown in geographically isolated areas within their potato-growing state or province. For example, 'Russet Burbank' seed potatoes are grown in a ring rot-free region in Idaho's Tetonian sector near Ashton, Idaho. 'Russet Burbank' seed potatoes are grown over 24,325 acres, which represents the largest of all seed-potato-growing regions of the United States and Canada. Assessments of the quality of the single-drop seed tubers (small, whole potatoes used as seed versus cut-drop potato pieces containing meristematic, lateral buds) are collected from each seed lot to be certified. Ring rot assessments of Idaho potatoes are made by inspection of young plants growing out from randomly selected tubers planted in a non-potato-growing region, such as in Oceanside, California, in late November. Certification is performed by the Idaho Crop Improvement Association. There is zero tolerance for ring rot.

Likewise, cotton free of angular leaf spot (bacterial blight), caused by *Xanthomonas axonopodis* pv. *malvacearum*, is grown for blight-free seed (mainly the cultivar Pima) in desert areas of Arizona that are naturally fenced in by high mountains. Cotton seeds can also be certified. They are treated with sulfuric acid to remove the cotton fibers and certified as angular leaf spot-free seeds. Young banana explants free of Moko disease, caused by *Ralstonia solanacearum*, are commercially propagated in the laboratory of Lembang Horticulture Research Institute in the Netherlands, and the certified plants are shipped to plantations (Stanfilco, a subsidiary of Dole Foods) in the Philippines (Sasson, 1993). Disease-free banana shoot tip cultures are used to maintain genetic stocks (Van den Houwe et al., 1995).

Certainly, plants showing visible symptoms are excluded. Quarantines aid in excluding the importation of disease into a disease-free region (see Legal Applications (Regulations) below).

Therapy

If economically feasible, plants infected by a given bacterial pathogen may be cured by several approaches. Infected plants can be subjected to high temperatures to kill the established pathogen. Both hot-water treatments and incubators with high ambient air temperatures have been used to treat infected stocks. Antibiotics, such as chlortetracycline, infiltrated into tree and vine crops have been used to suppress *Xylella fastidiosa* infections of shade and ornamental trees (Bentz et al., 2005). Similar infusion of agricultural-grade oxytetracycline into pear trees was used to control pear decline, caused by the pear decline phytoplasma (McIntyre et al., 1979). The pear decline phytoplasma belongs to the apple-proliferation (AP) group (Seemüller et al., 1998). Tetracycline derivatives have been injected into the trunks of palm and elm trees to treat lethal yellows diseases caused by phytoplasmas (McManus and Stockwell, 2001). Shoot tip culture of infected grapevines on agar media containing tetracycline removes *Xylella fastidiosa* (Hopkins, 1976). Treatment of vine cuttings in hot-water baths also eliminates *Xylella fastidiosa*. Meristem culture of bananas is a means to provide planting stock of this fruit crop that is free of Moko disease, caused by *Ralstonia solanacearum* race 2.

Protection

Protection is placing a barrier between the pathogen and the host. It may be either physical or chemical and it is applied in close proximity to the host. This principle is a major approach in disease control. Various chemicals have been used. Copper-based bactericides have been applied to seeds and plants for protection during peak periods of potential infection. Several clinically based antibiotics, such as derivatives of tetracyclines, streptomycin, and other aminoglycosides, have been used as protectants against such diseases as fire blight of pear and apple. These antibiotics

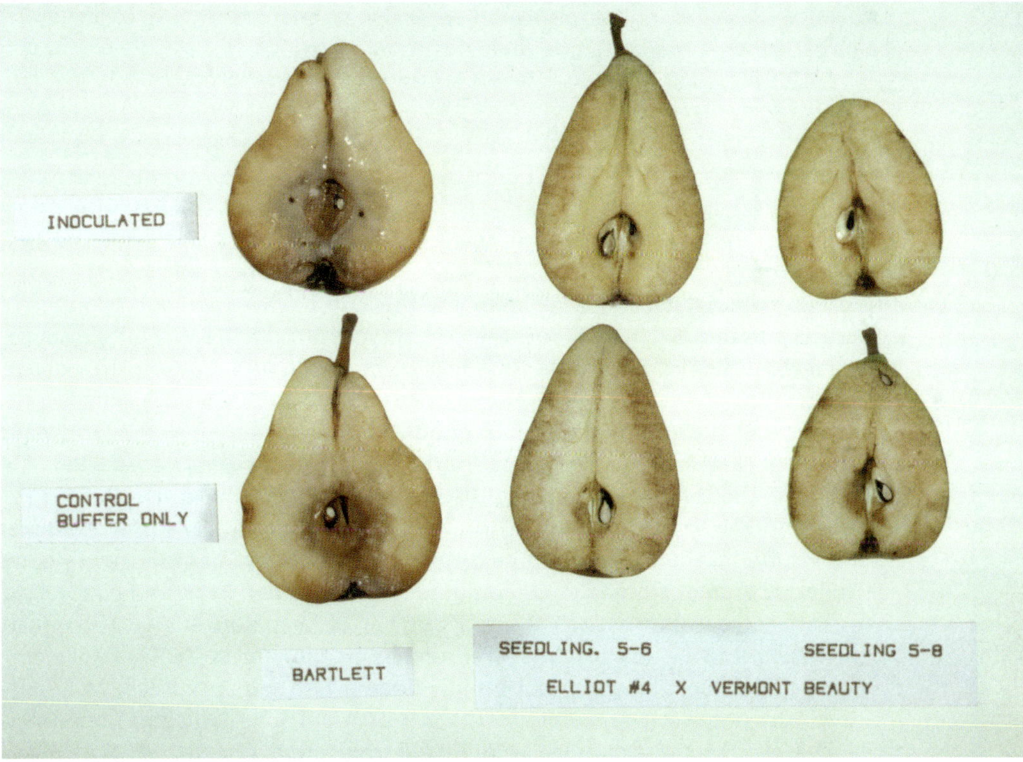

Fig. 11-2. Pear fruits assayed for resistance to *Erwinia amylovora*. The top row shows inoculations with *Erwinia amylovora* made on the susceptible cultivar Bartlett (left column) and on two hybrids (5-6 [middle column] and 5-8 [right column]) resulting from an 'Elliot #4' × 'Vermont Beauty' pear cross. The novel variety Elliot #4 was derived from a rootstock on the Elliot Ranch, California. The bottom row shows the control buffer only. (Photo by Kay Ryugo)

have also been used to a limited extent in therapy (see Therapy above). Sanitation and avoidance practices are also included under this control principle since clean environments and cultural practices constitute a form of protection.

This principle (sometimes referred as immunity) also applies to developing disease-resistant cultivars through conventional plant-breeding programs employing resistance genes from wild crop species bred into the horticulturally favored cultivar. Breeding for resistance to diseases caused by fungi and viruses has been successful. There are few examples of successful breeding for bacteria-borne diseases. Fire blight has been a target. A great deal of effort at the University of California, Davis, has been made to develop new pear cultivars that show tolerance or resistance to fire blight (Fig. 11-2). Coupled with classical genetic approaches is the development of resistance using molecular genetic strategies in identifying and incorporating specific genes or sets of genes that are known to confer resistance on the selected cultivar of a crop plant. This approach, often referred to as producing genetically modified organisms (GMOs), has been critically viewed by some environmental groups. Often their views are misguided and demonstrate misinterpretations of the use of genetics, since it is clear that genes from many different species are horizontally transferred and incorporated into diverse species and kingdoms through a natural process that has been in place for millions of years (Syvanen and Kado, 2002).

In addition to genetic approaches, and supported by arguments against the use of chemicals because of environmental concerns, prophylaxis by use of biological agents has gained popularity (see Biological Control below).

Legal Applications (Regulations)

Attempts to prevent the spread of pathogens across borders are made through mandated legal regulatory mechanisms. Border agricultural inspection stations and port embargoes are examples of instituting legal means, such as quarantines, to prevent the entry of specific plant diseases and pests potentially carried on specific crops, fruits, nuts, and seeds. Much of this approach is motivated politically rather than stemming from a true concern about introducing a plant pathogen. These legal mechanisms are used to reduce competition and ensure price control of a given crop.

Cases exist, however, in which economic disasters are prevented through quarantines, particularly when specialized crops are geographically isolated. Although sporadic instances of a given disease are observed, the absence of the insect vectors that spread a pathogen prevents epidemic outbreaks. For example, *Xylella fastidiosa*, the cause of alfalfa dwarf; Pierce's disease of grapevine; phony peach; leaf scorch of almond, oak, elm, mulberry, oleander, pear, and coffee; and citrus variegated chlorosis, as well as a number of latent infections, are spread by the xylem-feeding sharpshooter leafhoppers in the family Cicadellidae, subfamily Cicadellinae. The disease originally known as Anaheim disease (originating from the first occurrence of the disease in Anaheim, California) remained in check until the sharpshooters were introduced into California (Hewitt et al., 1942). In 1989, the glassy-winged sharpshooter, *Homalodisca coagulata* (Say), a large leafhopper species native to the southeastern United States, was introduced into southern California (Temecula, Riverside County) and proceeded to actively spread Pierce's disease in vineyards (Sorensen and Gill, 1996). Quarantines against the glassy-winged sharpshooter have been instituted in Alabama, Arkansas, California, Florida, Georgia, Louisiana, Mississippi, Missouri, North Carolina, South Carolina, and Texas. There are numerous wild plant species that harbor *Xylella fastidiosa* in the absence of visible symptoms (see Chapter 9). Such infected plants can easily pass through and into quarantined regions. Hence, the insect vector has been the quarantine target.

Biological Control

An alternative strategy to control bacterial plant diseases through the use of chemically based bactericidal agents, including antibiotics, is the biological control of bacterial plant diseases. Biological control exemplifies the principle of protection or prophylaxis. The principle of eradication through biological means is difficult to apply here. Biological approaches to eradicate established pathogens are often an unsuccessful approach to disease management because successful disease control by this method is highly dependent on the ability of the biocontrol organism to enter the site of infection and establish itself predominantly in areas where the pathogen is entrenched in the host and to inhibit further pathogenesis by the pathogen. In addition, the environment for propagation and establishment of the biocontrol organism can differ between regions and may not be optimal for its growth and maintenance in the environment in all regions where a crop is grown. The biocontrol organism must be equipped genetically to adapt to different environments, cultivars, locations, soil types, etc., since its usefulness in only one specific location would be commercially limiting.

A number of bacteria used as biocontrol organisms, such as *Pseudomonas* and *Bacillus* species, are known to produce various antibiotics, such as phenazines, 2,4-diacetylphloroglucinol, iturin A, luteolin, pyrrolnitrin, oomycin A, agrocin 84, herbicolin A, chaetomin, gliovirin, gliotoxin, and viridian. Other compounds produced are toxic volatiles, such as hydrogen cyanide, ammonia, ethanol, and methanol. Moreover, enzymes produced by the biocontrol organism can play a role. Cellulase, protease, β-gluconase, and chitinase are examples. Siderophores are also produced. Together, the effects of these compounds and enzymes are a significant component of the arsenal of the biocontrol organism.

Some successful examples of commercial disease control by biological means have been reported. For instance, the biological control of crown gall, caused by agrocinopine-producing strains of *Agrobacterium tumefaciens*, is achieved by using *Agrobacterium radiobacter* strains (Kerr, 1972; Kerr and Htay, 1974; Jones and Kerr, 1989). Also, *Pantoea herbicola* (previously called *Erwinia herbicola*; syn. *Pantoea agglomerans*) (Beer, 1981; Kearns and Hale, 1995; Kearns and Mahanty, 1998) and *Pseudomonas fluorescens* (Lindow et al., 1996) have been used to control fire blight of pear and apple, caused by *Erwinia amylovora*. *Pseudomonas fluorescens* and *Bacillus* species have also been used to control bacterial blotch of melon, caused by *Acidovorax avenae* subsp. *citrulli* (Santos et al., 2006). A Gram-positive bacterium prevented the bacterial blotch of cultivated mushrooms, caused by *Pseudomonas tolaasii* (Tsukamoto et al., 1998). *Pseudomonas fluorescens* has been used to curb the same disease (Herrera Estrella and Chet, 1998). In these examples, control of the disease was attributed, in part, to antibiotics and inhibitory secondary metabolites produced by the applied bacteria and, in part, by microbial competition. The severity and incidence of the disease apparently is proportional to the population size of the pathogen versus the amount of the applied biological-controlling microbe with respect to surface area (Lindemann et al., 1984).

In the biological control of fire blight, *Pantoea herbicola* prevented *Erwinia amylovora* from colonizing pear nectaries by the production of D-alanylgriseoluteic acid, whose production and elaboration by the biocontrol bacterium requires a 16-gene cluster (Giddens et al., 2002). In leaf scald of sugarcane, caused by *Xanthomonas albilineans*, *Pantoea dispersa* produces albicidin hydrolase, which detoxifies the toxin albicidin produced by the pathogen (Zhang and Birch, 1997). The hydrolase is a member of the serine hydrolase superfamily that includes serine proteases, lipases, and esterases (Derewenda and Derewenda, 1991; Nardini and Dijkstra, 1999). Albicidin hydrolase is encoded by the *albD* gene in *Pantoea dispersa*. *albD* is a candidate for the genetic engineering of sugarcane for resistance to leaf scald disease (Bower et al., 1996).

The effectiveness of biological control is compromised when there are differences in climate, microbial preferences, indigenous microbial strains, cultivars used, epiphytic strains, soil biosphere composition, etc. These conditions can differ appreciably and thus affect the establishment, growth, and localized dissemination of the biocontrol organism. These conditions can vary geographically. Biological control is most effective when seeds and vegetative propagules are used. Less inoculum of the biocontrol organism is required in these cases in comparison with field applications, in which large quantities of inoculum are required. For example, field control of pink disease of pineapple, caused by *Tatumella citrea*, in the Philippines requires huge amounts of fresh packed cells (*Pseudomonas* species) to be administered to 10,000 hectares of pineapple plants per month on a year-round basis. This requires large fermenters in full operation plus bacterial cell harvesting equipment and housing. Dosage per application is 100 g/acre, which is equivalent to 2.47 metric tons each month. Technical personnel to run the fermenters, plus the huge quantities of bacteriological medium that is continually prepared, adds to the cost of the operation (Kado, 1996). Taken together, the costs were too prohibitive for commercial operation. It was a lot cheaper to use insecticides to reduce insect vectors of the disease.

Precautionary care also must be taken to ensure that plasmids harbored in the biocontrol organism that encode one or more bactericidal compounds and their antidotes do not transfer to bacterial recipients in the surrounding environment. For example, plasmid pAgK84, which encodes the bacteriocide agrocin-84, is a conjugative plasmid and readily transfers under field conditions to other *Agrobacterium* species, including the pathogen *Agrobacterium tumefaciens* (Stockwell et al., 1996). To avoid this problem, *Agrobacterium radiobacter* harboring a nonconjugative agrocin-producing plasmid has been a good substitute (Jones and Kerr, 1989; Farrand et al., 1992).

Chemically Based Control

As early as 300 year ago, chemically based protection of seeds and seedlings from pests and fungi relied on poisonous compounds, such as mercury and arsenic (Fischer and Holton, 1957; Neergaard, 1977). Mercury-containing organic compounds, such as (2-carboxyphenylthio)-ethylmercury, are still used, mainly in Third World countries. Copper-based bactericides, such as cupric hydroxide, which are less toxic to mammals, have substituted for mercury. The copper substitutes appear to be effective in protecting seeds, foliage, and fruits against bacterial pathogens. The use of copper-based bactericides has its limitations. These materials are phytotoxic and can cause blemishes on foliage and fruits if applied when young leaves and fruits are developing. Timely applications are therefore required. Copper-based compounds have been applied to reduce the incidence of spot and speck diseases of tomato and bell pepper (Jones et al., 1991; Kousik and Ritchie, 1996; Flaherty et al., 2000). In California and Oregon, copper-based bactericides have also been used to reduce the incidence of walnut blight, caused by *Xanthomonas arboricola* pv. *juglandis*, and filbert blight, caused by *Xanthomonas arboricola* pv. *corylina* (Olson et al., 1976).

Owing to the typical mutation rate of one per million bacterial cells (10^{-6} mutations per gene per generation) even in stationary phase (Sniegowski, 2004), the use of single compounds in plant disease control will select bacterial cells of a pathogenic population that are resistant to the applied compound. This has been the case with *Xanthomonas vesicatoria* in tomato and bell pepper fields in Florida, Oklahoma, Georgia, and California (Gitaitis et al., 1985; Stall et al., 1986; Bender et al., 1990; Cooksey et al., 1990). Copper-resistant strains of *Pseudomonas syringae* pv. *tomato* have been found worldwide (Bender and Cooksey, 1986; da Silva and Lopes, 1995; Pernezny et al., 1995; Scheck et al., 1996; Alexander et al., 1999).

Besides chromosomal mutations leading to copper resistance, there are genetic determinants for preventing the damaging effects of copper ions in bacteria. At least four plasmid-borne genes are needed for copper metabolism (uptake, intracellular storage, transport, export, and regulation) (Brown et al., 1992). In *Pseudomonas syringae* pv. *tomato*, these genes are arranged as an operon (as *copABCD*) (Mellano and Cooksey, 1988). *Pseudomonas syringae* regulates the expression of this operon using a copper-induced, two-component signal transduction regulatory system that consists of CopS, a transmembrane signal transducer, and CopR, a transcriptional activator protein (Mills et al., 1993). Two-component, copper-mediated induction and regulation have also been observed in *Helicobacter pylori* (Waidner et al., 2005). In some instances, copper resistance genetic determinants are located on the chromosome, such as on the chromosome of *Xanthomonas arboricola* pv. *juglandis* (previously *Xanthomonas campestris* pv. *juglandis*) (Lee et al., 1994), and on the chromosome of *Pseudomonas syringae* (Lim and Cooksey, 1993). On the other hand, copper resistance is encoded by a highly conserved plasmid called pPT23D in many strains of *Pseudomonas syringae* pv. *tomato* (Bender and Cooksey, 1987; Cooksey, 1987).

Quaternary ammonium compounds have also been used as bactericides. Ammonium lignosulfonate, a by-product of wood pulp processing, together with potassium phosphate spray applications, were shown to reduce the severity of bacterial spot symptoms on tomato and pepper plants (Abbasi et al., 2002). Phosphate salts induce partial resistance in plants against fungal pathogens and likely aid in conferring resistance to bacterial spot.

Compounds that disrupt the bacterial membrane are likely to be good candidates for bacterial control in the future. Biodegradable household detergents and surface disinfectants, such as sodium 4-dimethylaminobenzenediazosulfonate (fenaminosulf) and 3-allyloxy-1,2-benzothiazole 1,1-dioxide (probenazole), in dilute form might be used in combination with antibiotics to enhance penetration of the bacterial cells of the pathogen. Detergents, including naturally occurring types such as rhamnolipids, enhance nondiscretionary uptake of bactericides. Rhamnolipids alter cell surface properties, increasing permeability (Al-Tahhan et al., 2000). However, certain types of rhamnolipids are produced by *Pseudomonas* species, such as *Pseudomonas aeruginosa* (McClure and Schiller, 1996) and *Pseudomonas fluorescens* (Hernandez et al., 2004), to enhance their virulence in humans and plants, respectively. The swimming and swarming motility of *Pseudomonas aeruginosa* is regulated by rhamnolipids (Caiazza et al., 2005). Rhamnolipids are also used to detach bacterial cells from biofilms (see Chapter 10). Biofilms help protect the pathogen from innate immune responses. Compounds that inhibit quorum sensing and reduce biofilm formation have been recently found. Azithromycin, for example, is effective in blocking the biofilm-forming ability of *Pseudomonas aeruginosa* (Hoffmann et al., 2007). This antibiotic is currently successfully used in curing pneumonia in human patients.

As mentioned in the above sections on Eradication, Therapy, and Protection, antibiotics such as streptomycin and tetracycline derivatives have been used in attempts to control bacterial plant diseases to a manageable level. For example, foliar applications of streptomycin on pear trees following periods of rain throughout the late spring and early summer effectively reduced the shoot blight phase of fire blight (Bonn and Morand, 1980). Oxytetracycline is also sprayed in pear orchards to curb the incidence of fire blight. The concentration of this agricultural-grade antibiotic is extremely high, at 1 mg/ml! Continued application of a single type of antibiotic, such as streptomycin, can enrich and select for plant-pathogenic bacteria resistant to the antibiotic (Stall and Thayer, 1962; Coyier and Covey, 1975; Young, 1977; Lacy et al., 1984; Shaffer and Goodman, 1985; Burr et al., 1988; Minsavage et al., 1990; Jones et al., 1991; Loper et al., 1991; Burr et al., 1993; Thomson et al., 1993; McManus and Jones, 1994). The appearance of antibiotic resistance in bacterial populations is a natural phenomenon that originates from the need for bacteria to survive harsh and often lethal environmental changes (see Chapter 2). For bacteria to survive the lethal effects of an antibiotic compound, the acquisition of horizontally transferred antibiotic resistance genes carried on mobile plasmids in different bacteria plays an important role in adaptive evolution (Syvanen and Kado, 2002). Replicative transposons in plasmids also help disseminate antibiotic resistance (Buii et al., 2007).

Recognition of emerging antibiotic resistance has propelled the search for novel strategies to combat bacterial infections of plants, animals, and humans. One strategy is to target the bacterial secretion systems that are needed to deliver effectors into the host cell (see Chapter 10). Small compounds akin to salicylidene acyl hydrazides have been shown to affect the type III secretion activity as well as its structure (Nordfelth et al., 2005; Veenendaal et al., 2009).

Crops Transformed for Resistance Against Bacterial Diseases

Genes that confer resistance (*R* genes) to specific bacterial plant diseases originate from various sources. Some *R* genes are present in naturally resistant wild relatives of a given crop plant. *R* genes from microorganisms or from insects and higher cells are alternative sources. The *R* genes may be isolated and cloned into a suitable plant transformation shuttle vector (for a list of shuttle vectors, see Komari et al., 2006). Using a shuttle vector, *R* genes are incorporated via *Agrobacterium*-mediated transformation or electrotransformation (Wu and Shimamoto, 1995) into the host genome of the susceptible cultivar and tested for expression of the resistance-conferring product and for basic resistance to a specific bacterial pathogen of interest. For example, the *R* gene *Pto* that confers resistance to bacterial speck, caused by race 0 of *Pseudomonas syringae* pv. *tomato*, has been introduced into several processing and some fresh-market tomato varieties (Oldroyd et al., 1997). Unexpectedly, the hybrid varieties heterozygous for *Pto* are not completely resistant to bacterial speck (Oldroyd et al., 1997). Interestingly, overexpression of a second plant resistance gene, *Prf*, results in conferring onto tomato plants an enhanced resistance to a number of normally virulent bacterial and viral pathogens (Oldroyd and Staskawicz, 1998). (*Prf* is a member of a superfamily of plant disease resistance genes that contain leucine-rich repeat and nucleotide-binding site protein motifs.) Because bacteria acquire resistance genes against eukaryotic defense responses through horizontal gene transfer events, single-gene products rarely confer complete resistance to a host plant attacked by the pathogen. Sets of genes or genetic islands have stronger impact on the recipient than do single genes (Kamoun and Kado, 1993; Syvanen and Kado, 2002).

Rice cultivars IR64 and IR72 were transformed with gene *Xa21* derived from an African (Mali) rice, *Oryza longistaminata* A. Chev. & Roehr. Gene *Xa21* encodes a receptor-like kinase that confers resistance against race 6 of *Xanthomonas oryzae* pv. *oryzae* (Song et al., 1995). The protein has 23 leucine-rich repeats and serine/threonine kinase autophosphorylating activity (Liu et al., 2002). A high degree of similarity between the *hrp* genes of *Xanthomonas oryzae* pv. *oryzae* and those of *Xanthomonas vesicatoria* and *Ralstonia solanacearum* suggests that *Xanthomonas oryzae* pv. *oryzae* injects harpin- and PopA-like (*hpa1*-encoded) effectors, along with avirulence proteins AvrXa10 and AvrXa7, through similar type III secretion machinery (Zhu et al., 2000). A monoclonal antibody-encoding gene prepared against Hpa1 and transformed into rice cultivars is likely to produce in vivo an antibody that interacts and quenches the activity of Hpa1. Loss of Hpa1 activity abolishes the pathogenicity of *Xanthomonas oryzae* pv. *oryzae* (Zhu et al., 2000). This is a viable strategy for the biological control of bacterial blight disease of rice since the monoclonal-transgene strategy is commercially available (Hein and Hiatt, 2002; Stoger et al., 2002).

The defense response of the host is controlled by a complicated network of finely tuned regulatory factors that include GTP-binding proteins, cytokinins, jasmonic acid, and salicylic acid (Sano and Ohashi, 1995). It becomes very clear that the incorporation of more than one novel gene for enhancing host resistance will be necessary. Genetically engineered plants with single-gene resistance introduced in the field are confronted by the natural selection of new virulent races of the pathogen. More than 40 *R* genes have been cloned and characterized. In the case of the *Oryza sativa–Xanthomonas oryzae* pv. *oryzae* pathosystem, 9 of the 30 *R* genes are recessively inherited (Iyer-Pascuzzi and McCouch, 2007). Recognition of the limitation in using single-gene-based resistance should therefore be taken seriously (Kamoun and Kado, 1993).

REFERENCES

Abbasi, P. A., Soltani, N., Cuppels, D. A., and Lazarovits, G. 2002. Reduction of bacterial spot disease severity on tomato and pepper plants with foliar applications of ammonium lignosulfonate and potassium phosphate. Plant Dis. 86:1232-1236.

Alexander, S. A., Kim, S. H., and Waldenmaier, C. M. 1999. First report of copper-tolerant *Pseudomonas syringae* pv. *tomato* in Virginia. Plant Dis. 83:964.

Al-Tahhan, R. A., Sandrin, T. R., Bodour, A. A., and Maier, R. M. 2000. Rhamnolipid-induced removal of lipopolysaccharide from *Pseudomonas aeruginosa*: Effect on cell surface properties and interaction with hydrophobic substrates. Appl. Environ. Microbiol. 66:3262-3268.

Andrews, J. H., and Harris, R. F. 2000. The ecology and biogeography of microorganisms on plant surfaces. Annu. Rev. Phytopathol. 38:145-180.

Azad, H. R., Davis, J. R., and Kado, C. I. 1985. Relationships between rhizoplane and rhizosphere bacteria and Verticillium wilt resistance in potato. Arch. Microbiol. 140:347-351.

Beattie, G. A., and Lindow, S. E. 1995. The secret life of foliar bacterial pathogens on leaves. Annu. Rev. Phytopathol. 33:145-172.

Beer, S. V. 1981. Biological control of fire blight. Acta Hortic. 117:123-124.

Bender, C. L., and Cooksey, D. A. 1986. Indigenous plasmids in *Pseudomonas syringae* pv. *tomato*: Conjugative transfer and role in copper resistance. J. Bacteriol. 165:534-541.

Bender, C. L., and Cooksey, D. A. 1987. Molecular cloning of copper resistance genes from *Pseudomonas syringae* pv. *tomato*. J. Bacteriol. 169:470-474.

Bender, C. L., Malvick, D. K., Conway, K. E., George, S., and Pratt, P. 1990. Characterization of pXV10A, a copper resistance plasmid in *Xanthomonas campestris* pv. *vesicatoria*. Appl. Environ. Microbiol. 56:170-175.

Bentz, J.-A., Huang, Q., and Jordan, R. 2005. Bacterial leaf scorch of shade trees. U.S. National Arboretum, U.S. Department of Agriculture, Beltsville, MD.

Bonn, W. G., and Morand, J. B. 1980. Fire blight of pear: Control of shoot blight phase with streptomycin. Can. J. Plant Pathol. 2:39-41.

Bower, R., Elliott, A. R., Potier, B. A. M., and Birch, R. G. 1996. High-efficiency, microprojectile-mediated cotransformation of sugarcane using visible or selectable markers. Mol. Breed. 2:239-249.

Brown, N. L., Rouch, D. A., and Lee, B. T. O. 1992. Copper resistance determinants in bacteria. Plasmid 27:41-51.

Buddenhagen, I. W. 1994. Moko disease. Pages 15-16 in: Compendium of Tropical Fruit Diseases. R. C. Ploetz, G. A. Zentmyer, W. T. Nishijima, K. G. Rohrbach, and H. D. Ohr, eds. American Phytopathological Society, St. Paul, MN.

Buii, D., Sorto, N., Trigueros, S., Newmark, J. S., Sherratt, D. J., and Tolmasky, M. E. 2007. Role of replicative transposons in plasmids' host range expansion: Implications in dissemination of antibiotic resistance. (Abstr.) Plasmid 57:231.

Burr, T. J., Norelli, J. L., Katz, B., Wilcox, W. F., and Hoying, S. A. 1988. Streptomycin resistance of *Pseudomonas syringae* pv. *papulans* in apple orchards and its association with a conjugative plasmid. Phytopathology 78:410-413.

Burr, T. J., Norelli, J. L., Reid, C. L., Capron, L. K., Nelson, L. S., Aldwinckle, H. S., and Wilcox, W. F. 1993. Streptomycin-resistant bacteria associated with fire blight infections. Plant Dis. 77:63-66.

Burrage, S. W. 1971. The micro-climate at the leaf surface. Pages 89-101 in: Ecology of Leaf Surface Micro-Organisms. T. F. Preece and C. H. Dickinson, eds. Academic Press, London.

Caiazza, N. C., Shanks, R. M. Q., and O'Toole, G. A. 2005. Rhamnolipids modulate swarming motility patterns of *Pseudomonas aeruginosa*. J. Bacteriol. 187:7351-7361.

Cameron, H. R. 1970. *Pseudomonas* content of cherry trees. Phytopathology 60:1343-1346.

Cooksey, D. A. 1987. Characterization of a copper resistance plasmid conserved in copper-resistant strains of *Pseudomonas syringae* pv. *tomato*. Appl. Environ. Microbiol. 53:454-456.

Cooksey, D. A., Azad, H. R., Cha, J.-S., and Lim, C.-K. 1990. Copper resistance gene homologs in pathogenic and saprophytic bacterial species from tomato. Appl. Environ. Microbiol. 56:431-435.

Coyier, D. L., and Covey, R. P. 1975. Tolerance of *Erwinia amylovora* to streptomycin sulfate in Oregon and Washington. Plant Dis. Rep. 59:849-852.

da Silva, V. L., and Lopes, C. A. 1995. *Pseudomonas syringae* pv. *tomato* resistant to copper in copper-sprayed tomato fields. Fitopatol. Bras. 20:85-89.

Davis, R. M., and Frate, C. A. 2006. Alfalfa bacterial wilt. UC IPM Pest Management Guidelines: Alfalfa. Univ. Calif. Agric. Nat. Resour. Publ. 3430.

Derewenda, Z. S., and Derewenda, U. 1991. Relationships among serine hydrolases: Evidence for a common structural motif in triacylglyceride lipases and esterases. Biochem. Cell Biol. 69:842-851.

Farrand, S. K., Wang, C.-L., Hong, S.-B., O'Morchoe, S. B., and Slota, J. E. 1992. Deletion derivatives of pAgK84 and their use in the analysis of *Agrobacterium* plasmid functions. Plasmid 28:201-212.

Fischer, G. W., and Holton, C. S. 1957. Biology and Control of the Smut Fungi. Ronald Press, New York.

Flaherty, J. E., Jones, J. B., Harbaugh, B. K., Somodil, G. C., and Jackson, L. E. 2000. Control of bacterial spot on tomato in the greenhouse and field with H-mutant bacteriophages. HortScience 35:882-884.

Gardener, B. B. M., and Weller, D. M. 2001. Changes in population of rhizosphere bacteria associated with take-all disease of wheat. Appl. Environ. Microbiol. 67:4414-4425.

Giddens, S. R., Feng, Y., and Mahanty, H. K. 2002. Characterization of a novel phenazine antibiotic gene cluster in *Erwinia herbicola* Eh1087. Mol. Microbiol. 45:769-783.

Gitaitis, R. D., McInnes, T. B., and Jones, J. B. 1985. A survey of Georgia and Florida for the presence of copper tolerant phytopathogenic bacteria. (Abstr.) Phytopathology 75:1288.

Gottwald, T. R., Hughes, G., Graham, J. H., Sun, X., and Riley, T. 2001. The citrus canker epidemic in Florida: The scientific basis of regulatory eradication policy for an invasive species. Phytopathology 91:30-34.

Handelsman, J., and Stabb, E. V. 1996. Biocontrol of soilborne plant pathogens. Plant Cell 8:1855-1869.

Hein, M. B., and Hiatt, A. 2002. Transgenic plants expressing assembled secretory antibodies. Scripps Research Institute. U.S. Patent 6,417,429.

Herder, K., and Turechek, W. W. 2006. Evaluating hot-water treatment as means for eradicating *Xanthomonas fragariae* in strawberry nursery stock. (Abstr.) Phytopathology 96:S47.

Hernandez, A. M., Suslow, T. V., Leloup, L., and Kado, C. I. 2004. A biosurfactant produced by *Pseudomonas fluorescens* promotes soft rot of processed vegetables. Plant Pathol. 53:596-601.

Herrera Estrella, A., and Chet, I. 1998. Biocontrol of bacteria and phytopathogenic fungi. Pages 263-280 in: Agricultural Biotechnology. A. Altman, ed. Marcel Dekker, Inc., New York.

Hewitt, W. B., Frazier, N. W., Jacob, H. E., and Freitag, J. H. 1942. Pierce's disease of grapevines. Calif. Agric. Exp. Stn. Circ. 353.

Hewitt, W. B., Houston, B. R., Frazier, N. W., and Freitag, J. H. 1946. Leafhopper transmission of the virus causing Pierce's disease of grape and dwarf of alfalfa. Phytopathology 36:117-128.

Hildebrand, D. C., and Schroth, M. N. 1964. Arbutin-hydroquinone complex in pear as a factor in fire blight development. Phytopathology 54:640-645.

Hirano, S. S., and Upper, C. D. 2000. Bacteria in the leaf ecosystem with emphasis on *Pseudomonas syringae*—A pathogen, ice nucleus, and epiphyte. Microbiol. Mol. Biol. Rev. 64:624-653.

Hoffmann, N., Lee, B., Hentzer, M., Rasmussen, T. B., Song, Z., Johansen, H. K., Givskov, M., and Hoiby, N. 2007. Azithromycin blocks quorum sensing and alginate polymer formation and increases the sensitivity to serum and stationary-growth killing of *Pseudomonas aeruginosa* and attenuates chronic *P. aeruginosa* lung infection in *Cftr*$^{-/-}$ mice. Antimicrob. Agents Chemother. 51:3677-3687.

Hopkins, D. L. 1976. Effect of tetracycline antibiotics on the development of Pierce's disease of grapevine. (Abstr.) Proc. Am. Phytopathol. Soc. 3:259.

Iyer-Pascuzzi, A. S., and McCouch, S. R. 2007. Recessive resistance genes and the *Oryza sativa–Xanthomonas oryzae* pv. *oryzae* pathosystem. Mol. Plant-Microbe Interact. 20:731-739.

Jones, A. L., Norelli, J. L., and Ehret, G. R. 1991. Detection of streptomycin-resistant *Pseudomonas syringae* pv. *papulans* in Michigan apple orchards. Plant Dis. 75:529-531.

Jones, D. A., and Kerr, A. 1989. *Agrobacterium radiobacter* strain K1026, a genetically engineered derivative of strain K84, for biological control of crown gall. Plant Dis. 73:15-18.

Jones, J. B., Woltz, S. S., Jones, J. P., and Porter, K. L. 1991. Population dynamics of *Xanthomonas campestris* pv. *vesicatoria* on tomato leaflets treated with copper bactericides. Phytopathology 81:714-719.

Kado, C. I. 1996. Pineapple pink disease research and control program. Annu. Rep. 1995-1996. Dole Foods, Inc., Polomolok, Philippines.

Kado, C. I., and Gardner, J. M. 1977. Transmission of deep bark canker of walnuts by the mechanical harvester. Plant Dis. Rep. 61:321-325.

Kamoun, S., and Kado, C. I. 1993. Genetic engineering for plant disease resistance. Pages 165-198 in: Advanced Engineered Pesticides. L. Kim, ed. Marcel Dekker, Inc., New York.

Kearns, L. P., and Hale, C. N. 1995. Incidence of bacteria inhibitory to *Erwinia amylovora* from blossoms in New Zealand apple orchards. Plant Pathol. 44:918-924.

Kearns, L. P., and Mahanty, H. K. 1998. Antibiotic production by *Erwinia herbicola* Eh1087: Its role in inhibition of *Erwinia amylovora* and partial characterization of antibiotic biosynthesis genes. Appl. Environ. Microbiol. 64:1837-1844.

Kerr, A. 1972. Biological control of crown gall: Seed inoculation. J. Appl. Microbiol. 35:493-497.

Kerr, A., and Htay, K. 1974. Biological control of crown gall through bacteriocin production. Physiol. Plant Pathol. 4:37-44.

Komari, T., Takakura, Y., Ueki, J., Kato, N., Ishida, Y., and Hiei, Y. 2006. Binary vectors and super-binary vectors. Pages 15-41 in: Agrobacterium Protocols (Methods in Molecular Biology), Vol. I, 2nd ed. K. Wang, ed. Humana Press, New York.

Kousik, C. S., Sanders, D. C., and Ritchie, D. F. 1996. Mixed genotypes combined with copper sprays to manage bacterial spot of bell peppers. Phytopathology 86:502-508.

Lacy, G. H., Stromberg, V. K., and Cannon, N. P. 1984. *Erwinia amylovora* mutants and *in planta*-derived transconjugants resistant to oxytetracycline. Can. J. Plant Pathol. 6:33-39.

Lee, Y.-A., Hendson, M., Panopoulos, N. J., and Schroth, M. N. 1994. Molecular cloning, chromosomal mapping, and sequence analysis of copper resistance genes from *Xanthomonas campestris* pv. juglandis: Homology with small blue copper proteins and multicopper oxidase. J. Bacteriol. 176:173-188.

Lelliott, R. A. 1967. Fireblight in England; its nature and its attempted eradication. EPPO Publ. Ser. A. No. 45-E.

Lim, C.-K., and Cooksey, D. A. 1993. Characterization of chromosomal homologs of the plasmid-borne copper resistance operon of *Pseudomonas syringae*. J. Bacteriol. 175:4492-4498.

Lindemann, J., Arny, D. C., and Upper, C. D. 1984. Use of an apparent infection threshold population of *Pseudomonas syringae* to predict incidence and severity of brown spot of bean. Phytopathology 74:1334-1339.

Lindow, S. E. 1991. Determinants of epiphytic fitness in bacteria. Pages 295-314 in: Microbial Ecology of Leaves. J. H. Andrews and S. S. Hirano, eds. Springer-Verlag, New York.

Lindow, S. E., and Brandl, M. T. 2003. Microbiology of the phyllosphere. Appl. Environ. Microbiol. 69:1875-1883.

Lindow, S. E., McGourty, G., and Elkins, R. 1996. Interactions of antibiotics with *Pseudomonas fluorescens* strain A506 in the control of fire blight and frost injury to pear. Phytopathology 86:841-848.

Liu, G.-Z., Pi, L.-Y., Walker, J. C., Ronald, P. C., and Song, W.-Y. 2002. Biochemical characterization of the kinase domain of the rice disease resistance receptor-like kinase XA21. J. Biol. Chem. 277:20264-20269.

Loper, J. E., Henkels, M. D., Roberts, R. G., Grove, G. G., Willett, M. J., and Smith, T. J. 1991. Evaluation of streptomycin, oxytetracycline, and copper resistance of *Erwinia amylovora* isolated from pear orchards in Washington State. Plant Dis. 75:287-290.

Malvick, D. K. 1988. Bacterial wilt of alfalfa. Univ. Ill. Ext. RPD No. 300.

McClure, C. D., and Schiller, N. L. 1996. Inhibition of macrophage phagocytosis by *Pseudomonas aeruginosa* rhamnolipids in vitro and in vivo. Curr. Microbiol. 33:109-117.

McIntyre, J. L., Schneider, H., Lacy, G. H., Dodds, J. A., and Walton, G. S. 1979. Pear decline in Connecticut and response of diseased trees to oxytetracycline infusion. Phytopathology 69:955-958.

McManus, P. S., and Jones, A. L. 1994. Epidemiology and genetic analysis of streptomycin-resistant *Erwinia amylovora* from Michigan and evaluation of oxytetracycline for control. Phytopathology 84:627-633.

McManus, P. S., and Stockwell, V. O. 2001. Antibiotic use for plant disease management in the United States. Online. Plant Health Progress doi:10.1094/PHP-2001-0327-01-RV.

Mellano, M. A., and Cooksey, D. A. 1988. Nucleotide sequence and organization of copper resistance genes from *Pseudomonas syringae* pv. *tomato*. J. Bacteriol. 170:2879-2883.

Miller, S. A., and Lewis Ivey, M. L. 2005. Hot water and chlorine treatment of vegetable seeds to eradicate bacterial pathogens. Ohio State Univ. Ext. Fact Sheet HYG-3085-05.

Mills, S. D., Jasalavich, C. A., and Cooksey, D. A. 1993. A two-component regulatory system required for copper-inducible expression of the copper resistance operon of *Pseudomonas syringae*. J. Bacteriol. 175:1656-1664.

Minsavage, G. V., Canteros, B. I., and Stall, R. E. 1990. Plasmid-mediated resistance to streptomycin in *Xanthomonas campestris* pv. *vesicatoria*. Phytopathology 80:719-723.

Nardini, M., and Dijkstra, B. W. 1999. α/β Hydrolase fold enzymes: The family keeps growing. Curr. Opin. Struct. Biol. 9:732-737.

Neal, J. L., Jr., Larson, R. I., and Atkinson, T. G. 1973. Changes in rhizosphere populations of selected physiological groups of bacteria related to substitution of specific pairs of chromosomes in spring wheat. Plant Soil 39:209-212.

Neergaard, P. 1977. Seed Pathology. Vol. I. Macmillan Press, London.

Nordfelth, R., Kauppi, A. M., Norberg, H. A., Wolf-Watz, H., and Elofsson, M. 2005. Small-molecule inhibitors specifically targeting type III secretion. Infect. Immun. 73:3104-3114.

Oldroyd, G. E. D., and Staskawicz, B. J. 1998. Genetically engineered broad-spectrum disease resistance in tomato. Proc. Natl. Acad. Sci. U.S.A. 95:10300-10305.

Oldroyd, G., Suslow, T. V., Staskawicz, B., and Miyao, G. 1997. *Pseudomonas syringae* pv. *tomato* infecting *Pto* gene-bearing cultivars carry functional *avrPto* in CA. (Abstr.) Phytopathology 87:S72.

Olson, W. H., Moller, W. J., Fitch, L. B., and Jeter, R. B. 1976. Walnut blight control. Calif. Agric. 30:10-13.

Pernezny, K., Kudela, V., Kokoskova, B., and Hladka, I. 1995. Bacterial disease of tomato in the Czech and Slovak Republics and lack of streptomycin resistance among copper-tolerant bacterial strains. Crop Prot. 14:267-270.

Rudolph, B. A. 1933. Bacteriosis (blight) of the English walnut in California and its control. Calif. Agric. Exp. Stn. Bull. 564:3-88.

Sano, H., and Ohashi, Y. 1995. Involvement of small GTP-binding proteins in defense signal-transduction pathways of higher plants. Proc. Natl. Acad. Sci. U.S.A. 92:4138-4144.

Santos, E. R., Gouveia, E. R., Mariano, R. L. R., and Souto-Mior, A. M. 2006. Biocontrol of bacterial fruit blotch of melon by bioactive compounds produced by *Bacillus* spp. Summa Phytopathol. 32:376-378.

Sasson, A. 1993. Asia: Biotechnology research and development overview. Australas. Biotechnol. 3:200-204.

Scheck, H. J., Pscheidt, J. W., and Moore, L. W. 1996. Copper and streptomycin resistance in strains of *Pseudomonas syringae* from Pacific Northwest nurseries. Plant Dis. 80:1034-1039.

Schuster, M. L., and Coyne, D. P. 1974. Survival mechanisms of phytopathogenic bacteria. Annu. Rev. Phytopathol. 12:199-221.

Seemüller, E., Marcone, C., Lauer, U., Ragozzino, A., and Göschl, M. 1998. Current status of molecular classification of the phytoplasmas. J. Plant Pathol. 80:3-26.

Shaffer, W. H., and Goodman, R. N. 1985. Appearance of streptomycin-resistant *Erwinia amylovora* in Missouri apple orchards. (Abstr.) Phytopathology 75:1281.

Singer, C. E., and Ames, B. N. 1970. Sunlight ultraviolet and bacterial DNA base ratios. Science 170:822-826.

Sleesman, J. P., and Leben, C. 1976. Bacterial desiccation: Effects of temperature, relative humidity, and culture age on survival. Ecol. Epidemiol. 66:1334-1338.

Sniegowski, P. 2004. Evolution: Bacterial mutation in stationary phase. Curr. Biol. 14:R245-R246.

Song, W.-Y., Wang, G.-L., Chen, L.-L., Kim, H.-S., Pi, L.-Y., Holsten, T. E., Gardner, J., Wang, B., Zhai, W.-X., Zhu, L.-H., Fauquet, C., and Ronald, P. C. 1995. A receptor kinase-like protein encoded by the rice disease resistance gene, *Xa21*. Science 270:1804-1806.

Sorensen, J. T., and Gill, R. J. 1996. A range extension of *Homalodisca coagulata* (Say) (Hemiptera: Clypeorrhyncha: Cicadellidae) to southern California. Pan-Pac. Entomol. 72:160-161.

Stall, R. E., and Thayer, P. L. 1962. Streptomycin resistance of the bacterial spot pathogen and control with streptomycin. Plant Dis. Rep. 46:389-392.

Stall, R. E., Loschke, D. C., and Jones, J. B. 1986. Linkage of copper resistance and avirulence loci on a self-transmissible plasmid in *Xanthomonas campestris* pv. *vesicatoria*. Phytopathology 76:240-243.

Stockwell, V. O., Kawalek, M. D., Moore, L. W., and Loper, J. E. 1996. Transfer of pAgK84 from the biocontrol agent *Agrobacterium radiobacter* K84 to *A. tumefaciens* under field conditions. Phytopathology 86:31-37.

Stoger, E., Vaquero, C., Torres, E., Sack, M., Nicholson, L., Drossard, J., Williams, S., Keen, D., Perrini, Y., Fischer, R., and Christou, P. 2002. Producing valuable pharmaceutical molecules in rice. Rice Genet. Newslett. 16:138.

Syvanen, M., and Kado, C. I., eds. 2002. Horizontal Gene Transfer, 2nd ed. Academic Press, San Diego, CA.

Thomson, S. V., Gouk, S. C., Vanneste, J. L., Hale, C. N., and Clark, R. 1993. The presence of streptomycin resistant strains of *Erwinia amylovora* in New Zealand. Acta Hortic. 338:223-225.

Tsukamoto, T., Shirata, A., and Murata, H. 1998. Isolation of a gram-positive bacterium effective in suppression of brown blotch disease of cultivated mushrooms, *Pleurotus ostreatus* and *Agaricus bisporus*, caused by *Pseudomonas tolaasii*. Mycoscience 39:273-278.

Van den Houwe, I., De Smet, K., du Montcel, H. T., and Swennen, R. 1995. Variability in storage potential of banana shoot cultures under medium term storage conditions. Plant Cell Tissue Organ. Cult. 42:269-274.

van Loon, L. C., Bakker, P. A. H. M., and Pieterse, C. M. J. 1998. Systemic resistance induced by rhizosphere bacteria. Annu. Rev. Phytopathol. 36:453-483.

Veenendaal, A. K. J., Sundin, S., and Blocker, A. J. 2009. Small-molecule type III secretion system inhibitors block assembly of the *Shigella* type III secretion. J. Bacteriol. 191:563-570.

von Mering, C., Hungenholtz, P., Raes, J., Tringe, S. G., Doerks, T., Jensen, L. J., Ward, N., and Bork, P. 2007. Quantitative phylogenetic assessment of microbial communities in diverse environments. Science 315:1126-1130.

Waidner, B., Melchers, K., Stähler, F. N., Kist, M., and Bereswill, S. 2005. The *Helicobacter pylori* CrdRS two-component regulation system (HP1364/HP1365) is required for copper-mediated induction of the copper resistance determinant CrdA. J. Bacteriol. 187:4683-4688.

Walker, T. S., Bais, H. P., Grotewold, E., and Vivanco, J. M. 2003. Root exudation and rhizosphere biology. Plant Physiol. 132:44-51.

Wu, C., and Shimamoto, K. 1995. Electroporation-mediated gene transfer to japonica rice protoplasts and regeneration of transgenic plants. Pages 93-98 in: Gene Transfer to Plants. I. Potrykus and G. Spangenberg, eds. Springer-Verlag, Berlin.

Yao, J., and Allen, C. 2006. Chemotaxis is required for virulence and competitive fitness of the bacterial wilt pathogen *Ralstonia solanacearum*. J. Bacteriol. 188:3697-3708.

Young, J. M. 1977. Resistance to streptomycin in *Pseudomonas syringae* from apricot. N. Z. J. Agric. Res. 20:249-251.

Zeidan, M., and Czosnek, H. 1994. Acquisition and transmission of *Agrobacterium* by the whitefly *Bemisia tabaci*. Mol. Plant-Microbe Interact. 7:792-798.

Zhang, L., and Birch, R. G. 1997. The gene for albicidin detoxification from *Pantoea dispersa* encodes an esterase and attenuates pathogenicity of *Xanthomonas albilineans* to sugarcane. Proc. Natl. Acad. Sci. U.S.A. 94:9984-9989.

Zhu, W., MaGbanua, M. M., and White, F. F. 2000. Identification of two novel *hrp*-associated genes in the *hrp* gene cluster of *Xanthomonas oryzae* pv. oryzae. J. Bacteriol. 182:1844-1853.

Future Prospects in Plant Bacteriology

12

Bacteria are very small, unicellular, haploid organisms that multiply exponentially, rapidly achieving vast populations. Consequently, the probability of mutation and acquisition of DNA by lateral gene transfer is high, thus providing a very efficient and flexible means of genetic variation.

With new data generated through genomics and metagenomic sequencing (which has provided information on the identity and diversity of bacteria in a variety of locations, including marine sediments, sea water, and soil), it has become clear that lateral gene transfer events, horizontal gene transfer events, or both, albeit at extremely low frequencies, have provided the means for adaptive evolution in plant–microbe interactions. Such events have apparently led to the generation of plant-pathogenic species from organisms that were previously specific to mammals or humans. Certain human-pathogenic members of the family *Enterobacteriaceae*, such as *Serratia marcescens* (Zhang et al., 2003; Bruton et al., 2004), and of the family *Pseudomonadaceae*, such as *Pseudomonas aeruginosa* (Plotnikova et al., 2000; Walker et al., 2004), have become capable of infecting plants, presumably by the acquisition of novel genes for plant infection through lateral gene transfer. Similarly, certain plant pathogens may have become human or animal pathogens through these same gene transfer mechanisms. For example, consider *Burkholderia cepacia*, which has crossed the animal/plant host boundary.

Interestingly, bacterial plant pathogens do not exploit host cell functions to create a replication niche inside plant cells, but rather they remain intercellular occupants. In the future, will some bacterial plant pathogens evolve to become intracellular, like their counterparts that infect animals?

Our research to date on the mechanisms of virulence and pathogenicity has helped to identify specific genetic targets for controlling the effects of pathogens on our domesticated plants. Future studies will further improve our understanding of pathogenesis, providing the critical information needed to protect crops against bacterial diseases.

Purpose for Studying Plant-Pathogenic Bacteria

The primary goal of research on plant bacterial pathogens is to formulate strategies for disease control. In-depth knowledge of the pathogen itself is essential for achieving the goal of disease control. Although plant pathologists might commonly believe that there is a race between the pathogen and its host or hosts, in actuality, it is the domestication of wild species of plants that has led to the spread of acute and chronic diseases. Wild species were highly tolerant to infection by their long association with nearly innocuous bacteria. Experience has demonstrated that antibacterial compounds, such as antibiotics and copper, provide control of bacterial diseases that is only temporary because of mutations, the horizontal acquisition of antibiotic and copper resistance genes by the pathogen, or both (Lai et al., 1977;

Bender and Cooksey, 1986; Burr et al., 1988; Minsavage et al., 1990; Sundin and Bender, 1996). This scenario is played out over and over because of the power of adaptive evolution in nature.

Novel strategies for disease control are essential to maintain a sustainable agriculture that feeds the world. Our best hope is to become completely acquainted with each important pathogen by learning the functions of each gene product that is used for pathogenesis and maintaining virulence. This understanding will eventually lead to the acceptance of genetically engineered plants for resistance to diseases, including those caused by bacteria. Some perspectives are given here as examples but space limits extensive and comprehensive discussions. Some relatively recent and proactive strategies for the near future are presented here.

Genetic Engineering of Endophytes as Protective Biologicals

Endophytic microorganisms and fungi are likely prospects for genetic manipulation to confer protection against specific pathogens on specific crop plants without genetically altering the horticulturally desired features of the plant itself. Endophytes can enhance host fitness, protect against insects and nematodes, enhance drought tolerance, and improve growth of the host (Malinowski and Belesky, 2000; Schardl et al., 2004).

Recombinant strains of *Leifsonia xyli* subsp. *cynodontis* have been developed to deliver biopesticides in corn and other crops. In corn, a recombinant strain of *Leifsonia xyli* subsp. *cynodontis* that produces the δ-endotoxin of *Bacillus thuringiensis* shows insecticidal activity against the European corn borer in corn plants inoculated with the recombinant strain (Dimock et al., 1993; Lampel et al., 1994). Concerns about the potential spread of the bioengineered bacterium by xylem-feeding insects were minimized by extensive studies showing that a wide variety of leaf-chewing and xylem-sucking insects are not normally vectors of *Leifsonia xyli* subsp. *cynodontis* (Barbehenn and Purcell, 1993). The market for genetically engineered corn producing the endotoxin, however, remains limited.

Endophytes could be genetically engineered to produce bactericides or virulence-disarming compounds that target, for example, the two-component sensory regulation system in a given plant-pathogenic bacterium. The two-component regulatory system consists of a histidine kinase and a cognate response regulator. In response to host-generated signals, the kinase phosphorylates its own histidine residue. The phosphate is then transferred to a conserved aspartate residue in the receiver domain of the cognate response regulator, whose conformational change facilitates binding to the promoter region of operons and initiates transcription. Targeting any one of the steps that normally activate operons involved in virulence would be a strategy for disease control.

Histidine kinase appears to be a suitable target candidate (Stephenson and Hoch, 2004). Histidine kinases have not been found in mammals (Stephenson and Hoch, 2004) but they are found in plants. *Arabidopsis thaliana* (L.) Heynh. and perhaps other members of the family Cruciferae possess histidine kinases but they appear to differ structurally from bacterial histidine kinases (Urao et al., 1999; Urao et al., 2000; Urao et al., 2001). Recently, thienopyridine (TEP) (= 3,6-diamino-5-cyano-4-phenyl-thieno[2,3-b]pyridine-2-carboxylic acid (4-bromo-phenyl)-amide) (Fig. 12-1) was found to be a specific inhibitor of bacterial histidine kinases (Gilmour et al., 2005). TEP belongs to a new class of competitive ATP inhibitors of bacterial histidine kinases and could serve as the starting compound for new inhibitors that specifically inhibit bacterial histidine kinases. Although TEP is chemically derived, it is plausible that, in the future, genes producing novel inhibitors of bacterial histidine kinases might be cloned into endophytes for their production as they reside in plants. TEP derivatives might also be used as a protective barrier against bacterial infection.

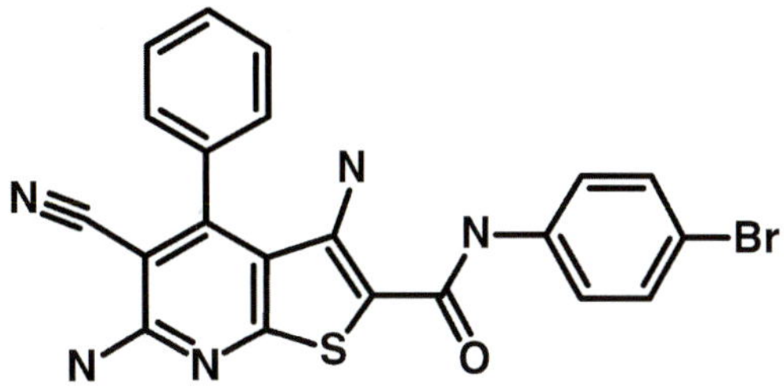

Fig. 12-1. Chemical structure of thienopyridine, a new class of competitive ATP inhibitors of bacterial histidine kinases.

Genetic Engineering of Pseudopathogens as Molecular Scarecrows

A large number of plant-pathogenic bacteria harbor plasmids (Kado and Liu, 1981; Coplin, 1989). Many plasmids resident in these bacteria harbor genes essential for virulence and pathogenicity. Removal of these genes or the entire plasmid or plasmid set renders the bacterial strain avirulent. The plasmid-free strain might be engineered to harbor the same plasmid or plasmid set with the virulence-conferring genes removed by selective deletions. The engineered plasmid would also be equipped with fertility inhibition genes that stringently prevent entry of like plasmids that might restore virulence. The engineered strain could be called a pseudopathogen and might be used as a competitor in the same ecological niche that the primary pathogen would occupy. For example, the rhizoplane or the phylloplane could be sprayed with the pseudopathogen, which would form a biofilm similar to that of the pathogen. It is recognized that part of the ecological success of bacteria that are incessantly exposed to variable environmental conditions is through their ability to grow as surface-attached biofilms, embedded in a protective extracellular matrix (Davey and O'Toole, 2000). Not only would the pseudopathogen occupy the same niche as that of the pathogen, it could also be equipped to produce a specific antibacterial compound that would be secreted in the niche.

Genetic Engineering of Crop Plants for Resistance to Bacterial Diseases

The basic premise of genetic engineering of crop plants for disease resistance is to facilitate and extend plant breeding without affecting the horticulturally desired properties of the plant. There are numerous possibilities for having crop plants produce a bactericide or an inhibitor of bacterial infection. Identification and characterization of various virulence effectors and their targeted receptors have helped identify potential targets for plant-produced antieffector compounds. Transgenic plants that are propagated vegetatively are the most desirable because they circumvent genetic variations.

Prevention of Bacterial Quorum Sensing

Chapters 6 and 10 described the use of bacterial lactonase to destroy the quorum-sensing signal compound *N*-acyl-homoserine lactone, which is needed for induction of virulence genes. Tobacco and tomato plants were genetically engineered to produce a lactonase derived from a *Bacillus* species. The transgenic plants showed resistance to soft rot disease, caused by *Erwinia carotovora* subsp. *carotovora* (now *Pectobacterium carotovorum* subsp. *carotovorum*) (Dong et al., 2000; Dong et al., 2001). Here is a case in which the lactone moiety of *N*-acyl-homoserine lactone was the target. Curiously, quorum-sensing signaling occurs on the cell surface of biofilms (rhizoplane and phylloplane) rather than at an internal location. One must assume that the lactonase in engineered tomato and tobacco is secreted to the external surface of the plant cells in order to be effective against the bacteria. Lysis of host cells by *Pectobacterium carotovorum* subsp. *carotovorum* likely would also release the enzyme. Lactonase genes derived from soil bacteria through metagenomic cloning have been shown to inhibit the biofilm formation of *Pseudomonas aeruginosa* (Schipper et al., 2009).

The antibiotic azithromycin has been used to block quorum sensing and alginate polymer formation in *Pseudomonas aeruginosa* (Hoffmann et al., 2007). Azithromycin might be useful in aiding the control of plant-pathogenic bacteria that use quorum sensing for pathogenesis. The antibiotic has proven successful in curing bacterial pneumonia in humans.

Inactivation of Bacterial Tyrosine Phosphatase

Another prospective target in the engineered host plant is the gene product that is responsible for altering host tyrosine phosphorylation. Protein modification by tyrosine phosphorylation is not generally found in prokaryotes (Foster et al., 1989). Protein modification by tyrosine phosphorylation is critical for the normal growth and development of eukaryotic cells. Plant proteins used in development and growth are precisely modified by concerted actions of protein tyrosine kinases and protein tyrosine phosphatases (Barizza et al., 1999; MacRobbie, 2002). Plant-pathogenic bacteria have evolved mechanisms that subvert the normal signal transduction pathways that modulate the structure and function of plant growth as well as the signal transduction functions involved in host resistance to bacterial infection. Consider *Pseudomonas syringae* pv. *tomato*, which secretes a 51.4-kDa protein tyrosine phosphatase (HopPtoD2) via the type III secretion mechanism (Espinosa et al., 2003; Espinosa and Alfano, 2004). HopPtoD2 suppresses the hypersensitive response, which is one of several activities that perturb the normal physiology of the eukaryotic cell, much like that of YopH, a 50-kDa tyrosine phosphatase secreted by *Yersinia* species (Guan and Dixon, 1990; Bliska et al., 1991). Although the precise role of HopPtoD2 remains to be determined, the action of tyrosine phosphatase is thought to suppress general defense responses that are activated upon recognition of an avirulence (Avr) protein (Espinosa et al., 2003). A gene product that degrades or inhibits the action of bacterial tyrosine phosphatase found through screening various bacteria can be expressed in transgenic plants. Clinical trials have used chemical inhibitors of protein tyrosine phosphatase, for example, 2,5-dihydroxy-*N*-methyl-*N*-nitrosoaniline (Imoto et al., 1993), sodium stibogluconate (Pathak and Yi, 2001), sodium vanadate (Posner et al., 1994), and trifluoromethyl sulfonyl and trifluoromethyl sulfonamide and benzylphosphonate derivatives (Huang et al., 2003).

Inactivation of Bacterial Protein Chaperones

Gram-negative plant-pathogenic bacteria have effectively adapted and used the type III and type IV secretion systems to deliver effector proteins and DNA (see

Chapter 10). Some effectors require the assistance of secretion chaperones for effective translocation through the secretion apparatus. The *Agrobacterium* Ti plasmid VirE2 protein assists T-DNA passage across the host cell membrane (Duckely and Hohn, 2003). This protein binds the transit single-stranded T-DNA (= T strand), assists in nuclear import (Tzfira et al., 2001; Ziemienowicz et al., 2001; Citovsky et al., 2004), and is facilitated by VirE3 (Lacroix et al., 2005). VirE2 possibly interacts with plant lipids since it binds sterols and sphingolipids to form channels (Duckely et al., 2005).

Transgenic tobacco and *Arabidopsis thaliana* (L.) Heynh. expressing VirE1 protein were shown to be less susceptible to *Agrobacterium*-induced tumor formation than were the wild-type plants (Szegedi et al., 2001; Humann et al., 2006). The idea of using VirE1 was to possibly block T-DNA transmission by preventing VirE2 from binding T strand since VirE1 forms complexes with VirE2, which then bind to lipids (Duckely et al., 2005). Although the exact mechanism of crown gall inhibition is unknown, the use of *virE1* may prove to be a useful transgene in the control of crown gall.

Designer Genes Encoding Cyclic, Amphipathic Bactericidal Peptides

Lower and higher animals produce antibiotic, amphipathic peptides containing 23–29 amino acid residues called magainins, cecropins, and defensins. In plants, defensins are found in seeds and peripheral cell layers of reproductive organs, flowers, and fruits and can be induced in other parts of the plant (Thevissen et al., 2003). These antimicrobial peptides are arranged in separate β-sheet domains held in place by intramolecular disulfide bonds between cysteine residues (usually three pairs). The antimicrobial activity resides in the hydrophobic membrane spanning domain of these peptides, with the hydrophilic residues providing the leakage pathway for water-soluble ions. Novel synthetic genes encoding artificial defensins have been used to generate transgenic tomato plants. They proved to be resistant to *Pseudomonas syringae* pv. *tomato* and *Ralstonia solanacearum* (Kado et al., 1995). Recently, a synthetic gene encoding a cecropin-like peptide was incorporated into transgenic tobacco plants that were shown to be resistant to *Pseudomonas syringae*, *Pseudomonas marginata*, and *Pectobacterium carotovorum* subsp. *carotovorum* (Zakharchenko et al., 2005). Interestingly, the artificial gene encoding a cecropin-like peptide increased resistance to late blight and white stem rot in transgenic potato (Zakharchenko et al., 2007).

It is quite apparent that synthetic genes can be designed and used to generate novel peptides with bactericidal activity. The specificity of these gene products is likely to be broad since the mechanism of cecropin activity is similar to that of detergents that degrade phospholipid-bearing membranes. Detergents and surfactants usually cause membrane leakage and loss of essential fluids of the bacterial cell.

RNA Interference or Posttranscriptional Gene Silencing

RNA interference (RNAi) is a biochemical process common to eukaryotes whereby double-stranded RNA (dsRNA) inhibits gene expression in a sequence-specific manner (Meister and Tuschi, 2004). Even though plants can use virus-specific nucleic acid molecules as systemic silencing signals that generate RNA-silencing defenses ahead of viral systemic invasion (Zamore, 2004; Matranga and Zamore, 2004), the use of dsRNA as the initiator of posttranscriptional gene silencing, usually called RNAi, has been successfully used against plant viruses (Waterhouse et al., 1998; Waterhouse and Wang, 2001). RNAi requires the production of dsRNA that is homologous to the gene that is targeted for silencing. The dsRNA is cleaved into smaller interfering RNA pieces (siRNAs) of about 21–25 nucleotides in length by an endonuclease called Dicer or RNase III (Elbashir et al., 2001). The siRNA forms a multiprotein effector complex called RISC (RNA-induced silencing complex) composed of helicases and nucleases. Within the RISC, siRNAs direct the cleavage of a single phosphodiester bond in target RNAs to which they are sufficiently complementary (Elbashir et al., 2001). The processed siRNA is then unwound and the RISC preferentially utilizes one of the RNA strands as a guide to target complementary RNA molecules for degradation. Figure 12-2 summarizes a simplified version of this degradative pathway. The relative stabilities of the 5' ends of the siRNA determine which strand of the siRNA duplex is utilized for mRNA targeting (Schwarz et al., 2003). The 3' region of the siRNAs contains the target cleavage while the sequences closer to the 5' end of the molecule are responsible for target binding (Schwarz et al., 2006). More potent and reproducible gene inhibition occurs with siRNAs than with the earlier antisense oligonucleotides approach. Research has revealed several gene-silencing pathways, such as microRNA (miRNA), *trans*-acting siRNA (tasiRNA), natural-antisense siRNA (natsiRNA), and repeat-associated siRNA (rasiRNA), of which the latter involves methylation (Eamens et al., 2008).

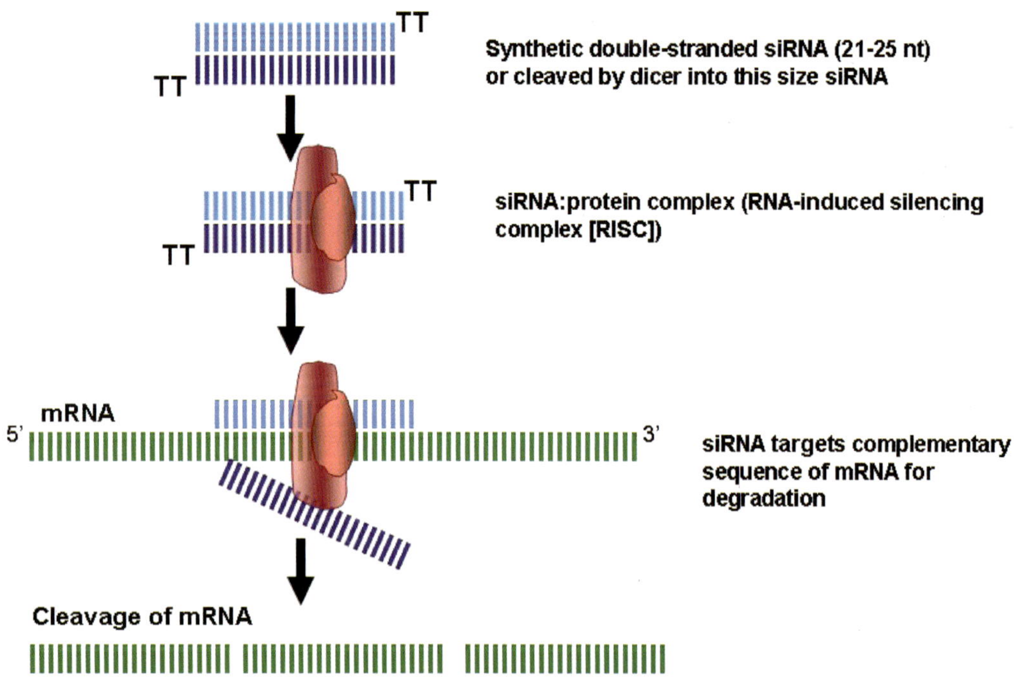

Fig. 12-2. A simplified version of a typical gene-silencing pathway that utilizes either double-stranded (ds) RNA or synthetic dsRNA of 21–25 nucleotides (nt) with 2-nt deoxyribosylthymine (dT) 3' overhangs. When either smaller interfering RNA (siRNA) is introduced into a cell, the molecule is bound by an RNA-induced protein complex (RISC). RISC scans the mRNA population of the cell and targets and binds to the mRNA bearing the complementary sequence. The helicase and nucleases of the RISC then degrade the mRNA.

The RNAi strategy has been used to reduce or prevent crown gall in transgenic *Arabidopsis thaliana* (L.) Heynh. and tomato (*Lycopersicon esculentum* Mill. var. *esculentum* Mill.) (Escobar et al., 2001). The targeted genes for the RNAi strategy were the *iaaM* (encodes tryptophan monooxygenase) and *ipt* (encodes isopentenyl transferase) oncogenes encoded by the T-DNA of *Agrobacterium tumefaciens* (described in Chapter 8). Escobar et al. transformed *Arabidopsis thaliana* and tomato with a transgene hpRNA (hairpin siRNA) construction designed to generate self-complementary *iaaM* and *ipt* transcripts (Escobar et al., 2001). The resultant transgenic plants retained susceptibility to *Agrobacterium* transformation but, in some tests, they were highly refractory to crown gall tumor formation.

Agrobacterium tumefaciens has been shown to suppress the plant defensive RNAi response against transcripts from the invading T-DNA as the crown gall tumor establishes (Dunoyer et al., 2006). The suppressing mechanism has not been identified. It is quite apparent that *Agrobacterium tumefaciens* is a highly sophisticated pathogen that evolved to deal with plant defenses, including the ability to block the hypersensitive response (Kado, 1994).

The natsiRNA is induced by *Pseudomonas syringae* pv. *tomato* infection and down regulates a *PPRL* gene that encodes a negative regulator of the RPS2 disease resistance pathway (Katiyar-Agarwal et al., 2006). Infection by this bacterium causes the accumulation of a novel, long siRNA (30–40 nucleotides in length compared with the typical 21–24 nucleotides) that silences a negative regulator of plant defense called AtRAP (Katiyar-Agarwal et al., 2007).

The RNAi strategy should prove extremely useful in generating transgenic plants bearing a gene encoding the dsRNA that targets and silences receptors of pathogenic effector proteins produced and injected by the bacterial pathogen. RNAi approaches have been directed to generating mainly transgenic plants with resistance to viruses. The exploitation of this approach is expanding to the control of plant diseases caused by bacteria and fungi as well as to the prevention of the damaging effects of nematodes and insects (Eamens et al., 2008).

Genomics

As stressed in this book, detailed knowledge of the genes used by the bacterial pathogen for virulence and pathogenicity will be greatly fortified by comparative genomic analyses as the database of genomic sequences continues to grow. Genomics and functional genomics will also clear the way for very precise taxonomic assignments and provide an excellent platform for developing concrete standards for speciation, whereby classical taxonomic approaches that have generated confusion and erroneous classification schemes will no longer be acceptable. The knowledge generated through genomics will provide valuable information as to the evolution of each pathogen and the time when each pathogen originated through molecular clock analyses that give more precise phylogenic assignments using appropriate filters (removal of incongruent data) (Clarke et al., 2002).

The sequencing of hundreds of bacterial genomes has shown that the percentage of genes devoted to regulatory function increases with genome size (Stover et al., 2000). On the other hand, obligate types of bacterial pathogens, such as phytoplasmas and spiroplasmas, have reduced genome sizes that are believed to have originated from larger genomes through ameliorative pressures in the host (Mira et al., 2001). Plant bacteriologists will be able to categorically characterize each common group of genomic species as the functions of all virulence and pathogenicity genes become known. Because members of the family *Enterobacteriaceae* have clinical relevance, the Enteropathogen Resource Integration Center (ERIC) provides genomics information on the pathogens of this family. ERIC resources are available through the web address www.ericbrc.org. Annotations of ERIC genomes are continuously updated. The annotations online are derived from ASAP (A System for community Annotation of Pathogens). ERIC also provides tools, such as MAUVE (a powerful multiple genome aligner and visualization tool) and GBrowse (Generic Genome Browser), for viewing individual sequence features to assist with research. The new generation of genome sequencing, such as nanopore sequencing methods, has reduced the cost of sequencing significantly. It is therefore feasible to sequence an entire bacterial genome for less than $1,000 U.S. dollars (Rusk, 2009).

Conclusion

The study of plant bacteriology has led to many novel findings that have benefited medicine and biotechnology as well as agriculture. The identification and characterization of plant bacterial virulence mechanisms and host–pathogen interactions have led to the identification of homologous mechanisms in human- and animal-pathogenic bacteria and have proven useful in clinical medicine (see Chapter 10). Precise knowledge of virulence gene functions will clear the path toward designing and developing bacterial plant disease control strategies.

Since 1970, our knowledge base of plant bacteriology has gone through extraordinary growth, largely due to federal granting agencies that have funded the fundamental work in this field. Comparative genomics has revealed many virulence gene cassettes situated in so-called pathogenicity islands and located on mobile elements either inserted in the chromosome or on a resident plasmid of the bacterial plant pathogen. The genes in the pathogenicity islands appear to be highly conserved, a finding that suggests bacterial pathogens evolved through the exchange or acquisition of novel genes useful in virulence and adaptation to particular hosts. Careful studies of the functions of the virulence gene products have revealed important strategies used by the pathogen to secure a niche in the host. It should be realized that these bacterial pathogens did not appear overnight but have long-standing relationships with the native wild hosts. Human domestication of useable wild plants has led to degenerate physiological responses by the domesticated crop plants when they encounter the pathogen. Chapters 2 and 9 have extensively discussed the evolution and adaptations of bacteria with plants.

Our future outlook on plant bacteriology is bright as we continue to unravel the mysteries of host–pathogen interactive mechanisms through the use of sophisticated analytical tools. Mother Nature has been very cryptic in the design of these mechanisms but, once revealed, they appear logical, quite simple, and understandable.

REFERENCES

Barbehenn, R. V., and Purcell, A. H. 1993. Factors limiting the transmission of a xylem-inhabiting bacterium *Clavibacter xyli* subsp. *cynodontis* to grasses by insects. Phytopathology 83:859-863.

Barizza, E., Schiavo, F. L., Terzi, M., and Filippini, F. 1999. Evidence suggesting protein tyrosine phosphorylation in plants depends on the developmental conditions. FEBS Lett. 447:191-194.

Bender, C. L., and Cooksey, D. A. 1986. Indigenous plasmids in *Pseudomonas syringae* pv. *tomato*: Conjugative transfer and role in copper resistance. J. Bacteriol. 165:534-541.

Bliska, J. B., Guan, K., Dixon, J. E., and Falkow, S. 1991. Tyrosine phosphate hydrolysis of host proteins by an essential *Yersinia* virulence determinant. Proc. Natl. Acad. Sci. U.S.A. 88:1187-1191.

Bruton, B. D., Mitchell, F., Fletcher, J., Pair, S. D., Wayadande, A., Melcher, U., Brady, J., Bextine, B., and Popham, T. W. 2003. *Serratia marcescens*, a phloem-colonizing, squash bug-transmitted bacterium: Causal agent of cucurbit yellow vine disease. Plant Dis. 87:937-944.

Burr, T. J., Norelli, J. L., Katz, B., Wilcox, W. F., and Hoying, S. A. 1988. Streptomycin resistance of *Pseudomonas syringae* pv. *papulans* in apple orchards and its association with a conjugative plasmid. Phytopathology 78:410-413.

Citovsky, V., Kapelnikov, A., Oliel, S., Zakai, N., Rojas, M. R., Gilbertson, R. L., Tzira, T., and Loyter, A. 2004. Protein interactions involved in nuclear import of the *Agrobacterium* VirE2 protein *in vivo* and *in vitro*. J. Biol. Chem. 279:29528-29533.

Clarke, G. P. P., Beiko, R. G., Raga, M. A., and Charlebois, R. L. 2002. Inferring genome trees by using a filter to eliminate phylogenetically discordant sequences and a distance matrix based on mean normalized BLASTP scores. J. Bacteriol. 184:2072-2080.

Coplin, D. L. 1989. Plasmids and their role in the evolution of plant pathogenic bacteria. Annu. Rev. Phytopathol. 27:187-212.

Davey, M. E., and O'Toole, G. A. 2000. Microbial biofilms: From ecology to molecular genetics. Microbiol. Mol. Biol. Rev. 64:847-867.

Dimock, M., Turner, J., and Lampel, J. 1993. Endophytic microorganisms for delivery of genetically engineered microbial pesticides in plants. Pages 85-97 in: Advanced Engineered Pesticides. L. Kim, ed. Marcel Dekker, Inc., New York.

Dong, Y.-H., Xu, J.-L., Li, X.-Z., and Zhang, L.-H. 2000. AiiA, an enzyme that inactivates the acylhomoserine lactone quorum-sensing signal and attenuates the virulence of *Erwinia carotovora*. Proc. Natl. Acad. Sci. U.S.A. 97:3526-3531.

Dong, Y.-H., Wang, L.-H., Xu, J.-L., Zhang, H.-B., Zhang, X.-F., and Zhang, L.-H. 2001. Quenching quorum-sensing-dependent bacterial infection by an *N*-acyl homoserine lactonase. Nature 411:813-817.

Duckely, M., and Hohn, B. 2003. The VirE2 protein of *Agrobacterium tumefaciens*: The Yin and Yang of T-DNA transfer. FEMS Microbiol. Lett. 223:1-6.

Duckely, M., Oomen, C., Axthelm, F., Van Gelder, P., Waksman, G., and Engel, A. 2005. The VirE1VirE2 complex of *Agrobacterium tumefaciens* interacts with single-stranded DNA and forms channels. Mol. Microbiol. 58:1130-1142.

Dunoyer, P., Himber, C., and Voinnet, O. 2006. Induction, suppression and requirement of RNA silencing pathways in virulent *Agrobacterium tumefaciens* infections. Nat. Genet. 38:258-263.

Eamens, A., Wang, M.-B., Smith, N. A., and Waterhouse, P. M. 2008. RNA silencing in plants: Yesterday, today, and tomorrow. Plant Physiol. 147:456-468.

Elbashir, S. M., Lendeckel, W., and Tuschi, T. 2001. RNA interference is mediated by 21- and 22-nucleotide RNAs. Genes Dev. 15:188-200.

Escobar, M. A., Civerolo, E. L., Summerfelt, K. R., and Dandekar, A. M. 2001. RNAi-mediated oncogene silencing confers resistance to crown gall tumorigenesis. Proc. Natl. Acad. Sci. U.S.A. 98:13437-13442.

Espinosa, A., and Alfano, J. R. 2004. Disabling surveillance: Bacterial type III secretion system effectors that suppress innate immunity. Cell. Microbiol. 6:1027-1040.

Espinosa, A., Guo, M., Tam, V. C., Fu, Z. Q., and Alfano, J. R. 2003. The *Pseudomonas syringae* type III-secreted protein HopPtoD2 possesses protein tyrosine phosphatase activity and suppresses programmed cell death in plants. Mol. Microbiol. 49:377-387.

Foster, R., Thorner, J., and Martin, G. S. 1989. Nucleotidylation, not phosphorylation, is the major source of the phosphotyrosine detected in enteric bacteria. J. Bacteriol. 171:272-279.

Gilmour, R., Foster, J. E., Sheng, Q., McClain, J. R., Riley, A., Sun, P.-M., Ng, W.-L., Yan, D., Nicas, T. I., Henry, K., and Winkler, M. E. 2005. New class of competitive inhibitor of bacterial histidine kinases. J. Bacteriol. 187:8196-8200.

Guan, K. L., and Dixon, J. E. 1990. Protein tyrosine phosphatase activity of an essential virulence determinant in *Yersinia*. Science 249:553-556.

Hoffmann, N., Lee, B., Hentzer, M., Rasmussen, T. B., Song, Z., Johansen, H. K., Givskov, M., and Holby, N. 2007. Azithromycin blocks quorum sensing and alginate polymer formation and increases sensitivity to serum and stationary-growth-phase killing of *Pseudomonas aeruginosa* and attenuates chronic *P. aeruginosa* lung infection in *Cftr*$^{-/-}$ mice. Antimicrob. Agents Chemother. 51:3677-3687.

Huang, P., Ramphal, J., Wei, J., Liang, C., Jallal, B., McMahon, G., and Tang, C. 2003. Structure-based design and discovery of novel inhibitors of protein tyrosine phosphatases. Bioorg. Med. Chem. 11:1835-1849.

Humann, J., Andrews, S., and Ream, W. 2006. VirE1-mediated resistance to crown gall in transgenic *Arabidopsis thaliana*. Phytopathology 96:105-110.

Imoto, M., Kakeya, H., Sawa, T., Hayashi, C., Hamada, M., Takeuchi, T., and Umezawa, K. 1993. Dephostatin, a novel protein tyrosine phosphatase inhibitor produced by *Streptomyces*. I. Taxonomy, isolation, and characterization. J. Antibiot. 46:1342-1346.

Kado, C. I. 1994. Anti-host-defense systems are elaborated by plant pathogenic bacteria. Pages 581-591 in: Molecular Mechanisms of Bacterial Virulence. C. I. Kado and J. H. Crosa, eds. Kluwer Academic Publ., Dordrecht, the Netherlands.

Kado, C. I., and Liu, S.-T. 1981. Rapid procedure for detection and isolation of large and small plasmids. J. Bacteriol. 145:1365-1373.

Kado, C. I., Buryanov, Y., and Huang, Y. 1995. Transgenic tomato expressing cyclic peptide 124 is resistant to *Pseudomonas syringae* pv. tomato. (Abstr.) Page 20 in: Proc. Conf. Genet. Eng. Resist. Pathogens. Fallen Leaf Lake, South Lake Tahoe, CA.

Katiyar-Agarwal, S., Morgan, R., Dahlbeck, D., Borsani, O., Villegas, A., Jr., Zhu, J.-K., Staskawicz, B. J., and Jin, H. 2006. A pathogen-inducible endogenous siRNA in plant immunity. Proc. Natl. Acad. Sci. U.S.A. 103:18002-18007.

Katiyar-Agarwal, S., Gao, S., Vivian-Smith, A., and Jin, H. 2007. A novel class of bacteria-induced small RNAs in *Arabidopsis*. Genes Dev. 21:3123-3134.

Lacroix, B., Vaidya, M., Tzfira, T., and Citovsky, V. 2005. The VirE3 protein of *Agrobacterium* mimics a host cell function required for plant genetic transformation. EMBO J. 24:428-437.

Lai, M., Shaffer, S., and Panopoulos, N. J. 1977. Stability of plasmid-borne antibiotic resistance in *Xanthomonas vesicatoria* in infected tomato leaves. Phytopathology 67:1527-1530.

Lampel, J. S., Canter, G. L., Dimock, M. B., Kelly, J. L., Anderson, J. J., Uratani, B. B., Foulke, J. S., and Turner, J. T. 1994. Integrative cloning, expression, and stability of the *cryIA(c)* gene from *Bacillus thuringiensis* subsp. *kurstaki* in a recombinant strain of *Clavibacter xyli* subsp. *cynodontis*. Appl. Environ. Microbiol. 60:501-508.

MacRobbie, E. A. C. 2002. Evidence for a role for protein tyrosine phosphatase in the control of ion release from the guard cell vacuole in stomatal closure. Proc. Natl. Acad. Sci. U.S.A. 99:11963-11968.

Malinowski, D. P., and Belesky, D. P. 2000. Adaptations of endophyte-infected cool-season grasses to environmental stresses: Mechanisms of drought and mineral stress tolerance. Crop Sci. 40:923-940.

Matranga, C., and Zamore, P. D. 2004. Plant RNA interference in vitro. Cold Spring Harbor Symp. Quant. Biol. 69:403-408.

Meister, G., and Tuschi, T. 2004. Mechanisms of gene silencing by double-stranded RNA. Nature 431:343-349.

Minsavage, G. V., Canteros, B. I., and Stall, R. E. 1990. Plasmid-mediated resistance to streptomycin in *Xanthomonas campestris* pv. *vesicatoria*. Phytopathology 80:719-723.

Mira, A., Ochman, H., and Moran, N. A. 2001. Deletional bias and the evolution of bacterial genomes. Trends Genet. 17:589-596.

Pathak, M. K., and Yi, T. 2001. Sodium stibogluconate is a potent inhibitor of protein tyrosine phosphatases and augments cytokine responses in hemopoietic cell lines. J. Immunol. 167:3391-3397.

Plotnikova, J. M., Rahme, L. G., and Ausubel, F. M. 2000. Pathogenesis of the human opportunistic pathogen *Pseudomonas aeruginosa* PA14 in *Arabidopsis*. Plant Physiol. 124:1766-1774.

Posner, B. I., Faure, R., Burgess, J. W., Bevan, A. P., LaChance, D., Zhang-Sun, G., Fantus, I. G., Ng, J. B., Hall, D. A., and Lum, B. S. 1994. Peroxovanadium compounds. A new class of potent phosphotyrosine phosphatase inhibitors which are insulin mimetics. J. Biol. Chem. 269:4596-4604.

Rusk, N. 2009. Cheap third-generation sequencing. Nat. Meth. 6:244-245.

Schardl, C. L., Leuchtmann, A., and Spiering, M. J. 2004. Symbioses of grasses with seedborne fungal endophytes. Annu. Rev. Plant Biol. 55:315-340.

Schipper, C., Hornung, C., Bijtenhoorn, P., Quitschau, M., Grond, S., and Streit, W. R. 2009. Metagenome-derived clones encoding two novel lactonase family proteins involved in biofilm inhibition in *Pseudomonas aeruginosa*. Appl. Environ. Microbiol. 75:224-233.

Schwarz, D. S., Hutvagner, G., Du, T., Xu, Z., Aronin, N., and Zamore, P. D. 2003. Asymmetry in the assembly of the RNAi enzyme complex. Cell 115:199-208.

Schwarz, D. S., Ding, H., Kennington, L., Moore, J. T., Schelter, J., Burchard, J., Linsley, P. S., Aronin, N., Xu, Z., and Zamore, P. D. 2006. Designing siRNA that distinguish between genes that differ by a single nucleotide. PLoS Genet. 9:1307-1318.

Stephenson, K., and Hoch, J. A. 2004. Developing inhibitors to selectively target two-component and phosphorelay signal transduction systems of pathogenic microorganisms. Curr. Med. Chem. 11:765-773.

Stover, C. K., Pham, X. Q., Erwin, A. L., Mizoguchi, S. D., Warener, P., Hickey, M. J., Brinkman, F. S., Hufnagle, W. O., Kowalik, D. J., Lagrou, M., Garber, R. L., Goltry, L., Tolentino, E., Westbrock-Wadman, S., Yuan, Y., Brody, L. L., Coulter, S. N., Folger, K. R., Kas, A., Larbig, K., Lim, R., Smith, K., Spencer, D., Wong, G. K., Wu, Z., Paulsen, I. T., Reizer, J., Saier, M. H., Hancock, R. E., Lory, S., and Olson, M. V. 2000. Complete genome sequence of *Pseudomonas aeruginosa* PAO1, an opportunistic pathogen. Nature 406:959-964.

Sundin, G. W., and Bender, C. L. 1996. Dissemination of the *strA-strB* streptomycin resistance genes among commensal and pathogenic bacteria from humans, animals, and plants. Mol. Ecol. 5:133-143.

Szegedi, E., Oberschall, A., Bottka, S., Oláh, R., and Tinland, B. 2001. Transformation of tobacco plants with *virE1* gene derived from *Agrobacterium tumefaciens* pTiA6 and its effect on crown gall tumor formation. Int. J. Hortic. Sci. 7:53-56.

Thevissen, K., Ferket, K. K., François, I. E., and Cammue, B. P. 2003. Interaction of antifungal plant defensins with fungal membrane components. Peptides 24:1705-1712.

Tzfira, T., Vaidya, M., and Citovsky, V. 2001. VIP1, an *Arabidopsis* protein that interacts with *Agrobacterium* VirE2, is involved in VirE2 nuclear import and *Agrobacterium* infectivity. EMBO J. 20:3596-3607.

Urao, T., Yakubov, B., Satoh, R., Yamaguchi-Shinozaki, K., Seki, M., Hirayama, T., and Shinozaki, K. 1999. A transmembrane hybrid-type histidine kinase in *Arabidopsis* functions as an osmosensor. Plant Cell 11:1743-1754.

Urao, T., Miyata, S., Yamaguchi-Shinozaki, K., and Shinozaki, K. 2000. Possible His to Asp phosphorelay signaling in an *Arabidopsis* two component system. FEBS Lett. 478:227-232.

Urao, T., Yamaguchi-Shinozaki, K., and Shinozaki, K. 2001. Plant histidine kinases: An emerging picture of two-component signal transduction in hormone and environmental responses. Sci. STKE 2001(109):re18.

Walker, T. S., Bais, H. P., Déziel, E., Schweizer, H. P., Rahme, L. G., Fall, R., and Vivanco, J. M. 2004. *Pseudomonas aeruginosa*-plant root interactions. Pathogenicity, biofilm formation, and root exudation. Plant Physiol. 134:320-331.

Waterhouse, P., and Wang, M.-B. 2001. Gene silencing as an adaptive defence against viruses. Nature 411:834-842.

Waterhouse, P., Graham, M., and Wang, M.-B. 1998. Virus resistance and gene silencing in plants can be induced by simultaneous expression of sense and antisense RNA. Proc. Natl. Acad. Sci. U.S.A. 95:13959-13964.

Zakharchenko, N. S., Rukavtsova, E. B., Gudkov, A. T., and Buryanov, Y. I. 2005. Enhanced resistance to phytopathogenic bacteria in transgenic tobacco plants with synthetic gene of antimicrobial peptide cecropin P1. Russ. J. Genet. 41:1187-1193.

Zakharchenko, N. S., Rukavtsova, E. B., Gudkov, A. T., Yukhmanova, A. A., Shkolnaya, L. A., Kado, C. I., and Buryanov, Y. I. 2007. Expression of the artificial gene encoding anti-microbial peptide cecropin P1 increases the resistance of transgenic potato plants to potato blight and white rot. Dokl. Biol. Sci. 415:267-269.

Zamore, P. D. 2004. Plant RNAi: How viral silencing suppressor inactivates siRNA. Curr. Biol. 14:198-200.

Zhang, Q., Weyant, R., Steigerwalt, A. G., White, L. A., Melcher, U., Bruton, B. D., Pair, S. D., Mitchell, F. L., and Fletcher, J. 2003. Genotyping of *Serratia marcescens* strains associated with cucurbit yellow vine disease by repetitive elements-based polymerase chain reaction and DNA-DNA hybridization. Phytopathology 93:1240-1246.

Ziemienowicz, A., Merkle, T., Schoumacher, F., Hohn, B., and Rossi, L. 2001. Import of *Agrobacterium* T-DNA into plant nuclei. Two distinct functions of VirD2 and VirE2 proteins. Plant Cell 13:369-384.

Index